PRACTICAL ONSHORE GAS FIELD ENGINEERING

PRACTICAL ONSHORE GAS FIELD ENGINEERING

DAVID A. SIMPSON, P.E.

ELSEVIER

Gulf Professional Publishing
An imprint of Elsevier

Gulf Professional Publishing is an imprint of Elsevier
50 Hampshire Street, 5th Floor, Cambridge, MA 02139, United States
The Boulevard, Langford Lane, Kidlington, Oxford, OX5 1GB, United Kingdom

British Library Cataloguing-in-Publication Data
A catalogue record for this book is available from the British Library

Library of Congress Cataloging-in-Publication Data
A catalog record for this book is available from the Library of Congress

ISBN: 978-0-12-813022-3

For Information on all Gulf Professional Publishing publications
visit our website at https://www.elsevier.com/books-and-journals

Publishing Director: Joe Hayton
Senior Acquisition Editor: Katie Hammon
Editorial Project Manager: Katie Chan
Production Project Manager: Sruthi Satheesh
Cover Designer: Victoria Pearson

Typeset by MPS Limited, Chennai, India

CONTENTS

PREFACE

When I started in Oil & Gas in 1980 as a "User Analyst" for Standard Oil (Indiana) (later to become Amoco Production Company and finally to merge with British Petroleum to become BP), the job was to represent users of (mainframe) computer programs in the process of developing applications for engineering and field use. To say that I was "fighting over my weight" would be a significant understatement. Everyone I talked to had vastly more knowledge of their job than I had of mine, but I was required to represent their needs and wishes to programing teams that barely knew what industry they were in. The only way that I could find to bridge that gap was to talk to people—a lot of people, in a lot of different disciplines. When I say "talk" I mean "listen." I learned fairly quickly that people like talking about themselves and their jobs to someone who displays interest. I spent a lot of time on the road in field and regional offices trying to find out what people need from the computer programs I was developing. As the programs ranged from a payroll application to a program that managed $200 million USD worth of pipe inventory to a program that allowed reserves estimators to store their results for aggregation, I talked to a wide range of people with a very wide range of skill sets. After 10 years of representing end-users in application development, I had a reasonably good handle on the business, the operations, and the industry.

In subsequent roles, I found myself managing a group responsible for representing the company's interest in wells operated by other companies. Part of that role (as the only engineer in the group) was to participate in estimating the coalbed methane reserves in the San Juan Basin. As I was signing the Reserves Change Form for several hundred wells representing a working interest proved reserves of hundreds of billion standard cubic feet of gas, I had pause. I wondered, "I'm saying that we can achieve a 125 psia reservoir pressure, but I have no earthly idea how to do that as our off-system delivery pressure is contractually 112 psia, and these wells cannot flow economic rates at a 13 psid pressure drop from deep in the reservoir to the compressor station."

The more I thought about it, the less confident I was that we could meet this abandonment pressure in this totally unknown asset. There was no analog in the world for the way that we expected the CBM field to release the adsorbed gas. We truly did not know what kind of surface pressures were required late in the life of a CBM field or what flow rates could be expected at those pressures. I was nervous enough to approach my boss and explain my concerns and ask for the opportunity to operate the company-operated portion of the CBM in the San Juan Basin and learn how to do it. After considerable management discussion, I assumed that role "in addition to other duties."

At the same time, I was working on my master's degree in fluid mechanics, and my master's thesis was on a topic in gas measurement. As that project progressed, the regional measurement engineer was transferred to another role without being replaced, so I stepped into the role of gas-measurement engineer once again "in addition to other duties"—it was very lucky that my staff in the outside operated group did not require much supervision.

After about 3 years of wearing all of these hats, the CBM operations reached peak production and began a 60% annual decline. That magnitude of decline was not anticipated in the production budget, and everyone became very concerned. I was "asked" to present the current state of the field to management. At the end of my bleak review, my boss asked the key question "what will it take to reverse this decline?" I had expected this question and had thought about it, so my answer was "there are five questions that no one in the world knows the answer to, I believe that if I have the space to work on this, I can come up with answers or work-arounds for them, but I can't tell you today what those answers will be or if they will have reasonable economics." He asked me what I needed in order to do that. I answered, "Leave me alone for four months." His response was a perfect "If you can list the problems and estimate a duration to solve them, then you've already started on the solution, see you in February." I spent the next 4 months in my basement developing what became "The POD" that is discussed in detail in Chapter 10, Integration of Concepts, and forms the basis of my understanding of the way that low-pressure/high-rate reservoirs perform. Developing and implementing this analysis and the projects that it spun off contributed to the breadth of experience that this book represents.

When I retired from BP in 2003 and started MuleShoe Engineering, I realized that 23 years with Standard Oil (Indiana) → Amoco Production Company → BP was really an apprenticeship for actually being an engineer. I've continued to learn in my consultancy, but the things I learned while working at a major were a significant portion of the price of entry to the world of consulting.

A recurring theme in my interactions with clients was the question "why doesn't this engineer know that?" As that question came up more and more often with new-hires, mid-career, and late-career engineers, I started wondering where a person would acquire the seemingly eclectic range of knowledge needed to be an effective facilities engineer. In 2008, I started teaching a 5-day class that I described as covering "everything that I wish I had known when I started my career." The class was well received, and I was frequently asked where someone could get more information on various topics covered in the course. Some of the sources could be used for multiple topics, but mostly it required multiple books or websites for each single topic. The course evolved over the years until by 2015 it felt fairly mature. At that point, the slowdown in the industry coincided with an even more severe slowdown in my consultancy, and I decided it was time to turn the course into the book you hold in your hands (or view on your screen).

Writing this book showed that the course was not nearly as mature as I'd thought, and there is one huge difference—in the classroom if I say something dumb, wrong, or confusing, the students can (and do) stop me and require me to fix it right then, but if I write something dumb, wrong, or confusing, there is no venue for clarification. Consequently, I've proofed the text a dozen times, imposed upon colleagues to review the individual chapters, and the editors at Elsevier have done a sterling job of finding my inconsistencies, but I am confident that there are errors and omissions that have escaped all of these reviews. For those, I apologize.

Although I sincerely appreciate the work of the various reviewers from six continents and thank them heartily, I won't list them here because I want it made clear that any problem or shortfall of this manuscript is mine, and those shortcomings should not fall to the people who kindly donated their time to making the book better than it would have been otherwise.

CHAPTER ZERO

Introduction

0.1 BACKGROUND

When I started working in Oil & Gas in 1980, there was a very long list of things that I wish I had known. In retrospect, more than anything else, I needed access to the terminology that specialists in various parts of the industry used. For example, when a Tool Pusher was installing a down-hole pump that I was experimenting on he asked me "do you want it set in tension or compression?" I did the "engineering thing" and made up an answer without anything to base it on. It turns out that I made the right choice, but coin tosses are a poor basis for an engineering decision. In this case I simply did not know exactly what he meant (I knew the terms, but I didn't know at that time how the terms applied). This book is an attempt to make the esoterica of the onshore upstream gas portion of the industry a bit more accessible to engineers whose remit is surface facilities.

Any book that purports to follow a hydrocarbon molecule from "squashed dinosaurs" to the ultimate burner tip faces the related dilemmas of "where do you start?" and "what can you assume the reader already knows?" Hopefully, it is safe to assume that anyone who picks up this book is comfortable with the concepts of algebra, but how about Newton's laws of motion? The general laws of thermodynamics? Kinematics? Organic chemistry? How much fluid mechanics can be assumed? There are no clear answers to these questions, but there are a few concepts that are used across all engineering disciplines in Oil & Gas that tend to get lost in the noise. For example, every part of Oil & Gas uses "volume flow rate at standard conditions." That idea is basic to reserves estimates, well-bore tubular requirements, pipe design, vessel capacity, and plant operations.

Everyone understands SCF (or SCm) don't they? Maybe not. In my master's thesis I used a volume flow rate at standard conditions to calculate a "velocity" that was central to my basic assertions. I defended this concept in front of my Thesis Committee and none of them even hinted that all of the conclusions contained in my thesis were absolutely invalid

Practical Onshore Gas Field Engineering.
DOI: http://dx.doi.org/10.1016/B978-0-12-813022-3.00021-3

because of this bonehead mistake. This experience leads me to the conclusion that the concept of volume flow rate at standard conditions is so ubiquitous in the industry that everyone assumes that everyone else understands it properly. Almost all of us are wrong about almost all of us. This chapter is focused on removing the veil of complacency from several of these ubiquitous concepts so we don't need to revisit them multiple times in the subsequent chapters. If you are uncomfortable with algebra you might want to go elsewhere for a refresher prior to proceeding.

0.2 FLUID TERMINOLOGY

When discussing the fluids we encounter in Oil & Gas, definitions often get blurred and many of the differences are of critical importance to selection and operation of surface equipment.

Crude oil: A mixture of naturally occurring chemicals which contain hydrocarbons and is liquid at reservoir conditions and remains liquid at temperatures somewhat above ambient at atmospheric pressure.

Condensate: A mixture of naturally occurring chemicals which contain hydrocarbons that are gaseous at reservoir conditions and liquid at ambient conditions. Condensate generally includes species of hydrocarbons including pentane (C_5H_{12}), hexane (C_6H_{14}), and occasionally heptane (C_7H_{16}) and octane (C_8H_{18}). Lighter hydrocarbon species can be present in condensate, but over time it will boil off. API gravity (see below) of condensate is generally 45–75°API.

Dry gas: A mixture of gases that does not contain hydrocarbon species that become condensate.

Note that all of these definitions start with "a mixture." This is because no reservoir has pure fluids, Mother Nature just doesn't work that way. The exact mix of fluids is quite variable from field to field and even from well to well within a field. A term like "West Texas Intermediate (WTI)," "North Sea Brent," or "Saudi Light" actually refer to a narrow range of specific gravities and API gravities (see later) and upper limits of contaminants.

Also note that "dry gas" does not mean that there is no liquid, it means that there is no marketable liquid hydrocarbons, there can be a considerable quantity of water in a "dry gas" field.

Natural gas liquids (NGL): Hydrocarbon species whose boiling point is near-ambient and cannot be reliably stored at atmospheric pressure in a vented vessel. NGL generally includes ethane (C_2H_6), propane (C_3H_8), butane (C_4H_{10}), and isobutane (C_4H_{10}).

Specific gravity: Specific gravity is a convenient way to represent a fluid's mass relative to a reference fluid.

- *Liquid*: The specific gravity of a liquid is the density of the liquid divided by the density of pure water at a reference temperature (usually 60°F or 15.6°C, but other temperatures are used) (i.e., 64.4 lbm/ft^3 (1000 kg/m^3)). If it is known that a liquid's density is 60% of the density of water, then you can input the density of water in units that are convenient for your current calculation and multiply it times 0.60. Liquid specific gravity is not an intrinsic property of the liquid and must be referenced to a specific temperature.
- *Gas*: Specific gravity of a gas is a ratio of the mass per mole of the gas divided by the mass per mole of air (28.9625 lbm/lb-mole (28.9695 gm/gm-mole)). Gas specific gravity is an intrinsic property of the gas and is not dependent upon pressure or temperature.

Wet gas: A mixture of gases that include hydrocarbon species that become liquid with pressure or temperature changes of a magnitude expected in normal gas production.

0.3 OILFIELD UNITS

Traditional oilfield units are based on the foot–pounds–second (FPS) system as bastardized by the industry. Primary oilfield units are:

- bbl → Barrel, used for all liquid measurements defined as 42 US gallons (159 L).
- SCF → Standard cubic feet, volume of gas stated as though it were measured at a reference pressure and temperature (see later).
- M → Roman numeral for 1000.
- MM → Multiplication of Roman numerals (i.e., in Roman numerals "MM" would be 2000, but in Oil & Gas "MM" is 1 million or 1000 × 1000).
- B → Abbreviation for "billion" or 10^9.
- T → Abbreviation for "trillion" or 10^{12}.

The prefix "M" and "MM" are only used in this book to mean "1000" and "1,000,000," respectively, with traditional oilfield units.

The SI units that this book will use are:

- m^3 → Cubic meter, only used in this book for physical volumes like liquid volumes and gas volumes at actual conditions (never for gas volumes at standard conditions).
- SCm → Standard cubic meters, volume of gas stated as though it were at a reference pressure and temperature (see later).
- k → Kilo, as a prefix it is the base value times 10^3.
- M→ Mega, as a prefix it is the base value times 10^6 (this book will try to avoid the use of "M" as an SI unit, but values like MPa are far too ubiquitous and useful to completely forego).
- G → Giga, as a prefix it is the base value times 10^9.
- T → Tera, as a prefix it is the base value times 10^{12}.
- P → Peta, as a prefix it is the base value times 10^{15}.
- E → Exa, as a prefix it is the base value times 10^{18}.

0.3.1 Unit Conversions

For conversions, in Tables 0.1 and 0.2, the units you have are in the first column. Go across and get the multiplication factor (Table 0.3).

0.3.2 g_c

Mass in the fps system is messy. The "official" mass unit is the "slug" which is designated as "the mass that would represent a weight of 32.174 lbf on the surface of the earth." Slugs tend to be a difficult concept for many people and calculation errors with mass in fps are so common that an informal unit called "pounds mass" and designated "lbm" was developed. One lbm weighs 1 lbf on the surface of the earth, which creates a

Table 0.1 Gas Unit Conversions

	SCF	MSCF	MMSCF	SCm	kSCm
SCF	1	1E − 3	1E − 6	0.0283	0.283E − 3
MSCF	1E3	1	1E − 3	28.32	0.0283
MMSCF	1E6	1E3	1	28,320	28.32
SCm	35.31	0.0353	35.3E − 6	1	1E − 3
kSCm	35,310	35.31	0.0353	1000	1

Note: In this book, when reference temperature and pressure are not stated, 14.73 psia (101.56 kPaa or 1.0156 bara) and 60°F (15.56°C) were used. If your project is expecting a different reference temperature or pressure, then multiply the above table by $\rho_{14.7/60F} \div \rho_{\text{required values}}$.

Table 0.2 Liquid Unit Conversions

	BBL	Gallon	L	kL = m^3
BBL	1	42	159	0.159
Gallon	23.8E − 3	1	3.79	3.79E − 3
L	6.29E − 3	0.264	1	1E − 3
kL = m^3	6.290	264	1000	1

Table 0.3 Temperature Unit Conversions

	°F	R	°C	K
°F	1	°F + 459.67	(°F − 32)*5/9	(°F − 32)*5/9 + 273.15
R	R − 459.67	1	(R − 491.67)*5/9	R*5/9
°C	°C*9/5 + 32	°C*9/5 + 491.67	1	°C + 273.15
K	K*9/5 − 459.67	K*9/5	K − 273.15	1

whole new package of confusion. The mathematical representation of mass and weight in the fps system is (all variables are defined in the "Nomenclature" section at the end of this chapter):

$$\text{lbf} = g \cdot \text{lbm} \Rightarrow 1 = g \cdot \frac{\text{lbm}}{\text{lbf}} \therefore g_c = g \cdot \frac{\text{lbm}}{\text{lbf}} = 32.174 \cdot \frac{\text{ft} \cdot \text{lbm}}{\text{s}^2 \cdot \text{lbf}} \tag{0.1}$$

Someone doing calculations in the fps system decides whether to include a g_c or not when they determine if the equation includes lbm and needs the results in lbf (e.g., a column of liquid has a mass and a density, but we frequently need the force that that column will exert at the bottom of the column in lbf per unit area), includes lbf and needs lbm, or has both and needs to cancel them.

This term has always been unique to the fps system, but recently the SI folks have been using kgf/cm^2 instead of kPa or bar so they will have to develop their own nonsense.

0.4 RESERVOIR FLUIDS

The Oil & Gas Industry exists to exploit in situ fluids in subterranean reservoirs. Any action that the industry takes that is not intended to facilitate that imperative is a waste of resources. Some fluid characteristics are so specific to a particular process, that it is appropriate to define those characteristics closer to the process (e.g., Chapter 7: Water Collection and Disposal seems like the appropriate place to discuss what to expect within the produced water that must be collected and disposed of). There are some general concepts that are appropriate for this chapter.

Do we need to consider safety of employees and the public? Absolutely, but we need to take those safety actions in ways that consider the fact that the only reason that these safety considerations exist is because we are trying to extract, transport, and process reservoir fluids into compounds that are useful to people.

Reservoir fluids are central to every appropriate action that any of us take within our roles in the industry. Consequently, the discussion of these fluids is appropriate in this portmanteau chapter rather than in Chapter 1, Gas Reservoirs.

0.5 LIQUIDS

The liquids that we deal with are produced water, oil, and condensate. The field of oil chemistry is vast and complex. Since this book is only intending to consider the issues of gas wells, crude oil as a detailed topic seems to be outside of the intended scope. Many gas wells produce both condensate and NGL, but it seems more appropriate to discuss those topics in Chapter 9, Interface to Plants.

0.5.1 Liquid Specific Gravity

The specific gravity of a liquid is (all variables are described in the "Nomenclature" section at the end of the chapter):

$$SG_{liquid} = \frac{\rho_{liquid\ at\ 60^{\circ}F}}{\rho_{water\ at\ 60^{\circ}F}} \tag{0.2}$$

If the specific gravity of the liquid is known, we don't have to care if we call the density of water 62.4 lbm/ft^3 or 1000 kg/m^3 to determine the density of the liquid. Defining the specific gravity of liquid relative to water also has the benefit of allowing someone to tell at a glance whether the liquid will float or sink in a body of water.

Density is not an intrinsic property of a liquid. Density will change (slightly for water, significantly for many hydrocarbon liquids) with changes in either pressure or temperature. Consequently, it is necessary to define specific gravity of a liquid at a reference temperature and pressure. For most engineering problems, the magnitude of the density changes with pressure or temperature is less than the uncertainty in the calculation (e.g., if you are estimating temperature $\pm 20°F$ ($\pm 11.1°C$) then a density uncertainty of $0.02\ \text{lbm/ft}^3$ ($0.32\ \text{kg/m}^3$) is unlikely to make a material difference in your total uncertainty).

One area where the actual temperature and pressure of the fluid is critical to fluid density and specific gravity is situations with volatile hydrocarbons. Small changes in either pressure or temperature can cause phase-change events including liquids flashing to a gas or gases condensing back to a liquid. These phase changes can easily make a material difference in the mass of fluid in your tank or pipeline. Dealing successfully with volatile hydrocarbon species is a specialized activity that is rarely central to upstream decisions and while it is important to understand that it happens, the mechanics of accounting for hydrocarbon phase change is beyond the scope of a book focused on gas.

0.5.2 API Gravity

"API gravity" is unit specific to commercial transactions dealing with crude oil and condensate. It is basically a normalized density, but it is presented as a temperature.

$$
\begin{aligned}
°\text{API} &= \frac{141.5}{SG_{\text{at } 60°F}} - 131.5 \\
SG_{\text{at } 60°F} &= \frac{141.5}{131.5 + °\text{API}}
\end{aligned}
\tag{0.3}
$$

The "at 60°F" designation is important since crude and condensate can change density rapidly with changes in temperature when you pass the boiling point of volatile hydrocarbon components.

The commodities market breaks the °API numbers down into:

- $°\text{API} > 31.1 \rightarrow$ Light crude oil and condensate
- $22.3 < °\text{API} < 31.1 \rightarrow$ Medium crude oil
- $10 < °\text{API} < 22.3 \rightarrow$ Heavy crude oil
- $°\text{API} < 10 \rightarrow$ Extra heavy crude oil (will not float on water)

WTI is defined as 39.6 °API, North Sea Brent is 38.06 °API, and Dubai Crude is 31 °API, the heavy crude from the Athabasca Oil Sands is around 8 °API. Shale oil from the Eagle Ford field tends to be 40–45 °API.

Refineries are designed for a very narrow range of °API and they can operate only as long as they can find a mix of oils that can be blended to match the design range. Trying to bring oil sands crude into a refinery designed for WTI requires a large amount of Eagle Ford (or similar) light crude to get the combined mix to 39.6 °API (and conversely, processing Eagle Ford crude requires some heavier oil to get down to the WTI target).

0.5.3 Barrel of Oil

King Richard III (1452–85) decreed that a "tierce" would hold 42 gallons (159 L). When oil was discovered in Titusville, PA, in 1859 and started the process of creating the Oil & Gas Industry, it was not known how oil would be packaged or sold since the internal combustion engines patented before that time were impractical (i.e., they did not compress the fuel/air mixture prior to combustion and developed inadequate power for most tasks) and the Titusville crude was not originally thought to be a motor fuel (the first Otto patent with compression wasn't until 1876).

Producers in Titusville looked at the various standard size vessels and found that a tierce full of oil weighed about 300 lbm (136 kg) which was about as much as two men could handle. They also found that 20 tierces would fit in a railroad car without significant wasted space. Based on these findings the state of Pennsylvania defined the unit of commerce for crude oil as a 42 gallon tierce (Historical, 1, 2016).

By 1901 when the Spindletop field was discovered in West Texas, industry in the United States was looking for a steel alternative to wooden barrels. In 1905, Henry Wehrhahn, an employee of Nellie Bly's Iron Clad company, patented a 55 gallon barrel for bulk liquids (Historical, 2, 2016). Most of the bulk movement of crude within the Spindletop field was by horse-drawn wagons on very primitive roads. The teamsters did not want to take the time to attach drum lids and they found that if they only put 42 gallons in a 55 gallon drum, the spillage was acceptable and the teamsters began painting a fill-line in the drums at the 42 gallon level. The teamsters were paid by the drum, not by hauled volume. Since you get what you measure, a different basis for payment certainly would have resulted in a different outcome, but they were counting drums.

Eventually the Spindletop experience was matched with the earlier Titusville experience and an oilfield barrel was defined as 42 gallons (159 L) in time to begin supplying bulk fuel to the limited mechanization that occurred during World War I.

The abbreviation for "barrel" is "bbl." There is an apocryphal story in the industry (Historical, 2, 2016, 2) that the abbreviation comes from the documented Standard Oil requirement that all barrels used for Standard's oil be painted blue and that the abbreviation stands for "Blue BarreL." Oil & Gas historians have uncovered 18th century bills of lading which referred to the quantity of many bulk liquids as being in "bbl" containers. The actual genesis of that abbreviation has been lost to antiquity, but it did not refer to the color of containers.

0.5.4 Liquid Hydrostatic Pressure

Since Oil & Gas wells are very long vertical conduits, we have to be concerned about the effects of stacking fluids vertically. For an incompressible fluid we see a linear gradient from the top of the column increasing to a maximum at the bottom. A gas gradient is a bit more complex. In both cases the cross-sectional area of the column has no impact on the pressure at the bottom, just the height.

0.5.5 Hydrostatic Gradient

For a liquid the pressure at the bottom of a column is:

$$P_{bot} = \rho \cdot g \cdot h + P_{top} = \rho_{water} \cdot SG_{liquid} \cdot g \cdot h + P_{top} \tag{0.4}$$

It is common to replace the density times gravity term with a pressure gradient term:

$$\text{grad} = \rho \cdot g = \rho_{water} \cdot SG \cdot g = \left(0.433 \cdot \frac{\text{psi}}{\text{ft}}\right) \cdot SG = \left(9.81 \cdot \frac{\text{kPa}}{\text{m}}\right) \cdot SG \tag{0.5}$$

This makes it simple, if you have a 20 ft (6.1 m) standpipe going into a tank, it is going to take at least 8.7 psig (59.7 kPag) for a pump to overcome before the water can enter the tank.

0.5.6 Liquid Compressibility

For most calculations involving liquid it is reasonable to assume that liquids are incompressible, but they are not actually incompressible. The term "bulk modulus" is used to indicate the amount of pressure required to lower the volume occupied by a specified mass of a fluid by 1%. For a gas the bulk modulus is very close to zero. For water it is 319,000 psi (2.2 GPa).

A column of liquid water 196 miles (314 km) high would cause the water at the bottom to be 1% denser than the water at the top, not a real concern. On the other hand, a cyclical hydraulic system that imparts 20,000 psig (137,900 kPag) on a string of 3-in diameter pipe running 16,000 ft into the ground and then releases the pressure to retract a plunger and then repeats will have to add nearly 5 gallons of water on the pressurization stroke and then have a place for that water to go on the depressurization step—a 3 gallon supply tank would run dry, a 6 gallon supply tank would likely hit a low-level alarm every step.

0.6 GAS

0.6.1 Gas Equation of State

There are a number of refinements of the classical ideal gas law (Wikipedia, 1) that is taught in college chemistry classes. The refinements are used in Oil & Gas primarily in high-pressure/high-temperature reservoir calculations and within plant processes. For field operations in gas, the complexities of most of these refinements do not result in different decisions from the decisions you would reach with a simple equation of state (EOS). Consequently, in this book, we will use:

$$\begin{aligned} P\cdot V &= n\cdot \overline{R}\cdot T \Rightarrow P\cdot V = m\cdot R_{gas}\cdot T \text{ (for an ideal gas)} \\ P\cdot V &= m\cdot R_{gas}\cdot T\cdot Z \text{ (for real gases)} \end{aligned} \tag{0.6}$$

As will be discussed later, the "Z" term is called "compressibility" and represents a gas deviation from ideal behavior. By convention, air is assumed to be an ideal gas ($Z = 1.0$) even though it does exhibit a slight deviation from ideal behavior which is typically well below the error introduced by measuring temperature and pressure (and gas composition for that matter).

0.6.2 Gas Specific Gravity

Gas density is not as convenient a term to use as liquid density, it is far too variable. Instead, in gas we use the ratio of the molecular weight of a gas to the molecular weight of air. This is an intrinsic value and it remains constant over the complete range of temperatures and pressures that a gas could be subjected to. The molecular weight of air is taken as 28.962 lb/lb-mole (28.962 gm/gm-mole).

Using molecular weight leads to an interesting observation. The universal gas constant is:

$$\overline{R} = 1545.3 \cdot \left(\frac{\text{ft} \cdot \text{lbf}}{\text{R} \cdot \text{lb-mole}}\right) = 8.314 \cdot \left(\frac{\text{J}}{\text{K} \cdot \text{mole}}\right) \tag{0.7}$$

The universal gas constant is rarely useful to a facilities engineer since we tend not to work in moles. The specific gas constant is more useful:

$$R_{gas} \equiv \frac{\overline{R}}{\text{MW}_{gas}} \Rightarrow R_{air} \equiv \frac{\overline{R}}{\text{MW}_{air}} \Rightarrow R_{gas} = \frac{R_{air} \cdot \text{MW}_{air}}{\text{MW}_{gas}} = \frac{R_{air}}{\text{SG}_{gas}} \tag{0.8}$$

Now, to get to a gas constant that we can work with in familiar units, we only have to know gas specific gravity and that:

$$R_{air} = 53.355 \cdot \frac{\text{ft} \cdot \text{lbf}}{\text{R} \cdot \text{lbm}} = 287.068 \cdot \frac{\text{J}}{\text{K} \cdot \text{kg}} \tag{0.9}$$

0.6.3 Gas Compressibility

Gas is quite compressible, and when you stack it vertically you find that the pressure exerted by the gas stacked above it changes via both linear and nonlinear mechanisms. The primary nonliner mechanism is called "compressibility" (symbol "Z") which is fundamentally a measure of the amount that a gas deviates from ideal gas behavior. Air is very nearly an ideal gas where $Z = 1.0$. Methane and CO_2 exhibit distinctly nonlinear (and often nontrivial) response to applied force.

Gas compressibility is clearly a function of gas composition. At 1000 psig (6700 kPag) and 60°F (15.6°C) sweet gas (see Section 0.6.6.7 for the description of "sweet gas," "sour gas," and "CBM") has $Z = 0.727$, sour gas has $Z = 0.791$, and CBM has $Z = 0.863$—a range of 17% relative to the average. Sweet gas at standard temperature and pressure (STP) has $Z = 0.99$ and at 2000 psia it is $Z = 0.59$. Ignoring deviation from ideal behavior is one of the most common sources of engineering failure in Oil & Gas.

Gas compressibility is not an intrinsic characteristic of a gas mixture and methods of determining it range from very difficult to quite easy and the results range from an excellent representation of reality to a poor representation of reality, and there is not much correlation between ease of calculation and quality of results. The main techniques to develop compressibility relationships are:

- Equation of state. These values tend to be consistent and reliable, but generally require very complex programming. Programs like REFPROP.EXE from the US National Institute of Standards and Technology (NIST), which is inexpensive and quite capable, and the HYSYS from Aspen Technologies, which is quite expensive and able to evaluate very complex chemical and flow relationships, are required to use the EOS method.
- Corresponding states. This method uses critical behavior to predict deviation from ideal behavior. The most commonly used method is the Hall–Yarborough equation. Hall–Yarborough provides good results for hydrocarbon gases, but the so-called acid gases (e.g., CO_2 and H_2S) cause it to provide numbers that deviate too much from the EOS methods. Some later researchers have developed adjustments that improve performance in acid gases. This "equation" is a complex multistage program that requires upward of 30 steps in an application like MathCad.
- Closed form. GPSA has a pair of equations that do a good job for many gases. For our sweet gas composition, both EOS and corresponding states show a distinct upturn in the compressibility value above 2000 psia (13,800 KPaa) that the GPSA equation dos not match. For our sour gas composition, up to about 1300 psia (8900 kPaa), the GPSA is closer to the EOS than corresponding states is. For our CBM gas composition, all three methods are close. The GPSA equations are (GPSA):

$$\begin{aligned}&\text{For pressure} < 145\ \text{psia}\\ &Z = \frac{1}{1 + 0.0002 \cdot P_{avg\_psia}}\\ &\text{For pressure} \geq 145\ \text{psia}\\ &Z = \frac{1}{1 + \dfrac{P_{avg} \cdot 3.444 \times 10^5 \cdot 10^{1.785 \cdot SG}}{T_{avg}^{3.825}}}\end{aligned} \tag{0.10}$$

The GPSA book shows the shift from one equation to the other to be 100 psia, but I find that value to have too much discontinuity. Allowing the low-pressure equation up to 145 psia (10 bara) reduces the discontinuity which is another support for using 145 as a break point in physical performance as will be discussed in Section 0.7.5. Of course both of these equations are empirical (see Section 0.9) and must be solved with pressure in psia and temperature in Rankine.

For pressures above about 2000 psia (13,800 kPaa), it is imperative that you use EOS method. For sweet gas at 8000 psia (55,100 kPaa) and 60°F (15.6°C) REFPROP shows $Z = 1.29$ and GPSA shows $Z = 0.42$.

0.6.4 Gas Gradient

Gas gradient is more complex than the liquid gradient discussed earlier.

$$P_{bot} = P_{top} \cdot \exp\left(\frac{0.01875 \cdot \mathrm{SG} \cdot \frac{\mathrm{h}}{\mathrm{ft}}}{\frac{T_{avg}}{R} \cdot Z_{avg}}\right) = P_{top} \cdot \exp\left(\frac{0.03418 \cdot \mathrm{SG} \cdot \frac{\mathrm{h}}{\mathrm{m}}}{\frac{T_{avg}}{K} \cdot Z_{avg}}\right) \tag{0.11}$$

This equation works out to a very small number that can often be ignored without impacting a decision. For example at 1000 m (3280 ft) the exponent term works out to 1.07 times pressure at the top so if the imposed pressure is 500 kPaa (73 psia) then the pressure at the bottom would be 535 kPaa (78 psia).

0.6.5 Gas Density and Atmospheric Pressure

Gas density is derived from the EOS and is defined as the ratio of the mass present divided by the volume it occupies.

$$\rho = \frac{m}{V} = \frac{P}{R_{gas} \cdot T \cdot Z} = \frac{P \cdot \mathrm{SG}}{R_{air} \cdot T \cdot Z} \tag{0.12}$$

An important special case is calculation of local atmospheric pressure. The density of air is a function of pressure and temperature. Air temperature is a linear function of elevation above sea level (ASL) (the slope of the line is 3°F/1000 ft (5.468°C/km). Weather impacts (i.e., current temperature, barometric pressure, precipitation, etc.) are insignificant relative to the weight of the entire column of air. If you know your local elevation, then you can determine local atmospheric pressure by (Wikipedia, 2):

$$P_{atm} = P_{std}\left(1 - \frac{0.003 \cdot h_{ASL}}{T_{std}}\right)^{\frac{g}{0.003 g_c \cdot R_{air}}} = 14.73\ \mathrm{psi}\left(1 - \frac{5.77 \times 10^{-6}}{\mathrm{ft}} h_{ASL}\right)^{6.248}$$

$$P_{atm} = P_{std}\left(1 - \frac{0.005468 \cdot h_{ASL}}{T_{std}}\right)^{\frac{g}{0.005468 \cdot R_{air}}} = 101.56\ \mathrm{kPa}\left(1 - \frac{18.94 \times 10^{-6}}{\mathrm{m}} h_{ASL}\right)^{6.248} \tag{0.13}$$

This equation is valid up to about 360,000 ft (110 km) ASL.

0.6.6 Fluid Characteristics

Mother Nature has allowed some fluids to be trapped within subterranean reservoirs. These fluids included natural gas (methane through about hexane), condensate, oil, sulfur compounds, CO_2, nitrogen, oxygen, and water. These mixtures are not very neat or tidy or in any way homogenous. For example, what we include as "water" may have total dissolved solids between 80 mg/L (ppm) and 400,000 mg/L with solids made up of nearly the entire periodic table of the elements. As facilities engineers we need to be able to accept whatever a reservoir might give us and to render it either a valuable commodity or a safely removed waste stream.

0.6.6.1 *Selected Properties*

Some of the fluid properties that are most commonly used in our industry are included in Tables 0.4 and 0.5. This data is extracted from the *GPSA Engineering Data Book* (GPSA) which is the one indispensable reference for all engineers in the Oil & Gas industry. Other authors have other recommendations for "the one indispensable reference," but for many years the only book on my shelf that was ever opened was GPSA (and I have all of reference books that I've ever heard described as "indispensable"). *GPSA Engineering Data Book* is far from perfect, but it is almost always "good enough."

0.6.6.2 *Adiabatic Constant*

Adiabatic is defined as "relating to or denoting a process or condition in which heat does not enter or leave the system concerned" (Webster). Adiabatic conditions are very common in many areas of our industry. In many other situations there is heat leaving the system (if you've ever laid your hand on a compressor discharge pipe, you will have experienced heat leaving the system very rapidly, into your hand), but the heat that enters or leaves is small relative to the system as a whole. The assumption that a process is adiabatic is often very useful.

Specific heat at constant pressure (c_p) is an intrinsic property that is included in Tables 0.4 and 0.5. Specific heat at constant volume is a derived property that is described by:

$$c_v = c_p - R_{gas} = c_p - \frac{R_{air}}{SG} \tag{0.14}$$

This allows the definition of an "adiabatic constant" as (sometimes the adiabatic constant is represented as the Greek letter "kappa" or "gamma,"

Table 0.4 Selected Properties (fps) (GPSA)

	Formula	MW	Boiling Point (°F) at Standard Pressure	SG (Relative to Air)	Net Heating Value (MMBTU/MSCF)	c_p (BTU/lb-°F)
C1—Methane	CH_4	16.043	−258.7	0.5539	0.9094	0.52669
C2—Ethane	C_2H_6	30.070	−127.5	1.0382	1.6178	0.40782
C3—Propane	C_3H_8	44.097	−43.7	1.5226	2.3149	0.38852
C4—Butane	C_4H_{10}	58.123	n: 31.1	2.0068	n: 3.0108	n: 0.39499
			i: 11.1		i: 3.0004	i: 0.38669
C5—Pentane	C_5H_{12}	72.150	n: 97.0	2.4912	n: 3.7069	n: 0.38825
			i: 82.1		i: 3.6990	i: 0.38440
C6—Hexane plus	Various	95.514	>100	3.2979	4.9898	0.36724
carbon dioxide	CO_2	44.010	−109.1	1.5196	0	0.19908
Hydrogen sulfide	H_2S	34.082	−76.5	1.1767	0.5868	0.23839
Air	$N_2 + O_2$	28.963	−317.6	1.0000	0	0.23980
Water vapor	H_2O	18.014	212.0	0.6220	0	0.44401
Water (liquid)	H_2O	18.014	212.0	815.7	0	1.00000

The data is also available in the SI version of GPSA.

Table 0.5 Selected Properties (SI)

	Formula	MW	Boiling Point (°C) at Standard Pressure	SG (Relative to Air)	Net Heating Value (MJ/SCm)	c_p (J/gm/°C)
C1—Methane	CH_4	16.043	−161.5	0.5539	33.88	2.205
C2—Ethane	C_2H_6	30.070	−88.6	1.0382	60.28	1.707
C3—Propane	C_3H_8	44.097	−42.1	1.5226	86.25	1.627
C4—Butane	C_4H_{10}	58.123	n: −0.5	2.0068	n: 112.18	n: 1.654
			i: −11.6		i: 111.79	i: 1.619
C5—Pentane	C_5H_{12}	72.150	n: 36.1	2.4912	n: 138.12	n: 1.626
			i: 27.8		i: 137.82	i: 1.609
C6—Hexane plus	Various	95.514	>38	3.2979	189.05	1.538
Carbon dioxide	CO_2	44.010	−78.4	1.5196	0	0.834
Hydrogen sulfide	H_2S	34.082	−60.3	1.1767	21.86	0.998
Air	$N_2 + O_2$	28.963	−194.2	1.0000	0	1.000
Water vapor	H_2O	18.014	100.0	0.6220	0	1.859
Water (liquid)	H_2O	18.014	100.0	815.7	0	4.187

there isn't anything approaching a standard nomenclature, this book will use "k"):

$$k = \frac{c_p}{c_v} = \frac{c_p}{c_p - \frac{R_{air}}{SG}} = \frac{c_p \cdot SG}{c_p \cdot SG - R_{air}} \tag{0.15}$$

0.6.6.3 Gas Mixtures

Mixtures of gases take on the properties of the sum of the components of the gas:

$$y = \Sigma y_i \cdot x_i \tag{0.16}$$

Using Eq. (0.16) you can determine the molecular weight of a mix of gases, e.g.:

$$\mathrm{MW} = \Sigma(\mathrm{MW}_i \cdot x_i) \tag{0.17}$$

This says that the total molecular weight of the mixture is equal to each component's molecular weight times that component's mole fraction. We generally refer to "mole fraction" as "volume fraction" because it is a function of the amount of space that each molecule takes up in a mixture. Volume fraction is calculated using a "partial molar volume" technique, but it is generally taken in our industry as being equal to mole fraction even though volume fraction deviates from mole fraction as compressibility varies from 1.0.

Mass fraction is the percent of the total mass represented by the mass of a given component:

$$\text{Mass fraction} = \frac{x_i \cdot \mathrm{MW}_i}{\Sigma(x_i \cdot \mathrm{MW}_i)} \tag{0.18}$$

Mass fraction is used much less often that volume fraction, and using mass fraction in most calculations will lead to wrong answers.

0.6.6.4 Including Water Vapor

Field gas always has considerable water vapor. As will be discussed in Chapter 3, Well Dynamics, any time a gas is in contact with a coherent gas/water interface, the water will evaporate until the relative humidity (RH) at the surface is 100%. A very small distance from the interface, the humidity will have decreased, but at the interface the gas is fully saturated with water vapor. Theoretical water vapor is usually determined from a chart called McKetta–Wehe which presents temperature on the x-axis and water content at 100% RH, with individual lines for pressures. If you get a water vapor content from this graph and call it "W" then:

$$\text{Mole fraction}_{water} = \frac{\frac{W}{\text{lbm/MMSCF}} \cdot 3.8068 \times 10^{-4}}{\text{MW}_{water} + \frac{W}{\text{lbm/MMSCF}} \cdot 3.8068 \times 10^{-4}} = \frac{\frac{W}{\text{lbm/MMSCF}} \cdot 3.8068 \times 10^{-4}}{18.0153 + \frac{W}{\text{lbm/MMSCF}} \cdot 3.8068 \times 10^{-4}} = \frac{\frac{W}{\text{mg/SCM}} \cdot 2.3765 \times 10^{-5}}{18.0153 + \frac{W}{\text{mg/SCM}} \cdot 2.3765 \times 10^{-5}} \quad (0.19)$$

This says that if water takes up 4% of the gas mixture, there is only 96% of the total remaining for the other components. In that case you would "normalize" the analysis by:

$$\text{Mole fraction}_i = (1 - \text{Mole fraction}_{water}) \cdot \text{Analysis mole fraction}_i \quad (0.20)$$

This adjustment is often trivial, but it can be significant in specific cases. For example, field gas gatherers have to provide pipeline capacity for water vapor and rightly want to be paid for it. To aid in the calculation of "wet gas volume," the gathering company wants to call the gas saturated at STP instead of at field conditions. Gas will rarely be at 100% RH, but that difference represents a small difference. The "at STP" assumption is more problematic. If we have field gas at:

- Flowing conditions of 250 psig (1724 kPag), 100°F (38°C), 100 MSCF/day (2.83 kSCm/day), "sweet gas" as defined later.
- Water content from the McKetta–Wehe chart (see Chapter 3: Well Dynamics) at flowing conditions is 187.4 lbm/MMSCF (3000 mg/SCM) or 0.39% RH.
- Water content from the McKetta–Wehe chart at STP is 3080 lbm/MMSCF (33,300 mg/SCm) or 6.11%.
- Annual sales at $4/MMBTU

- Dry: \$153,440
- Saturated at flowing conditions \$152,835 (−\$605)
- Saturated at STP \$144,070 (−\$9370)

Most gas producers find a cost of \$605 to move 1.4 MMSCF of water vapor to be reasonable, and nearly \$10,000 to move the same water content to be excessive.

0.6.6.5 Inherent Energy

When someone purchases a fuel they are interested in the amount of heat that can be extracted from it. One of the fundamental parameters of a gas is its "heat content" or "heating value." This term comes in two flavors (DOE):

- Gross heating value: The energy released by a specified volume of gas when a compound undergoes complete combustion with oxygen at standard conditions and then cooling the exhaust gases back to a reference temperature. This value can be measured in a laboratory.
- Net heating value: Theoretical value calculated by determining the gross heating value and deducting the latent heat of vaporization of the mass of water condensed on cooling. This value cannot be measured.

The net heating value while it cannot be measured is the number that represents how much heat you can extract from a fuel in real life. Methane has a net heating value of 909 BTU/SCF (33,800 kJ/SCm).

It is common in some companies to assign an arbitrary value to natural gas of 948 BTU/SCF (35,300 kJ/SCm) and then:

- 1 GJ $\approx$ 1 MSCF
- 1 TJ $\approx$ 1 MMSCF
- 1 PJ $\approx$ 1 BSCF

These conversions are generally not very accurate, but are often good enough for everyday work (e.g., if a well produced 200 GJ yesterday and 210 GJ today, then it is doing better).

0.6.6.6 Energy Equivalents

It has become common to use the term "BOE" to mean "barrel oil equivalent." The concept is to find a way to add a mass of gas to a mass of liquid on a basis that reflects the marketplace uses of the two types of fuels. With a large number of assumptions, you can get an average energy contained in a barrel of crude oil as 5.8 MMBTU (1700 kW-h). With an additional list of assumptions, "natural gas" can be said to contain 5.642 MSCF/BOE (0.1569 kSCm/BOE).

Several producers have gone the other way and reported "gas equivalent" in their annual reports. Converting the oil into a gas equivalent is obviously the reciprocal of converting gas in to a BOE. The arithmetic is simple, but it turns out to be beyond the ability of stock analysist and they tend to downgrade companies that try to go this route, not something that a rational board of directors will accept to make a point that they are "a gas company, not an oil company." Companies that have tried this have generally only published gas-equivalent numbers in their financial statements once or twice then gone back to BOE.

0.6.6.7 C6 Plus

Commercial gas chromatographs are a trade-off between cost and functionality. Most of them have the ability to identify hydrocarbon gases through pentane (C5), but lack the processing equipment necessary to distinguish heavier components one from the other. To accommodate not wanting to spend the money (and training and maintenance) on the more complex equipment, the industry has accepted that components heavier than C5 are generally a very small proportion of any real gas and can be represented by an "average" or "typified" value. The characteristics of C6 + can be very different from lab to lab. One lab developed the data in Table 0.6 from an extended analysis of the inlet to a specific plant on a

Table 0.6 Extended Analysis

	% in Sample	Normalized %	MW	SG	Net Heating Value (MMBTU/MSCF)	c_p (BTU/lb-°F)
n-Hexane	0.049%	12.1%	10.401	0.3591	0.4474	0.04662
Other hexanes	0.155%	37.7%	32.476	1.1213	1.9182	0.14330
Heptanes	0.090%	21.9%	21.966	0.7584	1.1180	0.08428
Octanes	0.037%	9.1%	10.410	0.3594	0.5282	0.03493
Nonanes	0.013%	3.2%	4.107	0.1418	0.2079	0.01225
Decanes +	0.015%	3.7%	5.257	0.1815	0.2656	0.01411
Benzene	0.023%	5.7%	4.425	0.1528	0.2034	0.01376
Toluene	0.017%	4.2%	3.858	0.1332	0.1789	0.01089
Ethylbenzene	0.001%	0.2%	0.261	0.0090	0.0122	0.00068
Xylenes	0.009%	2.2%	2.353	0.0813	0.1099	0.00642
Total	0.410%	100.0%	95.514	3.2979	4.9898	0.36724

specific day and have used it in millions of gas samples since then. They periodically reassess this value and find that the relative proportions of the various components have remained reasonably constant over time.

In this particular case, the lab determined that C6 + made up 0.410% of the total sample. The extended analysis of that 0.410% is normalized as discussed earlier. Multiplying the normalized percentage times an inherent characteristic of a component yields its contribution (e.g., the MW of *n*-hexane is 86.18 lbm/lb-mole, multiplying that times 12.1% gives you 10.401 lbm/lb-mole). Then adding the individual contributions gives you the effective value of the made-up component "C6 + ." If the C6 + is 0.410% of the total gas, then this typification of C6 + adds 0.39 lbm/lb-mole and 0.02 MMBTU/MSCF to the gas mixture.

It is important to note that if you are required to report "BTEX" (i.e., benzene, toluene, ethylbenzene, and xylene) to a regulatory agency, it is unacceptable to use a typified analysis to determine the BTEX emissions.

0.6.6.8 Examples of Gas Types

Throughout this book we will use examples that require a gas analysis (or at least a specific gravity or adiabatic constant). To reduce the need for the reader to guess what gas is being discussed, Table 0.8 gas mixtures will be used throughout the book. When an example says "sweet gas" then any intrinsic parameters needed will come from the "sweet gas" column of Table 0.7 (all columns use the C6 + characteristics above).

Using these mixtures, you can determine intrinsic properties for each gas mixture (Table 0.8–0.11).

Table 0.7 Gas Compositions Example

Component	Sour Gas Mole %	Sweet Gas Mole %	CBM (CSG) Mole %
C1	81.02%	76.58%	92.00%
C2	4.00%	13.86%	0.00%
C3	2.00%	6.21%	0.00%
i-C4	0.50%	0.56%	0.00%
n-C4	1.50%	1.67%	0.00%
i-C5	0.20%	0.22%	0.00%
n-C5	0.80%	0.89%	0.00%
C6 +	0.01%	0.01%	0.00%
H_2S	2.90%	0.00%	0.00%
CO_2	7.07%	0.00%	8.00%
Total	100.00%	100.00%	100.00%

Table 0.8 Sour Gas Example

	Mole Fraction (%)	MW	SG	NHV (MMBTU/MSCF)	c_p (BTU/lb/°F)	k
C1	81.02	12.9980	0.4488	0.7368	0.4267	1.0591
C2	4.00	1.2028	0.0415	0.0647	0.0163	0.0477
C3	2.00	0.8819	0.0305	0.0463	0.0078	0.0226
i-C4	0.50	0.2906	0.0100	0.0150	0.0019	0.0055
n-C4	1.50	0.8718	0.0301	0.0452	0.0059	0.0164
i-C5	0.20	0.1443	0.0050	0.0074	0.0008	0.0022
n-C5	0.80	0.5772	0.0199	0.0297	0.0031	0.0086
C6 +	0.01	0.0096	0.0004	0.0005	0.0000	0.0001
H_2S	2.90	0.9884	0.0341	0.0170	0.0069	0.0384
CO_2	7.07	3.1115	0.1074	–	0.0141	0.0914
	100	21.0762	0.7278	0.9265	0.4836	1.2921

Table 0.9 Sweet Gas Example

	Mole Fraction (%)	MW	SG	NHV (MMBTU/MSCF)	c_p (BTU/lb/°F)	k
C1	76.58	12.2857	0.4242	0.6964	0.4033	1.0011
C2	13.86	4.1677	0.1439	0.2242	0.0565	0.1654
C3	6.21	2.7384	0.0946	0.1438	0.0241	0.0702
i-C4	0.56	0.3255	0.0112	0.0168	0.0022	0.0061
n-C4	1.67	0.9707	0.0335	0.0503	0.0066	0.0183
i-C5	0.22	0.1587	0.0055	0.0081	0.0008	0.0024
n-C5	0.89	0.6421	0.0222	0.0330	0.0035	0.0096
C6 +	0.01	0.0096	0.0004	0.0005	0.0000	0.0001
H_2S	0.00	–	–	–	–	–
CO_2	0.00	–	–	–	–	–
	100.00	21.2984	0.7355	1.1731	0.4971	1.2732

Table 0.10 CBM Example

	Mole Fraction (%)	MW	SG	NHV (MMBTU/ MSCF)	cp (BTU/lb/°F)	k
C1	92.00	14.7596	0.5096	0.8366	0.4846	1.2027
C2	0.00	–	–	–	–	–
C3	0.00	–	–	–	–	–
i-C4	0.00	–	–	–	–	–
n-C4	0.00	–	–	–	–	–
i-C5	0.00	–	–	–	–	–
n-C5	0.00	–	–	–	–	–
C6 +	0.00	–	–	–	–	–
H2S	0.00	–	–	–	–	–
CO2	8.00	3.5208	0.1216	–	0.0159	0.1034
	100.00	18.2804	0.6312	0.8366	0.5005	1.3061

Table 0.11 Various Nondimensional Numbers

Nondimensional Parameter	Forces	Equations
Reynolds number	Inertial forces/viscous forces	$Re = \frac{\rho \cdot \vec{v} \cdot L_{characteristic}}{\mu}$
Weber number	Inertial forces/interfacial tension	$We = \frac{\rho \cdot v^2 \cdot L_{characteristic}}{\sigma}$
Mach number	Velocity/sonic velocity	$M = \frac{v}{v_{sonic}}$
Euler number	Pressure forces/inertial forces	$Eu = \frac{p_1 - p_2}{\rho \cdot v^2}$
Cavitation number	Distance from vapor pressure/ kinetic energy	$Ca = \frac{p - p_{vapor}}{\left(\frac{1}{2}\right) \cdot \rho \cdot v^2}$
Froude number	Flow inertia/body forces	$Fr = \frac{v}{\sqrt{L_{characteristic} \cdot g}}$
Strouhal number	Vortex shedding frequency/velocity	$St = \frac{f_{vortex\ shedding} \cdot L_{characteristic}}{v}$

0.7 TOPICS IN FLUID MECHANICS

This is not a fluid mechanics textbook. In a fluid mechanics course we have to pretend that $\delta v/\delta t$ has meaning even if we do not have an equation for the change in velocity with respect to time to differentiate, and even when nonlinear, second-order differential equation really doesn't have much usefulness to a practical engineer. The number of fluid mechanics equations that have a closed-form solution is miniscule, and few field engineers have access to (or the time to master) the tools of converting second-order, nonlinear partial differential equations into answers that can impact real-time decisions. In a fluid mechanics course we also spend considerable time proving that if we actually knew any useful equations for $\delta v/\delta \rho$ we could do such fine, fair, and wonderful things. As an academic exercise, fluid mechanics is a fascinating subject for study and I am glad that I've had the opportunity to study it both as an undergraduate and in graduate school, but not many practical tools have come directly from it.

This book is about doing a good-enough job of describing and understanding physical phenomena that we can reach useful, informed decisions. Consequently, most of this book will be empirical equations rather than closed-form equations, and for the few closed-form equations that exist we'll explore the simplifying assumptions that were required to get from the general to the specific. We are not going to derive much.

This book tries very diligently to be clear on terms. When the word "fluid" is used, you should expect to be talking about "a substance without a fixed shape that can persist under mild external forces," which would include gases, liquids, emulsions, and other non-Newtonian substances that do not have a fixed shape that will persist under mild external forces. When it says "liquid" it means a fluid that will exhibit a coherent gas–liquid boundary. When it says "gas" it means a fluid that will expand to the three-dimensional limits of a container. It is very common in artificial lift and deliquification (Chapter 3: Well Dynamics) to interchange the words "liquid" and "fluid" which is imprecise and often quite wrong. This book tries to avoid interchanging these words, but the usage in that field is so pervasive that it can be hard to identify where you are making that particular mistake. If I missed any places where I said "fluid" and meant "liquid" please forgive me and know that it was not intentional.

Virtually all of the liquids that nonengineers work with are Newtonian. A Newtonian fluid is a fluid that exhibits linear stress–strain relationships. This means that if a stress is applied to the fluid, regardless of starting stress, the fluid will exhibit a displacement proportional to the applied stress. In other words an unconstrained (i.e., it has some place to go when you apply stress) Newtonian fluid does not have the ability to withstand imposed stresses. A Newtonian fluid constrained in three dimensions will have a very different stress–strain relationship, but that is due to the constraining material, not inherent to the fluid.

Conversely, non-Newtonian fluids have a nonlinear stress–strain relationship. Examples of non-Newtonian fluids are things like toothpaste and paint where it takes more applied stress to start the fluid flowing than it takes to keep the fluid flowing after it starts. In Oil & Gas we have to be more concerned about non-Newtonian fluids than most engineers. Paraffin (when it is not a solid), fluids with high levels of paraffin or asphaltene, and many emulsions and foams exhibit nonlinear stress–strain relationships and predicting pressure drops (for example) of any of these fluids can be very tricky. For example, diesel has a very high paraffin content and at laminar flow rates and/or at lower temperatures will often be "hard to pump," in that it takes more energy to start it flowing than to keep it flowing. When you get into the turbulent region (see Moody diagram in Chapter 4: Surface Engineering Concepts), it starts acting like a Newtonian fluid and our standard equations effectively represent field data.

0.7.1 Statics

"Fluid statics" is the study of fluids at rest or in constant motion in the absence of viscous drag.

The basic equation of is shown in Eq. (0.21):

$$P_{bottom} = \rho \cdot g \cdot h + (P_{imposed} + P_{atm}) \tag{0.21}$$

If we are talking about the pressure at the bottom of a lake then $P_{imposed}$ is zero. If we are talking about the pressure at the bottom of a well-bore then $P_{imposed}$ is flowing tubing pressure (in gauge). In either case, you only add atmospheric pressure once. In fps units, density is in lbm/unit-volume and the answer has to be in lbf/unit-area so in fps units (Eq. (0.22)):

$$P_{bottom} = \rho \cdot \frac{g}{g_c} \cdot h + (P_{imposed} + P_{atm}) \tag{0.22}$$

While the force per unit area that a column of fluid exerts is only a function of height, the weight of the column is very much dependent on the volume of fluid in the column (Eq. (0.23)):

$$W = \rho \cdot g \cdot V \tag{0.23}$$

This can be important if you are looking at structural supports for above-ground piping or vessels. It is very important if you have to lift the column of liquid. For gases, it is less likely to be a major factor, and it is often reasonable to ignore the first term in Eq. (0.22) (if you don't ignore it, then you need a relationship for density vs pressure, see Chapter 3: Well Dynamics).

0.7.1.1 Buoyancy

Buoyancy is a static vertical force on a substance submerged in a fluid. If the mass of the displaced fluid is greater than the mass of the submerged substance, then the submerged substance will try to rise. This why a ship that weighs thousands of tonne can float, it displaces more liquid than it weighs. When we talk about droplets in Chapter 3, Well Dynamics, the size of the droplet will determine whether the liquid will be buoyant in the light gas. If the droplet is large then it displaces a smaller mass of gas than it contains and it falls. If the droplet is small, then it displaces a mass of gas that is near to the droplet mass and the droplet tends to remain in suspension for a time. Buoyant forces are often necessary to explain why things happen in facilities the way they do.

0.7.2 Dynamics

Fluid dynamics describes every possible influence on a flow stream, including:

- pressure;

- temperature;
- body forces (like gravity and magnetic attraction/repulsion);
- friction;
- rotation;
- compression and expansion.

Every force can be a function of time, point of reference, an intrinsic property of the fluid, or an extrinsic property of the fluid. When you put all the forces into all the proper relationships, the result shown in Eq. (0.24) is kind of messy.

0.7.2.1 Navier–Stokes Equation

French engineer Claude-Louis Navier first proposed a series of equations that represent fluid dynamics in 1822. In 1845, Irish mathematician-physicist George Gabriel Stokes redeveloped this equation that now carries the name "Navier–Stokes equation." One representation of the expansion of the Navier–Stokes equation is shown in Eq. (0.24).

$$\rho\cdot\left(\frac{\partial u_i}{\partial t}+u_j\cdot\frac{\partial u_i}{\partial x_j}\right)=\rho\cdot\Sigma F_{body}-\frac{\partial P}{\partial x_i}+\frac{\partial}{\partial x_i}\left(\mu\cdot\left(\frac{\partial u_i}{\partial x_j}+\frac{\partial u_j}{\partial x_i}-\frac{2}{3}\cdot\left(\frac{\partial u_k}{\partial x_k}\right)\cdot\delta_{i,j}\right)\right)$$

If gravity is the only body force, then expanding in Cartesian coordinates (z vertical) yields:

$$\rho\cdot\left(\frac{\partial u}{\partial t}+u\cdot\frac{\partial u}{\partial x}+v\cdot\frac{\partial u}{\partial y}+w\cdot\frac{\partial u}{\partial z}\right)=\rho\cdot g_x-\frac{\partial P}{\partial x}+\frac{\partial}{\partial x}\left[\mu\cdot\left(2\cdot\left(\frac{\partial u}{\partial x}\right)-\frac{2}{3}\cdot\left(\frac{\partial u_k}{\partial x_k}\right)\right)\right]$$
$$+\frac{\partial}{\partial y}\left[\mu\cdot\left(\left(\frac{\partial u}{\partial y}\right)+\left(\frac{\partial v}{\partial x}\right)\right)\right]+\frac{\partial}{\partial z}\left[\mu\cdot\left(\left(\frac{\partial u}{\partial z}\right)+\left(\frac{\partial w}{\partial x}\right)\right)\right]$$
$$\rho\cdot\left(\frac{\partial v}{\partial t}+u\cdot\frac{\partial v}{\partial x}+v\cdot\frac{\partial v}{\partial y}+w\cdot\frac{\partial v}{\partial z}\right)=\rho\cdot g_y-\frac{\partial P}{\partial y}+\frac{\partial}{\partial x}\left[\mu\cdot\left(\left(\frac{\partial v}{\partial x}\right)+\left(\frac{\partial u}{\partial y}\right)\right)\right]$$
$$+\frac{\partial}{\partial y}\left[\mu\cdot\left(2\cdot\left(\frac{\partial v}{\partial y}\right)-\frac{2}{3}\left(\frac{\partial u_k}{\partial y_k}\right)\right)\right]+\frac{\partial}{\partial z}\left[\mu\cdot\left(\left(\frac{\partial u}{\partial z}\right)+\left(\frac{\partial w}{\partial x}\right)\right)\right]$$
$$\rho\cdot\left(\frac{\partial w}{\partial t}+u\cdot\frac{\partial w}{\partial x}+v\cdot\frac{\partial w}{\partial y}+w\cdot\frac{\partial w}{\partial z}\right)=\rho\cdot g_z-\frac{\partial P}{\partial z}+\frac{\partial}{\partial x}\left[\mu\cdot\left(\left(\frac{\partial w}{\partial x}\right)+\left(\frac{\partial u}{\partial z}\right)\right)\right]$$
$$+\frac{\partial}{\partial y}\left[\mu\cdot\left(\left(\frac{\partial w}{\partial y}\right)+\left(\frac{\partial v}{\partial z}\right)\right)\right]+\frac{\partial}{\partial z}\left[\mu\cdot\left(2\cdot\left(\frac{\partial w}{\partial z}\right)-\frac{2}{3}\cdot\left(\frac{\partial u_k}{\partial z_k}\right)\right)\right]$$

(0.24)

Every term in Eq. (0.24) represents a function. Something like velocity in the "x" direction ("u" in Eq. (0.24)) will have an equation that is time dependent, another equation that is dependent on position down the "x" direction, another equation for the "y" direction, and a fourth equation for the "z" direction. All may be material. If you were able to resolve all of the variables into equations for each parameter for a specific flow, then it would likely be worthless for the next flow pattern you want to analyze.

Many people have worked on a closed-form solution to Navier–Stokes, and there is a $1 million USD bounty for anyone who can develop a general solution. In spite of many close calls (the latest in 2006 and 2014), the prize remains unclaimed.

There have been a few useful special cases that throw out the hard terms, but they tend to have so many simplifying assumptions that they lack universal applicability.

0.7.2.2 Bernoulli Equation

If you start with the assumption that the flow is:

- inviscid (i.e., viscosity and friction are zero);
- incompressible;
- irrotational (neither vorticity or rigid body rotation);
- reversible;
- isothermal;
- isentropic (flow not a function of position);
- adiabatic (no heat is gained from or lost to the environment);
- there is no work done on or by the fluid;
- no fluid is added or removed from the control volume being investigated.

Then you can reduce Eq. (0.24) to Eq. (0.25):

$$\frac{\rho \cdot v^2}{2} + \rho \cdot g \cdot h + P_{static} = \text{constant} \tag{0.25}$$

We've assumed that density is constant (i.e., the flow is incompressible), so this equation for two points in the same control volume (e.g., pipe, conduit, section of atmospheric air surrounding a moving body), then Eq. (0.25) becomes Eq. (0.26).

$$\begin{aligned} &\frac{\rho \cdot v^2}{2} + \rho \cdot g \cdot h + P_{static} = \frac{\rho \cdot v^2}{2} + \rho \cdot g \cdot h + P_{static} \\ &\frac{\rho}{2} \cdot (v_1^2 - v_2^2) + \rho \cdot g \cdot (h_1 - h_2) = P_2 - P_1 \end{aligned} \tag{0.26}$$

If you can pick a control volume where the two reference heights are equal, then Eq. (0.26) relates a static pressure change to the change in the square of velocity (the velocity term is referred to as "dynamic pressure" in Eq. (0.27)). This turns out to be quite useful. The square-edged-orifice flow-measurement equations (Chapter 5: Well-Site Equipment) started with Bernoulli's equation. This equation describes why airplanes can fly. It allows HVAC engineers to predict flow in ducts.

Bernoulli's equation is not Navier–Stokes. It cannot have friction, so it is limited to small control volumes. It can't have changes in density. It can't have rotation. It can't have heat loss/gain. If you satisfy all of the assumptions then it can yield amazing results. If you don't satisfy all of the assumptions then you get numbers.

Many of the current generation of fluid mechanics texts include an abomination called the "modified Bernoulli equation." This equation purports to take the Bernoulli equation and scab a "friction loss" term onto the end of it. Nonsense. You cannot derive the Bernoulli equation if you can't eliminate all of the terms involving friction, or flow changes with regard to density or position. The assumptions allow you to discard fully half of the terms in Navier–Stokes and pretty much all of the really hard ones. Then to come back after the fact and add a friction term is just foolishness.

0.7.2.3 No-Flow Boundary

The fluid molecules that touch a foreign substance (i.e., a solid or another phase of fluid or another species of fluid) must have the same velocity as the foreign substance. The easiest example of this is a fluid flowing within a pipe. The fluid will create a "boundary layer" of some nonzero thickness that will have a velocity gradient ranging (more or less as a straight line) from zero at the pipe wall to some small velocity at the edge of the boundary layer. The bulk of the flow will also have a velocity gradient from a minimum at the edge of the boundary layer to a maximum in the centerline of the pipe (Fig. 0.1).

The no-flow boundary may be (and is) physically small, but it is crucial to understanding flow. For example, without this concept, it would be impossible to explain why a flowing gas stream can transport a liquid droplet or how a jet pump works.

0.7.2.4 Similitude

Fluid mechanics has a concept called "similitude" that says that if you can define a ratio of forces that can be resolved to a nondimensional

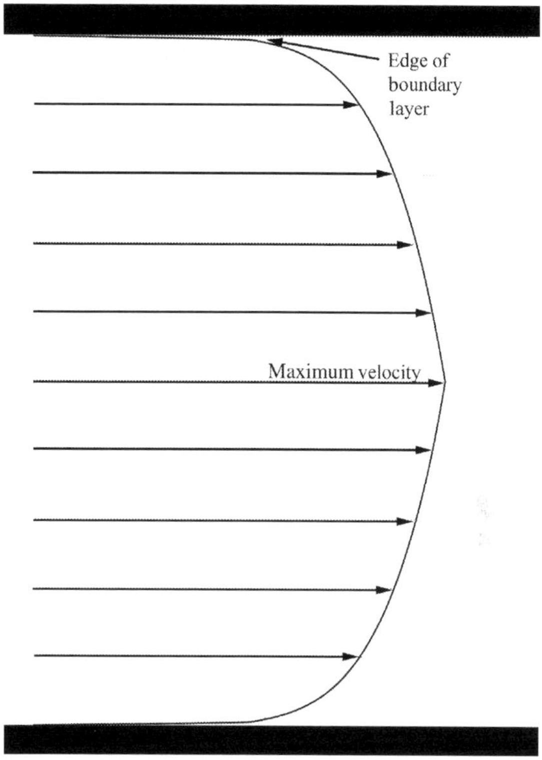

Figure 0.1 Turbulent flow profile.

parameter, then any system with the same value for that nondimensional parameter as a model of that system would have similar responses to external forces. It is generally required to have more than one of the nondimensional parameters match before conclusions based on a model can be thought to be reliably representative of a full system. Some of the more common nondimensional parameters are given in Table 0.11

0.7.3 Pressure and Temperature Measurement

Every subject we discuss in Oil & Gas has pressure and temperature lurking somewhere in the conversation. We cannot measure either one of them directly, but there are surrogates that provide very reliable indications of their values. The change in volume of many substances is linear with regard to temperature change over a specified range, so a mercury thermometer can be put into a calibrated glass tube and give us a simple instrument that reflects temperature, but we can't "measure" temperature.

Pressure is even harder to measure. The most common instrument is a Bourdon tube (Fig. 0.2) where a bent tube is connected to gears through a rack and pinion arrangement connected to a dial indicator. As pressure increases, the tube tries to straighten, rotating the dial toward a higher pressure and vice versa. This pressure indicator is only measuring a part of the total "pressure," the "static pressure," but there are two other components to total pressure ("dynamic pressure" and "hydrostatic pressure") that a Bourdon tube cannot react to.

The gauge in Fig. 0.3 is called a "test gauge." The thin silver circle that goes all the way around the gauge is a mirror. The end of the indicator needle is rotated to perpendicular with the face of the gauge and when you position your eye such that you cannot see a reflection of the needle, then you can read the proper pressure without a parallax error (i.e., a displacement in the apparent position of an object viewed along two different lines of site, a major cause of gauge-reading errors).

There is a rule for reading any gauge, but especially a test gauge: You can only interpolate to the midpoint between the smallest increments. So in the gage in Fig. 0.3, the pressure is 57 psig, not 57.1 or 57.2. If the pressure were a little higher then you would have to decide if it is 57,

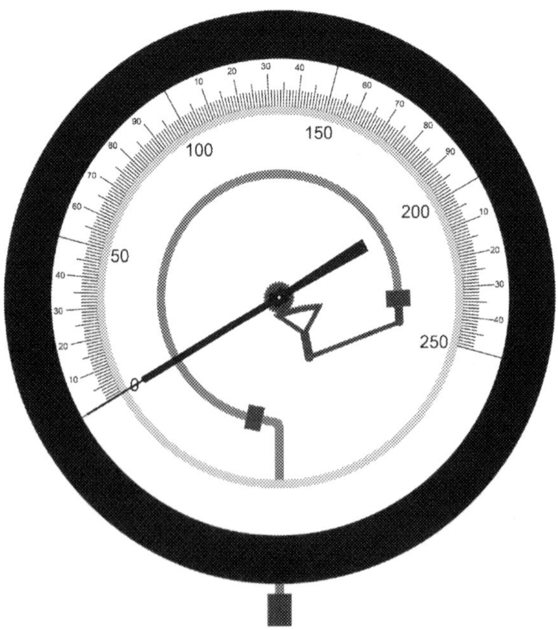

Figure 0.2 Bourdon tube pressure gauge.

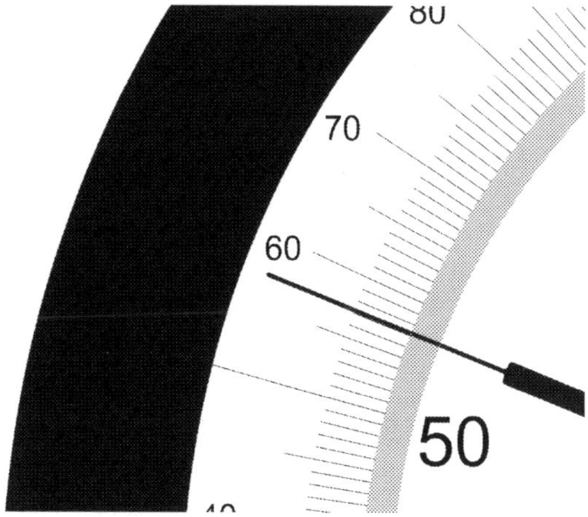

Figure 0.3 Analog gauge parallax.

57.5, or 58, but it is never 57.3. If you need to differentiate 57.3 from 57.4, then you need a gauge marked in 0.1 psi increments. Pressure gauges can only give you static pressure above atmospheric pressure (i.e., psig, kPag, or barg), not absolute pressure (i.e., psia, kPaa, or bara). Occasionally a clever measurement tech will shift the face of a gauge so that when it is depressurized it reads some value other than zero which is purported to be local atmospheric pressure—it almost never is actually local atmospheric pressure and this is a horrible practice.

0.7.4 Total Pressure

Total pressure is made up of three components:

$$P_{total} = P_{dyanamic} + P_{static} + P_{hydrostatic}$$
$$P_{total} = \frac{\rho_{flowingFluid} \cdot \vec{v}^{\,2}}{2} + P_{static} + \rho_{fluidMixAbove} \cdot g \cdot h \tag{0.27}$$

Within a system of incompressible fluids flowing at low velocities, total pressure must remain a constant over short distances. Systems flowing at a significant fraction of the speed of sound (i.e., something over 30% of sonic velocity) will not have a constant total pressure, but those cases are uncommon enough to generally start with the assumption that total pressure is constant over short distances.

For a system at rest in effectively still air, the $P_{hydrostatic}$ is local atmospheric pressure and the total pressure is $P_{static} + P_{atmospheric}$.

0.7.5 Pressure Continuum

In field operations it is common to talk about "high pressure" or "low pressure" and typically these terms are relative to some value that is clear in the speaker's mind, but not communicated in the current discussion. This often leads to suboptimal decisions or poor understanding of the issues.

While doing work on evaporation, three distinctly different regions on the plot were noted (Fig. 0.4).

It appears that above 145 psia (10 bara), this particular data could be reasonably represented as a straight line. It is also clear that below 43.5 psia (3 bara) the high-pressure straight line had no relationship to the data and in fact no straight line would do an adequate job of representing the data. Between these two points was a transition region that matched a straight line better than the low-pressure region, but still not very well. I found this to be intriguing and generated a number of plots including "entropy vs pressure," and "pipe performance vs pressure." All of the data reviewed could be divided into the three categories we saw with the evaporation data with the lines in the same places. The result was:

- >145 psia (10 bara) → High pressure. Equations developed for oil-fields tend to work very well, it is usually safe to ignore evaporation, and dehydration can be a reasonable process.

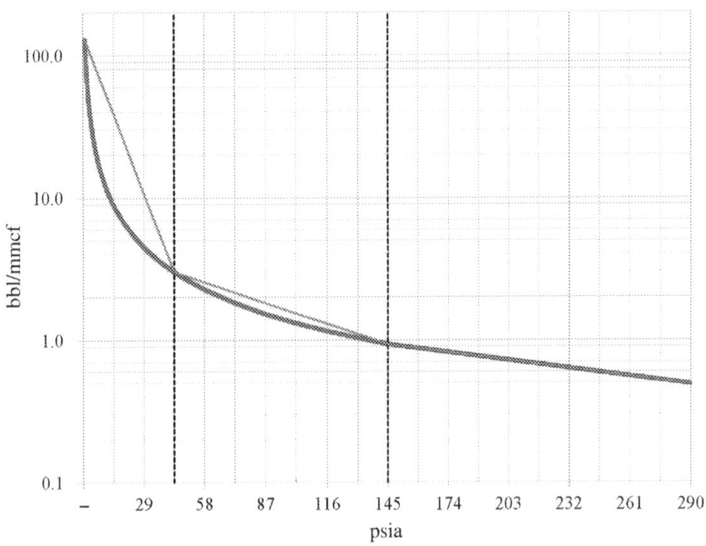

Figure 0.4 Water content vs pressure.

- 43.5 psia (3 bara) to 145 psia (10 bara) → Normal pressures. Error band on oilfield equations tends to increase, ignoring evaporation has higher risk, dehydration becomes less effective.
- < 43.5 psia (3 bara) → Low pressures. Traditional equations range from "mostly useless" to "misleading," evaporation represents a major challenge to all equipment performance and must be considered in any facilities decision, the amount of water in the gas makes dehydration too expensive to be considered (the problem is the mass of water that must be cooked out of the dehydration media in the recharge step becomes too large for reasonably sized equipment and fuel load, see Chapter 9, Interface to Plants, for a discussion).

There are other changes that can occur at very high pressure (probably above 10,000 psia (70,000 kPaa)), but environments with that magnitude of pressure should not be managed via rules of thumb.

0.8 STANDARD CONDITIONS

It is often useful to pretend that a gas is at a constant temperature and pressure as it moves through various processes. It is very important to keep in mind that the pressures and temperatures you are pretending the gas is at are simply a fantasy that does not control the physical forces and reactions on or of the gas. For example, if you calculate a gas velocity by dividing a volume flow rate at standard conditions by the flow area, you will get a number. You should be aware that the number you get will have velocity units, but will not represent any velocity that actually exists in the world. Consequently this book will never cancel ft^3 with SCF or vice versa. SCF is an imaginary unit and ft^3 is a physical unit.

The fantasy conditions that industry uses are called "standard temperature and pressure (STP)." This is an unfortunate term because the exact values for pressure and temperature in the fantasy world are anything but "standard." Table 0.12 has a sampling (far from complete, see Wikipedia, 3, for a longer list that is still far from complete) of some of the values that are regularly used for STP.

The reader might assume that I don't approve of reporting gas at standard conditions and nothing could be further from the truth. Using volume flow rates at standard conditions (for purposes of this discussion we

Table 0.12 Selection of STP Definitions

	Pressure	Temperature
Undergrad Chemistry Texts	14.696 psia (101.325 kPaa)	60°F (15.56°C)
Gas Measurement (USA)	14.73 psia (101.56 kPaa)	60°F (15.56°C)
EPA Reporting	14.696 psia (101.325 kPaa)	20°C (68°F)
NM and LA State Reporting	15.025 psia (103.59 kPaa)	60°F (15.56°C)
ISO	101.33 kPaa (14.696 psia)	0°C (32°F)
Gas Measurement (Europe)	100.0 kPaa (14.5 psia)	15°C (59°F)
Gas Measurement (Queensland)	101.325 kPaa (14.696 psia)	15°C (59°F)

will call that "SCF") is an elegant solution to the complementary problems of: (1) we do not have a way to measure mass flow rate (see Chapter 5, Well-Site Equipment, for a discussion); and (2) volume flow rate at actual conditions varies widely for the same mass flow rate at varying pressures and temperatures. Using SCF per unit time solves both of these problems nicely: (1) we have instruments that can infer a volume flow rate from parameters that we can measure; and (2) SCF remains constant (for a given mass flow rate) across any range of pressure and temperature. Using SCF is a fantastic solution to the requirement to do commerce at varying pressures (e.g., as a homeowner you purchase 100 SCF of gas delivered at a pressure of 15 psia, the molecules that made up your delivery left a processing plant at 1000 psia, and they might have left the gas well at 50 psia, it is the same number of molecules and it is the same 100 SCF in all three cases). SCF can be called "a surrogate for mass flow rate" because, like mass flow rate, a fixed number of gas molecules are contained in a given delivery unit (i.e., lbm or SCF).

Volume flow rate at standard conditions is a useful tool, as long as there is communication. Just because a government regulator defines "standard pressure" as 15.025 psia (103.59 kPaa) that only means that when you report to that bureau you have to pretend your gas was at 15.025 psia (103.59 kPaa) instead of pretending another value. Your delivery contract may specify that "standard pressure" is 14.73 psia (101.56 kPaa) (which is very common). At the same time you may be subject to extracting flow data from a corporate database that defines "standard pressure" as 14.5 psia (100 kPaa). It is incumbent on the user of flow data to know: (1) how the data is stored; (2) what fantasy value is required; and (3) how to convert from one to the other. The last item is pretty easy if you take care. We know that if we don't add or remove fluids from an enclosed system, mass flow rate must be the same everywhere. We also know that mass flow rate is product of volume flow rate and density, so use the relationship in Eq. (0.28):

$$\dot{m} = q_1 \cdot \rho_1 = q_2 \cdot \rho_2 \therefore q_2 = q_1 \cdot \frac{\rho_1}{\rho_2} \tag{0.28}$$

This says that to change from any pressure and temperature to any other pressure and temperature, divide the old mass flow rate by the new density. Looking at Eq. (0.28) you can see that many of the terms can be canceled (Eq. (0.29)).

$$q_2 = q_1 \cdot \frac{\rho_1}{\rho_2} = q_1 \cdot \frac{\left(\frac{P_1 \cdot SG}{R_{air} \cdot T_1 \cdot Z_1}\right)}{\left(\frac{P_2 \cdot SG}{R_{air} \cdot T_2 \cdot Z_2}\right)} = q_1 \cdot \frac{P_1}{P_2} \cdot \frac{T_2}{T_1} \cdot \frac{Z_2}{Z_1} \tag{0.29}$$

For any conversion from one version of STP to another, the compressibility will cancel. If the temperature is close enough it will also cancel. For nonfinancial purposes, this calculation can usually be reduced to Eq. (0.30):

$$q_2 = q_1 \cdot \frac{P_1}{P_2} \tag{0.30}$$

If we wanted to convert 100 MSCF/day from STP of 14.73 psia and 60°F to 14.5 psia at 59°F we would do the calculation in Eq. (0.31).

$$\begin{aligned} q_{15.025} &= 100 \cdot \frac{\text{MSCF}}{\text{day}} \cdot \frac{14.73 \cdot \text{psia}}{14.5 \cdot \text{psia}} \cdot \frac{519.63R}{518.63R} \cdot \frac{0.997533}{0.997510} = 101.393 \cdot \frac{\text{MSCF}}{\text{day}} \\ q_{15.025} &= 100 \cdot \frac{\text{MSCF}}{\text{day}} \cdot \frac{14.73 \cdot \text{psia}}{14.5 \cdot \text{psia}} = 101.586 \cdot \frac{\text{MSCF}}{\text{day}} \end{aligned} \tag{0.31}$$

You have to decide if your calculation can handle a 0.19% error or not (most can, some can't).

You cannot use SCF to calculate physical properties (such as velocity). To calculate physical properties of a gas flow, you must convert the volume flow rate from imaginary flow rates to actual, physical flow rates using the technique in Eq. (0.29) or (0.30) before starting on further calculations.

0.9 EMPIRICAL EQUATIONS

Empirical equations take advantage of coincidences in a physical phenomenon that can be exploited. For example, if you are interested in knowing how much water will be needed to fill a pipeline for a hydrostatic pressure test, you can calculate the volume by Eq. (0.32).

$$V = \left(\frac{\pi}{4}\right)\cdot \text{ID}^2 \cdot \text{Len} \tag{0.32}$$

Since we would typically source the water in "barrels," and this equation would yield inch^2 times feet and would need to some unit conversions (Eq. (0.33)).

$$V = \left(\frac{\pi}{4}\right)\cdot \text{ID}^2 \cdot \text{Len}\cdot \left(\frac{\text{ft}^2}{144\cdot \text{in}^2}\right)\cdot \left(\frac{\text{bbl}}{5.61\ \text{ft}^3}\right) \tag{0.33}$$

If we fix the length at 1000 ft, then the constants and unit conversions resolve to $0.9722 \approx 1.0$ and Eq. (0.33) can be approximated as Eq. (0.34).

$$V_{bbl\_per\_1000\text{ft}} \approx \left(\frac{\text{ID}}{\text{in}}\right)^2 \tag{0.34}$$

Notice a crucial element of empirical equations—they only "work" with exactly one set of units. Using Eq. (0.34), it is easy to estimate that you'll need 144 bbl (22.9 m^3) of water to fill 1000 ft (304 m) of 12 in (DN 300) ID pipe (Eq. (0.33) gives you 139.9 bbl (22.24 m^3), so Eq. (0.34) allows some extra for partially filled trucks and spillage), but what if you input millimeters and meters into the equation? The answer is not so simple (Eq. (0.35)).

$$V_{SI} = (304.8\cdot \text{mm})^2 = 92416\cdot ???? \tag{0.35}$$

Squaring the ID gives you a number, but what the heck does it mean? Let us look at an SI version of Eq. (0.33).

$$V = \left(\frac{\pi}{4}\right)\cdot \text{ID}^2_{mm}\cdot \text{Len}_m = \left(\frac{\pi}{4}\right)\cdot\left(\frac{304.8\cdot \text{mm}}{\frac{1000\cdot \text{mm}}{\text{m}}}\right)^2\cdot 304.8\cdot \text{m} = 22.24\cdot \text{m}^3 \tag{0.36}$$

If the length is fixed at 1000 ft (304.8 m) then you can collapse all of the non-ID terms into 0.0002394, but so what? This is not a memorable number and it is much easier to just use the actual calculation. In the field, a person has a chance of successfully squaring "12" and multiplying it by the number of 1000 ft increments in the line in their head. Most of us have no chance of being able to successfully determine how many 304.8 m increments there are, let alone squaring a three digit number and multiplying it by an awkward constant without a calculator. This equation in millimeters would be far worse than using the actual volume calculation.

This trivial example is important because Oil & Gas is full of empirical equations, and every one of them only works with exactly one set of units. An equation that works with 22% input as 22.0 will not work with 22% input as 0.22 (regardless of how "correct" 0.22 is and how "incorrect" 22.0 is). An equation that wants length in miles will not work with length in meters, kilometers, or feet. Often the units defy a logical explanation. Fine. Don't try to find logic, just use the equation with the units that take advantage of the coincidence.

Any equation that contains a numerical constant that is not a small integer or a physical constant (e.g., π or the ideal gas constant) is likely an empirical equation and requires a precisely defined set of units.

Another important telltale is when units don't cancel within a term that is then raised to a noninteger exponent (or a logarithm or included in a term that is used as an exponent). Look at Eq. (0.11) (reproduced as Eq. (0.37)).

$$P_{bot} = \frac{P_{top}}{\mathrm{psi}} \cdot \exp\left(\frac{0.01875 \cdot \mathrm{SG} \cdot \frac{\mathrm{h}}{\mathrm{ft}}}{\frac{T_{avg}}{R} \cdot Z_{avg}}\right) \qquad (0.37)$$

$$\mathrm{psi} = \mathrm{psi} \cdot \exp\left(\frac{\mathrm{ft}}{\mathrm{R}}\right)$$

There is no way to take an exponent of a "ft/R." That is a dead giveaway that this is not a closed-form equation, but an empirical equation.

REFERENCES

DOE. "Lower and higher heating values of fuels." Web. 14 April 2016.

GPSA. *GPSA Engineering Data Book.* (2004). F.P.S: GSAP.

Historical, 1. "History of the 42-gallon oil barrel—American Oil & Gas Historical Society." American Oil Gas Historical Society. 2015. Web. 14 April 2016.

Historical, 2. "Nellie Bly oil drum—American Oil & Gas Historical Society." American Oil Gas Historical Society. 2015. Web. 14 April 2016.

Webster (n.d.). Retrieved April 21, 2016, from http://www.merriam-webster.com/dictionary/biogenic.

Wikipedia, 1. "Equation of state." *Wikipedia.* Wikimedia Foundation. Web. 14 April 2016.

Wikipedia, 2. "Atmospheric pressure." *Wikipedia.* Wikimedia Foundation. Web. 14 April 2016.

Wikipedia, 3. "Standard conditions for temperature and pressure." *Wikipedia.* Wikimedia Foundation. Web. 14 April 2016.

NOMENCLATURE

Symbol	Name	fps Units	SI Units
c_p	Specific heat at constant pressure	BTU/lbm/R	J/gm/K
c_v	Specific heat at constant volume	BTU/lbm/R	J/gm/K
$f_{vortex\ shedding}$	The frequency that von Karman Streets are shed around a bluff body in flow	decimal	decimal
g	Gravitational constant	32.174 ft/s^2	9.81 m/s^2
g_c	Unit converter	32.174 ft-lbm/s^2-lbf	N/A (for now)
h	Height	ft	m
ID	Inside diameter	in	mm
k	Adiabatic constant (c_p/c_v)	None	None
Len	Length	ft	m
$L_{characteristic}$	Some length related to an analysis, can be OD, length, hydraulic diameter, etc.	ft	m
m	Mass	lbm	kg
$\dot{m}$	Mass flow rate	lbm/s	kg/s
MW	Molecular weight	lbm/lb-mole	gm/gm-mole
MW_{air}	Molecular weight of air	28.962 lb/lb-mole	28.962 gm/gm-mole
n	Number of moles		
P	Pressure	psia	Paa, bara
P_{vapor}	Vapor pressure of a component of a liquid	psia	Paa, bara
q	Volume flow rate at standard conditions	SCF/day	SCm/day
$\overline{R}$	Universal gas constant	1545.3 ft*lbf/R/lb-mole	8.314 J/K/mole
R_{air}	Specific gas constant for air	53.355 ft*lbf/R/lbm	287.068 J/K/kg
R_{gas}	Specific gas constant for any specified gas	R_{air}/SG_{gas}	R_{air}/SG_{gas}
SG	Specific gravity	None	None
v	Velocity	ft/s	m/s
v_{sonic}	Sonic velocity	ft/s	m/s
V	Volume	ft^3	m^3
$\dot{V}$	Volume flow rate at actual conditions	ft^3/s	m^3/s

(Continued)

(Continued)

Symbol	Name	fps Units	SI Units
W	Water vapor in a gas	lbm/MMSCF	mg/SCm
W	Weight	lbf	N
x	Mole fraction of a component in a mixture	None	None
y	Any intrinsic property of a element or compound (e.g., specific heat at constant pressure)	Various	Various
Z	Compressibility	None	None
ρ	Density	lbm/ft^3	kg/m^3
σ	Interfacial tension	cP	Pa × s

Subscripts

ASL	Above sea level
atm	Atmospheric
gas	Variable is referring to the gaseous phase
i	Array counter
liquid	Variable is referring to the liquid phase
SI	Metric version of equation
std	Indicates variable is at standard conditions
unit	Indicates variable may be in nonstandard units
water	Variable is referring to the characteristics of pure water at atmospheric pressure and 60°F (15.6°C)
1	Upstream conditions, or first state
2	Downstream conditions, or second state

UNITS

Symbol	Name	Type Unit	Equivalent Unit
bbl	Barrel (42 US gallons)	fps—volume	$0.1590\ m^3$
BTU	British thermal unit	fps–energy	1.055 kJ
°C	Celsius	SI–temperature	°C*9/5 + 32 = °F
cm	Centimeter	SI—length	0.394 in
cm^2	Square centimeters	SI—area	$0.155\ in^2$
dyne	Dyne	SI—force	2.248E − 6 lbf

(Continued)

(Continued)

Symbol	Name	Type Unit	Equivalent Unit
°F	Fahrenheit	fps—temperature	(°F − 32)*5/9 = °C
ft	Foot	fps—length	0.3048 m
ft^2	Square feet	fps—area	0.0929 m^2
ft^3	Cubic feet	fps—volume	0.02832 m^3
gm-mole	Gram mole (mass of one mole of gas)	SI—molar mass	lb-mole
in	Inch	fps—length	25.4 mm
in^2	Square inches	fps—area	0.000645 m^2
J	Joule	SI—energy	947.8E − 6 BTU
K	Kelvin (°C + 273.15)	SI—temperature	R*5/9 = K
kg	Kilogram	SI—mass	2.205 lbm
km	Kilometer	SI—length	3281 ft
lbf	Pounds force	fps—force	4.448 N
lbm	Pounds mass	fps—mass	0.454 kg
lb-mole	Pound mole (mass of one mole of gas)	fps—molar mass	gm-mole
mm	Millimeter	SI—length	0.0393 in
m	Meter	SI—length	3.281 ft
m^2	Square meters	SI—area	10.764 ft^2
m^3	Cubic meters	SI—volume	35.314 ft^3
N	Newton	SI—force	0.225 lbf
R	Rankine (°F + 459.67)	fps—temperature	K*5/9 = R
s	Second	all—time	
SCF	Standard cubic feet	fps—volume	0.283 SCm
SCm	Standard cubic meters	SI—volume	35.314 SCF

Unit Prefixes

M	fps—Thousand, oilfield prefix = 1000
MM	fps—Million, oilfield prefix = 10^6
B	fps—Billion, oilfield prefix = 10^9
T	fps—Trillion, oilfield prefix = 10^{12} (duplicates an SI unit, but both have the same magnitude)
k	SI—kilo, 10^3
M	SI—Mega, 10^6 (limited use in this book due to confusion with oilfield units)
G	SI—Giga, 10^9
T	SI—Tera, 10^{12}
P	SI—Peta, 10^{15}
E	SI—Exa, 10^{18}

EXERCISES

1. An upthrust fault connected three isolated reservoirs (Fig. 0.5) sometime in a past epoch and the sour gas mixed with the sweet and CBM gas. The geologists say the contribution to the reservoir was:
   - CBM—39%
   - Sweet gas—46%
   - Sour gas—15%

   Using the example gas analyses in this chapter, find:
   - H_2S fraction
   - Gas SG
   - Net heating value
   - Adiabatic constant (k)
2. We drill into the "mixed" formation in Fig. 0.5 and find the gas analysis in Table 0.13. Using the sample analyses in this chapter, what portion of the discovered gas must have come from each of the three formations?
3. The dark sections of Fig. 0.6 represent a 1.07 SG liquid. The intermediate section between point "2" and point "3" is trapped air. Assume that atmospheric pressure is 12.0 psia. Find:

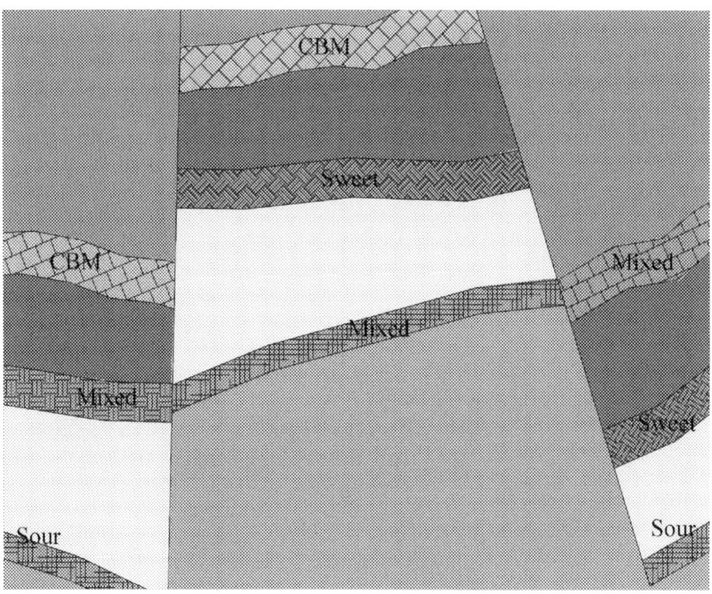

Figure 0.5 Downhole gas mixture.

Table 0.13 Gas Analysis for Mixed Formation

	Mole Fraction (%)	MW	SG	*k*
C1	85.0448	13.64	0.4711	1.1118
C2	5.2538	1.58	0.0545	0.0627
C3	2.3893	1.05	0.0364	0.0270
i-C4	0.2698	0.16	0.0054	0.0030
n-C4	0.8061	0.47	0.0162	0.0088
i-C5	0.1066	0.08	0.0027	0.0011
n-C5	0.4297	0.31	0.0107	0.0046
C6 +	0.0050	0.00	0.0002	0.0001
H_2S	0.4930	0.17	0.0058	0.0065
CO_2	5.2019	2.29	0.0790	0.0673
	100	19.75	0.6820	1.2929

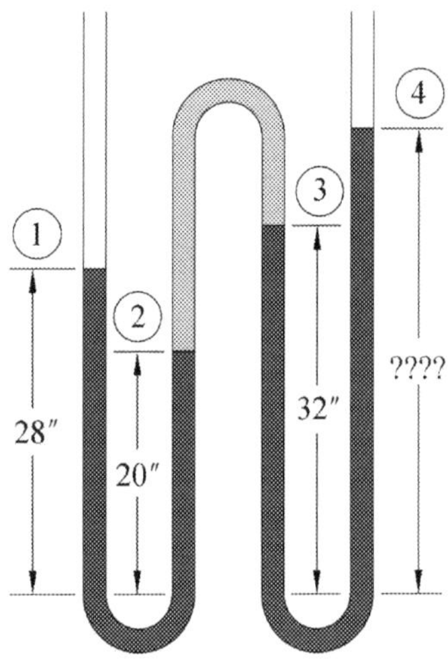

Figure 0.6 Manometer exercise.

- Height of forth leg
- Pressure in intermediate section

4. A pipe has a venturi in an airstream (Fig. 0.7). Pressure at P_1 is 100 psia (689 kPaa). Pressure at P_2 is 97 psia (669 kPaa). The air is isothermal at 80°F (26.7°C). Find:

- Change in velocity between the plane of instrument P_1 and P_2
- State the assumptions required to evaluate this flow

5. A company has gas production in several countries (Table 0.14) and their home country requires worldwide production to be aggregated

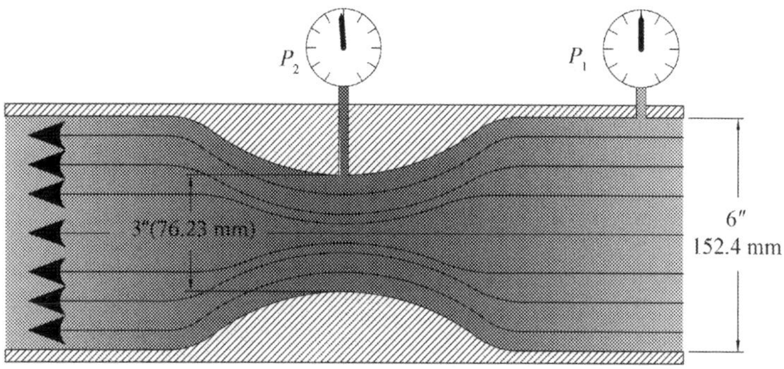

Figure 0.7 Venturi meter.

Table 0.14 Production by Company

Country	Pressure Base	Temperature Base	Reported Rate
Australia	101.325 kPaa	15°C	21,662 e3m^3/day
United Kingdom	101.5 kPaa	0°C	29.7 Mm^3/day
United States	14.73 psia	60°F	2.7 MMSCF/day

into a single volume stated in kSCm/day at 100 kPaa (14.5 psia) and 15°C (59°F). Find:

- Aggregated gas volume
- Error that would be exhibited if you assumed that every country used your home country's values for "standard"

6. While blowing down an air receiver, the air in the tail pipe reached a velocity of 1139 ft/s (347.2 m/s) at a pressure of 264 psia (1.82 MPaa) and a temperature of 80°F (26.7°C). If atmospheric pressure is 12 psia, what is the total pressure?

CHAPTER ONE

Gas Reservoirs

Simpson's First Postulate: Every activity, joint of pipe, piece of equipment, and facility should have the goal of maximizing reservoir profitability—any activity which ignores that goal is going to result in sub-optimum performance.

Contents

Practical Onshore Gas Field Engineering.
DOI: http://dx.doi.org/10.1016/B978-0-12-813022-3.00001-8

1.1 SOURCE OF HYDROCARBONS

The Oil & Gas industry has always considered hydrocarbons as being biogenic (i.e., produced by living organisms) and it is certain that some proportion of the hydrocarbons recovered from wells is biogenic.

However, the idea that hydrocarbons can be formed in the core of the earth and migrate through the mantle has been gaining a resurgence in acceptance. This abiotic (i.e., not derived from organisms) gas and oil would provide some unknown quantity of supplemental hydrocarbons that would have a potential for recovery and commercialization.

Exactly what that quantity would be requires some in-depth analysis.

1.1.1 Recoverable hydrocarbons explained

Recoverable hydrocarbons must first be hydrocarbons, then they must be located in a rock stratum that allows them to be stored. Finally, there must be some sort of containment to keep them from leaking out of the storage strata.

In the Oil & Gas industry, we call these entities "source rock," "reservoir rock," and "cap rock" (Fig. 1.1). Once a reservoir is delineated by having a source, a reservoir, and a containment, it requires a flow path to an outlet. Most of the time this flow path takes the hydrocarbons around the edges of the cap rock and they leak to the surface of the earth.

Occasionally, the cap rock has an effective seal on the reservoir. When the Oil & Gas industry comes along and drills a well and stimulates the reservoir, the flow path becomes a commercial transaction—this is the reason that the Oil & Gas industry exists.

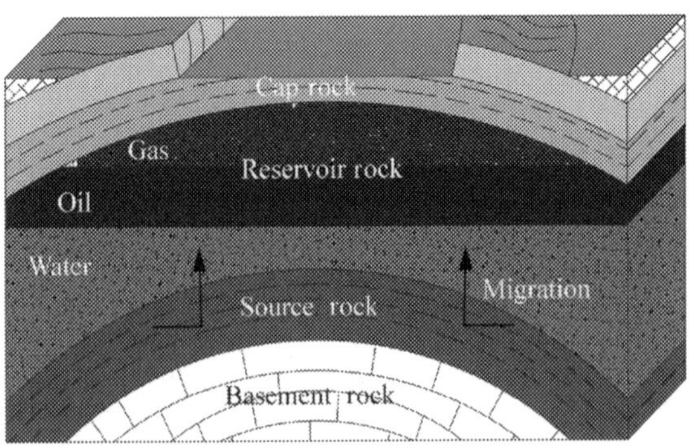

Figure 1.1 Source rock, reservoir rock, and cap rock.

1.1.2 Biotic hydrocarbons

The total amount of carbon that is tied up within living biological material on the earth is about 4 trillion metric tons (tonne, 4.4 trillion US tons). Something like 105 billion tonne of carbon are discarded by their living host each year (Wikipedia, 1, 2016)—the leaves of deciduous trees fall to the ground each autumn, some number of animals, plants, bacteria, and viruses die each year, all organisms emit some amount of solid, liquid, and gaseous waste, and so on.

This carbon will eventually become CO_2 and water vapor in an aerobic environment and mostly CH_4 in an anaerobic environment. The discarded biomass is approximately equally divided between land and sea.

Aerobic decomposition is exothermic and tends to be fairly rapid (i.e., from the onset of decomposition to a sterile, carbon-free mass takes weeks or months) and is largely free of the worst odors. Aerobic decomposition converts the carbon in the waste material to CO_2 and H_2O.

Anaerobic decomposition is endothermic, much slower and very smelly. It can take hundreds or thousands of years for an undisturbed anaerobic process to run to completion. The time required is largely a function of the energy input to the process. Anaerobic decomposition converts the carbon in the waste material to CH_4 and amounts of CO_2, and H_2O limited by available oxygen.

It is common for decomposing organic material to accumulate on the seafloor away from thermal vents and begin the decomposition process at very low energy input. While the mass is decomposing it sometimes happens that a storm event or a seismic event will cover the biomass with sand. Above the sand over many years, more organic material will collect that can eventually turn into shale which is one of the most common cap rocks.

Now we have a source, a reservoir, and a containment. Over millions of years, the methane product of decomposition can be converted through the application of heat and pressure into heavier hydrocarbons. The sealed volume can move upward or downward or it can tip (usually allowing the hydrocarbons to spill out of the reservoir, but not always) due to tectonic and seismic activity.

There is no competent theory to allow prediction of the proportion of biomass that will be subject to anaerobic decomposition. We can make some (arbitrary, but conservative) assumptions about the proportions:

- At least 1% of the biomass on land undergoes anaerobic decomposition (this includes human sanitary landfills, lakes, and wetlands, and in the stomachs of many animals and insects). One percent of half of 105 billion tonne per year is 0.525 billion tonne of carbon per year subjected to anaerobic decomposition.

- If we assume that 50% of the biomass in the sea is subjected to anaerobic decomposition, half of half 105 billion tonne per year is 26.25 billion tonne per year.
- With these assumptions, 26.775 billion tonne of carbon are converted to methane each year through anaerobic decomposition. Adding four molecules of hydrogen to every molecule of carbon gives us 35.7 billion tonne of methane per year.
- The density of methane is 0.042 lbm/SCF, so 35.7 billion tonne per year is 5 TSCF/day.
- World natural gas production (EIA, 2016) by the Oil & Gas industry in 2014 was 0.332 TSCF/day.

Virtually all of this biotic methane will escape to the atmosphere or be consumed by microbes, but some will be trapped by sediments. Again, there is no competent theory to allow a reasonably accurate method of determining the mix of trapped and escaped methane, it is a matter for conjecture, but it is unlikely that as much as 0.1% is trapped. A more conservative number might be 0.005%.

That would mean that something on the order of 2.5 BSCF (72 MSCm) per day would eventually be captured for possible future recovery. If this goes on for 360 million years, then the upper limit on the mass of hydrocarbons from biological sources stored in natural reservoirs is on the order of 6.9 Zettagrams (6.9×10^{21} g or 6.9×10^{15} tonne).

The US Geological Survey (USGS) estimates (Schmoker and Dyman, 2016) that as of the end of 2012, cumulative worldwide production (EIA, 2016) has been 1218 billion barrels of crude and condensate and 2438 trillion SCF of natural gas (192 petagrams, 192×10^9 tonne). The same source shows proved reserves of oil and gas to be on the order of 8 petagrams (8×10^9 tonne). This indicates that we have found, developed, and/or produced 200 petagrams (200×10^9 tonne) or 0.0029% of possible biotic hydrocarbons production.

These numbers seem to refute the idea that we have produced so much fossil fuel that it has to have another source. The five one-thousandths of a percent captured is a guess without much basis, but is unlikely to be as far off as the duration of 360 million years.

Some estimates have been made that assume seeps started within 100 years of the first organic waste being generated, so the duration of the seeps could be closer to 2 billion years. In other words, the conventional wisdom placing the source of all produced hydrocarbons as anaerobic decomposition remains plausible.

1.1.3 Abiotic hydrocarbons

The concept that petroleum and natural gas are formed by inorganic means has been proposed many times. One of the major theories was put forth by Thomas Gold in 1955 and expanded upon in the 1980s and 1990s. His theory was that elemental carbon and elemental hydrogen in the core of the earth at very high pressure and temperature would be adequate to facilitate the formation of hydrocarbons.

Simple hydrostatic pressure at 75 miles (120.7 km) below sea level would be about 170,000 psi (1,200,000 kPa) and the temperature of the earth's core is estimated to be about 10,800°F (6000°C).

DeRosa (2007) hypothesized that the presence of biological debris in petroleum products was the result of microbes feeding on the oil and gas, rather than a waste product of microbes feeding on organic material and this debris did not eliminate the possibility of abiotic hydrocarbons.

Although this theory and the competing theory that hydrocarbons arrived as space debris have both been discredited by modern science, we've seen this sort of group-think in other fields of inquiry recently and the collective opinion of "modern scientists" carries less weight every year.

A recent study by The Carnegie Institution's Geophysical Laboratory (Phys.org, 2016) was able to form ethane propane, butane, molecular hydrogen, and graphite (carbon) from methane at high temperature and pressure without any biological agents. They further found that the process was reversible, where ethane formed methane at the pressure and temperature of their experiment, which was an unexpected result.

1.1.4 Do abiotic hydrocarbons matter to the oil & gas industry?

When extracting hydrocarbons, we have no way to determine if they are abiotic or biotic in source. Nor do we have any economic reason to care. Abiotic hydrocarbons are certainly not required to support the volumes of petroleum products that have been recovered to date, but recent studies indicate that the scientific community could easily have been premature in rejecting the very concept of abiotic hydrocarbons.

At the end of the day, it really does not matter if the gas and oil that arrives at a processing facility came from "squashed dinosaurs" or from chemical reactions within the core of the earth—they still had to migrate from a "source" to a "reservoir" that is "capped." It doesn't hurt anything to call them "fossil fuels."

With the specificity of the required storage environment, nearly all abiotic hydrocarbons would have migrated to the surface without being trapped over geologic time just like biotic hydrocarbons did/do.

1.2 RESERVOIR ROCKS

As discussed in the last section, a hydrocarbon accumulation must have: (1) a source of hydrocarbons (source rock); (2) a means to store hydrocarbons (reservoir rock); and (3) a barrier to leakage (cap rock). Additionally the accumulation must either have a path for hydrocarbon migration or the ability to accept an imposed path for migration.

Historically we have thought about hydrocarbons being stored in some sort of void space. This space can be tiny like the spaces between grains of sand in sandstone or huge like the vugs and cracks in limestone. The public perception of "lakes of oil" underground, while not impossible, has never been discovered thus far.

Some of the unconventional gas and oil production is requiring that we rethink the concept that gas is only stored in void space. In coalbed methane (CBM) the preponderance of the gas is adsorbed to the surface of the coal. CBM reservoirs have "porosity" (see the following section) values far smaller than the historical minimum that was considered viable for a gas well. Shale reservoirs tend to have a mixture of organic material that includes adsorption sites and inorganic rock that has void space.

1.2.1 Porosity

Porosity is a measure of the void space in a rock.

$$\phi = 1 - \frac{V_{solidrock}}{V_{entirereservoir}} \tag{1.1}$$

Fig. 1.2 shows an impossible combination of sand grains stacked in such a way as to maximize the void space. Since nature is always tending

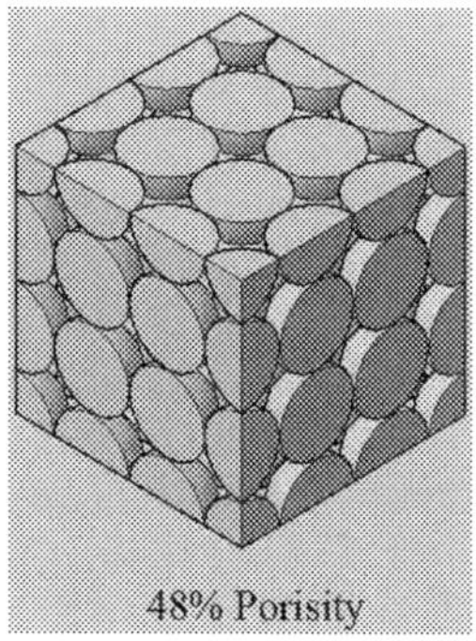

Figure 1.2 Stacked sand.

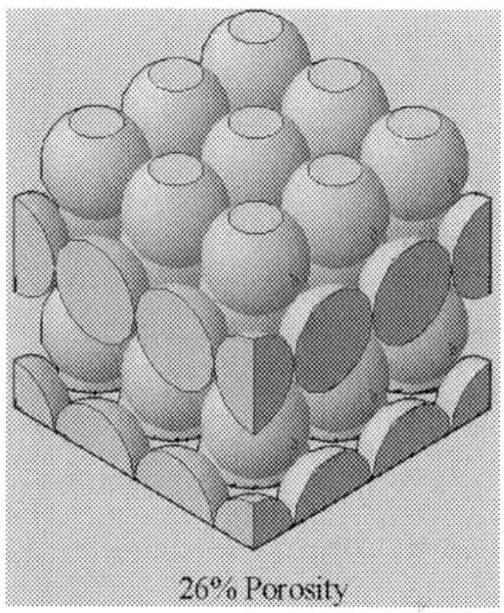

Figure 1.3 Stylized sand stacking.

toward minimizing potential energy, this theoretical maximum void space is impossible. Fig. 1.3 is a stylized representation of these perfectly spherical "sand grains" stacked in a more likely configuration. This shows that even with minimizing potential energy during consolidation, you can get as much as one-fourth of the reservoir volume in a sandstone as void space. With real sand grains not being perfectly sorted into the same size and not being perfectly spherical, porosity in sandstone is rarely much more than 20% and can be much less.

Limestones and other carbonate strata tend to have a very fine grain structure that has very low porosity, but water that is slightly acidic can scour vugs and channels in the matrix. These vugs can hold large quantities of foreign material, but must be interconnected to be useful as a reservoir rock. It is common for a carbonate reservoir to have as much as 50% porosity, and the majority of high production-rate reservoirs (e.g., the North Sea reservoirs) are carbonates.

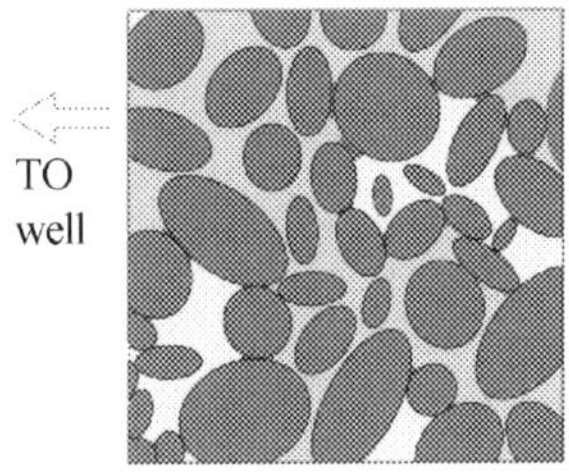

Figure 1.4 Porosity in sandstone.

Coal and shale reservoirs are called "unconventional" because they are. The porosity in coal beds tends to be around 0.2%, far too low to be an effective reservoir rock in conventional terms. This very low porosity is entirely contained within the cleat system and the material divided by the cleats has approximately zero porosity. In CBM reservoirs, the storage method is "adsorption" which is described in Section 1.6.2, not pore-volume storage.

Shale on the other hand is a mixture. Generally, 60% of the original gas in place (OGIP) and 100% of the original oil in place (OOIP) is stored in the pore volume of interbedded sandstones in the shale matrix. The remaining 40% of the OGIP is adsorbed to the surface of the organic material in the shale. As will be discussed in Section 1.6.3, this adsorbed gas is an effective source of energy to facilitate the movement of the pore-volume liquids.

Porosity can either be "connected" or "disconnected." In Fig. 1.4 the dark sections are connected porosity that allows flow from one space to the next. The light area shows disconnected porosity that does not communicate with other pore spaces. At initial discovery the pressure in the connected pore space and the disconnected pore space will be approximately equal. Over time and production, the pressure in the connected pores is depleted, creating a differential pressure across the media cementing sand grains and can cause this material to fail and produce the trapped hydrocarbons. Because of this ability to fail the trapping mechanism we generally treat all pore volume in sandstones as connected. Vuggy porosity and interbedded porosity are far more complex and beyond the scope of this discussion.

1.2.2 Permeability

Permeability is a measure of a reservoir's ability to flow. The field of flow through a porous media was first described by French Civil Engineer

Henry D'Arcy (1803–58). D'Arcy was addressing a surging problem with the aqueducts servicing Dijon, France. His solution was to fill the aqueduct with sand. When this technique solved the surging problem and still provided water to the city at an adequate rate, D'Arcy developed the equations that we currently use for flow through a porous media, primary among them is Darcy's Law (Wikipedia, 3, 2016).

$$q = \left(\frac{\kappa \cdot A}{\mu}\right) \cdot \left(\frac{\partial P}{\partial x}\right) \tag{1.2}$$

The permeability term is generally stated in miliDarcy (mD), but there are reservoirs where it is in Darcy and others where it makes the most sense to use nanoDarcy. The larger the permeability value, the easier it is for the fluids to get to the well-bore. Both coal and shale have a matrix permeability in the nanoDarcy range (up to a million times smaller than you would expect in a tight sandstone) which is the main reason that the unconventional reservoirs were not a significant portion of oil or gas production until technologies and strategies had matured to account for nearly zero permeability.

Relative permeability: Permeability is not one thing. It can have a different value if the fluid being used is water, crude oil, gas, or an emulsion. Typically we talk about "effective permeability" and "relative permeability" (or "Rel Perm") to a specific fluid. Absolute permeability is a measure of a reservoir's ability to facilitate the flow of the in situ mix of reservoir fluids and is measured in Darcy units. Effective permeability is the reservoir's ability to facilitate the flow of a single component of the reservoir fluid mix and is also in Darcy units. Relative permeability is a ratio of effective permeability to some base permeability such as absolute permeability, permeability to air, and permeability to the mix of reservoir fluids at a fixed set of reservoir conditions.

1.2.3 Hydrocarbon traps

The reservoir provides a storage method and a means to transport hydrocarbons within the reservoir, but hydrocarbons tend to be lighter than water so buoyancy will tend to cause them to migrate away from the center of the earth unless the reservoir is capped.

Traps fall into two broad categories: Structural and stratigraphic. Stratigraphic traps accumulate oil and gas due to changes in rock character. Stratigraphic traps commonly occur as "interbedded sand lenses"

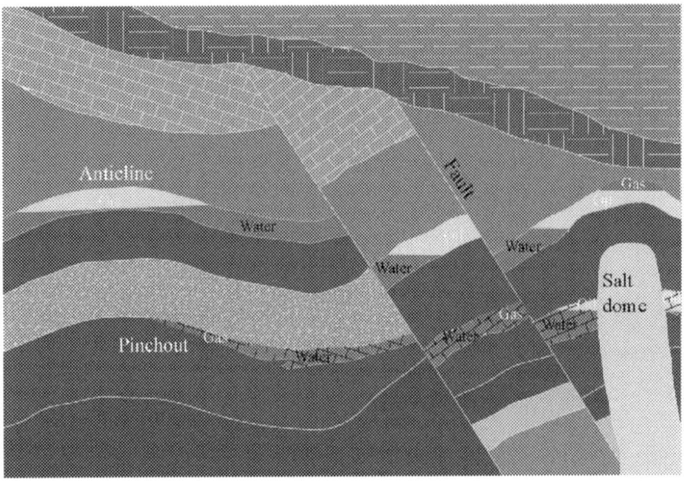

Figure 1.5 Some types of reservoir traps.

within a cap rock like shale. Stratigraphic traps can be very difficult to find or produce, but they can be quite prolific when found. In Fig. 1.5, the pinchout trap is stratigraphic. Stratigraphic traps can be difficult to find, they don't have the distinctive signature that you see for structural traps on seismic data. It is common to find a structural field, and then locate significant new reserves in step-out wells drilled into stratigraphic traps. The Eagle Ford Shale in South Texas is an example of a stratigraphic trap.

Structural traps represent a change to the strata after deposition. They generally show up on seismic as a discontinuity in the data caused by an abrupt change in rock density. Categories of structural traps are anticline, fault, and salt dome.

1.2.3.1 Anticline

An anticline is an arch of stratified rock in which layers bend downward in opposite directions from the crest (Webster, 2016). Anticline traps are most often formed by pushing from the edges, but they can be formed by compaction of the edges. The Jonah Field and Pinedale Anticline in southern Wyoming are examples of anticlines (WyoHistory, 2017).

1.2.3.2 Fault

When faulting is associated with displacement the reservoir strata can be displaced to a position where foreign strata cap the reservoir. The

Oklahoma City Field is an anticline that includes an example of a fault trap (Wikipedia, 5, 2017).

1.2.3.3 Salt Dome

Salt is less dense than most rocks, so over geologic time, salt plugs tend to rise which can create an anticline above it and a displacement cap on its flanks. Early wildcat oil drillers often found oil by looking on the surface for bumps that suggested a salt dome. The Spindletop field was formed around a salt dome (PRI, 2017).

1.3 RESERVOIR CONCEPTS

1.3.1 Reservoir temperature

As you move from the surface of the earth toward the center of the earth, you will encounter increasing temperatures. The core of the earth is very hot, the surface of the earth is quite temperate so it makes logical sense that the closer you get to the core the higher the temperature will get. Through the first few miles of depth, the temperature is a linear function of depth with a geothermal gradient of 0.0301 K/m (0.017 R/ft). Many things can cause local variations in this linear function, the most prevalent of them is the presence of fluids. A hydrocarbon reservoir can retain heat for millions of years so it is reasonably common for reservoir fluids to be warmer than the geothermal gradient would predict. You can take a first estimate at reservoir temperature (realizing that if any data contradicts this estimate then the estimate should be immediately abandoned) by:

$$T_{depth} = T_{surf} + 0.0301 \cdot \frac{\mathrm{K}}{\mathrm{m}} \cdot \mathrm{Depth} = T_{surf} + 0.017 \cdot \frac{\mathrm{R}}{\mathrm{ft}} \cdot \mathrm{Depth} \tag{1.3}$$

If you take surface temperature as 20°C (68°F) then a 600 m (1970 ft) well would have a reservoir temperature of 38°C (100°F). That is the rub. If you take surface temperature as 0°C (32°F) then reservoir temperature would be 18°C (64°F) and if you take surface temperature as 38°C (100°F) then reservoir temperature would be 56°C (133°F). But we assume that the reservoir is isothermal and we don't see wide swings in flowing wellhead temperature from winter to summer.

In fact, the current ambient temperature is only a factor in ground temperature down to the frost line which varies from a few inches to more than a meter down. Below the frost line, the ground temperature is generally around 20°C (68°F) and that is a reasonable value to use for T_{surf}.

1.3.2 Reservoir pressure

The "normal pressure gradient" is generally taken to be based on brine derived from seawater (i.e., SG = 1.07, pressure gradient = 0.465 psi/ft (10.52 kPa/m)). This is a useful value in assessing whether the reservoir is overpressured (i.e., the trap was sealed at a lower depth and geologic forces moved it from deep toward shallow), underpressured (i.e., either the trap was sealed at a more shallow depth and geologic forces moved it deeper or pressure was depleted through leakage), or normally pressured. Beyond that broad classification, a reservoir pressure based on hydrostatic pressure is reasonably worthless.

The only two methods to actually determine reservoir pressure are: (1) material balance and (2) pressure buildup test. A material balance requires knowledge of the shape and scope of the reservoir. We have to know the total void volume that we are accessing (i.e., both thickness and areal extent and connected porosity must all be known with reasonable confidence), which we never know in a conventional gas reservoir.

Pressure buildup tests can be effective at determining the reservoir pressure, but this can take a while. I did a statistical study of short-interval data in the Mesaverde formation in the San Juan Basin of Northern New Mexico in the United States and found that this tight gas formation would take something on the order of 20 years for wellhead pressure to reach average reservoir pressure. The short-interval data in that study showed that "connected porosity" seems to change based on differential pressure and over weeks and months the rate that the pressure is approaching an asymptote which represents reservoir pressure can change from minute to minute, which changes the apparent asymptote (reservoir pressure). In fact we almost never know what the reservoir pressure is in a conventional gas well. Data obtained while drilling can give you an indication, but local discontinuities, the nonhomogenous nature of reservoirs, and the removal of mass (through leakage, offset production, and production) make this number of limited value in most conventional reservoirs.

In unconventional gas reservoirs we can develop a reservoir pressure with considerably more confidence because the gas is stored as a part of the solid matrix rather than in pore spaces. This will be discussed under Sections 1.6.2 and 1.6.3.

1.3.3 Original gas in place

If you know how much stuff is stored within the reservoir, and you know how much stuff you have taken out of the reservoir, then you can have some confidence in predicting how much is left and what portion of that will be economically reasonable to recover. That recovery estimate is known as "reserves" which are a measure of the potential future value of the reservoir. Reserves tend to be the single value entity of any Oil & Gas producer and are a major factor in the valuation of a company.

Other than reserves estimating OGIP is not very useful after the well is on production.

1.3.4 Reservoir pressure versus gas in place overview

If you look at remaining gas in place (GIP) as a percentage of a well's OGIP and current reservoir pressure as a percentage of initial reservoir pressure (at the time the reservoir was first accessed at that location) (Fig. 1.6), then you can ignore the fact that one well may have tens of BSCF of OGIP and a well in the next field may have a fraction of a BSCF, or that one well may have an initial reservoir pressure over 10,000 psi (68,900 kPa) while a well in the next field may have 200 psi (1378 kPa).

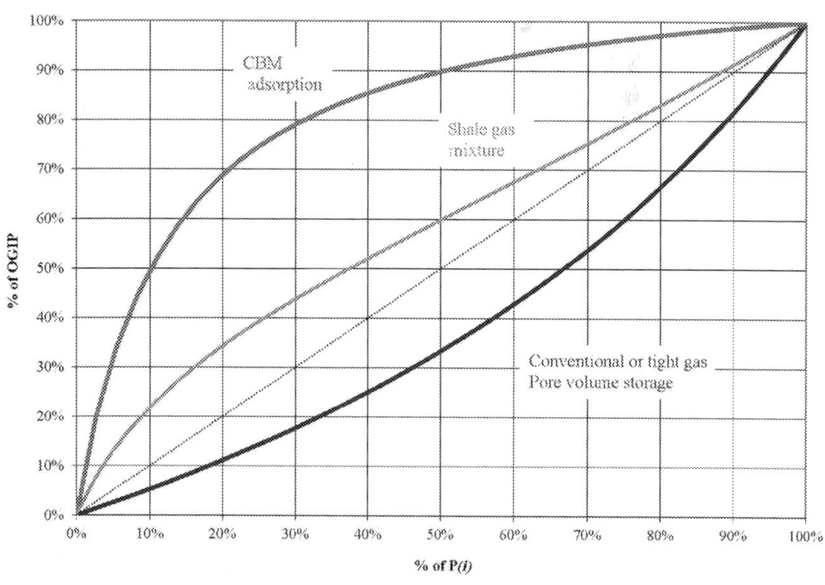

Figure 1.6 OGIP vs P_i.

Table 1.1 Reservoir type key points

	Conventional	Tight gas	CBM	Shale
Gas content (SCF/ton)	N/A	N/A	300–1000	50–400
Storage mechanism	Pore volume	Pore volume	Adsorption	Mixed
Ultimate recovery	40% of OGIP[a]	30% of OGIP[a]	95 + % of OGIP	70% of OGIP
Flow method	D'Arcy	D'Arcy	Channel	Channel
Permeability	>1 mD	10 μD to 1 mD	<10 nD	<10 nD to 10 μD
Porosity	20%–50%	0.5%–10%	<0.2%	0.1%–4%
Response to low pressure[b]	Minimal	Minimal	Excellent	Good
Liquid hydrocarbons	Usually	Usually	None	Variable
Water production	Generally low, not always	Low to none	Low to very high	High to very high
Water quality	Poor	Poor	Variable	Variable

[a]It is quite rare to ever know OGIP in conventional and tight gas fields. Claims that recovery is much higher than these values have not stood up to close scrutiny.
[b]"Low pressure" in this context is dropping flowing bottom-hole pressure much below half of current average reservoir pressure.

The dotted line in Fig. 1.6 is the equipotential line that represents percent of OGIP equal to percent of initial reservoir pressure. None of the types of reservoirs we are discussing comes close to that line, each for its own reasons that will be discussed as follows.

1.4 PRIMARY GAS-FIELD DISTINCTIONS

The distinctions between types of gas reservoirs are somewhat arbitrary, but there are some physical parameters that make understanding the distinctions easier (Table 1.1).

1.5 CONVENTIONAL GAS FIELDS

When the Oil & Gas industry was coming into being, gas did not have much of a market and was not considered a universally viable commercial product—its use required a pressurized infrastructure to get it from the well to processing to end-use and that infrastructure was slow to develop outside of new developing metropolitan and suburban areas. A national infrastructure was not even proposed until the rapid growth of suburbs after World War II created a significant demand. Consequently, the only reason to develop most gas fields in the first half of the 20th century was to recover the associated liquid hydrocarbons. Gas was often flared or vented in these fields. As gas started being piped into more homes and businesses, it finally reached the point where it was a valuable product on its own, but by that point the industry still realized more value from liquid associated with gas production than from gas production.

Conventional gas fields have the characteristic that they have fairly high permeability and fairly high content of associated hydrocarbon liquids.

1.5.1 Reservoir pressure versus OGIP conventional

If you recall the chart in Fig. 1.6, the conventional portion looked like Fig. 1.7. The line bows out below the 45 degree line. This is primarily due to the compressibility of the gas being pressure (and temperature, but

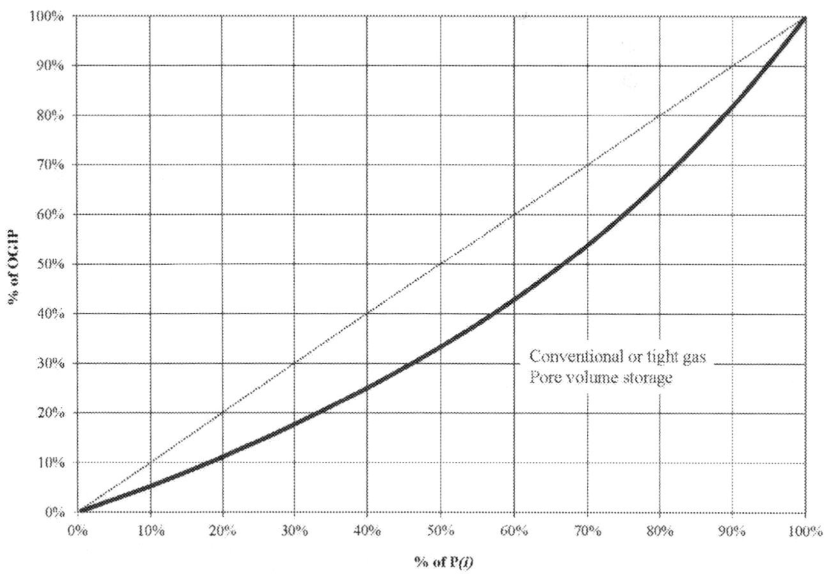

Figure 1.7 OGIP vs P_i for conventional gas.

temperature is assumed to be reasonably constant over the life of a field) dependent. The other factor is the incompressibility of the liquids which often leads to phenomena like "water block" where liquids fill the pore-to-pore passages and adhesive forces and surface tension are greater than the force available via differential pressure.

The bulk modulus of water is 319,000 psi (2200 MPa) which says that if your remove 1% of the liquid in a liquid-full portion of a reservoir, you need to lower the pressure in that portion by 319,000 psi (2200 MPa) which is not physically possible in any reservoir discovered to date. As you remove liquid from a liquid-full portion of the reservoir, pressure will begin to drop rapidly, liquids and gases in adjoining reservoir portions will flow into the portion being drained, when the pressure in adjoining portions is lowered to zero dP, flow will stop even if the section still has reasonably high pressure.

The primary ramification of the shape of the curve is that a step-change in pressure will almost never provide attractive increases in production and well-site compression in conventional fields tends to be viewed as a "rate acceleration" exercise rather than a "reserves adding" exercise. Rate acceleration generally has very poor economics in gas fields. When the reservoir pressure in a conventional field reaches about 20% of initial reservoir pressure, there is only about 10% of OGIP remaining, unlikely to be worth a large capital expenditure. Most conventional fields are abandoned at around 60%–70% of initial reservoir pressure, leaving 50%–75% of the OGIP in the ground. Fig. 1.7 represents part of the reason for walking away from these resources, the rest of the reason is that the pressure drop through the pore-to-pore pathways in the near well-bore are absolute-pressure dependent and there is no driving force within the reservoir to overcome this pressure drop.

As we will see in Chapter 3, Well Dynamics, while step-changes in flowing bottom-hole pressure are rarely effective in conventional fields, providing very constant flowing bottom-hole pressure (e.g., with well-site compression) can significantly increase ultimate recovery from conventional fields if we are willing to change our economic goals from rate acceleration to reserves adding.

1.5.2 Conventional gas

The flow mechanism from deep in a conventional reservoir toward a well-bore is described as "flow through a porous media," also known as

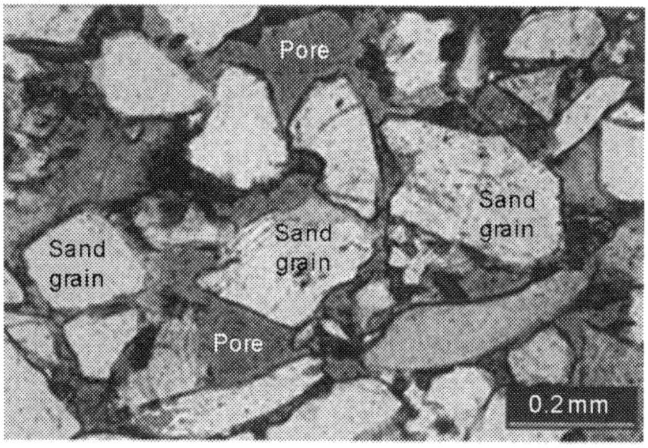

Figure 1.8 False color thin section of sandstone (Dollar Photo Club).

D'Arcy flow. This flow mechanism is characterized by very high pressure drops as reservoir fluids struggle to find pathways from one pore space to the next. Consequently, reservoir pressure can remain nearly constant for years on end. Effectively constant reservoir pressure is an important assumption in many of the equations that production engineers use to predict reservoir performance. While this is almost always a valid short-term assumption for conventional gas reservoirs from year to year, it needs to be periodically revisited to see if actual changes have altered the well's performance (Fig. 1.8).

1.5.3 Conventional reservoir materials

Conventional reservoirs are characterized by having reasonably large permeability and significant potential to extract liquid hydrocarbons. Further, the production is dominated by pore-volume storage, so porosity must be reasonably large to make a reservoir productive. Conventional gas is very similar to producing an oilfield and the techniques and technologies developed for oilfields can often be applied to conventional gas fields with little modification.

Carbonates: Calcium carbonate ($CaCO_3$) comes from dissolving animal bones and sea shells in slightly acidic water. As conditions change (temperature, pressure, and/or water chemistry), this $CaCO_3$ tends to precipitate out of the water. Under heat and pressure, the precipitate turns into limestone which has very low matrix porosity but tends to have significant void volume in cracks and vugs (i.e., a small unfilled cavity in a lode or in rock (Webster, 2016)). About 50% of the world's known reserves of oil are in $CaCO_3$ reservoirs (Schlumberger, 2107). The percentage of gas is much lower, but the highest rate gas fields have been carbonates.

Sandstones: Grains of sand tend to shift into the lowest possible potential energy. You can visualize this by thinking of a billiards table (see Figs. 1.2 and 1.3). Add pool balls to the table until there is no place for the balls to move to and then add some more. The added balls will find resting places touching multiple first-layer balls. Once you have multiple layers of pool balls stacked up, pour some small marbles (small enough to fit in the space between three pool balls touching) and the marbles will tend overtime to sift down to fill all the voids. This is what happens in a sandy beach, the sand grains are always trying to move toward the center of the earth and over geologic time reach the lowest potential energy state available.

With time and fluid movement, minerals tend to evolve out of the fluids to cement the sand grains into a sandstone, but still leaving significant void volume and pathways to allow fluids to move from pore to pore. Often these pathways are too small for rapid liquid movement but are adequate for gas movement. In this case the gas-to-liquid ratio puts the reservoir into a "gas" category rather than an "oil" category. A large number of the commercial conventional gas reservoirs in the world are sandstones.

Chalks: Chalk is a stable polymorph of carbonate with interbedded shales. The organic material in the shale tends to allow adsorption of gas in addition to the storage in cracks and vugs. As discussed in Section 1.6.2, this adsorbed gas provides an energy source for fluid movement of fluids stored in porosity. Chalks tend to have very low native permeability and must be extensively fractured through either stimulation or natural tectonic activity to be productive. Chalks make up most of North Sea and South Texas production.

1.6 UNCONVENTIONAL FIELDS

Unconventional gas is the stuff that the industry tended to avoid whenever there was anything else to develop. So far it is made up of tight gas, CBM, and shale (Fig. 1.9). Eventually hydrate mining and landfill gas will also be significant portions of this sector.

The industry tended to skip over the unconventional resources for four major reasons: (1) the permeability tends to be very low (Fig. 1.10);

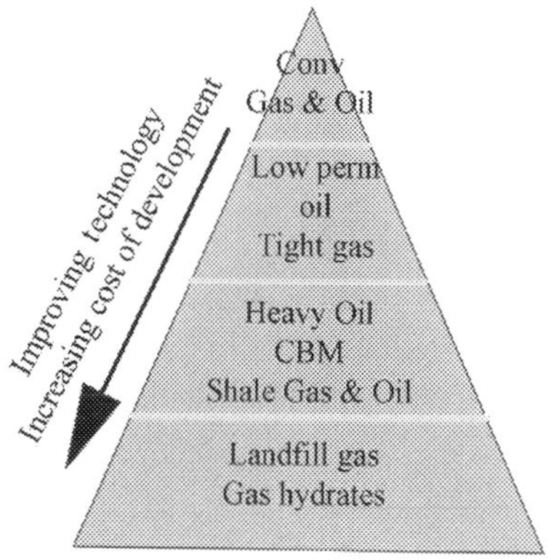

Figure 1.9 Reservoir development pyramid.

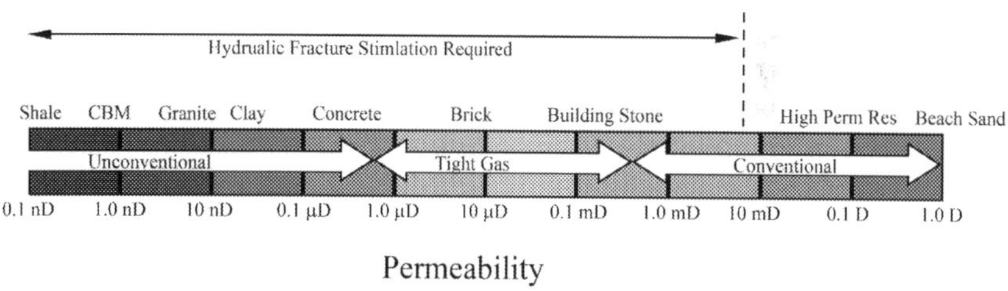

Figure 1.10 Permeability continuum.

(2) it often has no hydrocarbon liquids; (3) it often requires new or non-oilfield techniques to develop; and (4) it is often very expensive to develop and produce.

1.6.1 Tight gas

For many years tight gas was seen as "difficult conventional gas" in that the production fluids generally contained hydrocarbon liquids in about the same proportions that were seen in conventional gas, but tended to be significantly lower flow rates. Making money in tight gas required both a long attention span and a willingness to accept lower per-well production rates. The Mesaverde formation in the San Juan Basin was one of the first tight gas fields developed (in the 1950s and then infill drilled in the 1970s) and for many years it was the largest onshore gas field (by both total production and remaining reserves) in the US until the CBM development in the San Juan basin in the 1990s kept San Juan on top of the list while eclipsing the Mesaverde. The development of shale gas plays in the 2000s pushed San Juan off the bottom of the top 10 lists.

Tight gas was the beginning of awareness that gas is not oil, and that oilfield techniques and technologies might not be universally applicable. For example, in conventional gas, it was common to apply compression to lower flowing bottom-hole pressures in steps and quite rapidly. When we try that in tight gas, we get some increase in production for a very short time and then it stops—the permeability is so low that inflow generally cannot keep up with compression rate. We also found that applying small amounts of compression to slowly lower flowing bottom-hole pressure (on the order of 2–3 psi/month) tended to flatten decline even though flow rate increases were modest to nonexistent. Work in well-site compression in tight gas was really the first time that a company was able to confirm that adding compression to unconventional gas wells added recoverable reserves instead of simply accelerating the recovery rate from already booked reserves. This concept has been very important for CBM development and has been the key in tight gas economics.

1.6.2 Coalbed methane

In Fig. 1.8 we saw a thin section of sandstone where a scale of 0.2 mm = 1 in was adequate to see the pore space. A similar thin section from a coal bed (Fig. 1.11) would require 0.0001 mm = 1 in to see the pores—porosity is 1/2000th the size of pores in sandstone. If we expanded the scale on Fig. 1.8 to the scale on Fig. 1.11, every pore would appear infinite. Void space in CBM is so low and so isolated that permeability in CBM is essentially zero (it tends to be on the pD and nD scale), and there is so little void volume that there wouldn't be enough gas in the formation to be economic.

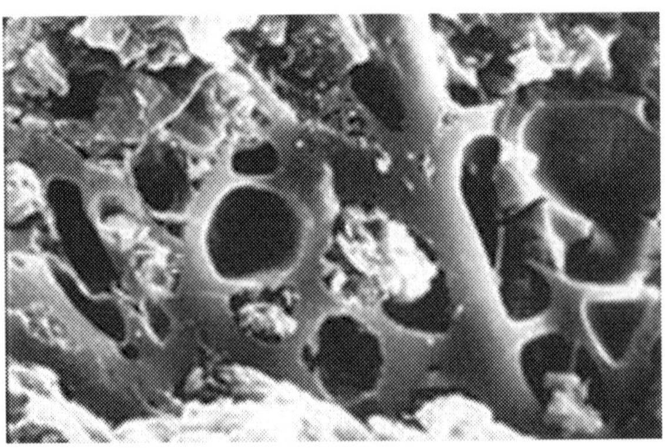

Figure 1.11 Thin section of coal, 0.0001 mm = 1 in (Dollar Photo Club).

CBM highlighted the difference between oilfield techniques and techniques appropriate for unconventional gas. The "pressure continuum" discussed in Chapter 0, Introduction, is a major piece of the difference. In conventional gas and most tight gas, all parts of the system are at pressures higher than transition from "normal pressure" to "high pressure" on Fig. 0.1 and at those pressures it is quite reasonable to assume that gas acts much like a liquid. In the low pressure part of that figure, the deviation from linear behavior becomes too large to ignore. The other major piece of the difference is that D'Arcy flow in nano-Darcy permeability is very nearly zero.

There are many excellent books on CBM Reservoir Engineering, the best I have found was written by Dr. John Seidle (2011). Of course, all of these books are written from the perspective of reservoir engineering for reservoir engineers. As a surface engineer it is difficult to reconcile actual uncertainty with the confidence that the authors show in their equations and models. For example, all of the books assign a value for "permeability" of the various CBM plays. They all use phrases like "permeability, which is primarily cleat permeability is ...". Permeability of an open conduit is infinite by definition. Permeability of the coal matrix is as close to zero as we can measure. This means that reservoir engineers have solved the age-old problem of using infinity in equations being undefined. They have defined "(infinity plus zero) divided by 2" as equal to "30 mD," now we know how to evaluate infinity. Having evaluated many reservoir models based on this new mathematical construct, I have to call "foul." These models tend to be slightly less able to predict future performance than the models used in climate science. I have had several discussions on this subject with Dr. Seidle and he smiles indulgently at this silly flangehead who has the audacity to say "D'Arcy is irrelevant to actual CBM performance." The following is based on my experience with actual well performance in several CBM fields located on four continents. In many regards it is very much at odds with the experts in CBM Reservoir Engineering, please keep that in mind as you review it.

Coal rank: Coal is ranked by the degree of alteration that occurs as coal matures. The ranks of coal that are commonly used are given in Table 1.2.

The "shininess" of coal is measured with a parameter called "vitrinite reflectance." Coal beds that have been economical for CBM have tended to be at least subbituminous or higher rank with vitrinite reflectance between 0.4% and 1.4% or "a bit dull" to "kind of shiny" (anthracite is usually around 6%).

Table 1.2 Coal rank

	Maturity	Color	Moisture	Heat capacity
Peat	Least	Dull brown	Wet	6 MBTU/lbm (14 MJ/kg)
Lignite	More	Dull black	Drier, still wet	7 MBTU/lbm (16.2 MJ/kg)
Bituminous	More	Almost shiny	Drier	11 MBTU/lbm (25.6 MJ/kg)
Anthracite	Most	Shiny black	Dry	14 MBTU/lbm (32.6 MJ/kg)

Coal cleats: According to Seidle, "Coals are unique, naturally fractured reservoirs with two mutually perpendicular fractures, both of which are perpendicular to the bedding plane" (Seidle, 2011). The dominant or "face" cleats tend to be connected and colinear. The secondary or "butt" cleats are perpendicular to the face cleats and generally terminate at the face cleats. The limited ability of an undisturbed coal bed to flow gas is due to limitations of the connectedness of the face cleats with the system of butt cleats.

Gas storage method: Since CBM does not have appreciable pore volume, it is normal to ask "then where the heck is the gas?" That is a fair question. The gas in a CBM field is "adsorbed" to the surface of the coal. Webster defines "adsorption" as "the adhesion in an extremely thin layer of molecules (as of gases, solutes, or liquids) to the surfaces of solid bodies or liquids with which they are in contact" (Webster, 2016). Adsorbed gases sit on "adsorption sites" on the surface of the coal and even with very small void spaces in the coal the total surface area is enormous. Adsorbed gas is held in place in CBM by the hydrostatic pressure of water filling the cleat system (rather than the cap rock necessary in conventional and tight gas reservoirs).

Adsorbed gas acts as part of the coal matrix and does not behave like a gas. As long as the molecules are attached to the solid, the gas equation of state is not valid for those molecules. In short $PV \neq nRT$. This concept is crucial to understanding exactly how much gas a CBM field can contain. A good tight gas field will recover about 0.5 BSCF/well. In the central "fairway" portion of the San Juan Basin CBM, an average well will recover about 12 BSCF/well.

Desorption of adsorbed gas requires a drop in pressure (i.e., the gas content of the thin film layer is a function of the confining pressure). Once the gas has left the coal surface, it has all of the characteristics of any free gas. This free gas has expansive energy that acts to "push" reservoir fluids toward the well-bore while with conventional and tight gas

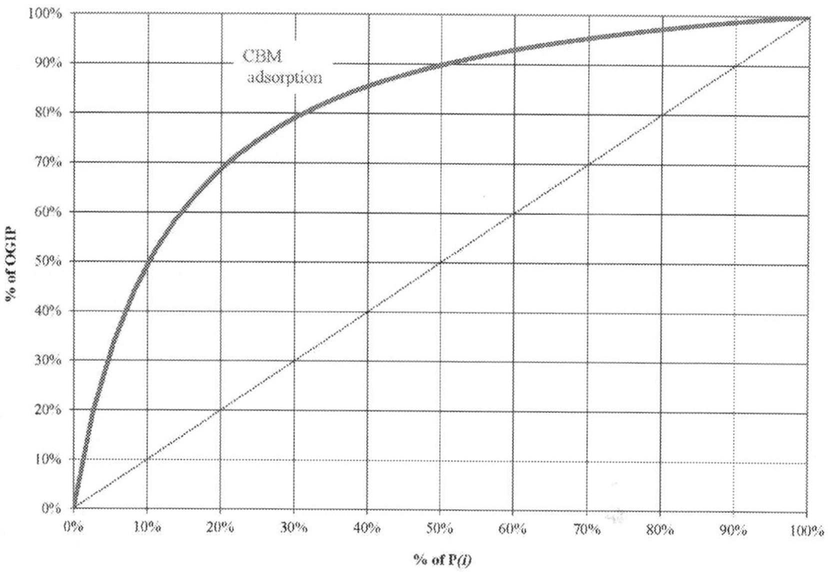

Figure 1.12 GIP vs pressure for CBM.

reservoirs all of the reservoir energy that will ever be available is in the gas within the pore volume at initial completion, when the first molecule of reservoir fluid is produced, the energy has been diminished (this phenomenon can be thought of in a conventional or tight gas reservoir as "pulling" the fluids to the well-bore). Fig. 1.12 shows the effect of this difference. In the particular well used for this graph, when reservoir pressure was 10% of initial pressure, there was still 50% of the OGIP still in the reservoir. This "pull vs push" analogy begins to explain why it is nearly always economic to apply compression to late-life CBM wells while it is rarely economic to apply compression to late-life conventional and tight gas wells (but there are some good reasons to apply well-site compression in midlife in conventional and tight gas reservoirs, see Chapter 3: Well Dynamics).

Fig. 1.12 is what is known as a "combined isotherm" in that each coal has a different preference for each species of gas present. It is common for a coal to have a greater ability to store CO_2 than methane, and over geologic time if there were enough CO_2 present then the methane would not be able to find adsorption sites. Since (as we discussed earlier) methane comes from anaerobic decomposition, the available oxygen that can become CO_2 is limited and CBM reservoirs generally contain fairly low levels of CO_2. Fig. 1.12 is the combination of a 100% CO_2 line and the 100% CH_4 line with the actual proportions of gases (i.e., in this case 82% CH_4 and 18% CO_2) (Fig. 1.14).

Water storage: Hydrostatic pressure in the cleat system is required for gas to remain adsorbed to the coal. Typically the entire cleat system is water-full under pressure. Since all gas flow must initiate at the coal surface within a cleat, the water in the face cleats that intersect the well-bore must be (nearly) completely removed before there are pathways for gas to flow away from adsorption sites. Producing this "thoroughfare water" is commonly known as the "dewatering period" and can last from days to months. At the end of the dewatering period, the well will still produce water, but the water/gas ratio will tend to approximately stabilize and remain constant for most of the remaining life of the well.

As will be discussed in Chapter 3, Well Dynamics, there is no economic benefit of removing more water than is naturally flowing from the butt cleats, and there is strong evidence that overpumping a well will actually reduce the ultimate gas recovery.

Langmuir isotherm: If you start with an assumption that reservoir temperature is approximately constant over time, then you can treat depletion as "isothermal." With that assumption you can develop an equation for OGIP based on the Langmuir adsorption model (Wikipedia, 4, 2016). One representation of OGIP using this model is:

$$\text{OGIP} = 0.031214 \cdot A \cdot h \cdot V_m \cdot y \cdot \rho_{lang} \cdot \frac{b \cdot P_i}{1 + b \cdot P_i} \tag{1.4}$$

All of the terms in Eq. (1.4) come from log and core analysis except drainage area. This term is often mandated by regulatory fiat. While the regulations yield a value that is quite removed from the conditions at any particular well, it can be useful to start with the mandated drainage area and refine it over time. This mandated value allows a first guess at OGIP. To get gas remaining, simply replace initial reservoir pressure with current average reservoir pressure.

You can solve Eq. (1.4) for pressure to allow prediction of reservoir pressure relative to cumulative production.

$$\overline{P} = \frac{\text{OGIP} - \text{Cum}}{0.031214 \cdot A \cdot h \cdot V_m \cdot y \cdot b \cdot \rho_{lang} - (\text{OGIP} - \text{Cum}) \cdot b} \tag{1.5}$$

Since flow within the CBM reservoir is channel flow, average reservoir pressure can be available at the wellhead within 48–72 hours of a well shut-in for high-productivity wells, weeks or months may be required in lower productivity wells. It is common to require field shut-in periods for compressor station and plant maintenance annually. I was very careful in the field I operated to manually shut in the wells (rather than letting the wells pressure up the gathering system) and to make sure that a calibrated transducer was communicating with the production database upstream of that shut valve (usually casing pressure measuring point). Once pressure stopped increasing (or it was clear that it was approaching an asymptote), we had current reservoir pressure and could tune drainage area to refine the OGIP value. Out of 65 wells, the first calibration was the only one necessary on 60 wells. The other five wells required a second calibration. Using the refined drainage area, we were able to develop a very good estimate of average reservoir pressure at any given time.

Another observation is that wells behave very differently during the various stages of their "lives." Plotting the remaining gas version of Eq. (1.4) helps define these life stages (Fig. 1.13).

The added lines are reflecting the observation that in the early and late parts of the isotherm, the relationship is approximately linear. Using that observation allows you to extend a line from 0,0 and another line

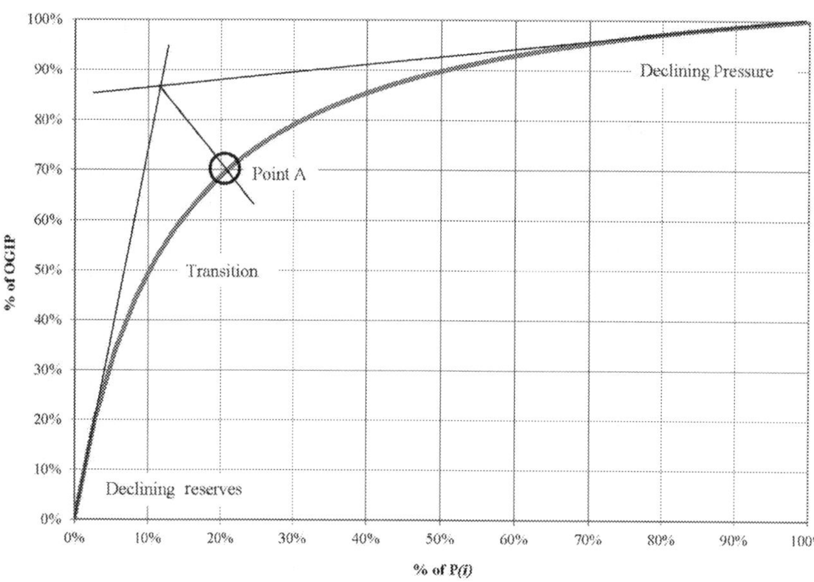

Figure 1.13 Annotated GIP vs P_i for CBM.

from 100,100 and where those lines cross, drop a line normal to the isotherm. The interpretation of these regions is:

- Declining pressure region
  - Pressure falls rapidly for small reductions in GIP.
  - Reservoir energy is very high and rapid production during this time can cause coal structures to fail and open larger channels which dramatically improves late-life production rates.
  - This period can be reasonably short (in the well in Fig. 1.13 it lasted 7 months).
  - Wells in this region do not respond to very low pressures but sometimes deliquification methods improve production.
  - Wells in this region respond very well to flowing bottom-hole pressures that are approximately equal to the pressure at point A.
- Declining reserves region
  - Pressure changes very slowly while GIP changes relatively quickly.
  - This period can be very long (in the well in Fig. 1.13 this period started at 11 years and at the time of this writing it has been on production for 27 years and is still making commercial volumes).
  - The only way to recover commercial volumes in this stage is with very low flowing bottom-hole pressures.
- Transition region
  - This is the change from "easy" to "difficult" for operations.
  - At the beginning of the period most wells don't need deliquification and few wells need compression.
  - At the end of the period all wells need both a coherent deliquification strategy and very low wellhead pressures.
  - This period is often fairly long (in the well in Fig. 1.13 it lasted 10 years) and following reservoir pressure with bottom-hole pressure during this period is very important.
- Point A
  - Prior to point A the target flowing bottom-hole pressure should be somewhere around the point A value (in the Fig. 1.13 well that was 350 psig (2413 kPa)).
  - As pressure approaches point A the flowing bottom-hole pressure needs to drop to maintain the proper relationship between average reservoir pressure and flowing bottom-hole pressure to optimize commercial production.

Every CBM reservoir has a "sweet spot" in flowing bottom-hole pressure that optimizes ultimate recovery. In the San Juan Basin Fairway

CBM wells that number was a flowing bottom-hole pressure of one-half of average reservoir pressure (discovered through many trials and even more errors). These observations allowed a prediction of what pressure was going to be needed in the future. The isotherm analysis does not have a rate or time component, but as part of our normal business activities we regularly do production forecasts. The forecast has an implied cumulative production, so matching the forecast with the isotherm it is a reasonable thing to develop pressure drawdown schedules to allow budgeting compression and gathering system upgrades early enough to allow for budget cycles.

CBM contamination: As discussed earlier, decomposition of organic material with inadequate oxygen (not necessarily zero oxygen) produces methane, and limited amounts of water and carbon dioxide. CO_2 tends to be a better fit for the adsorption sites in a coal bed than CH_4, and the coal will preferentially accept the CO_2 to the exclusion of CH_4. Each of these gases can be represented on an isotherm as seen in Fig. 1.14. The top line shows the isotherm for 100% CO_2, the bottom line shows the isotherm for 100% CH_4. The line in the middle shows the combined

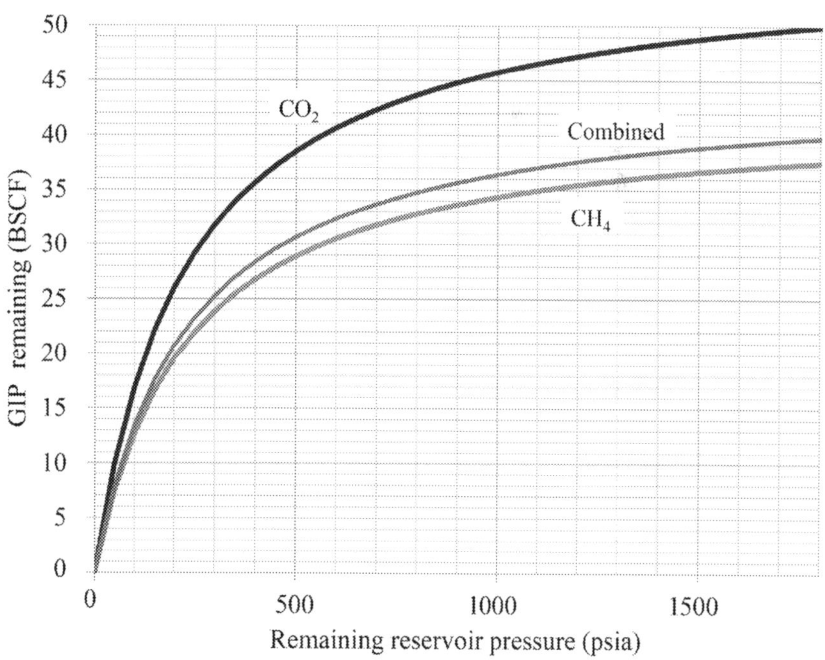

Figure 1.14 Carbon dioxide and methane isotherms.

isotherm for the actual mix of gases in this particular well in this particular formation (i.e., 82% CH_4 and 18% CO_2).

Determining the actual in situ mixture of gases can be a challenge. When you produce some gas, you leave adsorption sites vacant to be filled by gas flowing from deeper in the cleat system, but the sites prefer to attach CO_2 over CH_4. The near-well-bore coals rapidly replace methane with carbon dioxide. When the well in Fig. 1.14 was first produced the sales gas contained just under 6% CO_2, 12 years later the gas stream was up to 23% CO_2. The carbon dioxide in the sales gas passed through the 18% that we think was in the formation at the beginning without even pausing. This seems to support the idea that the coal is acting as an "activated" coal filter to strip CO_2 from the reservoir fluids and over time as the saturation of CO_2 on the near-well-bore coals increases toward some limit (well above 18%) it begins evolving off of the coal more rapidly.

Using this theory a mass balance on the San Juan Basin fairway was developed and it was determined that the "missing" CO_2 from the early production would be balanced with the "excess" CO_2 from the later production at a sales concentration of slightly less than 30%. That has proven to be a reasonable estimate as late-life CO_2 concentration has leveled off and begun declining back toward the original 18%. Computer models (Seidle, 2011) looking at homogenous desorption pressure predict that CO_2 will increase toward 100% at complete drawdown. This is not what operators have seen.

CBM completions: A conventional completion installs production casing across the productive seams, cements it into place, then perforates the production casing and cement sheath, and stimulates the formation. Coal has very low mechanical strength and putting a concrete hydrostatic gradient (about twice the gradient imposed by water) on the coal always causes a considerable amount of concrete to break its way into the target formation, often doing irreparable damage to the productivity of the well.

To avoid this damage, it is good practice to set production casing above the coal and cement it to surface. Then the coal is left either open hole or with an uncemented liner installed across the coal.

It is common for coalbeds to consist of multiple coal seams. If we leave the entire interval uncemented (good practice), then control of stimulations becomes more difficult. If you are doing a hydraulic fracture stimulation, you have no way to control which seam(s) will get any frac fluid at all and what proportions of the job will go to what seams will always remain a mystery.

Statistical analysis of upward of 1000 hydraulic fracture stimulations in various CBM basins around the world has been puzzling. There does not seem to be any type of stimulation that has a better track record than any other type. Proppant type and proppant volume do not correlate with success in any basin. Any given stimulation type that provides good results in one well may give poor results in the next well and vice versa.

Coal miners have always talked about coal being "self-healing." This means that any inclusion in the coal matrix will eventually become part of the coal matrix and will be indistinguishable from the surrounding matrix. This does not bode well for the ability of injected proppant to successfully prop open flow paths in the coal.

There is a weak correlation between carrier volume and success. Frac jobs that include a lot of liquid (or foam) carrier and just enough "proppant" to enhance the abrasive characteristics of the fluid have tended to have a higher incidence of success, likely due to scouring vertical channels in the coal matrix that are too large to "heal."

The very best results have been achieved by "cavitating" the wells. This technique either uses surface compressed air or reservoir pressure to build up high pressure in the near well-bore and then rapidly drop the pressure to "surge" the well. These surges tend to fracture large sections of coal and the high velocity flow carries much of the broken coal to surface. The pressurize/surge cycles are repeated until the coal has "stopped flowing" (i.e., solid coal to surface has diminished). This process requires the coal to have very low mechanical strength and it simply doesn't work in most coals, but since production improvements up to a factor of 40 times have been reported, where it works it is foolish to avoid.

Horizontal wells are the flavor of the week in Oil & Gas and many operators have tried horizontal wells in CBM with varied (mostly poor) success. Successful frac jobs in horizontal coals have been rare, open-hole horizontal wells have a very high incidence of collapse, and cemented casing tends to damage the coal. "Success" in horizontal CBM wells is extremely difficult to gauge. For the most part the operator is comparing a terrible completion process (like drilling the coal with mud, setting production casing across the coal, and cementing to surface) with a less terrible completion process (like air drilling a horizontal/vertical pair and leaving the horizontal drill pipe in the well as a liner, perforating the liner, and doing a hydraulic fracture stimulation on the entire lateral) and they see improvements over the terrible offsets. Does that mean that horizontal wells in CBM fields is a good idea? I don't believe that it has been proven that horizontal wells have better results than vertical wells drilled and completed with good practices.

Flow within a CBM reservoir: CBM has few of the characteristics of conventional reservoir rock: all flow is in cracks and channels with approximately infinite permeability; matrix permeability is so low that it is effectively zero; and the "reservoir" is capped with hydrostatic pressure of liquids in the cleat system instead of a cap rock. This makes conventional reservoir-performance tools largely ineffective because reservoir pressure changes far faster than the tools expect; the void volume is quite variable as gas evolves from the coal surface; and the drainage area can be very asymmetrical and very large (dependent on cleat and fracture geometry even more than conventional wells are).

In fact the actual flow matches a pipe-flow model much more closely than a D'Arcy-flow model.

Key CBM learnings: Coal is very sensitive to formation damage:

- Mud drilling the formation has a high risk of the mud destroying the productivity of a CBM well
- Cementing across the coal generally results in significant reduction of production capacity
- Hydraulic fracture stimulation gives very different results from similar stimulations in conventional or shale reservoirs.

There is a maximum containing pressure required to prevent desorption, production requires being below that containing pressure, and that containing pressure changes rapidly with time:

- Above the maximum pressure the gas will mostly stay on the coal.
- Below the maximum pressure the gas will desorb.
- Significantly below the maximum pressure degrades the efficiency of the flow conduits, often to the point that the pressure at the desorption sites is higher than it would be at a higher wellhead pressure.
- Very low pressures in the early days will often increase water rate and total water at the cost of gas rate and ultimate recovery.

In short, you have to treat CBM as a unique opportunity, it isn't just mushy sandstone.

1.6.3 Shale

The first commercial well in the United States was a Shale Gas well in Fredonia, New York in 1821 (30 years before the first oil well in Pennsylvania). It was 27 ft (8.32 m) deep in the Devonian shale. Flow was

Figure 1.15 Photomicrograph of shale section (Dollar Photo Club).

so low that output was only suitable for gas lights and little liquid moved at those flow rates.

The photomicrograph in Fig. 1.15 shows that the shale matrix is made up of sand, quartz, organic material (e.g., peat, coal, kerogen, and bitumen), and trash. This mixture of materials results in shale being very much nonhomogenous, much stronger than coal, and has mixed storage methods.

Production from shales has evolved from a very minor portion of the mix of US gas and oil production at the beginning of the 21st century to the dominant position in US gas and oil production in 2016 production. Much of the growth in production has been in states that have not historically been Oil & Gas producers, and consequently production reporting in the shale is quite spotty and reported production is so much less than consumption that the people responsible for publishing this data have resorted to "we produced *x*, we consumed 130% of *x* so the missing gas must have come from _____." Not a great way to do reservoir engineering, but you work with what you have. Individual companies have good data, but it is kept at the level required for reporting and financial settlement, which is often at the lease level instead of the well level. When we say "in 2015 the Marcellus Shale produced 17.8 BSCF/day," that number likely came from looking at plant and compressor station throughput. When we say "in 2015 the Barnett Shale averaged 4.379 BSCF/day" we are talking about individual-well volumes measured on high-quality gas measurement equipment and consistently reported to the Texas Railroad Commission.

Storage method: Shale is messy. In CBM we saw microstructures we called "cleats" which created surface area for adsorption. In shale the inclusions of sand, peat, quartz, etc. lead to, if not homogenous, then a heterogeneous mix of largely homogeneous cells of widely varying sizes. The cells can range from cubic inches to cubic yards, and the relationship of one cell to its neighbor may be significant or it may be largely absent. Nothing as organized as a cleat system exists. Also the shale matrix is much stronger than the coal matrix as we discussed earlier. Permeability is close enough to zero to make shale an excellent and very common cap rock.

The mixed nature of shale provides a range of storage mechanisms. The organic material in the shale contains adsorption sites for storing adsorbed gases. The sand provides locations for liquid hydrocarbon storage. It is typical for 60% of the OGIP and 100% of the OOIP to be in the pore volume. This leaves a very important 40% of the OGIP adsorbed to the coal-surface deep in the reservoir. Desorbing this gas provides a motive force to move the liquids stored in the pores. The storage can be demonstrated by a combined OGIP vs pressure graph.

Fig. 1.16 implies that early-life production will be D'Arcy-flow dominant and late-life will be desorption dominant. With permeability so low, D'Arcy flow is limited to very short distances. Consequently when we

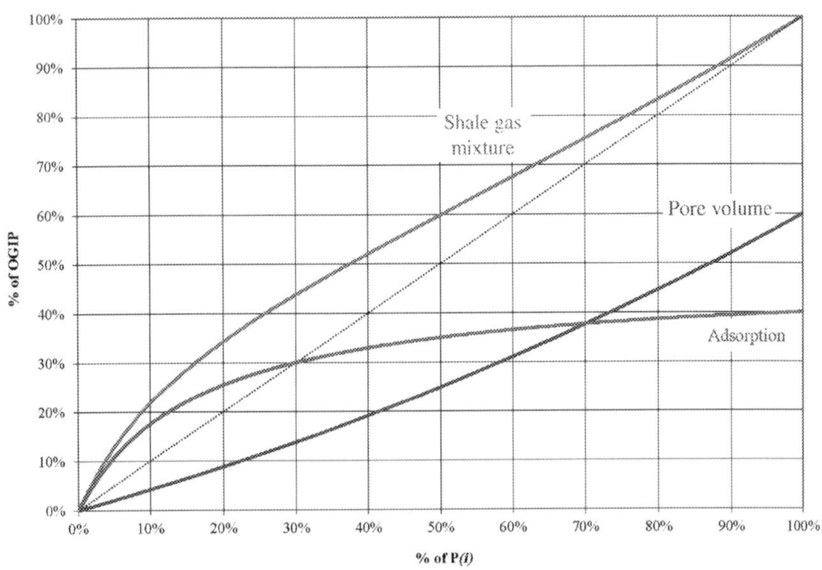

Figure 1.16 OGIP vs P_i, for shale.

drill vertical shale wells, the tendency is for them to produce for a short time and then fall off rapidly. To combat rapid decline, the industry has learned to vastly increase the well-bore access to the reservoir by drilling horizontal wells with very long laterals and precisely controlling stimulation energy to create cracks deep into the reservoir over the entire length of the lateral. This technique results in vast volumes of shale that are "a short distance" from a flow channel with approximately infinite permeability.

Hydrocarbon liquid-rich fluids: The fluids are as much a mixture as the lithography. We find thermogenic gases (i.e., primarily methane that has resulted from heavier hydrocarbons modified by heat and pressure), biogenic gases (i.e., gases from contemporary biological activity), and a range of hydrocarbon liquids and water. Many of the shale plays have been categorized as "liquids-rich" gas wells. Other plays are "dry gas." Still others are "gassy oil plays." Regardless of how they are categorized, multiphase flow and flashing/condensing hydrocarbons create a challenging flow problem. It is common for cooling in the well-bore to create paraffin and hydrate problems in tubing and production equipment. Keeping oil tanks hot to minimize paraffin accumulation generally results in excessive flashing of natural gas liquids (NGL)-weight hydrocarbons (and accompanying lost sales and environmental exceedances).

Processes and procedures to deal with liquids-rich environments will be discussed in the well-site and gathering chapters. For a reservoir discussion, it is enough to know that liquids-rich fluids are common in shale plays and that the reservoir is going to give you what the reservoir is willing to give you.

Shale mechanical characteristics: Shale plays tend to be very thick. It was long thought that shales thinner than 300 ft (90 m) would not be commercial. Some wells in the Marcellus Shale encountered individual productive intervals over 1000 ft (300 m) thick. This thickness suggests that the wells could easily have very long productive lives (upward of 100 years in some of the plays). As technology evolves to optimize completions in the very thick shales, it will likely become economic to develop thinner shales, opening up many millions of additional acres of potential shale development.

Shale tends to be quite hard and brittle. This characteristic suggests that hydraulic fracture stimulation should have the dual benefit of breaking a lot of shale with the applied pressure and the ability to prop open the fractures with proppant is quite effective.

Shale conclusion: Shale gas is still in its early stages. We don't:

- know how the wells will respond over time;
- know how the water rates will change with time;
- have a clear strategy for what pressures will be required over time;
- know how we are going to do midlife and late-life deliquification.

Some of this information may require 30–50 years to develop. Beware of the idea that you can design a shale field once and the facilities will last forever—as an industry we will make just about as many mistakes in shale as we made in CBM, but hopefully they will be different mistakes.

1.7 RESERVOIR DEVELOPMENT

As a reservoir moves from a "prospective resource" to "proved and developed reserves," it has to move through several stages of evaluation, data gathering, evaluation, test drills, evaluation, and market assessments (Fig. 1.17). The current economics are considered at each stage to determine what the project is potentially worth. For onshore gas-field

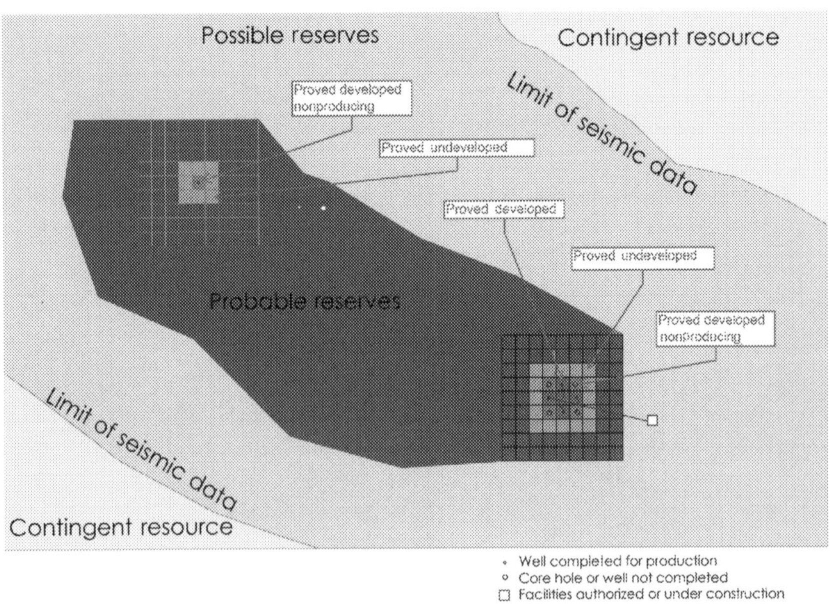

Figure 1.17 Types of resources.

development in a place with reasonable access to oil-field workers and adequate takeaway pipeline capacity, this can be a reasonably contained and quick process. For something like the Marcellus shale where the developers had to create an employment, processing, and takeaway infrastructure, it can take decades.

1.7.1 Types of resources

The steps that a new field goes through from an idea to a resource are: (1) prospective resource; (2) contingent resource; (3) possible reserves; (4) probable reserves; and (5) proved reserves. Each of these terms has a specific definition, the only one that impacts a company's net worth under Securities and Exchange Commission (SEC) guidance is the last one.

Prospective resource: At this stage work has been limited to a geologist studying maps and possibly visiting the site to evaluate rock type, surface indications of faults, or domes. The company may or may not even own mineral rights in the area.

Contingent resource: Once the geologists convince their management that there is enough evidence on the ground to be intrigued, the company starts spending money. Frequently this money is spent on things like seismic, aeromag surveys (i.e., searching magnetic anomalies from airplanes), studies of existing infrastructures, assessing mineral ownership, and engineering guesses about the cost to develop. All of this work is followed by the first of many economic evaluations, generally called "scoping economics," this stage defines the field as "discovered."

Possible reserves: If the scoping economics meet the company hurdle rates (most don't), then the evaluation continues with some amount of drilling-rights acquisition, drilling core holes, more detailed infrastructure assessment, and the second economic analysis, generally called "preliminary economics." It is common at this point to begin quietly acquiring mineral rights which may or may not ever become a producing entity. This stage and the previous two stages represent "exploration."

Probable reserves: If the preliminary economics indicate that the play might have merit, then work moves to probable reserves. During this stage it is common to drill pilot wells, do a detailed infrastructure feasibility analysis, and run detailed economics. At the end of this step, a memo version of reserves can be placed on the books, but these "indicated additional" reserves do not add to the financial position of the company. This stage represents "appraisal."

Proved reserves: When you are drilling production wells, building pipelines and processing facilities, and have committed the money to make gas and/or oil sales possible, then the company can book reserves that do actually affect the value of the company. The rules for booking "reserves" are very detailed and quite complex and have considerable variation from country to country. In general the details vary, but the central issue is constant—money. Your field does not have to actually be on production, but you need to have let contracts, scheduled work, and committed funds to field development before you can book any reserves. The key element in determining the magnitude of the asset you are going to create is that the hydrocarbons represented by the booking must be: (1) able to be developed with existing technology using already acquired acreage and (2) be worth more than the cost of recovery. We could not have booked reserves for the Marcellus Shale in 1940 because: (1) the expected price of gas was approximately zero; (2) horizontal drilling technology did not exist; and (3) hydraulic fracture stimulation had not yet been invented. At that time, while the industry knew that there was a lot of gas to be recovered, these resources would have failed both the "existing technology" and the "economic viability" tests.

Proved reserves are further broken into the following:

- Proved undeveloped: resources under acreage that has been leased, has been evaluated, but does not yet have funds committed to drill the wells and/or develop the infrastructure. In many countries this category cannot be reported to regulators and/or potential investors.
- Proved developed, nonproducing: leased acreage, some number of wells drilled, infrastructure development underway (funds committed). This is a financial asset.
- Proved developed: leased acreage, wells drilled (not necessarily all of them), some amount of infrastructure completed enough to put the acreage on production.

1.8 CONCLUSION

This chapter was not intended to turn facilities engineers into reservoir engineers; it was intended to help you master the concept that there is little certainty in the calculations and predictions of the quantity, pressure, temperature, and fluid composition of the fluids that a facilities engineer has to design for.

You need to think about building facilities that are extensible for the case where the reservoir engineering prediction is one-fourth of actual. You need to think about building facilities that are still profitable if the reservoir engineering prediction is four times actual. No set of actual conditions will ever exactly match your design parameters, and no one is ever going to apologize for that. Often they won't even be close. No system you ever build will be suitable for a 16-fold variation in well performance. It can't be done. The only solution is to think about extensibility. For example, you can use a 20-in trunk that was designed for 60 MMSCF/day for 10 MMSCF/day if you understand that the fluid dynamics in the line are closer to the fluid dynamics of a separator than a pipeline and build a pigging capability to remove the liquid that will inevitably collect. In the same vein, if you terminate your 4-in flow line into an 8-in tee on the trunk, then you can economically extend the capability of the flow line without having to shut the trunk down for the new tie in if the expected 3 MMSCF/day well comes in at 15 MMSCF/day. We'll talk about the details of both of these scenarios (and how to deal with them) in later chapters.

REFERENCES

EIA. U.S. Energy Information Administration—EIA—Independent Statistics and Analysis. (n.d.). Retrieved April 22, 2016, from https://www.eia.gov/totalenergy/data/annual/showtext.cfm?t = ptb0402.

DeRosa, N. (2007, September 15). Black Gold: Thomas Gold's deep hot biosphere and the deep-earth theories of the origin of petroleum. Retrieved June 13, 2016, from http://www.metaresearch.org/publications/bulletin/2007issues/0915/Mrb07cp5.asp.

Phys.org. Hydrocarbons in the deep Earth? (2009, July 26). Retrieved May 05, 2016, from http://phys.org/news/2009-07-hydrocarbons-deep-earth.html.

PRI. Spindletop, Texas, Paleontological Research Institute. Retrieved January 3, 2017, from https://www.priweb.org/ed/pgws/history/spindletop/spindletop.html.

Schlumberger. Carbonate reservoirs. Retrieved January 3, 2107, from http://www.slb.com/services/technical_challenges/carbonates.aspx.

Seidle, J. (2011). Fundamentals of Coalbed Methane Reservoir Engineering. Tulsa, OK: PennWell.

Schmoker, J.W., & Dyman, T.S. (n.d.). A review of previous USGS World Energy Assessments. Retrieved May 2, 2016, from http://energy.cr.usgs.gov/WEcont/chaps/RV.pdf.

Webster (n.d.). Retrieved April 21, 2016, from http://www.merriam-webster.com/dictionary/biogenic (other dictionary references replace "biogenic" with the referenced term).

Wikipedia, 1. Biomass (ecology). (n.d.). Retrieved April 22, 2016, from https://en.wikipedia.org/wiki/Biomass_(ecology).

Wikipedia, 3. Darcy's law. (2016, March 23). Retrieved May 17, 2016, from https://en.wikipedia.org/wiki/Darcy's_law.
Wikipedia, 4. Langmuir adsorption model. (n.d.). Retrieved June 20, 2016, from https://en.wikipedia.org/wiki/Langmuir_adsorption_model.
Wikipedia, 5. Oklahoma City Oil Field, Retrieved January 3, 2017 from https://en.wikipedia.org/wiki/Oklahoma_City_Oil_Field.
WyoHistory. The Jonah Field and Pinedale Anticline: A natural-gas success story. Retrieved January 3, 2017, from http://www.wyohistory.org/encyclopedia/jonah-field-and-pinedale-anticline-natural-gas-success-story.

FURTHER READING

DOE. "Lower and higher heating values of fuels." Web. April 14, 2016.
The Flood geology of oil. (2011, February 01). Retrieved June 13, 2016, from https://ageofrocks.org/2011/02/01/the-flood-geology-of-oil/.
King, G. "Hydraulic fracturing 101: What every representative, environmentalist, regulator, reporter, investor, university researcher, neighbor and engineer should know about estimating frac risk and improving frac performance in unconventional gas and oil wells", SPE 152596, 2012 Hydraulic Fracturing Technology Conference, The Woodlands, TX, February, 6–8, 2012.
Kutcherov, V. G., & Krayushkin, V. A. (2010). Deep-seated abiogenic origin of petroleum: From geological assessment to physical theory. Reviews of Geophysics 48 http://dx.doi.org/10.1029/2008RG000270, ISSN 8755-1209.
Oil On My Shoes. The original petroleum geology site! (n.d.). Retrieved May 17, 2016, from http://www.geomore.com/oil-and-gas-traps/.
Wikipedia, 2. Abiogenic petroleum origin. (n.d.). Retrieved May 05, 2016, from https://en.wikipedia.org/wiki/Abiogenic_petroleum_origin.

NOMENCLATURE

Symbol	Name	fps units	SI units
A	Area	ft^2	m^2
b	Langmuir shape factor	1/psi	
Cum	Cumulative production	SCF	
g	Gravitational constant	32.174 ft/s^2	9.81 m/s^2
g_c	Unit converter	32.174 ft-lbm/s^2-lbf	N/A (for now)
h	Height	ft	m
ID	Inside diameter	in	mm
Len	Length	ft	m
m	Mass	lbm	kg
$\dot{m}$	Mass flow rate	lbm/s	kg/s
MW	Molecular weight	lbm/lb-mole	gm/gm-mole
MW_{air}	Molecular weight of air	28.962 lb/lb-mole	28.962 gm/gm-mole

(Continued)

(Continued)

Symbol	Name	fps units	SI units
n	Number of moles		
OGIP	Original gas in place	SCF	
P	Pressure	psia	kPaa, bara
$\overline{P}$	Average reservoir pressure	psia	
P_i	Initial reservoir pressure	psia	
q	Volume flow rate at standard conditions	SCF/day	SCm/day
$\overline{R}$	Universal gas constant	1545.3 ft*lbf/R/lb-mole	8.314 J/K/mole
R_{air}	Specific gas constant for air	53.355 ft*lbf/R/lbm	287.068 J/K/kg
R_{gas}	Specific gas constant for any specified gas	R_{air}/SG_{gas}	R_{air}/SG_{gas}
SG	Specific gravity	None	None
T	Temperature	R	K
Z	Compressibility	None	None
v	Velocity	ft/s	m/s
V	Volume	ft^3	m^3
V_m	Coal gas content	SCF/ton	
$\dot{V}$	Volume flow rate at actual conditions	ft^3/s	m^3/s
γ	Mineral matter free mass fraction	Fraction	
κ	Permeability	D	D
μ	Viscosity	cP	cP
ρ	Density	lbm/ft^3	kg/m^3
ρ_{lang}	Rock density	gm/cc	
σ	Interfacial tension		
$\partial P/\partial x$	Change in pressure per distance unit	psi/ft	kPa/m

Subscripts

ASL	Above sea level
atm	Atmospheric
gas	Variable is referring to the gaseous phase
liquid	Variable is referring to the liquid phase
SI	Metric version of equation
std	Indicates variable is at standard conditions
unit	Indicates variable may be in nonstandard units
water	Variable is referring to the characteristics of pure water at atmospheric pressure and 60°F (15.6°C)
1	Upstream conditions, or first state
2	Downstream conditions, or second state

EXERCISES

1. What is the difference between biotic hydrocarbons and abiotic hydrocarbons and why should we care?
2. In a carbonate reservoir, the absolute permeability is 160 mD. The effective permeability to oil is 128 mD, effective permeability to water is 170 mD, and effective permeability to gas is 300 mD. What is the relative permeability to oil, water, and gas as compared to absolute permeability?
3. Is a pinchout a structural or stratigraphic trap?
4. What classification of reservoir would have "adsorption" as a significant contributor to OGIP?
5. If a reservoir is located at 16,000 ft (4877 m) below ground level has an initial reservoir pressure of 7135 psig (49.2 MPag), would it be called "overpressured" or "underpressured" and why?
6. A company's annual report lists several categories under the heading "potential resources" including prospective resource, contingent resource, possible reserves, probable reserves, and proved reserves, which of these is considered a current financial asset?

CHAPTER TWO

Well-Bore Construction (Drilling and Completions)

Contents

Practical Onshore Gas Field Engineering.
DOI: http://dx.doi.org/10.1016/B978-0-12-813022-3.00002-X

The discipline of "drilling" is thought by many to be the very heart of the Oil & Gas industry. The mantra goes something like "without drillers, you don't have a way to access the reserves, so we are the most important." It is interesting that the same could be said for every discipline (e.g., "without surface facilities you couldn't get the product to market"), but isn't. Drilling is inherently dangerous work that requires extreme discipline and focus. It is more common for a new engineer to start as a driller and spend their entire career as a driller than for a new engineer to spend a career in any other discipline (e.g., it is not at all uncommon for an engineer to move between production, reservoir, and facilities several times during a career).

The language of drilling is somewhat less accessible to outsiders than the language specific to other disciplines, and the drilling community is considerably less forgiving of errors in terminology than other disciplines. For example, I once saw a driller ask a receptionist on her first day in the industry "where is my wire?" When she asked "what wire?" he turned his back on her and walked off without explaining that he was looking for the "daily drilling wire" report from a drilling rig. This was an extreme example, but not knowing what a "blowout preventer" (BOP) is or not knowing that a "tower" is what the rest of the world calls a "shift" (i.e., a scheduled work period) is unlikely to encourage a driller toward cooperation.

"Drilling" ends with a wellhead on top of a hole in the ground. When the drilling rig is gone, a "completions rig" is brought in to run any logs, perforate the casing, and/or install tubing. This chapter includes both drilling and completions.

As a facilities engineer, I was never asked to do a casing design, mud design, or bit selection, but knowing that these things have to take place helped me to understand the constraints that the drillers put on a construction schedule.

2.1 DRILLING ENVIRONMENTS

Potential oil and gas reservoirs exist in every environmental subclass on earth. We have drilled on the ice in the North Slope of Alaska and Siberia, in most of the world's deserts, in deep (deeper every year) oceans, swamps, and on top of high mountains.

2.1.1 Onshore

Onshore drilling rigs typically have the benefit of space that few of the offshore operations have. Looking at onshore rigs makes it easier to visualize the systems that must be present for any drilling operation. Much of the onshore drilling equipment is truck-portable and can be moved from one site to another with minimal disassembly. Reassembly at the next well is a very well-defined process that tends to go in a predictable sequence. Some environments and reservoirs are not suitable for drilling rigs optimized for rapid disassembly/reassembly and these rigs are constructed in place and a rig move is a major and very expensive event.

Onshore rigs are typically classified "singles," "doubles," or occasionally "triples." This refers to the height of the derrick. A single can fit one joint of pipe between the traveling block and the rig floor. A joint of pipe is averages 31.6 ft (9.65 m) long. This means that to trip pipe on a Kelly drive single rig you have to stop at every joint to disconnect the Kelly, add the next joint, and reconnect the Kelly. A double rig has room to handle two-joint sections loaded in the pipe rack and has to stop drilling to add pipe half as often. For deep wells this results in a substantial time savings. For very deep wells a triple is cost-justified. Each step-up in lifting-length adds substantial cost per day for rig rental while reducing the number of days required to reach a target depth.

2.1.2 Offshore

Seventy percent of the earth's surface is currently covered by water. That number has ebbed and flowed over geologic time, and the coastlines have changed dramatically many times. We find marine deposits in the Rocky Mountains, and swampy deposits in deep water, but finding any given type of reservoir is more likely (70% vs 30%) in the ocean than on land. Consequently, offshore drilling, while significantly more difficult and expensive than onshore drilling, has been a significant segment of Oil & Gas drilling activities.

2.1.2.1 Fixed platform

In water up to about 1500 ft (450 m), it is often cost effective to attach a platform to the seafloor. These permanent structures are used for drilling, production, living quarters, and limited warehousing. Fixed platforms have long been the staple of offshore reservoirs in shallow water.

Wells drilled from fixed platforms typically have their wellhead above the ocean surface accessible from the platform.

2.1.2.2 Jack-up rigs

Jack-up rigs are transported with their legs jacked up to the minimum submergence and floated like a (very top heavy) barge. When the rig reaches the target location, the legs are jacked down (with additional sections added as needed) to the seafloor. Each leg can be individually adjusted for variations in the seafloor. Jack-up rigs are limited to fairly shallow depths, typically less than 400 ft (120 m), by the number of leg sections that can be transported. When a well is complete, a subsea wellhead is set and the jack-up rig is moved to the next well.

2.1.2.3 Semi-submersible rigs

Semi-submersible rigs have "legs" that are variable-buoyancy chambers that can be flooded or evacuated to keep the surface of the platform level and stable in unstable seas. The legs are evacuated to move the rig (creates a very high center of gravity that makes them unstable in heavy seas). When they are on location, the legs are flooded to the point that the surface of the platform is slightly higher than the final target elevation. Then multiple anchors are put out near each of the legs and set in the seafloor. The rig position control is a marvel of engineering in that the buoyancy of each leg and the tension on each anchor cable is independently adjustable on a continuous basis to keep the platform stable, level, and in place while a drill string is run to the seafloor and the well is drilled. A string of pipe that long has some flexibility, but not enough to handle the magnitude of swell or ocean current or storm winds, so the buoyancy and anchor tension system must be very robust and flexible to keep from shearing the drill pipe (or later production risers if applicable).

Semi-submersible rigs have been used in water up to 10,000 ft (3050 m) deep. They are used for both drilling and production.

2.1.2.4 Drillships

As you can see from Fig. 2.1, one significant difference between drillships and semi-submersible rigs is that drillships don't use anchors to hold them in place. To combat the forces of wind, current, and waves, they need very flexible propulsion systems that can compensate for surge (movement in the direction of the centerline), sway (rotation about the centerline), and yaw (rotation about some moving vertical line through the centerline) (Fig. 2.2). The propellers on drillships are independently powered

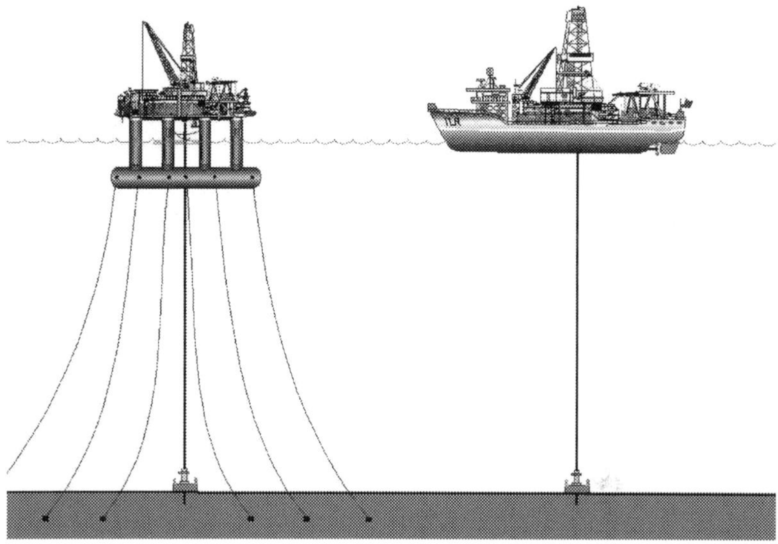

Figure 2.1 Comparison of semi-submersible (left) and drillship (Wikipedia, 1 Drillship). *Source: /Drillship (https://en.wikipedia.org/wiki/Drillship).*

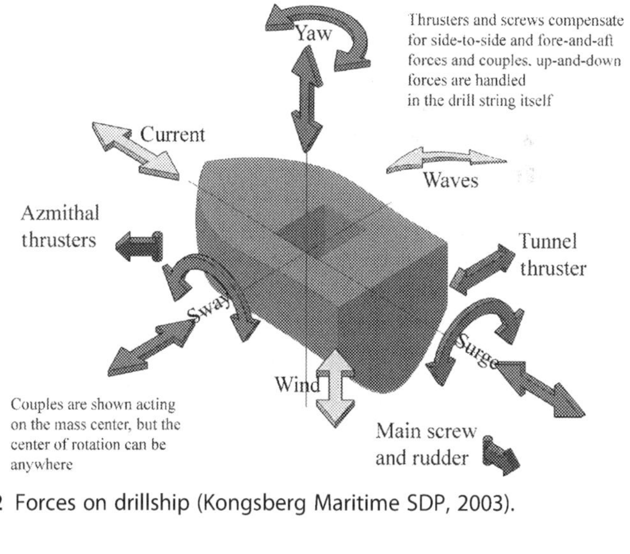

Figure 2.2 Forces on drillship (Kongsberg Maritime SDP, 2003).

via large electric motors and each screw can be rotated through 360 degrees to keep the ship in position. Less can be done about up and down movement, so the drill string has compensators to keep from translating vertical movement of the ship to the bit.

Drillships can be used in similar water depths as semi-submersibles. They can either be just drilling or they can stay on location after drilling is completed and become FPSO (floating production, storage, and off-loading) facilities (when used as an FPSO they put out anchors to reduce the precision required of the positioning system).

2.2 RIG COMPONENTS

Any drilling rig must have: (1) power, (2) the ability to raise and lower drill string; (3) the ability to rotate a bit; and (4) the ability to circulate drilling fluids. For onshore drilling the components look something like Fig. 2.3. Offshore and extreme onshore (i.e., rigs configured for very deep total depth or very challenging well fluids) rigs will have their components laid out differently, but they all have to accomplish these four functions.

Offshore rigs will additionally have subsea and/or subsurface pressure control and will tend to have multiple wells drilled from one surface "location" to try to keep costs as low as possible (still very high). An offshore drilling operation must also have the ability to transfer bulk materials from barges/supply ships onto the drilling facility.

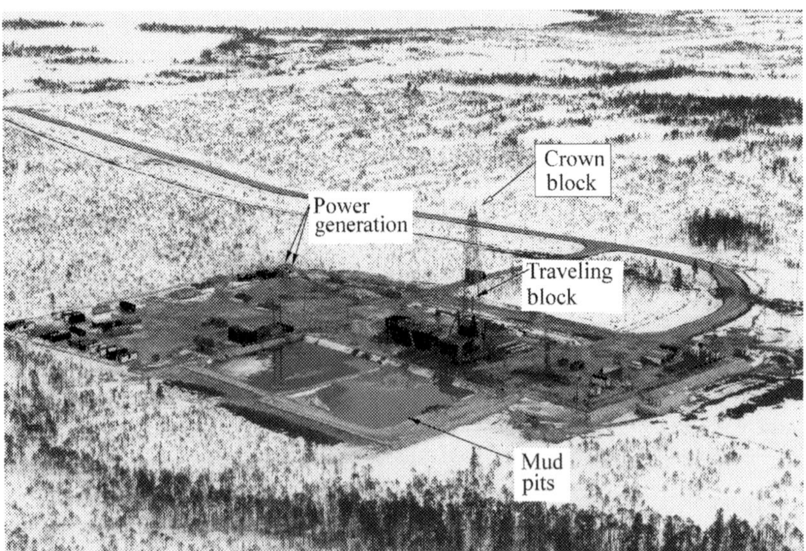

Figure 2.3 Land drilling rig components.

2.2.1 Power systems

The power is usually provided by several diesel engine powered generators in parallel. Loss of power during a critical drilling evolution can be exceedingly dangerous, so drilling rigs tend to have multiple redundant generators. Rigs can either be primarily electric or primarily hydraulic. The only major difference is that for an electric operation there will be a control switch house to distribute the power load and for a hydraulic operation there will be an additional (electrically driven) hydraulic pump to transfer the energy to the various systems.

2.2.2 Lifting Systems

The derrick carries the weight of the drill string, holds the crown block (Fig. 2.4), and holds pipe upright on Kelly drive rigs. The motive force for the lifting systems is the draw works which have a limited ability to store wireline. Excess wireline is stored in a spool off the dead line anchor, so if the rig needs to change from running one joint at a time to running several joints at a time, the extra wireline comes off of the storage.

The fast line runs from the draw works to the first pulley on the crown block. The wireline then runs down to the first pulley on the traveling block, then back up to the second pulley on the crown block, etc. There are typically five or more pulleys on the crown block and four or more

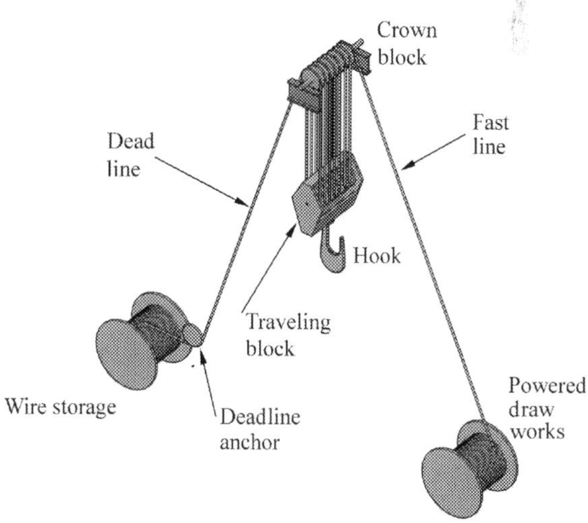

Figure 2.4 Lifting systems.

pulleys on the traveling block that results in the traveling block moving at one fourth the speed of the fast line (for a four active pulley system) and the force exerted by the hook being four times the force on the draw works. The wireline goes over the last pulley on the traveling block and down to the dead line, the dead line anchor, and the wire storage spool to anchor the wireline. Weight on bit measurement is determined at the dead line anchor (a critical parameter for drilling operations).

The hook on the traveling block holds the swivel which supports the drill string. The traveling block does not rotate.

2.2.3 Rotating systems

Historically, all energy needed for drilling a well was applied from the surface and the entire drill pipe, collars, and bit were all rotated together. For vertical wells this leads to good, predictable, and repeatable results. As the target location gets farther from being directly under the drilling rig, it becomes more difficult to rotate from the surface. For one thing, starting a deviation from vertical requires setting a down-hole mandrel (called a "whipstock") to point the bit away from straight down. For another thing, the drilling path was largely controlled by adjusting the weight on the bit. More weight on the bit tended to make a straighter hole while increasing the torque (and therefore the energy cost) required to drill while less weight on the bit tended to allow the bit to deviate so as to take the path of least resistance. The balance between too much weight and too little weight always was based on the Driller's "touch" on the rig. It often results in a straight hole, but not always.

To be able to drill high angle and horizontal wells, more control is required, so rotating the bit from the surface is ineffective for these sorts of wells.

2.2.3.1 Rotating from surface

For many decades, all drilling was done by rotary table rigs. Top-drive rigs are becoming more popular as materials capability improves because it allows fewer threaded connections to be made up each step (e.g., each joint of drill pipe in a "single" drilling rig).

Rotary table: In rotary table drilling, a section of the rig floor is a rotating table. Within the rotating table is a "master bushing" that will rotate with the rotating table and has a profile that will hold a "Kelly bushing" that can be lifted out of the master bushing. In the center of the Kelly bushing there is a rectangular or hexagonal hole that holds the "Kelly" and allows the Kelly to move up and down, but prevents it from rotating independently from the Kelly bushing. One end of the Kelly is attached to the "Swivel" on the traveling block and the other end is attached to the drill pipe.

When the bit has advanced to the end of Kelly travel, the drill string is lifted until the Kelly bushing is lifted from the master bushing and the top thread on the drill pipe is exposed. "Slips" (i.e., a tapered mandrel that prevents the drill pipe from falling when the Kelly is released) are installed in the master bushing and the Kelly is unscrewed. Now a joint of pipe can be added to the drill string and the Kelly can be attached to the new joint. At this point the driller lifts the string to allow the removal of the slips and the string is lowered back to the bottom of the hole, the Kelly bushing engages in the master bushing and the bit can be rotated again.

Top drive: It is easy to see that running two joints of drill pipe at a time (i.e., a "double") would require significant additional height on a derrick with a Kelly drive, in fact it would change the working height from around 70 ft (21 m) to over 140 ft (42 m). For very deep wells drilling from a single with a Kelly would take a very long time. To facilitate deeper wells, the industry developed "top drive". A top-drive rig puts an electric or (more common) hydraulic motor on the traveling block hook and attaches the drill pipe directly to the motor. The rig still has a rotary table and still uses slips in a master bushing to hold the pipe for adding or removing drill pipe from the string, but it lacks a Kelly or a Kelly bushing. A double top-drive rig would have the crown block about the same elevation as a single Kelly drive rig. A triple top-drive rig would only be about 30 ft (9.1 m) taller. A 15,000 ft (4570 m) vertical drill string is made up of about 500 joints of drill pipe and drill collars. With a Kelly drive single, the last trip into the well will have required setting slips 1000 times and making/unmaking threads 1500 times. The same job with a top-drive double would require setting slips 250 times and making/unmaking threads 500 times for a significant time savings.

2.2.3.2 Rotating in directional holes

Top-drive rigs have limited ability to control where the bit is or where it is going at any given time. For vertical wells this limited control is generally adequate. For high-angle and horizontal wells it is far from adequate. To get the control needed it was necessary to stop rotating the entire drill string and just rotate the drill bit (Fig. 2.5). This was accomplished by using a progressive cavity pump (PCP, see Chapter 3: Well Dynamics) in a unique configuration. Normally the rotor on a PCP is driven by a motor, but pumping fluid into the PCP will tend to rotate it. Using this characteristic, we are able to pump mud down the drill string, into the pump and out the bit. As the mud goes through the PCP it forces the rotor to spin, driving the drill bit through a transmission. A bent portion of the string points the drill bit slightly off of the drill pipe centerline. By rotating the string, the bit can be pointed in any direction.

Rotary Steerable Bit (RSS) technology was introduced by BP at the Wytch Farm (United Kingdom) extended-reach wells in the 1990s (PetroWiki Rotary Drilling). RSS technology is more complex than conventional steerable methods, but results in better hole control. For those interested in learning more about RSS, the reference given here contains a good description.

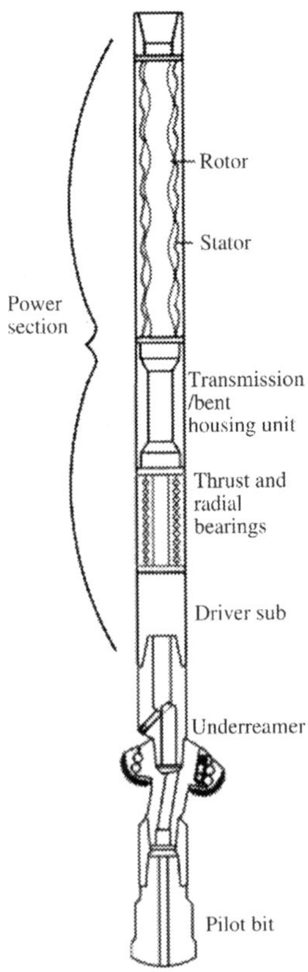

Figure 2.5 Mud motor sub.

2.2.4 Drill string

The down-hole portion of a drill string is generally similar for any type of drive. The very bottom of the string is a drill bit. Selection of drill bits is a very technical matter, four general types of drill bit are shown in Fig. 2.6. For every general type of bit there are hundreds of variations that are continuously changing as materials science evolves. A trip to replace a broken or dull drill bit typically costs many times the cost of the bit itself. The general types are listed as follows:

- Tricone bits are the most common historically, they are quite expensive and are used with mud.
- Polycrystalline diamond compact (PDC) bits are much less expensive than tricone bits and about as effective, they are taking over as the most common drill bit.
- Air hammer bits are used without mud, and have an internal mechanism to hammer the bit against the rock using airflow through the drill string that is then used to carry drill cuttings up the well.
- Coring bits are used to extract intact sections of target formations for laboratory analysis.

Above the bit you will have some combination of: (1) a down-hole motor (only for directional wells); (2) measurement while drilling (MWD) tools; (3) drill collars; and (4) drill pipe. Most of these are obvious (i.e.,

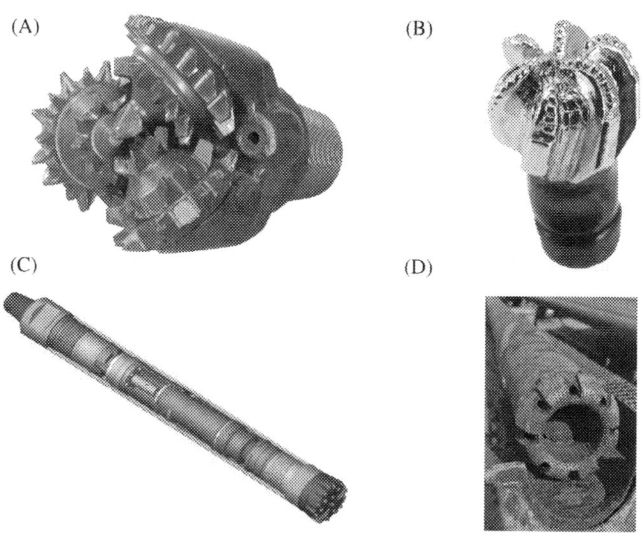

Figure 2.6 Types of drill bits: (A) tricone bit; (B) PDC bit; (C) air hammer bit; (D) coring bit. *Source: Purchased from Adobe Stock Photos.*

drill pipe) or discussed elsewhere (i.e., down-hole motor and MWD). Drill collars are the portion of this string that is not discussed elsewhere. Drill collars are the length of a joint of drill pipe (but come in various lengths to fine-tune weight on bit), they are hollow to allow drilling fluids to pass, very stiff, and much heavier than drill pipe. Adding drill collars above the bit will tend to keep a colinear force on the bit even if the drill pipe is not straight in the well (putting a very long string of pipe in excessive compression will cause the pipe to act like a rope in compression and will wander about the hole, but using inadequate compression will allow the bit to follow the path of least resistance). Since the drill rig cannot "push down," drillers adjust the number of drill collars to get the weight on the bit to the value desired.

2.2.5 Circulation systems

A 10-in (254 mm) diameter hole, 16,000 ft (4880 m) long has a volume of 8700 ft^3 (247 m^3) or a cube 21 ft (6.4 m) on a side. A major concern while drilling a well is removing that volume of pulverized rock and dirt from the hole. Removing the rock and dirt is done by the drilling fluid. The functions of the drilling fluid include: (1) lubricate, cool, and clean the bit; (2) control reservoir pressure; and (3) remove, transport, and release cuttings.

The circulation system needs to have a way to separate the drill cuttings from the mud for reinjection of the mud. This is done in a device called a "shale shaker" on the return line that mechanically removes the drill cuttings and sends the mud back to be reinjected in the well. The chemistry and density of the mud is frequently evaluated to ensure that the required properties are maintained.

2.2.5.1 Drilling fluids

Drilling fluids are selected for compatibility with the expected formation and the requirements for pressure maintenance. We generally assume that the reservoir pressure will be close to the pressure exerted by a column of brine (specific gravity of 1.07 lbm/ft^3 (17.1 kg/m^3)). In drilling terms this is called 8.92 ppg (pounds per gallon (1.07 kg/L)). If the company desires to drill a well "overbalanced" (i.e., with the drilling mud tending to flow into the formation instead of formation fluid tending to flow into the well-bore), then the driller will use a mud that is heavier than 8.92 ppg. Conversely if they want to drill "underbalanced" then they will mix a mud lighter than 8.92 ppg.

Drilling fluid will either be "water based," "oil based," or "gaseous."

Water-based drilling fluid: You start with water and add a drilling additive. *World Oil* magazine publishes an annual list of drilling additives and the list is about 2000 elements. Categories of additives include the following:

- Clays. Bentonite is the most common, but there are others that are used based on compatibility with the target formation. Clay is added to increase the mud weight.
- Viscosity control. These chemicals are generally called "friction reducers."
- Shale stabilization. Shale often has hydrophilic (i.e., "water loving") clay which can swell when wet. We add potassium chloride (KCl) to the water to reduce the affinity of the clay for the water.
- Corrosion control
- Biocides
- pH control
- Defoamers
- Emulsifiers
- Fluid loss reducers
- Floculants. These chemicals (primarily calcium) are added to help the drill cuttings clump for easier removal.

Any, none, or all of these additives will be present in the drilling fluid on any given well. There are times that call for drilling with water without any additives at all.

Oil-based drilling fluid: You start with a petroleum product (usually diesel, kerosene, fuel oil, or blends of these). Add chemicals from the categories discussed under "Water-based drilling fluid" earlier. Oil-based mud will do a better job of cleaning reservoir chemicals from the down-hole equipment, less likely to swell shales, improved lubrication, and it allows a higher penetration rate (more weight on bit).

Spills of oil-based mud carry with them an increased environmental risk. If oil-based mud gets into an aquifer the impact is greater than water-based mud. Finally, it is difficult to identify when/if reservoir oil has entered the reserve pit (a good thing) with oil-based drilling mud. Use of oil-based drilling fluids is becoming rare due to environmental concerns.

Gaseous drilling fluid: The use of gaseous drilling fluid is usually called "air drilling" and includes using high-pressure air, mist (i.e., water sprayed into air), and various foams. Air drilling is used to alleviate concerns about lost circulation of drilling mud into the formation (which can

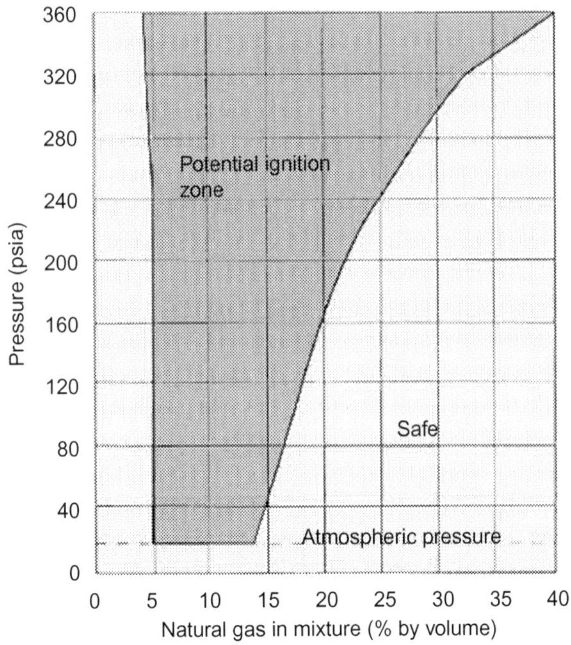

Figure 2.7 Methane LEL vs pressure.

damage the formation) and to allow formation fluids to be produced while the formation is being drilled.

Often, gas is saturated with water on the surface to: (1) minimize evaporation of reservoir fluids; (2) improve lubricity at the bit; and (3) suppress combustion. Air drilling is done with a hammer bit, so you create an interesting environment down-hole by injecting air, adding the potential for sparks by banging on solid rock with a steel bit, and (hopefully) bringing in formation hydrocarbons. That sounds like a perfect description of a completed fire triangle. To make matters worse, raising the pressure of the mixture increases the explosive range (Fig. 2.7). We reduce fire hazard by: (1) relying on the down-hole mixtures being "too rich to burn"; (2) replace air with nitrogen, CO_2, or methane; or (3) add enough water mist to quench any fires that start. The last option is the most common.

2.2.5.2 Pressure control

Pressure at the drill bit while mud drilling is a simple hydrostatic gradient.

$$P_{drill} = \rho_{mud} \cdot g \cdot h \tag{2.1}$$

The density of the drilling mud controls the pressure at the bit:

- $P_{drill} > P_{res} \rightarrow$ mud will flow into the reservoir. This is called "overbalanced drilling" and can result in formation damage, but keeps reservoir fluids off the rig floor.
- $P_{drill} < P_{res} \rightarrow$ reservoir fluid will flow into well-bore. This is called "underbalanced drilling" and results in a "kick" of reservoir fluids to the surface.
- $P_{drill} = P_{res} \rightarrow$ no net flow, and very unpredictable results.

Drillers adjust the density of the mud to manage hydrostatic pressure at the bit.

Reservoir pressure is a function of overburden pressure at the depth where the reservoir was sealed. The reservoir can move up or down over geological time. Mud weight is designed for expected pressures, but surprises are very common. If reservoir pressure is higher than the hydrostatic pressure on the bit, you can get formation fluids flowing into the well-bore and see a kick on the surface. Fluids in the kick can be saltwater, liquid hydrocarbons, or gas. This flow can be excessive which will result in blowing the mud out of the well and presenting the driller with a "blowout" or an uncontrolled well. The first line of defense against unpleasant surprises is mud weight. The second line of defense is the "BOP." BOPs are designed to perform specific functions:

- Annular BOP (Fig. 2.8), numbers refer to detailed design within the patent—hydraulic pressure is used to squeeze an elastomer around the pipe in the hole.
- Hydraulic Rams (letters refer to Fig. 2.9).
  - **a)** Blind rams—rams have a straight profile and are designed to seal with no pipe in well.
  - **b)** Pipe rams—each ram has a semicircle profile that is the same size as the drill pipe to seal around the drill pipe
  - **c)** Shear rams—rams have a cutting profile designed to cut through the drill pipe.

When reservoir fluids reach the surface as coherent volumes (as opposed to being dissolved in the drilling fluid) it is called a "kick." Not all kicks are as dramatic as a blowout. Often the first evidence of a kick is that the mud flow rate increases or pit volume increases or you see an oil sheen on the reserve pit. Sometimes you will see gas bubbles coming out of the mud (called "gas cut mud"). There are three general categories of kick control:

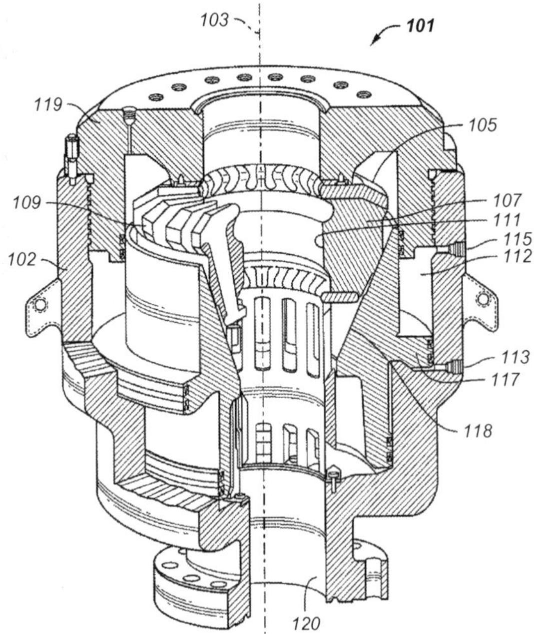

Figure 2.8 Annular BOP. *Source: Khandoker, S. (n.d.). Patent US20080023917—Seal for blowout preventer with selective debonding. Retrieved August 01, 2016, from https://www.google.com/patents/US20080023917.*

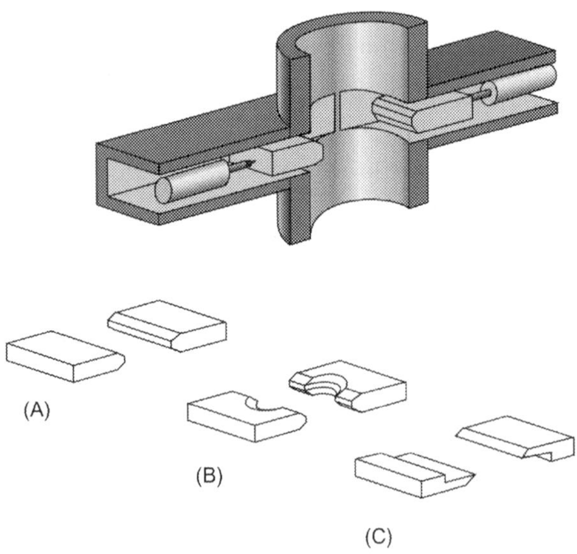

Figure 2.9 BOP rams. (A) blind rams; (B) pipe rams; (C) shear rams. *Source: Wikipedia, 2 Blowout Preventer. (n.d.). Retrieved August 01, 2016, from https://en.wikipedia.org/wiki/Blowout_preventer.*

- Driller's method—with the bit on the bottom, circulate reservoir fluid out of the hole with the pump at half speed. Raise mud density to kill weight. Circulate kill-weight mud.
- Engineer's method—with the bit on the bottom raise the weight of the drilling fluid to kill weight. Circulate kill-weight mud.
- Volumetric method—with the bit off the bottom you can equate change in mud pit volume to a change in annular pressure. As gas kick comes in, pressure increases. Reduce high casing pressure by bleeding off a volume of liquid to allow controlled expansion. When pressure stabilizes, trip pipe to the bottom and use one of the other methods.

  BOP stacks are activated when the fluid methods are ineffective.

2.3 HOLE TOPOLOGY

The traverse of a drill bit through the earth must be mapped in three dimensions. To define a bottom-hole location, you need to determine the following:

- Surface location—latitude, longitude, and elevation above sea level of the rig floor (historically designated as "Kelly bushing elevation" or KB, even if it is a directional well and the Kelly is absent).
- Kickoff point—this is the point below the surface location where the bore purposely leaves the latitude and longitude of the surface location.
- Azimuthal (compass) direction—this is the direction you would need to go from the surface location to stand over the drill bit.
- Build rate—this is a unique term to drilling. It is usually expressed in degrees per 100 ft (degrees per 30 m) and it is a rate of change of direction from the vertical. For example, if a tangent to the bit trajectory was 25 degrees from vertical at 5000 ft and you had a 3 degrees/100 ft build rate then at 5100 ft a tangent to the bit trajectory would be 28 degrees from vertical. The higher the build rate, the higher the "dog leg severity" and the more difficult the well will be to operate.
- Inclination angle—the angle from vertical of the tangent that results at the end of the build.
- True vertical depth (TVD)—the depth of the bit without regard to trajectory. It is measured from the surface location elevation so if the

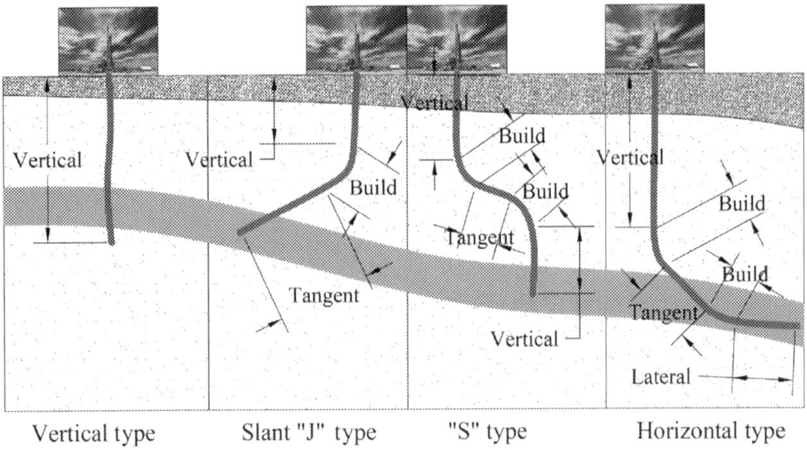

Figure 2.10 Hole shape.

bit runs laterally under a large hill you do not add the height of the hill to the surface location elevation to get the TVD.

- Measured depth—basically this is the length of the drill pipe.
- Horizontal displacement—distance along the azimuth from the surface location to a point above the drill bit.
- Lateral—the length of pipe running more or less horizontally and tangential to the last build section.

Wells generally have one of four shapes as shown in Fig. 2.10.

- Vertical type—the preponderance of the nearly 2 million Oil & Gas wells in the world are vertical type. When we say "vertical type" instead of "vertical" it is because no well is actually vertical, they all have some amount of deviation caused by rate of penetration, weight on bit, and natural discontinuities in the rock being drilled through. The vertical type well was drilled without any purposeful intention of deviating from vertical.
- Slant "J" type—these wells are "kicked off" using a device called a "whipstock" that creates a pathway for the bit to move off of vertical. We use slant "J" wells to reach specific bottom-hole locations that cannot be reached with a vertical well (e.g., the target location is under the grounds of a school). They are also used to create a sidetrack to move away from bottom-hole location that was damaged.
- "S" type—at times there are reasons to enter a formation with a vertical bore, but circumstances make drilling a vertical-type well undesirable or impossible. For example, when we started drilling the CBM wells in

the San Juan Basin, the concept of "cavitation" (see later) had not yet been invented so all of the early wells were cased and frac'ed (see later). The industry quickly realized the benefit of the new technique and many wells were reentered, sidetracked, and the new hole was cavitated. We then learned that the coal was so weak that entering it at a high-angle slant "J" configuration would not allow the hole to support itself and the holes in those wells collapsed. This was corrected by drilling "S" type wells and entering the coalbed with a vertical hole.

- Horizontal type—formations with low permeability have limited ability to flow reservoir fluids from the reservoir into the well-bore. In many of these formations you can increase the recovery by drilling horizontally through the formation and then opening the entire length of the lateral to the well-bore.

We choose to deviate from vertical wells for a number of reasons. We discussed inaccessible surface locations and sidetracking earlier. We can also use one of these general types of wells to access multiple bottom-hole target locations from a single drilling location (i.e., multiwell pad or platform).

Control of directional wells has gotten so advanced in recent years that it is common to be able to hit a target circle less than 1 ft (305 mm) in diameter from hundreds or thousands of feet away. This control makes it possible for one well to intersect the well-bore on another well from a considerable offset distance. For example in the event of a blowout with damaged casing, you can drill a "relief well" to intersect the damaged well to pump cement to kill the blowout.

2.4 WELL-BORE TUBULARS

The size designation of all Oil Country Tubular Goods (OCTG) is based on outside diameter, weight per foot, material strength, and end finish (e.g., type of threads). Nominal 2-in (50 DN) tubing is called "2-3/8." One very common inside diameter is 1.996 in (50.7 mm) which is designated "4.7 lbm/ft" (it doesn't have to make sense, you just have to recognize it). The pipe grades come in two parts, first there is a letter then a number (e.g., J55 or N80). Each letter designation refers back to a specific metallurgy specified in API 5CT. The number is the nominal

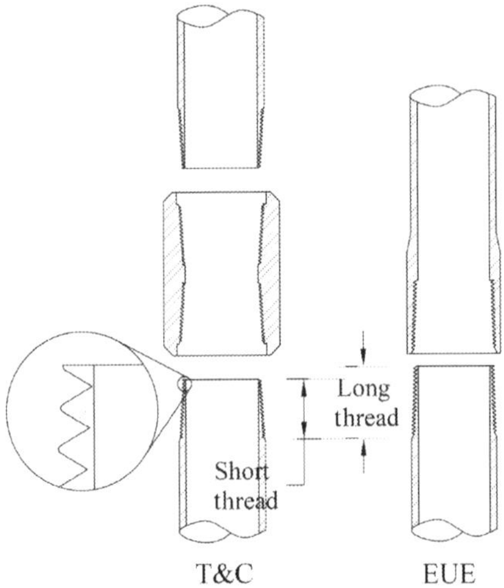

Figure 2.11 OCTG end connections.

yield strength of the steel ("55" is 55 ksi (379 MPa) yield strength, and LS125 would be a non-API composition specific to the Lone Star Steel subsidiary of US Steel with a yield strength of 125 ksi (862 MPa)). The end finish would be things like T&C (threaded and coupled), EUE (external upset end, meaning that the OD of the female threads is greater than the OD of the rest of the pipe, see Fig. 2.11), Buttress (a specialty thread used for increased strength), etc.

Tubing, liners, and casing all come in nominal 30 ft (9.14 m) lengths.

We put pipe into wells to support both the drilling activity and the production activity. During drilling the casing carries the BOP, prevents well-bore collapse, contains and conducts the drilling fluids, and guides logging/completion tools to target locations.

During production the well-bore tubulars contain reservoir fluids and keeps them out of up-hole formations, carries the weight of wellhead equipment, and reduces corrosion risks.

Well-bore tubulars include casing, liners, and tubing. Casing is always cemented in place. Liners may or may not be cemented in place. Tubing is not cemented in place.

2.4.1 Casing/Liners

The various strings of casing or liners (Fig. 2.12) in a well are each there for a reason. Some strings are there to protect the environment from the drilling/production fluids while some strings are installed to protect

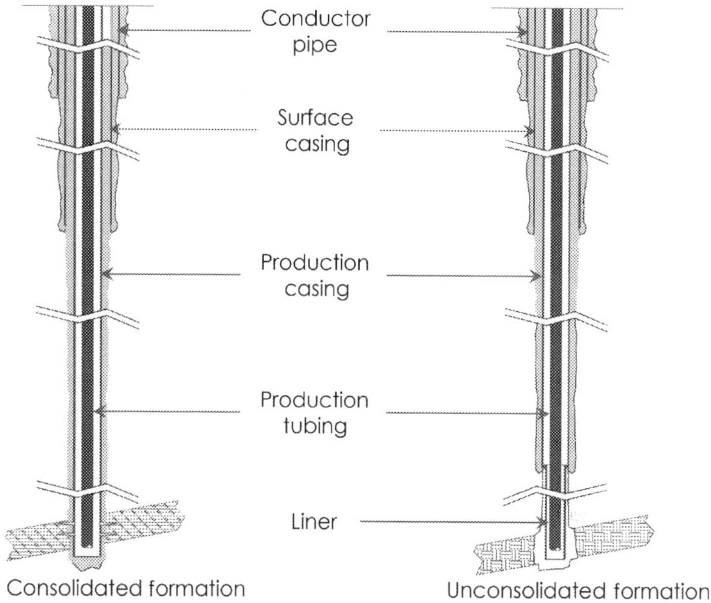

Figure 2.12 Down-hole tubulars.

formations above the target zone from excessive hydrostatic pressure and/or to prevent losing drilling fluids into up-hole formations. For example, many CBM plays lie on top of conventional and tight gas plays. As we've seen, the coal is not mechanically strong and can be fractured with relatively low imposed pressure. To keep from losing drilling fluids into the coal it was common when drilling through the coal to either set casing across it or to significantly lighten the mud density or both.

2.4.1.1 Casing design

Designing casing is a technical subject that must consider many factors. Fig. 2.13 describes the broad outlines of the process. The "boundary conditions" include the strength of the various rock strata, pore pressures in the strata, mud weights, geology, directional well plan, drilling fluid design, and the expected corrosiveness of reservoir fluids.

The "jewelry" being considered includes logging tools, testing equipment, production equipment, and contingencies. Production equipment is an important consideration. If the expectation is that the well will eventually have an electric submersible pump (ESP, see Chapter 3, Well Dynamics) and the expected liquid volumes from the formation require a pump with a 6-in (152 mm) outside diameter then casing or liners smaller

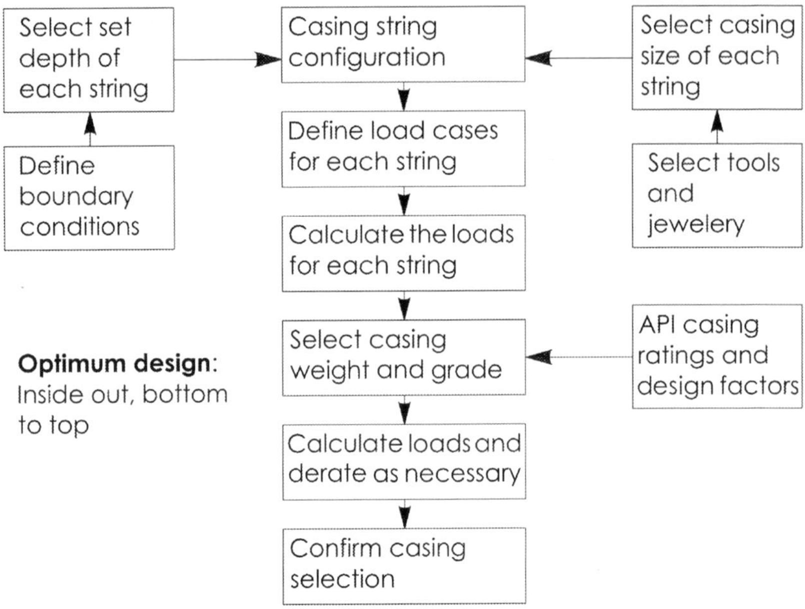

Figure 2.13 Casing design flowchart.

than 6-in would require changing the production plan to equipment that will fit (and will likely be more expensive or be unable to pump required volumes).

The load cases include drilling mud weight, string weight, and reservoir pressure. When doing load calculations you have to consider internal and external axial loads, net burst, and collapse loads.

2.4.1.2 Cellar

Often the first step in drilling an onshore well is to dig a "cellar." The cellar is a box that provides space below the rig floor for the BOP and casing hanger(s) and holds the drilling fluid returns while drilling conductor pipe and surface casing. Cellars can be lined with just about anything from prefabricated cement boxes to wood-lined holes or holes lined with corrugated metal culvert material.

2.4.1.3 Conductor pipe

When you look at historical well records, you see a "spud date." That is the date that the conductor pipe was started into the ground. If the well will have a cellar it is dug and lined prior to spudding the well. In unconsolidated surface formations, the conductor pipe can be "spudded" (hammered) into the ground, in other formations conductor pipe is drilled with the drill cuttings and mud return being accumulated in the cellar. Conductor pipe is set deep enough to allow later strings to be started in consolidated rock. Depths of 40–300 ft (12.2–91.4 m) are common. The conductor pipe is cemented to surface. The primary purpose of conductor pipe is to allow mud returns to be directed to the pits. Conductor pipe is not load bearing.

2.4.1.4 Surface casing

Surface casing is run through the conductor pipe and set deeper than any potential aquifers. It is not uncommon for companies to use a different drilling fluid for surface casing than for the rest of the well because contamination of a drinking water supply with oil-based mud (for example) would be exceedingly difficult to mitigate and would be potentially hazardous to the public. Once surface casing is set, it is cemented to surface and it is common to run cement bond logs (CBL) (see later) on the surface casing to ensure that it is protecting up-hole formations from subsequent drilling activities. Surface casing is therefore both to protect the aquifers and to prevent the aquifers from introducing uncontrolled water into the drilling mud and changing its properties. Surface casing is load baring, and has a flange at the top to hold the casing hanger for later intermediate and/or production casing (Fig. 2.14).

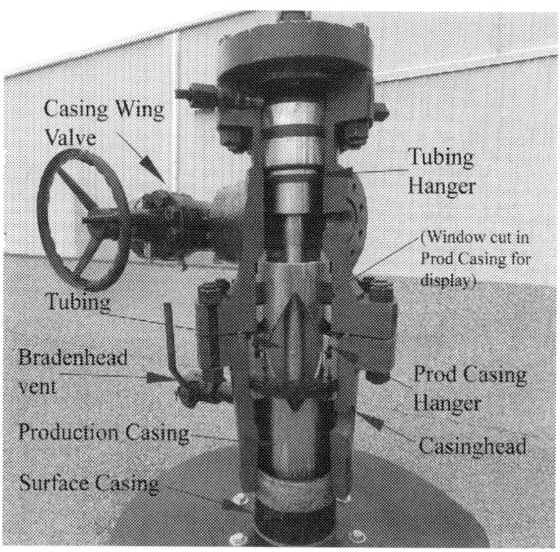

Figure 2.14 Wellhead.

The area between the surface casing and the intermediate casing is known as the "bradenhead." Pressure in the bradenhead is the primary indication of a poor or failed cement job on the intermediate or production casing. The bradenhead pressure is closely monitored during stimulation activities and periodically during the life of the well.

2.4.1.5 *Intermediate casing*

Intermediate casing (not shown in Fig. 2.14) is run inside the next larger casing (either surface casing or an up-hole section of intermediate casing). It is used to protect up-hole strata from the weight of mud required for pressure control in the target formation. Intermediate casing is hung from a casing hanger at surface and is installed past troublesome zone(s). There can be multiple intermediate strings on very deep or very complex wells. Intermediate casing is usually cemented to surface or above the casing shoe on the next larger string, but it can be un-cemented in certain situations.

2.4.1.6 *Production casing*

Production casing is the final casing run and is hung from a casing hanger on the surface. In conventional wells the production casing is generally set below the target zone, but it can stop above the target zone if it is desirable to change drilling fluid for penetrating the target zone. Production casing must be able to withstand full wellhead shut-in pressure, full bottom-hole pressure, and any mud or workover kill-fluid weight. It is common to flow reservoir fluids up the casing (or casing/tubing annulus) so the production casing must be compatible with expected reservoir fluids. Production casing is cemented to surface.

2.4.1.7 *Liners*

If part of the well-bore tubulars is hung from some point in the well below the wellhead, then it is called a "liner." The top of the liner is set inside a casing string and a liner can be cemented or un-cemented. People run liners that are unperforated (i.e., simply joints of pipe), slotted (slots are filled with some friable material that is supposed to be knocked out by dropping a tool in the well), or pre-perforated (which requires killing the well to run or remove).

2.4.1.8 Wellhead

Once the surface casing is set, a flange to attach the casinghead to the surface casing is attached either by threads, butt weld, or socket weld. Fig. 2.14 shows many of the typical components of a wellhead. If there were intermediate casing string(s) then each string would have its own casinghead and casing hanger. As you can see, the production casing ends in the tubinghead and is sealed from the production tubing by the tubing hanger assembly. The wing valve allows the annulus to be connected to a sales line. The wellhead is installed in stages. The largest casinghead to be used is bolted onto the surface flange after the surface pipe is cemented in the well (but before the cement plugs have been drilled out). The BOP is bolted onto the casinghead for the next drill section. When the largest production casing is hung from the first casinghead and cemented in place (but before the cement plugs have been drilled out), the BOP is removed and the next casinghead is installed, the BOP is reinstalled, and the cement plug is drilled out. In this manner the wellhead is installed in stages as the well progresses.

2.4.1.9 Tubing

Tubing is not normally put in the well by the drilling rig and will be discussed later.

2.5 CEMENTING

Cement is put into the well to provide zone isolation and segregation, corrosion control, formation stability, pipe strength improvement, to prevent migration of fluids into the pipe-to-dirt annulus, and to provide structural support for the casing string. The conductor pipe and surface casing are always cemented to surface. Intermediate strings are sometimes cemented to surface, but are often only cemented to a few joints above the casing shoe of the next bigger pipe. Production casing is normally cemented to surface. When it is not cemented to surface it is difficult to prove that produced fluids have not migrated into potable water aquifers.

In general terms cementing is either "primary" or "remedial." In primary cementing the cement exits the tubulars through the lower end. For remedial cementing the cement generally leaves the tubulars through holes cut into the pipe for the purpose of placing cement into intermediate location(s).

The majority of the cements used in oil and gas wells are based on Portland cement (limestone) made up of at least two-thirds (by mass) calcium silicate (Ca_2SiO_4 and Ca_3SiO_2). "Cement" in this application is not "concrete" because there is no sand or gravel in the mix. Properties of the cement slurry depend on its components and additives (things like accelerators to make it cure faster, retardants to make it cure slower, friction reducers, etc.). The fineness (and consistency) of the grind and impurities in the water impact both the cure time and ultimate strength of the cement.

The chemistry of cementing is complex and allows for a wide range of properties of the slurry and of the cured pour. The industry has been refining groupings of mixes to accomplish specific tasks in specific environments since American Society of Testing and Materials (ASTM) first published the specifications for five types of cement in 1940. The American Petroleum Institute (API) issued API Code 32 in 1948 (it later became API RP 10B in 1952) to define properties and suitable environments for the five classes (increased to six classes in 1952). In 1953 the document was upgraded from a "Recommended Practice" to a "Standard" (API Std 10A). In 1972 the "Standard" became a "Specification" (the highest designation in the API list of document types, specifications are generally very proscriptive, mandatory, and don't leave much room for interpretation). In 2000 ISO published API Spec 10A as ISO 10426. API Spec 10A has eight classes of cement designated "A" through "H" (a ninth class designated "J" has been in use, but it is not described in API Spec 10A).

Classes A, B, and C are classed as "construction cements" and have the characteristics of ASTM C150 cement types. These are only suitable for shallow wells (down to 6000 ft (1830 m)) and are rarely used down-hole (tend to be used where Classes G and H are not available to meet a drilling schedule or the job is such low pressure/temperature that widely available local ASTM Type V cement can be substituted for a much lower cost).

Classes D, E, and F were designed to be specific to high-pressure and -temperature environments (deeper than 6000 ft (1830 m)), but were abandoned in the 1980s when it was found that the desired properties were more economically achieved with additives introduced to the mix on site.

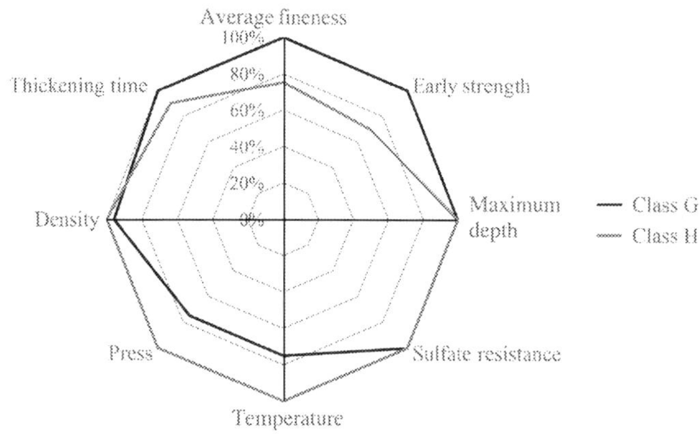

Figure 2.15 Cement comparison.

API Spec 10A describes the chemistry of the various classes, and Classes G and H have identical definitions. Looking at other sources, you find that Class H is a slightly coarser grind and takes a bit less water to form a slurry (resulting is slightly higher slurry density). There is a considerable published performance data for Class G and H cements, and a very wide range of results for any given parameter. Fig. 2.15 is an attempt to synthesize all of these divergent sources into a single picture, and it should be taken with considerable skepticism.

Something like 80%–90% of cement jobs in Oil & Gas use Class G cement. Both Classes G and H can be used with additives like accelerators and retarders.

Class J was developed for deep, high-temperature/high-pressure applications and is not included in API Spec 10A.

Additives are generally categorized as:

- Accelerators: purpose is to reduce cement setting time and speed up the development of compressive strength. Primarily used in shallow, low-temperature wells.
- Retarders: purpose is to extend cement setting time to allow enough time for slurry placement in deep wells.
- Extenders: purpose is to make the cement go farther and are used to reduce the density when cementing weak formations where full-weight cement would break down the formation.
- Weighting agents: purpose is to increase cement density for cementing high-pressure reservoirs which might become unstable if slurry density is too low.

- Dispersants: purpose is to reduce the viscosity of cement slurry and ensure good mud removal during placement.
- Lost-circulation material: purpose is to reduce the amount of cement lost to weak formations.
- Special additives: these chemicals include antifoam agents, corrosion control, etc. and are used for designated purposes.

The quality of water required varies widely depending on the cement performance required. There are examples of using seawater, brackish water, and of course potable water. Each water quality requires different additives and results in different properties. The properties (especially density) are checked periodically during the cement job.

2.5.1 Mixing

Cement is mixed with either jet mixers or batch mixers. Jet mixers are used for almost all large jobs and combine cement, water, and additives in a single pass operation. Batch mixers allow closer control of properties in critical, small jobs. You can extend the open time of batch jobs by rotating the mix in large tanks, but you can still only mix a limited amount at one time.

Bags of dry cement are cut open on the installed blade on the cutting table and emptied into the hopper. Water is pumped from the supply source into the jet pump (see Chapter 3: Well Dynamics for a discussion of jet pump operations and capabilities) that pulls on the dry cement in the hopper. The sizes of the nozzles in the pump ensure the ratio of water to cement. After the divergent nozzle in the jet pump, there is a port to pump in any additives that are required. The slurry empties into the slurry tub which serves as the suction vessel for the slurry pump.

Optimum ratio of water to cement is a compromise. Free water results in separation occurring at the top of a long column or creating pockets in the sheath. Pockets contribute to annular gas leakage and other annular flow problems.

2.5.2 Placing cement

The initial cement job fills the annular space between the casing and the hole from the casing shoe to the surface or a point several hundred feet above the zone that must be isolated. The first cement job is called primary cementing and its success is absolutely critical to the success of subsequent well control and completion operations.

Primary cementing either places a sheath of cement the entire length of the casing string (called "cementing to surface") or to a point some distance above the casing shoe of the next largest casing. The quality of the cement job is evaluated with various tools (see later) and if portions of the job do not meet the quality requirements then the job can be repaired using a technique called "squeeze cementing."

2.5.2.1 Primary cementing

The bore hole must be larger than the pipe or the pipe will not go in. The entire bore hole is full of drilling fluid when the casing is run in the hole. All of the drilling fluid (both inside and outside of the casing) must be removed from the well before the primary cement job will be effective. To accomplish this, a hollow "bottom plug" made of resilient material that conforms to any irregularities in the casing and wipes the casing wall is put in the well and chased with cement. Often a fluid is run on top of the mud and below the bottom plug to help the bottom plug remove all of the mud from the casing walls. When the bottom plug reaches the bottom of the casing (called the "casing shoe"), it stops and an internal rupture disk fails and allows the cement to flow through the hollow core of the bottom plug and into the annulus. At the calculated end of the volume required, a solid core "top plug" is run to push the cement out of the casing. When the top plug is against the bottom plug, pressure increases abruptly and the driller knows that the top plug is on the bottom. Both plugs can then be drilled out.

Filling the entire annular space between the bore hole and the casing with cement is crucial to the functions of a primary cement job. If the pipe is close to the borehole on one side (Fig. 2.16) then it is common for the cement to fail to displace the mud. Since the mud is not designed to perform the functions of a primary cement job, this can be a major problem. Off center pipe is corrected using centralizers.

In a normal cement job, you have the full hydrostatic pressure of the column of cement from a given formation to surface. When a weaker seam is in the cemented interval this pressure can be too high, and cement can flow into the formation. When there is concern about the mechanical strength of a given formation, the driller can use down-hole tools to do the job in two stages. The first stage is conventional and fills the annulus from the bottom. When the first stage cement should have reached the location of the ports for the second stage, stop, set a packer,

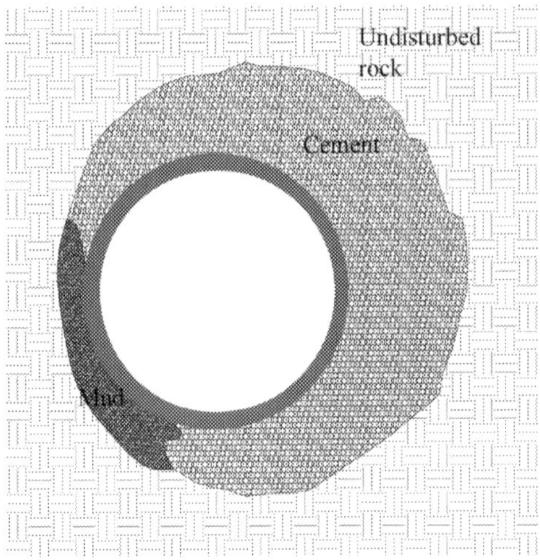

Figure 2.16 Uncentralized cement job.

and reposition a down-hole fitting to open ports in the production casing to allow the cement to enter the annulus from the location of the ports and rise to surface with much lower hydrostatic pressure. When the cement job is finished the ports are closed to maintain the steel/cement isolation system in that location.

2.5.2.2 Remedial cementing

The repair job is referred to as squeeze cementing. In a squeeze job, cement is forced into the zone through perforations, ports in tools, holes in casing produced by corrosion, or through the clearance between casing-overlap liners or strings. Although squeeze cementing has become commonplace, it is expensive and its use can be minimized through improved primary cementing procedures.

It is used to repair a primary cement job, repair a casing failure, or to seal off depleted zones to allow other zones to produce. A squeeze job requires perforating the casing (see later for a discussion on perforating) and setting a tool to force the cement through the perforations.

Squeezes are either done at high pressure or low pressure. High-pressure squeezes expect to fracture the formation and fill the voids developed with cement and a large fluid loss is expected. Low-pressure squeezes have a more controlled amount of cement and are not expected to fracture the formation.

2.5.2.3 Liner cementing

Cementing a liner requires special equipment and techniques to obtain a seal in the close clearances between the liner and the casing string. The types of jobs are "Modified Circulation Job" which is similar to a squeeze, and "Puddle Cement Technique." Under the puddle technique you spot cement in the open hole with drill pipe, and then put the liner in the cement "puddle." Puddle jobs are generally less effective and are only used with short liners.

2.5.3 Cement evaluation

The cement job is the primary assurance that groundwater will not contact well-bore tubulars and the secondary assurance that reservoir fluids will not contact groundwater. There are many things that can go wrong with a cement job, and the industry has developed extensive processes to predict the effectiveness of a cement job.

2.5.3.1 Pressure test

Once cement has set, the casing hanger is installed under the BOP, then a drill bit that will fit through the casing is run to drill out the plugs and the casing shoe and slightly into the formation below the casing shoe. At this point the casing is tested under pressure to ensure that a leak-tight cement job has been obtained. If the pressure test does not hold, the cement job is evaluated for the various problems discussed later and problem areas are squeezed and the casing is retested.

2.5.3.2 Temperature log

The action of curing cement is significantly exothermic. For something like 12–24 hours after a cement job, a simple down-hole temperature indicator can show you the top of the cement, but it cannot indicate how well the cement has filled the voids or bonded to the pipe/borehole.

2.5.3.3 Radioactive log

Sometimes radioactive tracers can be added to the cement slurry to identify where the cement is and isn't using a radiation counter. This works best in areas with low background radiation.

2.5.3.4 Cement bond log

This is basically a sonic (acoustic) tool run on a wireline. It indicates quality of the cement bond both to the pipe and to the borehole. The distance between the transmitter and receiver is about 3 ft (1 m) and the time it takes for a signal to reach the receiver and the amplitude of the returning signal indicate bond quality.

2.5.4 Drilling wrap-up

During drilling, various ancillary activities may-or-may-not, must-or-must-not take place. For example, it is becoming common on deviated wells to run logging tools immediately above the drill bit to assess the down-hole environment in real time. This is called "logging while drilling (LWD)" or "MWD" and is discussed later. If caliper logs (Fig. 2.17) are run, they have to be run while the drilling rig is on the well because they have to be run in the uncased hole. Logging is generally considered a post-rig-down activity since having the drilling rig on location while running logs, perforating, or running tubing is expensive when these activities can be done from less capable (and cheaper) equipment. It is common to set the tubinghead and tubing master valve(s) with the drilling rig, but installing the rest of the "Christmas tree" of valves and interconnections is usually left for a workover rig.

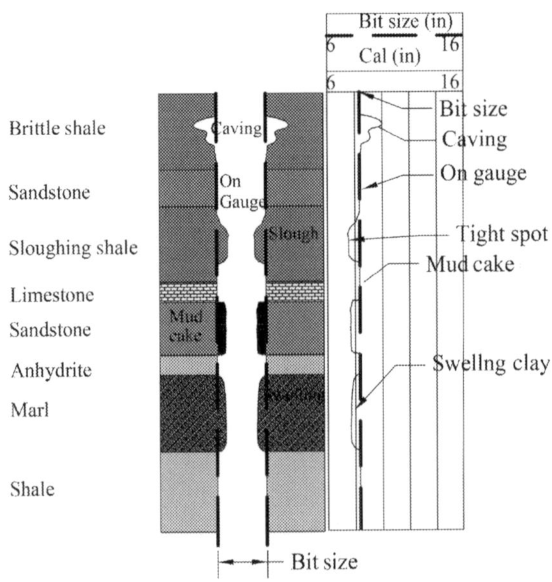

Figure 2.17 Mock-up of caliper log.

2.6 LOGGING

Logging tools can be used to estimate the formation density, porosity, permeability, type, natural fracture orientation, pressure, fluid (oil, gas, water) content and proportions, and temperature. Logging tools can also gauge the inside diameter of the raw well-bore. Logging tools are generally either electrical, sonic, or nuclear.

2.6.1 Electrical

Conventional electrical logs are run on a wireline and are capable of real-time data acquisition on either the trip into the well or the trip out of the well. These rigs have equipment to provide a precise location of the tools being run so multiple logging suites can be depth-indexed.

2.6.1.1 Caliper

A caliper tool (Fig. 2.17) can be thought of as central body with some number of spring loaded arms that have instrumentation to capture how far each arm is from the central body. The arms tend to be installed opposite of each other so that the central body position need not be considered. The tool is lowered into the open hole with the arms retracted against the central body. On the bottom of the hole, the arms are released and the hole profile is assessed as the tool travels up the well.

Tools with four or fewer arms give you an average inside diameter of the hole. Tools with more than four arms can also give you ovality of the hole which can be a good indication of formation anisotropy (e.g., the rock has a different stress tolerance in one direction from the stress tolerance in the other direction) which can be important for perforation orientation.

2.6.1.2 Acoustic (Sonic)

These tools measure a formation's ability to transmit seismic waves to allow the evaluation of porosity, lithology (or rock texture), and cement condition. Transit time decreases with increasing porosity. Sonic tools can be run in either open holes or cased holes.

2.6.1.3 Spontaneous potential

This tool measures the voltage potential difference between the formation and a ground wire to surface. Surface ground is a zero baseline potential, and the reading from the formation can deflect in either the positive direction or the negative direction. A potential difference is created by salinity difference between the bore-hole and the reservoir, streaming potential (i.e., an electrical current which originates when an electrolyte is driven by a pressure gradient through a channel or porous plug with charged walls), or electrochemical invasion.

2.6.1.4 Resistivity

The presence of liquids in a strata changes the resistance to current flow. Oil and saltwater are conductors. Rock and pure water are insulators. Resistivity logs are used in all formations to assess pay thickness. This is the primary tool for identifying pay thickness in CBM wells since pure coal is a very effective insulator and solid or liquid inclusions lower the insulation value. A coalbed full of water has a distinctive resistance signature.

2.6.2 Nuclear

Nuclear tools are used to evaluate naturally occurring radioactive material (NORM) and the formation response to imposed radiation sources.

- Natural gamma—this tool does not have an inherent radioactive source and is measuring the radiation given off by the strata (certain rock types have considerable native radiation). It uses the same technology as a Geiger counter.
- High-energy gamma—this tool bombards the formation with gamma rays and measures the returns after Compton scattering and photoelectric absorption. Compton scattering is an inelastic scattering of a photon by a free charged particle, usually an electron. It results in a decrease in energy (increase in wavelength) of the photon (which may be an X-ray or gamma ray), called the Compton effect. Part of the energy of the photon is transferred to the scattering electron. It is used to measure rock density.
- Compensated neutron (neutron porosity)—neutrons interact with anything in their path. Oil and water give off hydrogen under a neutron flux. The tool measures the free hydrogen to determine porosity. In addition to being effective in conventional formations, this tool is useful in identifying coal seams. While coal actually has a very low porosity, the neutron porosity log shows a very high porosity—an indication that unconventional reservoirs still do need an unconventional approach.

2.6.3 Logging while drilling

Logging while drilling (LWD or Measurement while drilling (MWD), the terms are used interchangeably) is a general term to describe systems and techniques for gathering down-hole data while drilling without the requirement to remove drill pipe from the well. LWD offers similar functionality as wireline logging with differences in data quality, resolution, and/or coverage. LWD tools are basically large, instrumented drill collars. In general, the practice is to put the "most important" tools closest to the bit. LWD tools can go anywhere that the bit can go, so they tend to be significantly more effective in extended-reach-lateral horizontal wells. They send their data to the surface using either pulses into the mud or signals superimposed on the drill pipe. Because the environment on the rig floor is so noisy, two-way communication is not possible and communication from the tool to the surface is very slow (1–2 bits of information/second) (Table 2.1).

2.6.4 Production logging tools

Production logging is used for producing wells to gather dynamic performance data used to optimize production or manage fluid profiles for injection wells. The data is used to identify producing zones and the fluids from them; detect leaks, low-pressure zones, or problem zones; calculate flow rate contributions from individual zones; determine cross flow between zones; and profile injection fluids. Production engineers call for production logging when they want to profile production/injection, diagnose excessive water production, or diagnose mechanical problems like blocked tubing. Common tools include the following:

Table 2.1 Wireline Logging vs LWD

Wireline	LWD
Small, light, delicate	Big, heavy, tough
Since the 1930s	Since the 1970s
Easy communications both ways	Limited communication up, none down
Powered through cable	Uses batteries and mud turbine
Delays drilling while running	Transparent to drilling
Limited to about 60 degrees from vertical	Can go where the bit can go
Susceptible to poor hole condition	Not concerned by hole condition

- Quartz gauges—measure and record down-hole pressure and temperature.
- Casing collar locator—uses magnets to locate collars, completion items, and perforations.
- Capacitance—used to determine fraction of water in a liquid mixture.
- Flowmeters/spinners—used to determine flow rate into well-bore (far more effective with perforations that with un-cemented completions).

2.7 PRODUCTION COMPLETIONS

Production tubing is run into the well to facilitate production. Fig. 2.18 shows some of the options that are seen with tubing. It can be "packed" (i.e., a seal mechanism is installed in the tubing/casing annulus to isolate the annulus from the reservoir) or "unpacked," it can be a single string or multiple strings, it can be large or small.

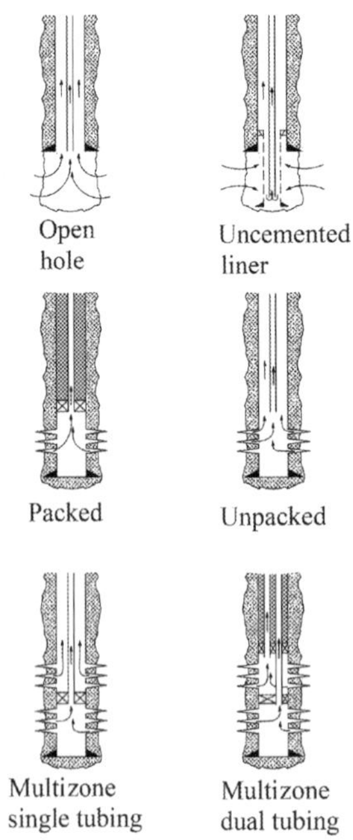

Figure 2.18 Type of completions.

2.7.1 Tubing

Historically, "tubing" has been defined as "any well-bore tubular smaller than 4-in nominal." This definition is outrageously limiting and serves no tangible value. In common usage, "tubing" is "any down-hole string which is hung off from surface, not cemented in place, and whose inside diameter is open to reservoir pressure." Tubing can be any size, and set at any depth.

Both size and set depth should be chosen based on the needs of the reservoir. This is an area that the Oil & Gas industry gets wrong more often than any other. Competing forces of "it is just tubing, we can save money by picking a single size/quality and sticking with it" and "we always set tubing below the perfs" and "our policy is to always set a packer above the producing formation" cause the wrong tubing to be placed in the wrong location too often. In Chapter 3, Well Dynamics, we will talk at length about how to select a tubing size and the costs/benefits of various set-depth and packer strategies.

2.7.1.1 Stick tubing

For tubing we can change the tubing many times over the life of a well. Stick tubing is OCTG as described earlier.

2.7.1.2 Coiled tubing

Coiled tubing comes in a long, continuous length wound on a spool. It is straightened prior to pushing it into the well-bore. If it is removed, it can be coiled back onto the spool. This pipe is bendable, yet stiff, and can be pushed into the lateral on a horizontal well. It is increasingly used for drilling, stimulations, logging, perforating, and fishing (i.e., retrieving dropped tools and pipe in well-bores). It is also used in production operations.

Spool sizes typically run from 2000 to 15,000 ft (610–4570 m) with diameters from ¾ to 4½ in (19–212 DN) and can be made of various grades of carbon or stainless steel.

2.7.2 Completion options

Depending on the reason for drilling a well, the well test, and/or logging results, we have the options of:

- plug and abandon the well,
- suspend the well as a future or possible production well,
- complete the well as a production or injection well.

Completing the well includes perforating any pipe installed across target formation, sand control decisions (rare in unconventional gas), production packer installation, installing the "Christmas tree" on the wellhead, and installing the tubing string.

The boxes with the "x" in them in Fig. 2.18 indicate packers that isolate the tubing/casing annulus from reservoir pressure and any corrosion risks posed by reservoir fluids. Packers are still reasonably common in single zone oil wells and are generally used in multizone gas or oil wells where the operator chooses not to (or is not allowed to) comingle the zones down-hole. The reasons for not being allowed to comingle zones down-hole generally relate to either regulatory proscriptions or to the different zones having different ownership.

2.7.2.1 Open-hole completions

Generally for open-hole completions, the production casing is set above the target zone and cemented into place. Further drilling extends the well-bore into the target seam(s), and the extended hole is left open.

Open-hole completions have the following advantages: the entire pay zone is open to the well-bore; perforating is not required; well can be easily deepened; risk of formation damage from cementing is avoided; and costs are reduced.

The risk of hole-collapse is very real in open-hole completions, so production engineers are reluctant to run down-hole pumps or even production tubing into an open-hole. If there are multiple seams in the open-hole section, there is no effective way to control which seams will receive any stimulation or cleanup treatments.

2.7.2.2 Uncemented liner completions

The risk of hole-collapse can be reduced by running a liner into the uncemented portion of the well. This completion type is very common in weak formations like CBM or unconsolidated conventional formations. The liner is hung from the foot of the production casing and is either pre-perforated or perforated in place.

Pre-perforated liners have holes in them. This can be a problem if you are running (or retrieving) a liner into (or from) a live well. As the perforations clear the stripping rubbers on the wellhead, you have a direct path from the reservoir to the rig floor and many ignition sources. Consequently you have to kill the well with some fluid prior to running a perforated liner. Since many operators go far out of their way to avoid ever putting a foreign liquid on a formation like CBM, this is a pretty major limitation in many operations. To get around this, operators either run unperforated pipe and perforate it in place or run "slotted liners" (see later), both of these solutions solve the problem of running the liner into the well, but neither solves the problem of pulling a liner. There does not seem to be a universal method recommended by the industry for pulling a perforated liner.

There are a number of "slotted" liners on the market where the liner is pre-perforated, but the perforations are filled with a friable material. This allows the liner to be run without killing the well and once in place the perforations can be opened by dropping a tool into the liner to knock the plugs out of the slots.

Neither open-hole nor lined-hole completions are effective on wells that have multiple pay-zones in a single completion interval that each require a specific stimulation method. This is because the zones are never homogenous and one of them will tend to break first and take the entire stimulation while the other zones will not be stimulated as much as designed.

2.7.2.3 Cemented casing completions

The most common completion has production casing run to the bottom of the well-bore and cemented in place. After cementing (and logging), a successful well will be perforated through the casing and the cement sheath into the formation.

Cemented casing completions offer lower risk to rig personnel, better control of inflow, and the possibility of selectively stimulating zones. Disadvantages include restricting the flow path from the reservoir to sales, expense, and risk of cement damage to formation.

Packed: Packed completions isolate the tubing/casing annulus from the reservoir pressure and fluids. They are seldom warranted from a corrosion or pressure viewpoint and in answer to the question "why is there a packer in this well?" the usual answer is "just because that is the way we do it." Any activity in Oil & Gas that is only done because of "company policy" or "tradition" is an activity that has not had adequate recent scrutiny. When you look hard at why people run production packers in single-completion (or down-hole comingled) wells, the real answer should have been "the company had a bad issue of corrosion in the casing on a well (in another field on another continent) in 1963 and we contaminated a potable water aquifer and we are never doing that again." This is a perfectly good reason to evaluate the reservoir fluids and metallurgy choices in this well. It is not a good reason to install a packer. This is more common in oil wells than gas wells, but gas wells do still get hit with company policies that try to make them into oil wells.

Unpacked: The tubing/casing annulus is a crucial reservoir management tool in gas wells (and is very useful in oil wells). As we'll see in Chapter 3, Well Dynamics, we can compare flowing tubing pressure to shut-in casing pressure to determine a specific well's ability to lift liquid with gas flow. We can only do this if the annulus is open to the wellhead. One of the questions that should always be asked in a production-equipment audit is "why is _____ in the well?" One of the first items to ask about is production tubing. The usual answer on gas wells is "for liquids management." That is a reasonable answer, but the next question should be "how does it provide that?" and generally results in a 1000 mile stare and a bit of kicking rocks around in the dirt, but rarely a considered answer. Production tubing and the tubing/casing annulus are tools of reservoir management only as long as they have a function and that function is being exploited. It cannot be exploited with a packer in the well.

Multizone single tubing. It is common to treat a well-bore as an asset and use the same hole in the ground for more than one producing formation. If the various formations in a well have different mineral ownership (which is common in several parts of the United States, but rare outside of the United States) then you have to keep track of what money is owed to the owners of each formation as you start selling gas or oil. You can accomplish this by putting a packer in the well above the lower formation and below the upper formation. This allows the lower formation to flow up the tubing and the upper formation to flow up the tubing/casing annulus. The deeper zone has a reasonable chance of responding to mechanical pumping, but the logistics to trying to pump the annulus precludes pumping it and the upper zone will typically be abandoned if it starts having liquid-loading problems.

Multizone dual tubing: To allow a multizone well-bore to have full access to the tools of production-management like mechanical pumps, you have to run a second tubing string to service that zone. The upper zone often has an upper packer as shown in Fig. 2.18, but that packer is not strictly necessary.

Both types of multizone wells can allow the operator to make the flow/no-flow decision for each zone independently. Completing a well-bore that has multiple independent zones as a down-hole commingled well (i.e., there is no isolation between the zones) can allow a lower-pressure zone to act as a "sink" or "thief zone" taking production from the stronger seam. When this happens, it is common to isolate the zones with a packer.

Horizontal: Horizontal wells are reasonably recent, and the industry is still trying to work out how to balance costs with capabilities. Some of the techniques that have been tried have included:

- Open hole—this option sets production casing at the end of the build section and drills an open hole to the lateral's extent. With this technique there is no way to stimulate specific sections of the formation and the risk of hole-collapse is too high.
- Uncemented liner—you can address hole-collapse with an uncemented liner, but you still have no control over where a stimulation will go (since the formation is open outside of the pipe). People try to address the need to stimulate sections of the reservoir by setting "external casing packers (ECP)" to allow them to put a temporary plug in the liner to try to direct a stimulation job to specific portions of the reservoir. ECP completions have not proven to be terribly effective.
- Cemented casing/liner—the pipe is run to the toe of the lateral and cemented in place and perforated. This technique is the normal method of completing most horizontal shale wells because they require multistage stimulation procedures and this completion technique maximizes flexibility.

2.7.3 Perforating

For a conventional well, we set heavy-wall production casing across the target formation and then sheathe it in cement to make sure that no reservoir fluids can get into the well-bore. This is useful for keeping explosive liquids and gases away from the rig floor, but it is ineffective for production. To make a cased well available for production, you have to make holes in the casing and cement. Big holes. Many of them. Deep. Deeper than you could get with a drill bit. We use explosive charges to make them.

The goal of perforating is to provide a clean flow-channel between the producing formation and the well-bore while causing minimal damage to the producing formation. The ultimate test of the effectiveness of a perforating job is well productivity. Productivity depends on perforation length, shot density, angular phasing (i.e., the direction that the holes go relative to each other), perforation diameter, and how much debris is left behind. Each of these dependencies is further dependent on external factors like the formation characteristics and the importance of each changes from well to well.

2.7.3.1 Gun types

Perforating guns are usually made up of a number of shaped charges, but there are some bullet guns for specialized applications. There are many types of shaped-charge guns that can be used and they fall into about three categories:

- Retrievable hollow gun—the charge is secured in a steel tube which is sealed against hydrostatic pressure. When the charge blows, the tube can expand and sometimes get stuck in the well and have to be milled out. This is rare and these guns leave minimal debris in the well.
- Nonretrievable (expendable) gun—made up of individual sealed cases, each containing a charge. Sealed cases are made of a frangible material such as aluminum, ceramic, or cast iron. The case is blown into small pieces that remain in the well.
- Semiexpendable gun—charges in sealed cases are secured on a retrievable wire carrier or metal bar. This configuration reduces the debris left in the well and increases the ruggedness of the gun.

2.7.3.2 Conveyance methods

We deploy perforating guns either with a wireline or on a tubing (or coiled tubing) string. When wireline deployment is used, it can either be done before or after the installation of tubing. "Thru casing" jobs are run before the tubing is installed and allow for the largest perforating gun that can fit through the casing. "Through tubing" jobs are run through the tubing and out the end and are limited in the diameter of perforating gun that can be run, and this method allows the casing above the gun to be isolated with inflatable packers to increase the force of the blast. Since it is ineffective to push on a rope, perforating in horizontal and high-angle wells often cannot be done with wireline-conveyed deployment. The weight of the gun is also limited by the carrying capacity of the wireline.

Tubing-conveyed perforating allows casing-conveyed gun sizes to be used with considerably improved control of placement, isolation from the casing above the tool, larger perforated intervals can be shot because the tubing has much more strength in tension than a wireline, better isolation of reservoir pressure from the surface, and you can push tubing into extended laterals. The downside of tubing-conveyed perforating is that it is more expensive to pull the tubing string than to spool up a wireline. Using coiled tubing has all of the benefits of both wireline and tubing-conveyed and is becoming very common.

2.8 STIMULATIONS

We saw in Chapter 1, Gas Reservoirs, that unconventional gas reservoirs tend to have very low permeability which translates into extreme difficulty for one unit volume of rock to communicate with the next unit volume that there is a new place for the reservoir fluids to go. The purpose of stimulating a well is get the word out to as much of the reservoir as possible that there is a new place for the reservoir fluids to go. If we look at a continuum of acceptability of stimulations, we see (in order of technological acceptance by management):

- Cased, cemented, and hydraulically fractured
- Open hole (or uncemented liner) and hydraulically fractured
- Open hole (or uncemented liner) and "cavitated"
- Open hole (or uncemented liner) and a "Mississippi Frac"

Virtually all conventional gas or oil wells are cased, cemented, and at least evaluated for a hydraulic fracture stimulation (and most wells are frac'd). Our industry understands this process very well and has a good handle on troubleshooting frac problems and a good understanding of suitable parameters and risks. The other techniques have none of this comfort-making history.

2.8.1 Hydraulic fracture stimulation

Applying a force to a rock that is greater than the forces holding the rock together will result in broken rocks. That force can be applied as fluid pressure instead of a solid pushing against a solid. If you break rock with a fluid that is filled with chunks of stuff, then the chunks might hold the fractures open to allow fluids to flow through them for extended periods of time.

Though the first well that was stimulated using modern hydraulic fracturing was Stanolind Oil & Gas (later to become Standard Oil (Indiana) then Amoco then BP) Kipper Gas Unit No. 1 in the Hugoton field in Kansas in 1947, the history actually goes back to the Civil War. In 1862 during the battle of Fredricksburg, VA, Union Col. Edward Roberts found that shooting an artillery shell into a body of water had an effect that could fracture rocks that drained a canal and opened a previously closed path of attack (Manfreda, 2015). In 1865 Roberts received US Patent No. 59,936 that described using a "torpedo" to improve water flow in an artesian well. This process was very successful even without added proppant (other than fragments from the iron casing of the torpedo) and Robert's company quickly moved to stimulating oil wells for \$100–200 USD and 1/15th royalty on the production. The first modern frac (i.e., using hydraulic pressure and proppants rather than explosives) used 1000 gallons of gelled gasoline (commonly known today as napalm) and sand into a limestone formation followed by a chemical to break the gel. This process did not significantly improve production, but did spark efforts to improve the process. In the 1960s Pan American Petroleum (a subsidiary of Standard Oil (Indiana) at that time) used walnut shells in gelled water to successfully stimulate wells in Stevens County, OK, and the industry rapidly adopted this process (with other proppant materials, Standard Oil (Indiana) held a patent on using walnut shells). This Stevens County process was technologically identical to a multistage frac job performed in 2016 in the Marcellus Shale.

When talking about hydraulic fracture stimulation you have to consider carrier fluid, proppant, and additives. The carrier fluid can be slick water (i.e., water with friction reducers added), potable water (city water or river/lake water, with or without additives), produced water from other wells, cross-linked gels, or foams. The original patent for this technique specified ground up walnut shells for proppant, but the industry quickly evolved to raw sand, coated sand, ceramic balls, and other materials with a strong resistance to crushing force.

The list of chemicals added is very long and is determined for each frac job. Most jobs with water add a powder made from guar beans from India or Pakistan to the water. This material is used in cooking and can quickly turn water into a thick gel with a very large capacity for carrying proppant. After the guar is specified, the engineer looks at the need for corrosion control chemicals, friction reducers, chemicals to stabilize swelling clay, chemicals to try to bind coal fines, etc. Every job will be unique.

At the end of the day, the goal is to convert at least a portion of the reservoir from radial D'Arcy flow to linear channel flow. Any success at this significantly lowers the differential pressure required to flow reservoir fluids into the well-bore for production.

For horizontal or vertical wells in very thick formations, stimulating an extended interval with one job will always provide inconsistent results—if any part of the extended interval is weaker than the rest then it will take a disproportionate amount of the stimulation fluids (maybe all of them). To combat this on wells with casing cemented into the hole over an extended interval, we do a "staged stimulation." In other words the tool has inflatable packers above and below the frac nozzles that seal most of the perforations from the stimulation pressure. At the end of a stage, the packers are released and the equipment is moved to the next part of the perforated interval. Wells in the shale with extended laterals can require upward of 100 stages.

Open-hole and uncemented liner wells are an entirely different issue. It is actually impossible to have any control over where the frac fluids will go in an open-hole completion, and an uncemented liner does not provide any mitigation. In both cases, the fluid will take the path of least resistance and most of the formation will not be stimulated at all. There is no way to prevent this.

To further reduce the effectiveness of hydraulic fracture stimulation in CBM, coal is not very mechanically strong. Coal miners have always talked about coal being self-healing so that something inserted into the coal matrix will quickly be integrated into the coalbed and will be hydraulically indistinguishable from the coal matrix. A grain of sand just gets swallowed up by the coal. Thousands of CBM/CSG wells have been frac'd around the world and the results are absolutely random. Procedures applied to a CBM well that then produced at a high rate were applied to future CBM wells that provided very low production rates. The data is random enough to make it look like frac jobs in CBM can either do harm or do nothing at all. It doesn't look like they ever do significant good and often do significant harm.

2.8.2 Open-hole cavitation

When universally discouraging results were seen from conventional completions with hydraulic fracture stimulation of the wells in the central "fairway" portion of the San Juan Basin, the operators started looking for viable alternatives. It was discovered that in these particular wells the coal matrix was so weak that a small differential pressure would cause the coal to fail and increase the surface area of the well-bore and improve production. It was found that by increasing the pressure in the well-bore (either

Figure 2.19 CBM cavitation.

through adding air or allowing reservoir pressure to build up) and then slamming open a surface valve to allow a rapid depressurization of the well-bore to a flare (Fig. 2.19), that the pressure a few inches into the reservoir would blast the coal into small pieces that would flow to the surface like a fluid. Repeated cycles of this pressurization, rapid depressurization cycle could be effective in creating a large surface area for the reservoir to flow into the well-bore. Wells treated with this process often increased their production by a factor of 20–40 over conventional completions. It only works on wells with very specific rock quality, but where it works the results are impressive.

Factors that contribute to the success of a cavity completion include: (1) strength (friability) of the coal must be low; (2) coal seams need to be fairly thick; (3) there must be a well-developed cleat system; (4) the coal must have a low ash content; and (5) the coal must have high in situ stress. Few of these parameters can be successfully evaluated pro forma and in new fields (or new areas of existing fields) my recommendation to companies is to try to cavitate and if the coal doesn't flow, to set a liner and see what the well will do, the data on CBM does not support substituting hydraulic fracture stimulation for wells that won't cavitate.

There is no way to know how many cavitation cycles will be required for a given well, or how long each flow back will take. Companies that evaluate their Driller's performance based on cost and adherence to a schedule have a great deal of difficulty committing to a process with an indeterminate duration and costs directly tied to the duration. Drillers rarely have a stake in well performance, so they often say "We can meet your schedule with a conventional completion and come in under budget, we've been doing conventional completions for many decades, what is the problem?" Actually, "the problem" is that conventional completions result in anemic performance and significant gas left in the ground at abandonment. One company has five cavitated CBM wells that average 12 MMSCF/day and 500 cased, cemented, and frac'd wells making an average of 220 MSCF/day. The last well that they "caviated" they simply ran a tool to under-ream the formation to a slightly bigger diameter and ran casing and cemented it—this well was labeled a "caviation failure" and ended the cavitation program. Now that company's wells don't make much gas but the drillers meet their schedule and budget every single quarter.

2.8.3 Mississippi clean-out

When talking about out-of-the-box thinking, there are few concepts farther out there than a Mississippi Frac. This technique has been used in isolated spots within many fields and was marginally successful in some unconsolidated conventional formations. It has been tried a couple of times in CBM and the results of those wells are very promising.

This process is fairly simple:

- Drill to the top of the coal, set production casing, and cement it. This section should have 7-in (175 DN) or larger casing.
- Stage enough produced water to do the stimulation (it could easily take 50,000 bbl (8000 m^3) and possibly more).
- Pressure up the well-bore to significantly above the parting pressure of the coal (call it 5000 psig (34.4 MPa)) with the produced water.
- Start drilling while maintaining high-pressure water flow through the bit. Expect a lot of lost circulation as the water pulverizes a considerable volume of coal.
- When you reach your target depth immediately run a liner to bottom and tubing down to the middle of the most productive formation, and rig down.
- Put the well on production as soon as the rig leaves.

Initial production should be expected to be very low as you have changed the hydraulics of the reservoir for a considerable distance into the formation. Within a few days, the water in the casing should drop enough to allow desorption in the near well-bore and production to start. In most CBM wells where this has been tried, the new channels dewatered within a few weeks and did not need mechanical deliquification.

The benefit of this process is that production at 6 months tends to be significantly higher than conventionally drilled offset wells, drilling costs tend to be 40%–70% of conventional drilling/stimulation while being as friendly to the drilling schedule and budget as conventional jobs. In this technique drill cuttings serve the proppant function and no additional chemicals are introduced.

2.9 CONCLUSION

My intention in this section is to demonstrate that wellheads are connected to stuff and that stuff did not magically appear. Drillers will tell you that without them, you wouldn't have anything to produce. You need to tell drillers that without us they wouldn't have any reason to drill. It is to all work together or none of it works.

REFERENCES

Adobe Stock Photos. (n.d.). Retrieved July 25, 2016, from https://stock.adobe.com/.

Britannica Rotary Drilling. (n.d.). Retrieved July 19, 2016, from https://www.britannica.com/technology/oil-rig/images-videos/A-land-based-rotary-drilling-rig/113917.

Crain, E. R. (n.d.). Crain's Petrophysical Handbook—Cement Integrity Logs—Part 1—CBL/VDL. Retrieved August 09, 2016, from https://www.spec2000.net/07-cementlog1.htm.

Khandoker, S. (n.d.). Patent US20080023917—Seal for blowout preventer with selective debonding. Retrieved August 01, 2016, from https://www.google.com/patents/US20080023917.

Kongsberg Maritime SDP. (November 2003). "Operator Reference Manual—Dynamic Positioning System," Rel. 5 Update 1.

Manfreda, J. (2015). The real history of fracking. Retrieved August 20, 2016, from http://oilprice.com/Energy/Crude-Oil/The-Real-History-Of-Fracking.html.

PetroWiki Rotary Drilling. (n.d.). Retrieved July 19, 2016, from https://www.britannica.com/technology/oil-rig/images-videos/A-land-based-rotary-drilling-rig/113917.

PetroWiki, 2 Cement Bond Logs. (n.d.). Retrieved August 09, 2016, from http://petrowiki.org/index.php?title = Cement_bond_logs.

Weatherford Centralizers. (n.d.). Retrieved August 08, 2016, from http://www.weatherford.com/en/products-services/well-construction/cementing-products/centralizers#.

Wikipedia, 1 Drillship. (n.d.). Retrieved July 15, 2016, from https://en.wikipedia.org/wiki/Drillship.

Wikipedia, 2 Blowout Preventer. (n.d.). Retrieved August 01, 2016, from https://en.wikipedia.org/wiki/Blowout_preventer. Egmason—Own Work.
Wintershall—Drilling. (n.d.). Retrieved July 25, 2016, from https://www.wintershall.com/technology-innovation/drilling.html.

FURTHER READING

Atlas Copco Quantum Leap Hammers. (n.d.). Retrieved July 25, 2016, from http://www.rockdrillsales.com/products/hammers/quantum-leap-hammers/.
Webster (n.d.). Retrieved April 21, 2016, from http://www.merriam-webster.com/dictionary/biogenic (other dictionary references replace "biogenic" with the referenced term).

NOMENCLATURE

Symbol	Name	fps Units	SI Units
g	Gravitational constant	32.174 ft/s^2	9.81 m/s^2
g_c	Unit converter	32.174 ft-lbm/s^2-lbf	N/A (for now)
h	Height	ft	m
ρ	Density	lbm/ft^3	kg/m^3

Subscripts

drill	At the drill bit
mud	Drilling fluid
res	Reservoir

EXERCISES

1. "Underbalanced drilling":
   a. Means drilling with less fluid weight that expected reservoir pressure
   b. Means drilling while expecting reservoir fluids to blow onto the rig floor
   c. Is used to reduce the likelihood that drilling fluids will enter the reservoir
   d. All of the above

2. Drilling mud is:
   a. A combination of water and soil in precise proportions
   b. A joint compound used to seal threads on drill pipe
   c. Not used while air drilling
   d. None of the above
3. Logging while drilling is characterized by:
   a. Good communication from surface to tool
   b. Limited to less than 60 degrees from vertical
   c. Slow data speeds
   d. Delaying drilling while running tools
4. Which of the following is least likely to be found in a CBM well?
   a. Conductor string
   b. Surface casing
   c. Intermediate casing
   d. Production casing
5. Nuclear logging tools are used for:
   a. Monitoring the mud to prevent the build-up of naturally occurring radioactive material (NORM)
   b. Measuring the diameter and shape of a borehole
   c. Quantifying lithology, density, and porosity
   d. Measuring the permeability of a formation
6. A well is being drilled with 9.18 lbm/gal (1.1 kg/L) water-based mud. How much pressure does the mud apply to the formation at 3000 ft (914 m) below ground level?

CHAPTER THREE

Well Dynamics

Simpson's Second Postulate: A mistake implemented (and corrected) promptly has a much better chance of success than a perfect decision made after months of sober deliberation

Corollary: If you don't implement (and analyze) mistakes, then you never improve

Contents

Practical Onshore Gas Field Engineering.
DOI: http://dx.doi.org/10.1016/B978-0-12-813022-3.00003-1

3.1 ROLE OF SURFACE PRESSURE IN WELL PERFORMANCE

At a given reservoir pressure and well condition, bottom-hole pressure (BHP) controls the rate of inflow to a well. BHP is the pivot point in a balance between reservoir flow and flow to surface. This says that there is no way to develop a single equation to model flow from deep in the reservoir to sales, the flow regimes are too different.

In the reservoir we are talking about phase behavior (both interference and phase change), constantly changing flow paths, and many small flow paths. In the well-bore we are concerned about flowing against gravity, phase interference more than phase change, a single constant flow path. All of these factors are (to a greater or lesser extent) functions of pressure and temperature.

A change in surface pressure cannot ever be translated intact to the reservoir. We often see reductions in surface pressure resulting in a significant increase in fluid friction in the well-bore and the converse is also common. The net result of a pressure change can even be the reverse of desired results. There are many examples of rapidly decreasing surface pressure actually lowering the efficiency of the flow conduits so much that flow rates decrease.

As we learned in Chapter 1, Gas Reservoirs, this balance between pressure and overall flow efficiency does not always (or even often) come down on the side of an extremely high-pressure differential between the reservoir and flowing BHP. Losing track of reservoir pressure is a prime cause of long-term underperforming wells. In the coalbed methane (CBM) example in Chapter 1, Gas Reservoirs, the flowing BHP wanted to be about half of reservoir pressure to optimize the recovery rate and ultimate recovery. Every reservoir will be different and few reservoirs will lose pressure as fast as CBM, but every reservoir will have a characteristic BHP that will optimize well economics.

3.1.1 Pressure consistency

Flow in a reservoir is complex and often tortuous. Fluids are the ultimate opportunist. They will always go from higher pressure toward lower pressure. When you shut a well in, the fluids from deep in the reservoir move toward the lower pressure in the near-well-bore until the differential pressure is less than the obstacles to flow (primarily fluid friction and phase blocking). This "repressurizing" will continue until the differential pressure is negligible. I did a statistical analysis of one tight gas formation (the Mesaverde in the San Juan Basin) and concluded with hourly data on over 2000 shut-in periods longer than 2 weeks each of 300 wells that the average reservoir pressure in the Mesaverde would stabilize at the well locations in something on the order of 20 years. This analysis was very eye-opening in showing wells that had been shut in for longer than a year still increased shut-in casing pressure a measurable amount (albeit quite small) from 1 hour to the next. The database had hourly data for 13 months, and the study took 4 months, so the longest pressure build up in that dataset was 17 months on four wells. All were still showing progress toward reaching an asymptote in the last hour of data.

Searches for the earliest use of the term "Flush Production" found a 1969 joint report by the US Bureau of Mines and El Paso Natural Gas Company on the nuclear stimulation called "Project Gas Buggy" where 24 kT nuclear explosions were used to stimulate two wells (Atkinson et al., 1969), and the term has come to mean "high initial flow rates immediately subsequent to ending a shut-in period." Many people have put forth theories as to why flush production occurs, but none of them really gets to the essences of the phenomena. Essentially, the near-well-bore volume repressurizes, while production is halted which increases the differential pressure between the reservoir adjacent to the well-bore and the BHP. Higher differential over a short distance increases the flow rate (and the friction, but the distances are short) and higher static pressure improves flow efficiency. Fig. 3.1 shows this effect. During the period marked "start-up," the pressure gradient in the near-well-bore is very steep and there is considerable driving force for fluids to move from the reservoir to the well-bore. Over time this pressure gradient shifts lower and the driving force from the reservoir to the BHP declines. If you can provide a steady pressure for a very long time (months), then you can move the pressure gradient down to the "long-term steady press" line. This line represents a significant increase in the volume of the reservoir accessed (i.e., the last place that there is an effective differential pressure between the average reservoir pressure and the near-well-bore volume) increases dramatically.

It is common wisdom (and therefore suspect) that "compression in tight gas is ineffective." The reason for this old-saw is that people insist on treating compression in gas reservoirs as "rate acceleration" and not

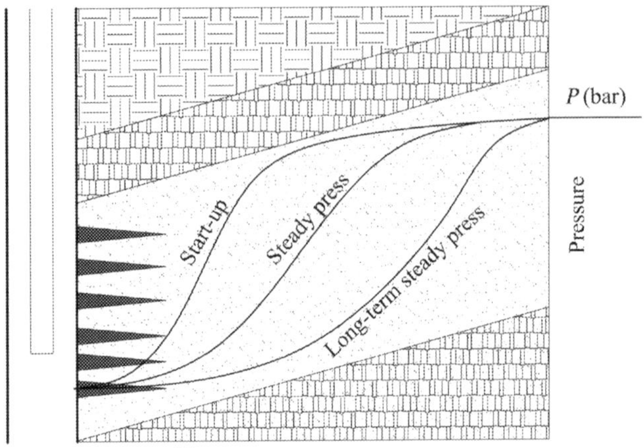

Figure 3.1 Pressure traverse.

"reserves adding." In times of flat to increasing product prices, rate acceleration economics are poor. Even in times of decreasing product prices, the rate acceleration economics are rarely as good as putting the capital in the Bond market. The common wisdom is a case of improper expectations—if we expect a big rate increase in tight gas from compression, we will be disappointed, but if we expect flatter decline and longer well lives then there is a very good chance of exceeding expectations.

Tight gas performance is a vivid example of the value of pressure consistency, but keeping pressure in a narrow range around a rationally developed target pressure is even more important in CBM and Shale. In all of unconventional gas, constant pressures are more important to ultimate recovery than a specific pressure.

The goal of everything we do on a well-site or in the downstream facilities from selection of well-bore tubulars to deliquification method to gathering system pigging schedule should be focused on the target of maintaining consistent, steady pressure within a range that we have defined based on a rational analysis of reservoir characteristics.

Fig. 3.2 shows an interesting well. It is a cased, cemented, and frac'd gas well with a production packer above the producing zone. This well began loading up, and deliquification options were nonexistent with the packer installed. The operator decided that he could increase the recovery by "riding the flush" with periodic shut-in periods followed by rapid flow to facilitate the gas carrying the liquid out of the hole. They were finding that their intermitting process was resulting in

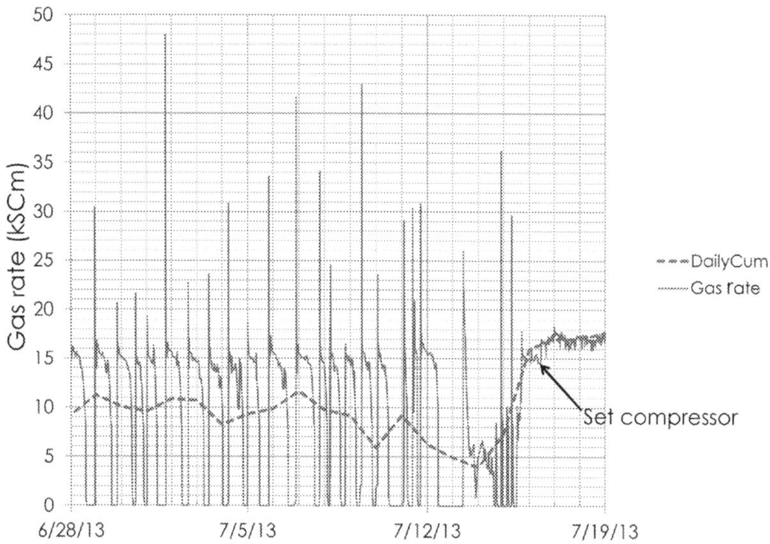

Figure 3.2 Change well production technique.

progressively shorter flush periods with lower peaks resulting in decreasing daily cum production every week. Finally they decided to abandon that technique and installed a compressor. The compressor only lowered the average flowing tubing pressure by about 7 psig (50 kPa), but the daily flow rate increased from about 8 kSCm/day (283 MSCF/day) to 17 kSCm/day (600 MSCF/day). Shortly after the end of this data, the compressor failed, but the replacement compressor was able to return the well to steady production at the elevated rate. This well is just one of many vivid examples of steady pressure outperforming widely varying pressure.

3.1.2 Water vapor

Water vapor is ubiquitous. We see it as "humidity." It makes our iced drinks and air conditioner coils sweat. But how much water vapor is represented by humidity? At sea level the water content of 100% relative humidity (RH) air at 95°F (35°C) is 2507 lbm/MMSCF (40,158 mg/m^3). That is 300 gallons (1137 L) of water in a million SCF (28 kSCm) of dry air. One million SCF of air would be needed to burn 358 gallons of gasoline. That means that on a hot day in Manila you would be putting as much water vapor as fuel into your vehicle's engine. OK, water vapor can be a big volume, but what is it? The human mind wants to think of a molecule as a "small thing," but we have trouble grasping how small. Liquid water droplets can be described as:

- Coarse spray—0.007–0.039 in (0.2–1.0 mm) diameter. A course spray would have a terminal velocity as a raindrop of 7 ft/s (2.1 m/s) so it would tend to fall flowing at target velocity within a production separator.
- Fine spray—0.0039–0.007 in (0.1–0.2 mm) diameter. Terminal velocity as a drizzle is 0.5 ft/s (0.152 m/s) so it would tend to be buoyant at separator velocities.
- Mist—0.002–0.0039 in (0.051–0.1 mm) diameter. Terminal velocity as fog is 0.05 ft/s (0.015 m/s) so it would tend to be buoyant in still air for a time.
- Aerosol—0.000039–0.002 in (0.001–0.050 mm) diameter. Terminal velocity as cloud is 0.003 ft/s (0.9 mm/s) so it will tend to be buoyant in still air for a long time.

An aerosol has very small droplets. A water-vapor molecule has a diameter of 16×10^{-9} in (38×10^{-6} mm). To put that into scale, if you took the smallest aerosol droplet (0.001 mm) and blew it up to the size of the earth, a water-vapor molecule would be 4 ft (1.2 m) diameter. On the same scale a methane molecule would be 6.4 ft (1.95 m) diameter. This says that if you were to devise a filter to separate methane from water it, would have to stop all of the methane molecules to allow the water vapor through. Mechanically separating gas from water vapor is simply not economically possible.

Water vapor and the residue of evaporation are an important factor in troubleshooting many problems with both downhole and surface equipment that are all too often misdiagnosed.

3.1.3 Evaporation

Whenever there is a coherent gas/liquid interface, the liquid will evaporate until the gas touching the interface is at 100% RH. The amount of water vapor that is represented by 100% RH varies with the temperature and pressure of the gas. In 1958, researchers J.J. McKetta and A.H. Wehe published a chart that describes the amount of water vapor that can be accommodated in a hydrocarbon gas at a specified temperature and pressure (McKetta and Wehe). The correlations underlying this chart have been published by ASTM International (formerly American Society for

Testing and Materials) as ASTM D-1142-95 (reapproved 2006). Generally available versions of the McKetta–Wehe (GPSA, fig. 20.4) are very useful in describing the process, but can be difficult to use. I've used the data from ASTM to generate Figs. 3.3 and 3.4. Fig. 3.4 is not much easier to

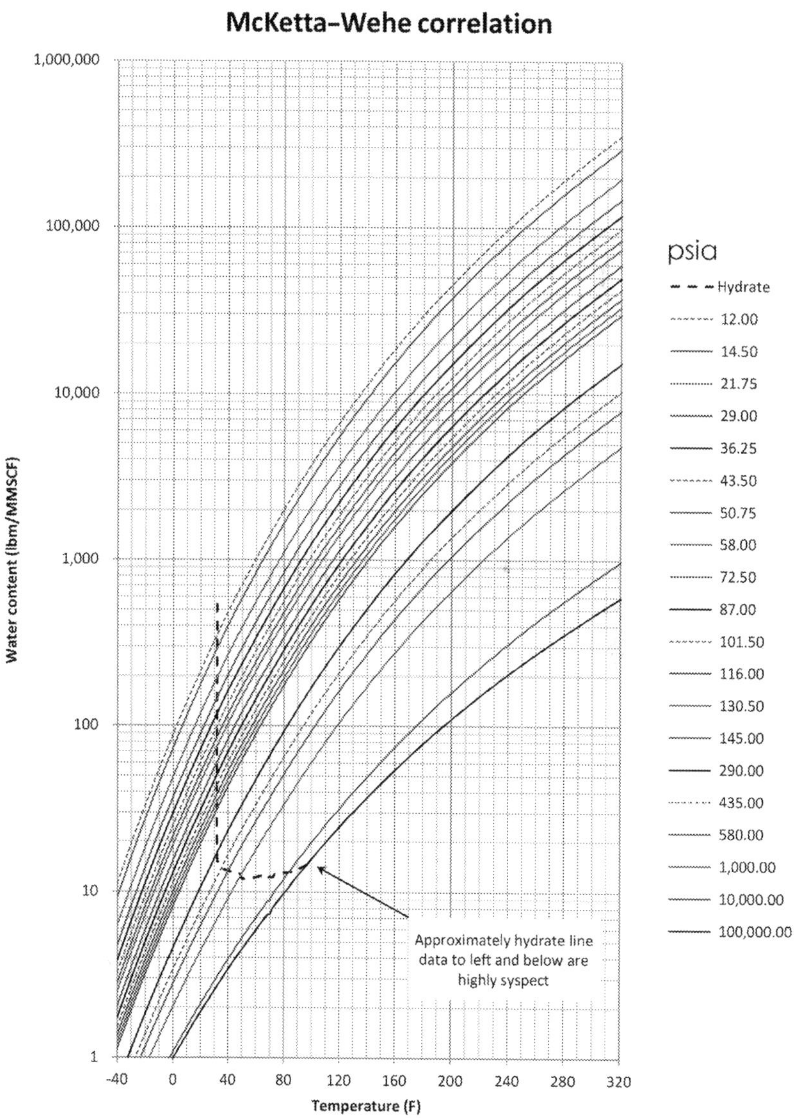

Figure 3.3 Mcketta–Wehe correlation.

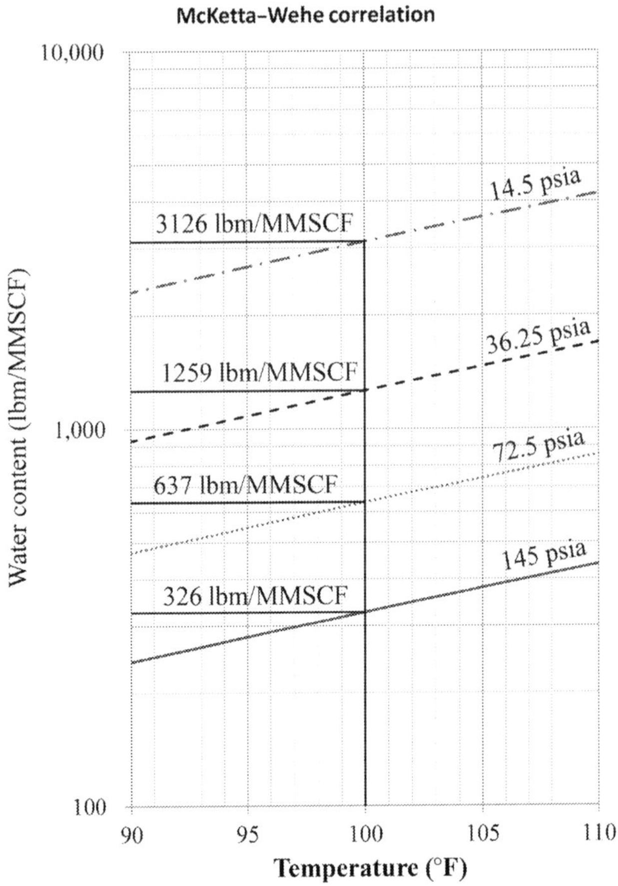

Figure 3.4 Selected water content points.

read than GPSA (Figure 20.4), but it is a high enough resolution that it blows up nicely (at E-Size (A0) it is reasonably useful).

As pressures decrease, the water content increases dramatically. Fig. 3.4 shows an extract from Fig. 3.3. Holding temperature constant at 100°F (37.78°C) the data shows that halving the absolute pressure approximately doubles the fully saturated water-vapor content (the SI version of this chart has mg/SCM on the Y-axis, this makes the 14.5 psia (100 kPa) value 50,704 mg/SCm and the 145 psia (1 MPa) value 5222 mg/SCM.

Fig. 3.3 can be used to estimate the water content of a gas:

- Enter the chart with a temperature
- Go up to a pressure

- Go across to a water content
- For example: 120°F (49°C) and 58 psia (400 kPa) is 1410 lbm/MMSCF (22590 mg/SCm).

Fig.3.3 can also be used to estimate a dew point of a gas:

- Enter with a water content
- Go across to a pressure
- Go down to a dew point
- For example: 7 lbm/MMSCF (110 mg/SCm) at 1000 psia (6900 kPa) is 32°F (0°C).

While Fig. 3.3 is effective for any hydrocarbon gas with reasonably pure water, the numbers must be corrected for acid gases (e.g., CO_2 and H_2S) and for highly saline water (every 10,000 mg/L is 1% by volume, the adjustments start at 1%). The corrections are described in GPSA *Engineering Data Book* in Chapter 20 (Dehydration) if you ever find that you need to apply them.

3.1.4 Phase-change scale

We often see solids deposited inside of compressors, separators, control valves, around packing leakage in control valves, and hanging off of threads in piping. This is often arbitrarily called "calcium carbonate" and occasionally carbonate scale inhibiters and carbonate dissolving chemicals are put into lines without ever getting a materials analysis. When this doesn't work, operators are often at a complete loss as to what to do next. When they call me, I ask for a water analysis. It is rare in unconventional gas for there to be more than a trace of calcium in the water, calcium carbonate is a problem in chalks and limestones, not in sandstone, coal, or shale.

If the solids are not calcium carbonate then what are they? When a barrel of water with 1% (10,000 mg/L) dissolved solids (often called total dissolved solid ("TDS")) evaporates, it leaves 3.5 lbm (1.6 kg) of solids behind. More TDS and each barrel will leave more solids behind. A barrel (159 L) of pure water weighs about 350 lbm (159 kg), so according to Fig. 3.3 at 145 psia (10 bara) and 100°F (37.8°C) every MMSCF would have a barrel of water and in 10,000 mg/L water would leave 3.5 lbm (1.6 kg) of solids behind. The scale we see in unconventional gas is almost never a supersaturated precipitant like we see in many conventional reservoirs, it is actually a dissolved solid that only plates out when the water evaporates.

Figure 3.5 Example of nahocolite.

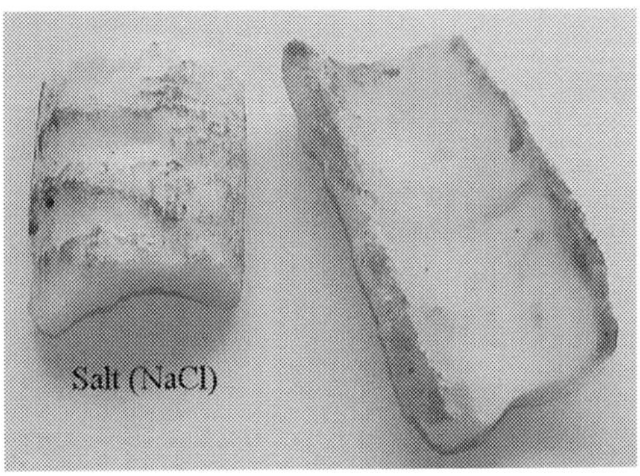

Figure 3.6 Example of NaCl salt.

What is left behind in phase change depends on the makeup of the water. Usually, the primary component is either nahocolite (Fig. 3.5) or table salt (Fig. 3.6). When the primary anions are bicarbonates (HCO_3^-) then a sodium (Na^+) cation will form nahocolite which is granite hard and barely soluble in strong acid. The fitting in Fig. 3.5 was a "half mule-shoe" (i.e., a tubing fitting with a 45-degree bevel on the end to allow tubing to be run into an open hole and if it hangs up on a ledge, rotating the tubing will cause it to slide off where a square cut end would try to drill) that had been in a well for less than 1 hour before flow cut off. The tubing was pulled and the fitting was brought to me for diagnosis. It was my first experience with this vile stuff and considerable research and lab work were required to find out what it was, let alone what to do about it. It was eventually determined to be slightly soluble in strong acid.

When the primary anion is chloride (Cl^-), then with sodium it will form table salt. Table salt is quite soluble in hot water, but getting hot water to the location of the accumulation can be a challenge. The example in Fig. 3.6 formed in the tubing/casing annulus of a well (2-3/8 inside 4-1/2 casing) and was over 30 ft (9.1 m) long, it came out of the well in much smaller sections.

No one wants to deal with these phase-change solids after they have clogged up the works, so we look to the chemical industry to "solve" this problem. They can't. "Scale" chemicals are readily available to treat precipitants from continuous-phase liquid systems. The key terms in the last sentence are "precipitants" and "continuous-phase liquid systems." The phase-change solids are not precipitants, they are the residue of evaporation and adding antiprecipitation chemicals to evaporating water simply adds mass to the deposit while doing nothing to prevent it. Any chemical additive must be transported to a specific location in order to be effective. In a liquid, scale chemicals will be dissolved in the liquid and will go wherever the liquid goes. In a gas the chemicals can sometimes be carried as an aerosol, but in a very short time, collisions between the droplets will create droplets too big to be buoyant and they will fall to the bottom of the pipe, and stay there. This is the crucial thing to understand, any remedial chemicals introduced into a piping system must have a transport mechanism, and gas rarely has the flow energy required to deliver the remedial chemicals to where they are needed. When we dump scale inhibiter (or corrosion inhibiter, paraffin reducer, etc.) into flow lines they will almost never be transported to where they are needed and will usually accumulate in inconvenient locations. Remedial chemicals in gas systems rarely do more good than harm.

Since evaporation is a statistical process that follows the laws of thermodynamics eventually, but not necessarily instantaneously, it is very difficult to predict where the salt will form. In Fig. 3.5, we can conjecture that the transition from reservoir flow to the well-bore dropped pressure significantly and began accelerating evaporation and the high-velocity gas accumulated at the end of the tubing and collected the nahcolite. In Fig. 3.7, about 1 L (1.06 quart) of water droplets came out of the suction scrubber per day for less than 2 weeks and when the droplets hit the hot suction valves the water flashed. In Fig. 3.6, I have been unable to come up with a theory to explain how the deposition started in the annulus,

Figure 3.7 Salted reciprocating compressor valves. *Source: Personal collection of Dejan Ivanovich.*

once it had partially blocked the flow, then a differential pressure could have shifted the stream down the McKetta–Wehe chart, but how did that differential pressure start? I don't know, and that is my answer all too often. Why a salt block forms in a specific location has proven to be very difficult to assess or (more importantly) to predict. If we could determine why salt forms in one place but not in the adjacent, seemingly identical, location then we might be able to create conditions that force salt formation in a convenient location rather than in the typical inconvenient location. I have searched for that elusive "trip point" for two decades with zero success.

3.1.5 Hydrates

A "clathrate" is "relating to or being a compound formed by the inclusion of molecules of one kind in cavities of the crystal lattice of another" (Webster), and a "hydrate" is "a substance formed when water combines with another substance" (Webster), so the stuff we call "hydrate" is more properly called "clathrate," but using the proper term will impede communication in this industry so at best we can call it "clathrate hydrates" or simply "hydrates" because communication is too rare and valuable to forego in the name of "accuracy." If we can agree on what we are going to call something, then "hydrate" works as well as anything. In essence the most common hydrate crystal (called "Type 1") is made up of two copies of the left-hand cage and six copies of the right-hand cage in Fig. 3.8. This structure is an appropriate size for CO_2 and methane, and each crystal will hold 46 water molecules and 8 methane or CO_2

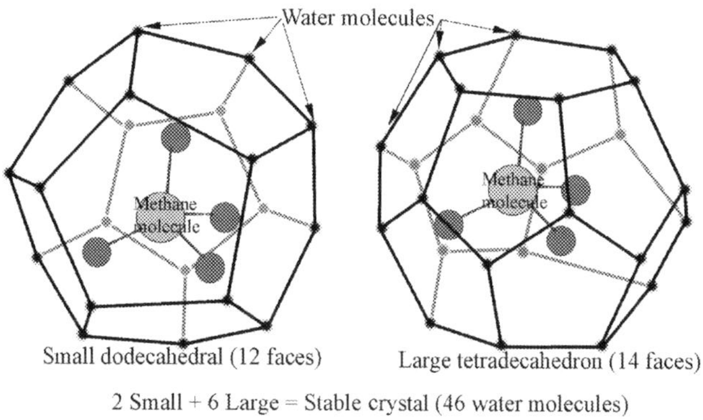

Figure 3.8 Hydrate cage.

molecules. Other structures require different numbers of different shaped structures to hold other size molecules (e.g., "Type H" hydrate will use 34 water molecules to hold a single molecule of butane or heavier gas).

Hydrates form at elevated pressure and (somewhat) depressed temperature. If the gas in question contains CO_2 or H_2S then the curves can shift somewhat, but the shift will rarely lead you to different predictions. You can see that for something like 59°F (15°C) gas, pure methane will be at risk of forming hydrates at any pressure above 2000 psia (14 MPaa), but for a heavier gas that pressure will be less (e.g., for 0.8 SG gas at 59°F (15°C) you have to be concerned about hydrate formation at 500 psia (3.45 MPaa).

We see significant hydrate formation below the seafloor and in the permafrost. Deposits in far northern Alaska have been estimated to be more stored energy than drilling fossil fuels have uncovered. The seafloor off of Japan also has very large hydrate deposits, and the Japanese government has been funding significant research in how best to monetize this asset.

The other location of hydrate freezes is not as useful. We see hydrate freezes in well flow lines and gathering systems that can cause significant flow problems. Usually the hydrate formation in piping happens near a flow restriction that can facilitate Joule–Thomson cooling to drop the temperature as the pressure drops.

Many field operators carry methanol (CH_3OH) or glycol ($C_2H_6O_2$) to help them "break freezes." The problem with this is that while methanol in liquid water will significantly lower the freeze point, methanol on a hydrate will not do anything chemically. If the methanol is warmer than the hydrate-formation point, then it will melt the hydrate and the water from the melted hydrate will have "antifreeze" properties, but you could do this as effectively with warm water. Hydrate freezes are generally best dealt with by depressurizing and adding heat. Chemicals are just expensive, nasty, dangerous, and ineffective.

The presence of liquid water in a mostly water-vapor environment or the presence of water vapor in a mostly liquid water environment will tend to suppress hydrate formation.

3.2 PREDICTING FLOW RATES

As an industry, we spend an inordinate amount of time predicting flow rates for building budgets, estimating reserves, setting company and national policy, and operational evaluations of well performance. By and large we do a reasonably horrible job of predicting the future. The US Energy Information Administration (EIA) has a whole staff that does nothing but try to predict the future of supply and demand for various energy commodities around the world. They look at all the data that is reported to any state agency and apply the most advanced statistical tools in the world to prepare their "Annual Energy Outlook." Fig. 3.9 is data from the 2008 report, updated for the 2013 data. Data to the left of the

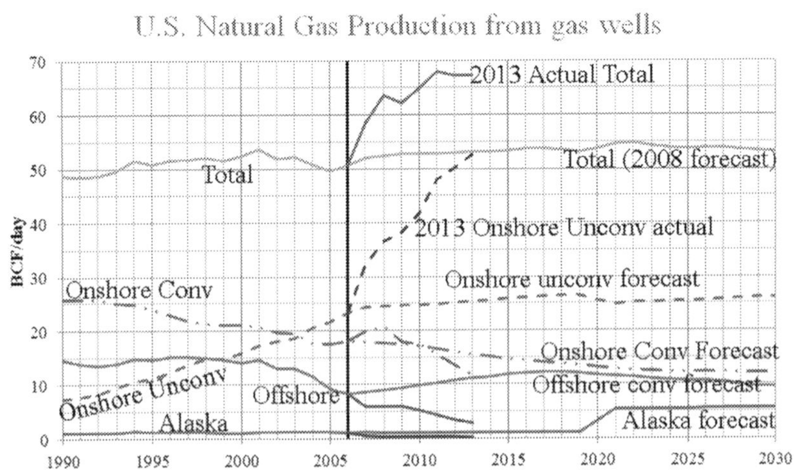

Figure 3.9 EIA annual energy outlook (2008, updated in 2013) (EIA).

vertical line (2006) is historical data. Data to the right of the vertical line (in the 2008 version) was the US gas forecast through 2030. It looked in 2008 like the US gas production was both flat and unable to meet any increased demand. Five years later (with 7 more years of production data) we were not stagnate at around 43 BSCF/day (1.218 GSCm), but had increased 55% to 67 BSCF/day (1.897 GSCm/day) in spite of the fact that offshore conventional production not only did not increase, but it decreased 75%. This is not to say that the EIA forecasters are bad at their job, they are actually some of the best in the world at this exercise—their job is forecasting and most engineers in Oil & Gas spend as little time as humanly possible at this impossible task. Remember, if computers and mathematical algorithms could predict the future, then there would be no mystery in either the stock market or weather forecasting. In 2008, no one really expected the sea change that has been shale gas. In 2008, the 10 largest gas fields in the United States were all conventional, CBM, or tight gas. The Barnett Shale made the top 100 fields in 2008, but no other shale play was even close. By 2013, the top 10 list had eight shale gas plays on it. That is probably the reason the 2008 prediction was so bad, but about once a decade there is a game-changing event in the industry that makes previous forecasts look foolish.

Having shown an example of unsuccessful forecasting, I have to say that reservoir engineers and production engineers are not allowed to say "It's too hard" and skip forecasting, they have to do it, and they have to do the best job they can.

3.2.1 Bureau of mines method

The starting point for most conventional prediction methods is the *Bureau of Mines Method* (first proposed by the US Bureau of Mines) and also known as the "backpressure equation":

$$q = c_p \cdot (\overline{P}^2 - P_{bh}^2)^n \tag{3.1}$$

Eq. (3.1) assumes that the reservoir pressure is constant (and known) over time, and the sum of all the flow resistance included in c_p is constant both with time and with drawdown. It seems to work reasonably well for conventional gas if you have a good guess for average reservoir pressure (not very common), have assessed c_p and n for each well (even less common), and have a good method to calculate flowing BHP (sometimes) since average reservoir pressure tends to change year-to-year instead of the day-to-day changes that are common in CBM.

Using Eq. (3.1) for unconventional wells leads us to some very bad conclusions. First, the reservoir pressure changes too fast to allow it to be called "constant" for a forecasting period. Second, the near-well-bore conditions are changing very rapidly over the life of the well and cannot be considered either "constant" or "characteristic." Finally, the equation does not accommodate inclining production. To try to overcome these limitations, a project called the POD (Simpson 1) looked at what might be included in the flow constant (c_p). Even though it is clear that the conventional "radial diffusivity" model was describing a flow regime that did not exist in CBM, it might just be close enough:

$$c_p = \frac{k \cdot h}{1422 \cdot \mu \cdot Z \cdot T \cdot \left(\ln\left(\frac{r_e}{r_w}\right) - \frac{3}{4} + \text{Skin} \right)} \tag{3.2}$$

All of the terms in Eq. (3.2) vary with both time and drawdown, but if we assume that most of those changes are second-order effects and can likely be ignored, then we can look to rock mechanics to dump all of the variability into the "skin" term. Ian Palmer and John Mansouri were doing work at Amoco's Tulsa Research Center while the POD team was doing our work and they provided an equation that was a precursor to what became the well-known Palmer–Mansouri model.

$$\text{Skin} = -13.42 + 4.92 \cdot \phi_0^{-1.67} \cdot \left(\frac{c}{b}\right)^{-0.438} \cdot (\overline{P} - P_{bh})^{0.2636} \tag{3.3}$$

All of the parameters in Eq. (3.3) are fixed at initial values except drawdown which comes out of an ongoing material balance in conjunction with a production forecast. Inserting Eq. (3.3) into Eq. (3.2) and combining constants yields:

$$c_p = \frac{k \cdot h}{1422 \cdot \mu \cdot Z \cdot T \cdot \left(\ln\left(\frac{r_e}{r_w}\right) - 14.17 + 4.92 \cdot \phi_0^{-1.67} \cdot \left(\frac{c}{b}\right)^{-0.438} \cdot (\overline{P} - P_{bh})^{0.2636} \right)} \tag{3.4}$$

Inserting Eq. (3.4) into the backpressure equation (Eq. (3.1)) with the nonlinearity term ("n") fixed at 1.0, seems to do a better job in rapidly declining reservoir pressure reservoirs. Fig. 3.10 shows the impact that this refinement of the backpressure equation can have. The two most

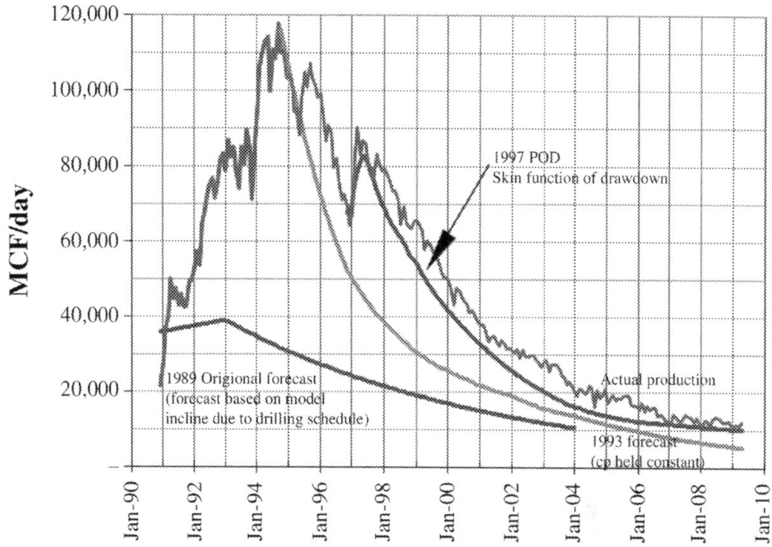

Figure 3.10 Group of 46 San Juan Basin Fairway CBM wells (Simpson 1).

encouraging features of this data are: (1) at 12 years both the actual production decline and the forecast have flattened out; and (2) the forecast was actually able to reflect the uplift from the POD project (one of the limitations of the backpressure equation is an inability to reflect periods of inclining production). It is interesting that the 1989 original forecast was done using a comprehensive compositional reservoir model and the most expensive tool did the worst job.

3.2.2 Inflow performance relationship

Inflow performance relationship (IPR) analysis is very well developed for oil wells. It is less well developed for conventional gas, but there are tools that mostly do an acceptable job. J.V. Vogel (1968) presented an easy to use IPR relationship for oil wells as:

$$\frac{q_{0(press)}}{q_{0(\max)}} = 1 - 0.2 \cdot \left(\frac{P_{bh}}{\overline{P}}\right) - 0.8 \cdot \left(\frac{P_{bh}}{\overline{P}}\right)^2 \tag{3.5}$$

M.K. Fetkovich adapted Eq. (3.5) to Eq. (3.1) (using graphical techniques to evaluate c_p to yield:

$$\frac{q_{0(press)}}{q_{0(\max)}} = \left[1 - \left(\frac{P_{bh}}{\overline{P}}\right)^2\right]^n \tag{3.6}$$

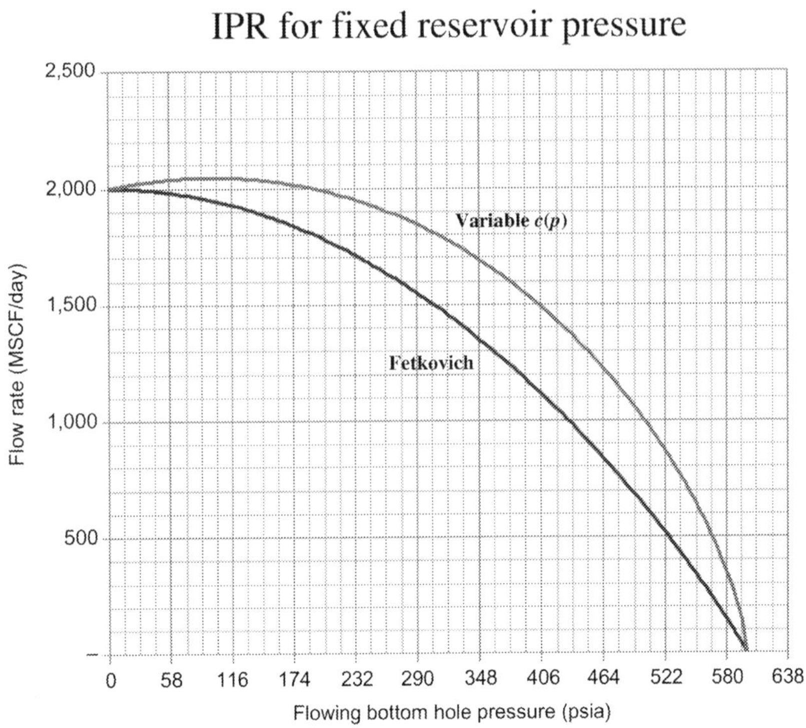

Figure 3.11 IPR relationship.

Comparing Eq. (3.6) using graphical methods to determine a fixed c_p to using Eq. (3.4) to determine c_p is plotted on Fig. 3.11. This shows that some pressure above zero can result in exceeding the absolute open-hole flow that is a cornerstone of the IPR model (and raising pressure resulting in increasing flow rate has been observed in many unconventional wells).

IPR analysis can be momentarily useful for selecting a flowing BHP at a given reservoir pressure, but efforts to use it for forecasting unconventional well performance have not resulted in useful forecasts.

3.2.3 Decline curve analysis

The other major approach to production forecasting is using historical time-series rates to anticipate future rates. This "decline curve analysis" first has to determine the physical extents of the reservoir. If the reservoir is "infinite" (meaning that the borders of the reservoir are beyond the scope impacted by the well), then you can assume that the decline will be exponential and you should plot historical production on a semilog scale.

If the reservoir is bounded within the scope impacted by the well, then you can assume a hyperbolic relationship and should plot historical production on a log–log scale.

Once you have plotted historical production on the appropriate scales, you can draw a best-fit straight line through the data and extend it beyond the limits of the historical data. The assumption is that historical production represents a characteristic performance of the reservoir and future performance will be dominated by the same forces as the forces that dominated historical production.

There are a number of computer programs on the market that will use statistical best-fit methods to match a line to the historical data and capture the data points on the extrapolation. Several of these will allow the computer to generate forecasts for a group of wells without human intervention. The best among them allow a person to review the graphs and include/exclude data points that are patently not representative. Using this feature with knowledge of curtailment periods can significantly improve the usefulness of the graphs.

Fig. 3.12 shows a series of forecasts that were made based on decline curve on a particular San Juan Basin well. We had 2 years of production

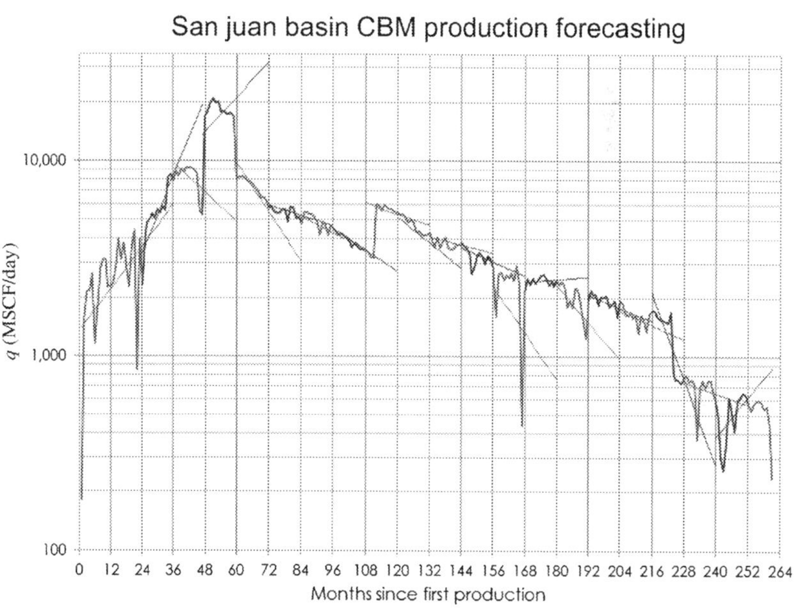

Figure 3.12 Decline curve forecasting.

history prior to doing our first decline curve analysis (this exercise was directed at 1-year budgets so each forecast is limited to 12 months). You can see from 14 forecasts represented on this graph that we got it really wrong far more often than we got acceptable results (1 month the actual was 181% of forecast, another month the actual was 12% of the forecast). Probably half the time we were within ±10%, but the other half required considerable explaining after the fact.

3.2.4 CBM method

If IPR and decline curve analysis don't work very well in unconventional gas wells, what is left? Computational fluid dynamics reservoir models have resulted in the worst forecasts provided by the industry. Nodal analysis procedures (not covered explicitly in this book) tend to rely heavily on IPR techniques and/or decline curve analysis and have been of limited value in unconventional fields.

We found in the POD and operating the POD wells (Simpson 1) that any analysis that treats an unconventional reservoir as D'Arcy flow through a porous medium to provide poor results. Remember Fig. 1.11 (repeated here as Fig. 3.13), that the permeability of unconventional reservoirs to be markedly less than the cement we use to contain the gas, approaching the permeability of steel pipe. As permeability dips below about 10 mD, equations based on D'Arcy's relationships become less and less representative of observed conditions. The only way to successfully produce unconventional resources is to create channels and other pipe-like pathways within the reservoir. In shales we do this with massive frac jobs. In CBM we facilitate the coal failing in a manner that creates flow channels. Flow in unconventional reservoirs is represented much more effectively by pipe-flow or channel-flow equations than by flow-through-a-porous-media equations.

If you know reservoir pressure and temperature, flowing BHP, and gas rate at a given time then you can use one of the pipe-flow equations

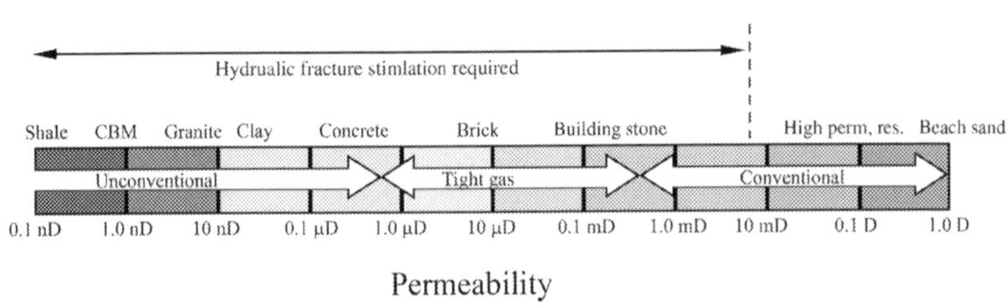

Figure 3.13 Permeability continuum.

discussed in Chapter 4, Surface Engineering Concepts, with an arbitrary pipe size (the POD used 4-in (DN 100) pipe) to determine a pipe length that matches the flow rate and pressures (this value will tend to remain constant over the life of the well).

The steps to use this pipe length to predict production are as follows:

1. Calculate reservoir pressure at the next time step by adding volume at current production rate to cumulative production and using the isotherm relationship from Chapter 1, Gas Reservoirs.
2. Assume a constant surface pressure and calculate flowing BHP at current rate:
   a. Calculate a new inflow rate (using pipe-flow equations) based on future reservoir pressure and calculated BHP.
   b. Use the new rate and constant surface pressure to calculate revised BHP.
   c. Calculate a new rate and repeat "2" until change is acceptably small from one step to the next.
3. Use the "final" flow rate from 2c to recalculate future reservoir pressure and repeat 2 until the change from step to step is acceptably small. This process is truly a pain, but results seem to stand up over time.

3.3 FLUID LEVELS

Flowing BHP is made up of three components: (1) flowing tubing (or flowing casing) pressure; (2) fluid friction in the well-bore tubulars; and (3) hydrostatic pressure from any standing fluids (Fig. 3.14). Since BHP is the key hand-off between reservoir performance and well-bore performance, being able to assess fluid levels in the well-bore is key to understanding inflow performance. The environment downhole is tumultuous and no condition exists for more than a few seconds. Determining BHP requires a way to assess each of the components of pressure.

Finding a practical method to assess well-bore fluid levels has proven to be more difficult than people expect it to be. The most common methods of finding the level are: (1) downhole pressure gauge; (2) comparing tubing pressure to casing pressure; (3) running a pressure "bomb"; and (4) sonic fluid shots.

Pressure "bombs" represent an interesting data gathering exercise. It is rare for data from nature to fall on a straight line. Downhole pressure

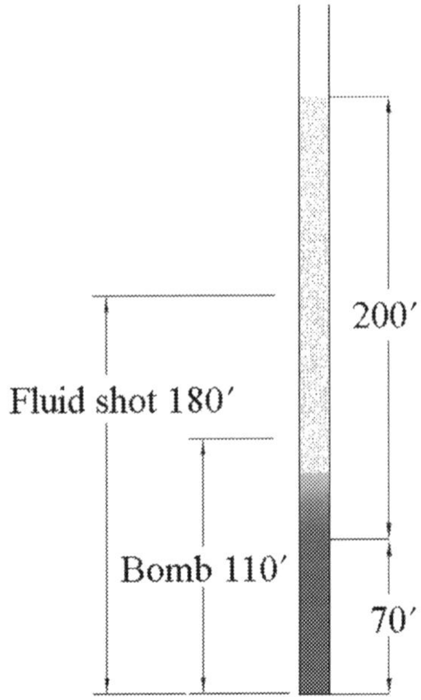

Figure 3.14 Well-bore fluid levels.

recorders represent that perfect dataset every time (if it doesn't, then the data collection process had an error). The slick-line operator has more control as he pulls the tool out of the well than when he's going in, so the typical process is to lower the tool to the bottom and wait a specified period of time to let it stabilize. Then the operator pulls the tool up the hole pausing at specified points to allow the data to be registered to specific points. The data up to the fluid level will lie on a straight line with a gradient of 0.433 psi/ft (10.02 kPa/m) times the water-specific gravity from the bottom of the well to the top of liquid water (the nearly horizontal line added to Fig. 3.15, represents 70 ft (21.3 m) of liquid), then there will usually be a section of froth that does not have a linear relationship between depth and pressure (about 200 ft (61 m) in Fig. 3.15), and finally a nearly linear section of just gas (the nearly vertical line added to Fig. 3.15) to surface. As you will see later in this chapter, the gas section is not actually a straight line, but the magnitude of the change (3 psi (21 kPa) over 2600 ft (793 m)) is so low that the curvature gets lost. Where the two superimposed lines cross (2890 ft (881 m)) defines the effective fluid height—in this case 70 ft (21.3 m) of water, and 40 ft (12.2 m) equivalent water from the 200 ft (61 m) of froth which imposes the same backpressure as 110 ft 33.5 m) of

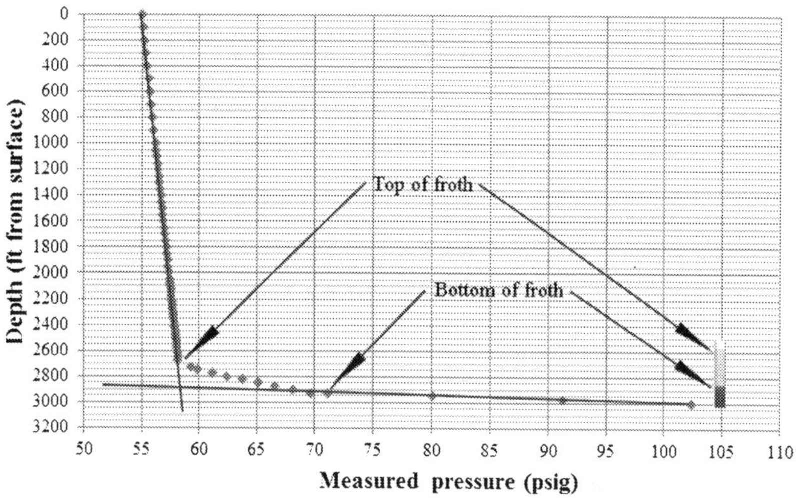

Figure 3.15 Data from pressure bomb.

liquid water. Pressure bombs are useful, but they require at least partial disruption of flow to run the tool in the well, and often the operator cannot guarantee his tool position if it is moving within flow. This is a prime case of the act of observing changing what is being observed.

Permanent downhole pressure gauges do not have the problem of moving within flow, but they also could not distinguish the 200 ft (61 m) of froth from 40 ft (12.2 m) of liquid and there are times where that distinction is important. This is sometimes corrected by placing several gauges in the well, but that can be very expensive and failed gauges can lead to incorrect decisions.

3.3.1 Tubing vs casing pressure

When the tubing/casing annulus is open to the reservoir (Fig. 3.16) and not flowing, then tubing pressure is controlled by delivery pressure alone, and casing pressure is controlled by BHP alone. The key uncertainty is how much pressure is lost to friction in the tubing. Keep in mind that the tubing pressure and the casing pressure at the entrance to the tubing must be exactly equal to each other since they are the same number taken at the same point.

The relationship between flowing tubing pressure, static casing pressure, and BHP is a function of both fluid level and friction losses. If there

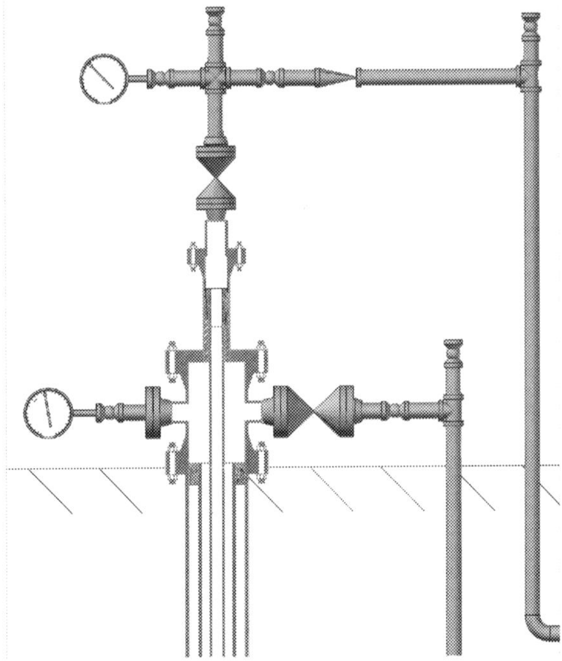

Figure 3.16 Wellhead.

is no liquid standing in the well-bore and the tubing/casing annulus is not flowing, then the difference between tubing pressure and casing pressure is:

$$\Delta P_{tubing\ friction} = (P_{\text{casing}} + P_{gas\ gradient}) - P_{tubing} \approx P_{\text{casing}} - P_{tubing} \tag{3.7}$$

As we saw earlier, with surface pressures under about 500 psig (3.5 MPag) gas gradient is generally small enough to be safely ignored in this calculation. If there is a liquid level in the well, then you need to add a term:

$$\Delta P_{tubing\ friction} = (P_{\text{casing}} + [\rho \cdot g \cdot h]_{water} + P_{gas\ gradient}) - P_{tubing} \approx P_{\text{casing}} + [\rho \cdot g \cdot h]_{water} - P_{tubing} \tag{3.8}$$

If the tubing/casing annulus is open to flow, then the relationship gets a lot more complex as described later in this chapter.

3.3.2 Sonic fluid shots

The workhorse for determining the height of liquids in oil wells and gas wells with nonflowing tubing/casing annulus is the Echometer system from a Texas company by the same name. This device connects to the casing wing valve (i.e., at the location where the pressure gauge is installed in Fig. 3.16) and sends a sound burst down the annulus. When the sound hits things like tubing collars or a liquid level, it reflects the sound back to the sensor on the surface. By counting tubing collars, it is reasonably an accurate method to determine a liquid level relative to the surface. By knowing the configuration of well-bore tubulars it is trivial to turn that liquid level into a height above the end of the tubing.

When there is a calm liquid surface the Echometer system provides very acceptable information to compute flowing BHP. This system has been used successfully in thousands of sucker rod pump (SRP) installations in oil fields (with no casing flow) to optimize pump performance and fluid shots have been used successfully to correct many pump-performance problems.

When there is flow in the casing, the situation gets more complex. A liquid on top of a flowing gas source will experience some amount of agitation. This agitation causes some amount of the surface of the liquid to splash and foam (see Fig. 3.14). This foam can rise to a considerable height, and as a compressible fluid, the weight of the column is unpredictable. When you shoot a sound burst into it, you will get a return, but there is no way to differentiate a return from a random point in the foam stack from an actual liquid level. In the example in Fig. 3.14 the Echometer system overstated the BHP by 70 ft (21.3 m) which overstates the BHP by 30 psi (207 kPa) which may or may not have a material impact on well performance or downhole pump performance.

Companies spend a lot of effort and money "shooting fluid levels" in pumping gas wells (with annular gas flow) only to find that the results from one shot are materially different from a second shot 2 minutes later. My usual recommendation is to discontinue shooting fluid shots altogether in pumping gas wells, but it is rarely followed. Companies see that they are getting unambiguous liquid levels and making decisions based on those clear liquid levels. Companies that experience frustration with the nonrepeatability of the tool generally stop doing multiple shots and simply "average" a single shot. Fluid shots have been a significant factor in pump failures in pumping gas wells because they tend to report higher-than-actual fluid levels that can result in a pump operating outside of its performance envelope.

3.4 VERTICAL MULTIPHASE FLOW

All other things being equal, gas will tend to flow at a higher velocity than a liquid in the same stream. At the gas/liquid interface there is a "no-flow boundary" that requires that either the gas is slowed to the speed of the liquid, or the liquid is accelerated to the speed of the gas, or some combination. In vertical flow,

- gas velocity will tend to drag liquid up the hole;
- buoyant forces will tend to lift the liquid up the hole, but
- gravity will tend to push the liquid down the hole.

The major variables are droplet size (bigger drops increase the impact of gravity and decrease the impact of buoyancy), droplet shape (spherical droplets require less energy to transport than oblate spheroids require), and gas velocity (higher velocity allows drag plus buoyant forces to exceed gravity).

3.4.1 Flowing gas gradient

Static gas gradient was shown as Eq. (0.11). With a liquid column, the static fluid gradient becomes:

$$\begin{aligned} P_{bh} &= \frac{P_{surf}}{\text{psia}} \cdot \exp\left(\frac{0.01875 \cdot \text{SG}_{gas} \cdot \dfrac{(h_{bh} - h_{liq})}{ft}}{\left(\dfrac{T_{avg}}{R}\right) \cdot Z_{avg}} \right) + \rho_{liq} \cdot g \cdot h_{liq} \\ &= \frac{P_{surf}}{\text{kPaa}} \cdot \exp\left(\frac{0.03418 \cdot \text{SG}_{gas} \cdot \dfrac{(h_{bh} - h_{liq})}{m}}{\left(\dfrac{T_{avg}}{K}\right) \cdot Z_{avg}} \right) + \rho_{liq} \cdot g \cdot h_{liq} \end{aligned} \tag{3.9}$$

(Note that this equation uses the convention of dividing the variables by the required units to show that the input data must be converted to the designated unit prior to using it.) If you add friction, then the dynamic pressure drop equation becomes:

$$S=\frac{0.0375\cdot \mathrm{SG}_{gas}\cdot \frac{h_{td}-h_{liq}}{\mathrm{ft}}}{\frac{T_{avg}}{R}\cdot Z_{avg}}=\frac{0.06836\cdot \mathrm{SG}_{gas}\cdot \frac{h_{td}-h_{liq}}{\mathrm{m}}}{\left(\frac{T_{avg}}{K}\right)\cdot Z_{avg}}$$

$$P_{bh}=\left[\left(\frac{P_{tbg}}{\mathrm{psia}}\right)^2\cdot e^S+\left(\left(\frac{\rho_{liq}}{\frac{\mathrm{lbm}}{\mathrm{ft}^3}}\right)\cdot\left(\frac{g}{g_c}\right)\cdot\left(\frac{h_{liq}}{\mathrm{ft}}\right)\right)^2+f_m\left(\frac{q}{\frac{\mathrm{MSCF}}{\mathrm{day}}}\cdot\frac{T_{avg}}{R}\cdot Z_{avg}\right)^2\left(\frac{e^S-1}{1500\cdot\left(\frac{d_{eff}}{\mathrm{in}}\right)^5}\right)\right]^{0.5}$$
$$=\left[\left(\frac{P_{tbg}}{\mathrm{kPaa}}\right)^2\cdot e^S+\left(\left(\frac{\rho_{liq}}{\frac{\mathrm{kg}}{\mathrm{m}^3}}\right)\cdot(g)\cdot\left(\frac{h_{liq}}{1000\cdot\mathrm{m}}\right)\right)^2+f_m\left(\frac{q}{\frac{\mathrm{SCm}}{\mathrm{h}}}\cdot\frac{T_{avg}}{K}\cdot Z_{avg}\right)^2\left(\frac{e^S-1}{1.2\times10^{-6}\cdot\left(\frac{d_{eff}}{\mathrm{mm}}\right)^5}\right)\right]^{0.5} \tag{3.10}$$

As will be discuss in Chapter 4, Surface Engineering Concepts, "f_m" is the "Moody friction factor" which is typically used for liquid flow.

3.4.2 Tubing flow vs casing flow

In tubing flow, the effective diameter (d_{eff} in Eq. (3.10)) is tubing inside diameter, but annular flow is somewhat more complicated. A simple area calculation converted back to a diameter seems logical:

$$A=\left(\frac{\pi}{4}\right)\cdot\left(\mathrm{ID}_{csg}^2-\mathrm{OD}_{tbg}^2\right)$$
$$d_{eff}=\sqrt{\left(\frac{4}{\pi}\right)\cdot A}=\sqrt{\left(\frac{4}{\pi}\right)\cdot\left(\frac{\pi}{4}\right)\cdot\left(\mathrm{ID}_{csg}^2-\mathrm{OD}_{tbg}^2\right)}=\sqrt{\left(\mathrm{ID}_{csg}^2-\mathrm{OD}_{tbg}^2\right)} \tag{3.11}$$

Another logical approach seems to be the "wetted perimeter" method that is used for liquid flow in a channel:

$$\begin{aligned} d_{eff}&=\frac{\text{Wetted area}}{\text{Wetted perimeter}}=\frac{4\cdot\left(\frac{\pi}{4}\right)\cdot\left(\mathrm{ID}_{csg}^2-\mathrm{OD}_{tbg}^2\right)}{\pi\cdot\left(\mathrm{ID}_{csg}-\mathrm{OD}_{tbg}\right)}\\ &=\frac{\left(\mathrm{ID}_{csg}+\mathrm{OD}_{tbg}\right)\cdot\left(\mathrm{ID}_{csg}-\mathrm{OD}_{tbg}\right)}{\left(\mathrm{ID}_{csg}-\mathrm{OD}_{tbg}\right)}=\left(\mathrm{ID}_{csg}+\mathrm{OD}_{tbg}\right)\end{aligned} \tag{3.12}$$

The first "4" in the equation results from the equation being for liquid and including the included angle from the center to the surface of the liquid, for a gas the volume is completely filled.

When we look to *Petroleum Engineering Handbook*, we see that (Bradley and Gipson, 1987, pp. 34–29, for derivation) you need to account for the extra surface area (no-flow boundary) relative to the flow area:

$$d_{eff} = [(\mathrm{ID}_{csg} + \mathrm{OD}_{tbg})^2 \cdot (\mathrm{ID}_{csg} + \mathrm{ID}_{tbg})^3]^{1/5} \qquad (3.13)$$

If we have 2-3/8, 4.7 lbm/ft, J-55, EUE tubing (OD is 2.375 in (60.33 mm)) inside 7-in, 28 lbm/ft, N-80, T&C casing (ID is 6.214 in (157.8 mm)), then these three approaches would give you:

- Intuitive—5.724 in (145.8 mm) (8% higher than petroleum method)
- Wetted perimeter—3.839 in (97.51 mm) (21% lower than petroleum method)
- Petroleum method—5.298 in (134.6 mm)

The intuitive method seems like it is likely close enough, until you realize that Eq. (3.10) (and most fluid flow equations) uses this number to the fifth power which introduces a 50% error out of the box.

When doing this calculation it is conventional to use the OD of the tubing body and ignore any restrictions due to the thread area. This is the technique I have used in many wells and was able to match actual data to calculated data very closely.

3.4.3 Annular flow in pumping wells

In the example in the last section we see that the flow area of the tubing/casing annulus is typically much larger than the flow area in the tubing (the inside diameter of the tubing in that example is 1.996 in (50.7 mm) which results in an area of 3.13 in^2 (20.19 cm^2) where the effective diameter of the annulus was 5.298 in (134.6 mm) which results in an area of 22.05 in^2 (142.3 cm^2)—7 times larger). This results in significantly lower velocity in the annulus than in the tubing. Any liquid that might be carried into the annulus is generally moving so slowly that the gas lacks the flow energy to carry the droplets and they will tend to fall against flow.

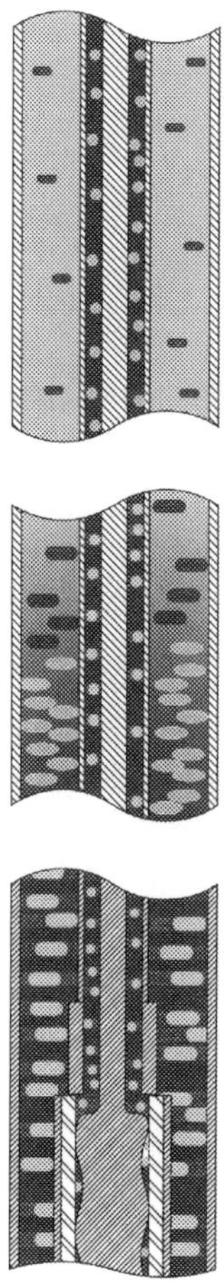

Figure 3.17 Annular flow.

In Fig. 3.17 you see the result of this. In the bottom of the annulus you have gassy water (which is made worse in the presence of well-bore components like tubing anchors). Somewhere up the hole you see the mix change to a watery gas. Finally as you move further up the annulus you see the mixture has changed to a largely dry gas. At the same time any gas that entered the tubing through the pump must stay in the tubing.

The gas in the tubing takes up physical volume. The mass flow rate of gas will be constant from the top of the tubing to the bottom, but the space required for that mass changes dramatically. Above some point as you descend into the well, the density of the gas will have increased to the point where the pump begins to see flow that is very similar to single-phase liquid.

The impact of this phenomena is best described with an example.

- Pump type—progressing cavity pump (PCP) which is positive displacement (PD) (see Section 3.5.3.6)
- Set depth—600 m (1969 ft)
- Separator pressure—360 kPag (52 psig)
- Pump design capacity—90 m^3/100 rpm (566 bbl/100 rpm)
- Speed at start of test—210 rpm (should be pumping 189 m^3/day (1189 bbl/day))
- Torque at start of test—37% of max (torque is a function of discharge pressure and flow rate)
- Water production—65 m^3/day (408 bbl/day) which was 34% of design for this rpm.

After gathering initial conditions, we added backpressure to get flowing tubing pressure up to 2200 kPag (320 psig). The results were surprising (Table 3.1).

As you go up the tubing, you see the gas expanding with decreasing pressure and taking up more and more space (Fig. 3.18), which leaves less and less space for the water to flow (the density of the water does not change appreciably with changes in pressure). Consequently the water

Table 3.1 Backpressure example

	Before	Expected after	Actual After
Flowing tubing pressure	358 kPag (52 psig)	2210 kPag (320 psig)	2210 kPag (320 psig)
Theoretical pump discharge pressure	6250 kPag (906 psig)	8090 kPag (1174 psig)	
Actual pump discharge pressure	2660 kPag (386 psig)		7050 kPag (1022 psig)
Water rate at 210 rpm	65 m^3/day (408 bbl/day)	65 m^3/day (408 bbl/day)	185 m^3/day (1164 bbl/day)
Water rate at 80 rpm			60 m^3/day (377 bbl/day)
Torque (percent of max) at 210 rpm	37%	57%	78%
Torque (percent of max) at 80 rpm			61%
Efficiency	34%		82%

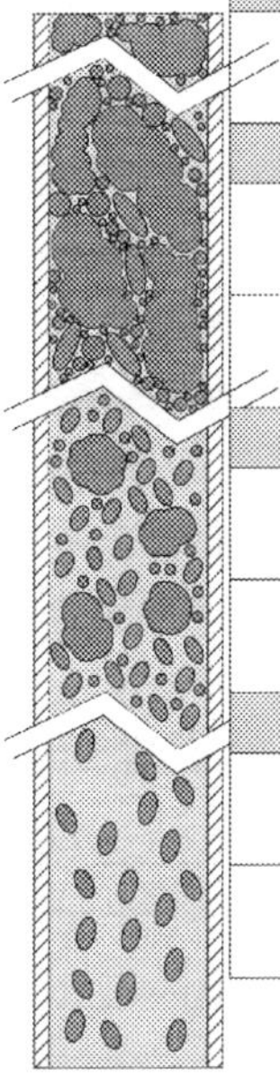

No back pressure		With back pressure
358 kPag (52 psig)	Wellhead Pressure	2206 kPag (320 psig)
	One joint of tubing down	
358.6 kPag (52.01 psig)	Pressure	2282 kPag (331 psig)
47 m/s (155 ft/s)	Water velocity	0.5 m/s (1.6 ft/s)
	30 Joints down	
758 kPag (110 psig)	Pressure	4550 kPag (660 psig)
0.9 m/s (2.8 ft/s)	Water velocity	0.3 m/s (0.9 ft/s)
	60 Joints down	
2661 kPag (386 psig)	Pressure	8094 kPag (1174 psig)
0.4 m/s (1.4 ft/s)	Water velocity	0.15 m/s (0.5 ft/s)

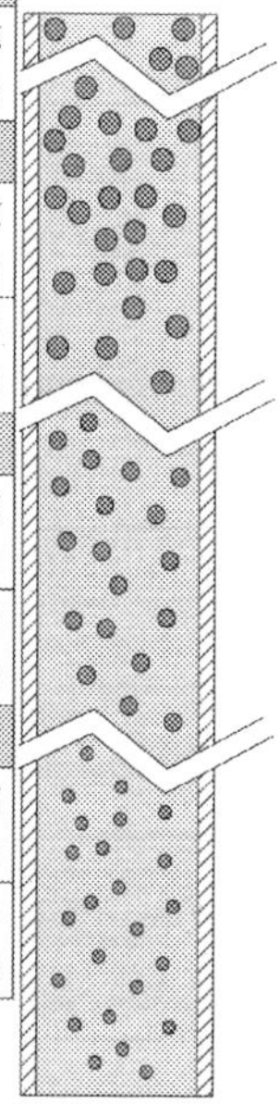

Figure 3.18 Tubing pressure traverse.

must go progressively faster to maintain a constant mass flow rate in a smaller portion of the total space available.

In this example, the gas took up 33% of the space available in the pump and in the bottom of the tubing, which reduces the capacity of the pump by one-third from the outset. Since a PCP (or really any pump designed for single-phase liquid) is designed to be full of liquid, having that much gas in the pump significantly increased the slippage of fluids from late in the pump back toward the suction, likely accounting for the rest of the decrease in efficiency. After adding backpressure, the amount of space in the pump that the gas occupied was about 15% of the available volume which significantly reduced the other sources of inefficiencies.

This experiment has been repeated on a number of PCPs and several SRP with the same results every time—having a flowing tubing pressure above about 145 psia (10 bara) on pumping gas wells always increases the pump efficiency and moves the pump closer to the pump curve. For wells with surface pressures above 145 psia (10 bara), adding backpressure seems to have minimal effect.

3.5 GAS WELL DELIQUIFICATION

Since 1970s natural gas has gone from being a waste product that was only accessed when there was a good chance of accompanying liquid hydrocarbons, to a primary, sought-after product. A strong consumer market, the memory of recent high prices, and the potential for the return of high prices are making producers rethink the terms "abandonment pressure" and "economic limit." Fields are reaching their original abandonment pressure while the wells are still profitable and operators are asking the question "what do you have to do differently to operate these gas wells at low pressures and lower flow rates?" The primary answer in many cases is "deliquification." To make sure we are talking about the same activity let me define terms (Simpson 2):

- *Artificial lift*: application of external energy to lift a *commercial product* from reservoir depths to the surface
- *Deliquification*: application of energy to remove an *interfering liquid* to enhance gas production

The key difference is that in artificial lift it matters where and in what condition the liquids reach the surface, but for deliquification the liquids just need to stop interfering. For example, we will see later that evaporation can be a reasonable deliquification method, but it would obviously be an artificial lift failure.

3.5.1 Gas well life cycle

Within a given reservoir and well-bore characteristics:

- Flow rate into the well-bore is a function of differential pressure between the reservoir and the well-bore.
- Flow rate out of the well is a function of the differential pressure between the bottom and top of the well-bore.
- The ability of the flowing gas to drag liquid along with it is a function of the gas velocity which is a function of the flow rate, pressure, and tubing size.

Pressure and flow management is the primary task of production teams. The formation provides the energy for flow. The rate of flow is limited by the resistance to flow (i.e., the sum of friction and imposed backpressure).

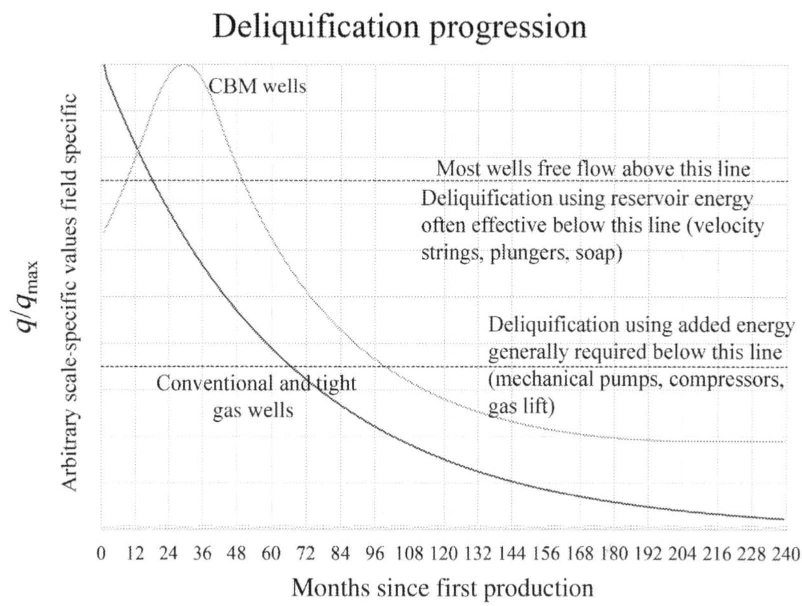

Figure 3.19 Deliquification progression.

Early in a well's life (Fig. 3.19) most wells are able to flow on their own if you let them. In some fields and with some operators, this option is off the table since the initial build-out includes a downhole mechanical pump. In many reservoirs there is a distinct possibility that pumping a well from day 1 will reduce ultimate recovery, but the data is ambiguous enough to prevent certainty. In CBM, the wells need to desorb the gas within the major channels to the well-bore, and there is a period on most CBM wells that production inclines as these channels convert themselves from "gas storage" to "fluid transmission." The data that I've reviewed indicates (at least to me) that this conversion is not aided by setting pumps early, but many operators disagree.

The middle section of Fig. 3.19 is important because the wells often need help, but for the most part they can get by with inexpensive options like plungers, velocity strings, or tubing flow control (see later). These methods are less expensive because they are mechanically simple while not requiring the input of external energy for their operation.

At some point in the well life, reservoir energy is no longer adequate to lift liquids to surface. This is an economic cusp for most wells—is the flow rate high enough to cover the cost of added energy and contribute to capital recovery? If not, then the well will have to be plugged and abandoned. If the flow rate is high enough, then deliquification equipment is installed.

3.5.2 Deliquification using reservoir energy

Liquids are typically not buoyant in gases, so to get liquids to flow vertically upward against gravity requires the conversion of energy to work. The source of energy for this work can either come from the reservoir or can be brought in from the outside. Free-flowing wells use reservoir energy to lift the liquid. "Deliquification using reservoir energy" implies human activities to reduce waste of reservoir energy in the energy conversion to translate the liquids from the bottom of the well to the top.

3.5.2.1 Critical flow

R.G. Turner et al. from Baker Oil Tools and his colleagues published the term "critical flow" in the November, 1969, issue of *Journal of Petroleum Technology*. This paper showed that the liquid volume reaching the surface was a function of gas velocity, which in turn is a function of interfacial tension and fluid density. "Critical velocity" is that flow rate of fluids moving vertically up a pipe that is just sufficient to drag produced liquid with the gas. This velocity will vary from well to well, from time to time, and from nominal pressure to nominal pressure depending on the quantity of liquid that needs to be lifted, the water droplet size and shape, and the conduit size.

Many other researchers have built on this concept with new interpretations of Turner's data and occasionally with new datasets. The magnitude of "critical velocity" and the method for determining it continues to be a source of heated academic debate, but it is certain that at some increasing velocity, the ability of a gas to transport liquids efficiently changes from "not effective" to "effective."

There are dozens of versions of the critical flow equation, and they mostly vary the constant part of Turner's basic equation. There is considerable confusion around this equation. The paper said that interfacial tension should be input in dyne/cm, but it appears in intermediate steps that they used lbf/ft. When you resolve the discrepancies, Turner's equation is:

$$v_{Turner} = 1.912 \cdot \frac{\text{ft}}{\text{s}} \cdot \left(\frac{\left(\frac{\sigma_{gas\ liq}}{\frac{\text{dyne}}{\text{cm}}} \right) \cdot \left(\frac{\rho_{liq} - \rho_{gas}}{\frac{\text{lbm}}{\text{ft}^3}} \right)}{\left(\frac{\rho_{gas}}{\frac{\text{lbm}}{\text{ft}^3}} \right)^2} \right)^{1/4} \tag{3.14}$$

Turner's work was done at quite high pressure and experiments show that water droplets in this pressure regime tend to be approximately spherical. In 1991, Steve Coleman et al. (1991), working with researchers from Exxon using a lower pressure dataset reached the same conclusion as Turner, but with a different constant:

$$v_{Coleman} = 1.593 \cdot \frac{\text{ft}}{\text{s}} \cdot \left(\frac{(\sigma_{gas\ liq}) \cdot (\rho_{liq} - \rho_{gas})}{\rho_{gas}^{\ 2}} \right)^{1/4} \tag{3.15}$$

Coleman explained that the lower constant was indicated by the observation that at lower pressures the water droplets tended to flatten out into an oblate spheroid. A sampling of the literature showed several minor tweaks to both Turner and Coleman, but not really anything major until M. Li and his colleagues at the State Key Lab of Oil/Gas Reservoir Geology and Exploitation at Chengdu University of Technology in China (Li et al.). This research looked at a revised droplet geometry that produced droplets shaped like a torus with a membrane filling the inner diameter. This configuration appears to be much more efficient and Li's version of the equation is:

$$v_{Li} = 1.322 \cdot \frac{\text{ft}}{\text{s}} \cdot \left(\frac{(\sigma_{gas\ liq}) \cdot (\rho_{liq} - \rho_{gas})}{\rho_{gas}^{\ 2}} \right)^{1/4} \tag{3.16}$$

The term in the root works out to ($\text{ft}^{3/4}/\text{s}^{1/2}$) so this is obviously an empirical equation and you must use the units shown in Eq. (3.14).

Interfacial tension (σ, also known colloquially as "surface tension") between water and gas can be estimated by:

If temperature $> 280°F$, then:

$$\sigma_{gas\ liq} = \sigma_{280} = \left(75 - 1.108 \cdot \left(\frac{P_{tbg}}{\text{psia}}\right)^{0.349}\right)$$

If wellhead temperature $\leq 74°F$, then:

$$\sigma_{gas\ liq} = \sigma_{74} = \left(53 - 0.1048 \cdot \left(\frac{P_{tbg}}{\text{psia}}\right)^{0.637}\right) \tag{3.17}$$

Otherwise:

$$\sigma_{gas\ liq} = \left(\frac{280 - \frac{T_{tbg}}{°\text{F}}}{280 - 74}\right) \cdot (\sigma_{74} - \sigma_{280}) + \sigma_{280}$$

Interfacial tension for condensate is about two-thirds of these numbers. If both are present then using water is more conservative.

There were some differences in the conclusions of the papers:

- Turner's work was done at high pressures (mostly over 1200 psi (8.3 MPa)), and concluded that below about 100 bbl/MMSCF (0.56 m^3/kSCm) critical flow was not a function of water volume flow rate.
- Coleman's work was done at pressures from 50 to 500 psia (345–3450 kPa) and concluded that below 220 bbl/MMSCF (1.24 m^3/kSCm) critical flow was not a function of water volume flow rate.
- Li's work was done at around 350 psia (2410 kPa) and while the paper was silent on water flow rate where the equation was valid, it implied that this research supported Coleman's water rates.

Many people try to make practical use of these equations using flowing tubing pressure without making the effort to translate that to BHP. With surface pressures in the "high-pressure" region (i.e., above 145 psia (100 kPa)), this is likely conservative and calculated values will be higher than actual critical velocity (usually a good outcome that wastes some energy to friction, but provides better assurance of keeping the well unloaded). There is anecdotal evidence that this equation does not represent reality at pressures less than 45 psia (310 kPaa) and will lead to bad decisions at these pressures.

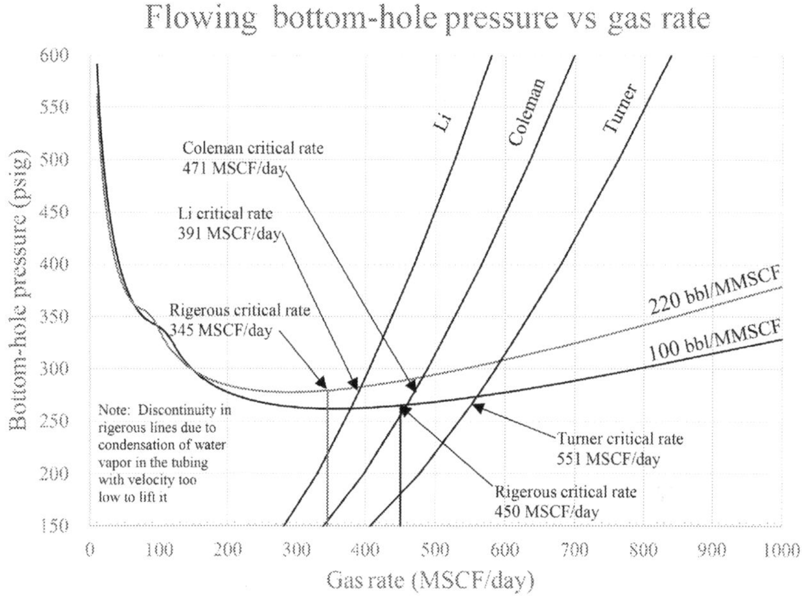

Figure 3.20 Flowing BHP vs flow rate.

Fig. 3.20 compares Turner, Coleman, and Li at 150 psig (1034 kPag) flowing from a 3000 ft (914 m) of 2-3/8 tubing (1.996 in (50.7 mm) inside diameter tubing) well to a rigorous multiphase model that includes evaporation and condensation. At zero flow rate, the model shows BHP equal to surface pressure plus a column of liquid consistent with reservoir pressure. As gas flow rate increases, the gas tries to lift the liquid column, which lowers the apparent weight of the liquid column. At the minimum, the lifting force is near a maximum, while not yet paying too high a price in friction. From that minimum, increasing flow rate increases fluid friction and therefore BHP. On the 220 bbl/MMSCF (1.24 m^3/kSCm) line, Li would require 13% more flow rate to reliably lift water and Coleman would require 37% more flow. On the 100 bbl/MMSCF line Turner requires 22% more flow. All of these differences require that more reservoir energy be expended than is absolutely necessary.

If Turner and later theories lose reliability with low-pressure operations, what can you do? Fig. 3.21 shows a nontheoretical, field method for determining an actual well's critical flow rate. This method is based on the fact that as long as the tubing and annulus are in easy communication, pressure in the annulus will contribute to the flow rate. When the end of the tubing is immersed in water, this easy communication

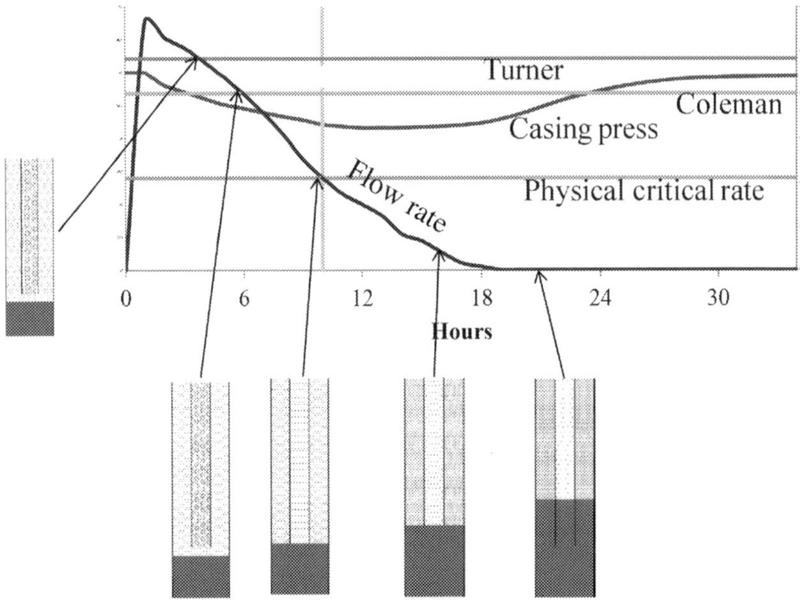

Figure 3.21 Nontheoretical method to determine critical rate.

stops. Obviously, as long as the well is efficiently lifting liquids, liquid will not build up in the well-bore. This method takes advantage of the fact that the volume of the tubing/casing annulus is finite and as you use the stored energy you will be able to see pressure drop on the annulus. The process that one field has used with very good effect has been:

- Shut in the well until tubing and casing pressure equalize. If the wellhead is equipped to allow the spaces to be equalized on the surface, make certain that these equalizer valves are shut.
- Set up the automation (or stand-alone data logger) to record data at 1 minute intervals. If the automation system does not allow capturing short-interval data (and what good is an automation system that does not allow engineering analysis?) then use a data logger. Reliable, short-interval data is crucial.
- Start the well flowing to sales. On the data log, you will be able to see casing pressure decreasing toward line pressure. If casing pressure reaches separator pressure then this technique cannot be used and you are limited to theoretical methods. Most of the time, the casing pressure will stop decreasing at a value well above separator pressure.

- At the point where casing pressure stops decreasing, note the time. It is usually a good idea to back up about an hour from the point that casing stops decreasing and call the flow rate at that point "critical flow." There is no need to allow the well to continue to flow while loading as is shown in Fig. 3.21.

One experienced operator runs this test anytime nominal line pressure changes more than 15 psig (103.4 kPag) or once per quarter whichever happens first. They input the critical flow into their automation computer and anytime the well falls below the nontheoretical critical rate they shut the well in with automation for a predetermined period. It is difficult to tell the benefit of this regimented approach, because every well that they have done the test on has significantly increased production rate above forecast, which has caused forecasts to be revised upward pretty much every quarter and the old data is not readily available. They seem quite happy with the process.

3.5.2.2 Velocity string

Realizing that a well's flow rate at (or above) critical velocity is a function of the velocity and the size of the tubing leads people to investigate matching tubing size with the well's ability to flow. While any tubing string in a well is a "velocity string," we tend to think in terms of smaller than 2-3/8 (DN 50) tubing as velocity strings. The technique of reducing tubing size to use reservoir energy to lift liquids has a very definite shelf life. As a well continues a natural decline, any size tubing will eventually be too big for the reservoir's inflow to be above critical in the tubing.

As tubing gets smaller, friction increases at an even faster rate. In Chapter 4, Surface Engineering Concepts, we'll see that the friction factor is a function of Reynolds number which has pipe ID in the numerator, reducing Reynolds number can put you on a very steeply increasing section of the friction curve so the impact on BHP of reducing tubing size both reduces the flow area and increases the friction factor.

Wells with aggressive velocity strings are very unforgiving. Some observations are as follows:

- If the inflow rate increases, increases in friction losses can dampen the inflow.
- If the inflow rate decreases slightly, you can drop below the critical rate and load up very quickly (1 bbl (159 L) of water equals 1030 ft

(314 m) in 1 in (DN 25) which will add 500 psig (3.4 MPag) to the BHP).
- A cold section in the well-bore can condense water vapor and upset the balance in a near-critical well.
- Small-diameter velocity strings preclude both plungers, pumps, and swabbing.
- When using aggressive velocity strings it is a good idea to keep the tubing/casing annulus valve shut.

3.5.2.3 Tubing flow controller

As people started accepting Turner's theories, an obvious (after the fact) next step is to manage tubing flow by augmenting it with casing flow. The basic principle is that as long as tubing flow is above critical, then flowing the tubing/casing annulus does not harm the well's flowing capacity while improving profitability. The problem was that technology had not caught up with the concepts when people first started trying to implement this. My first attempt is shown in Fig. 3.22. This well was experiencing liquid loading problems flowing 2.7 MMCSF/day (76.5 kSCm/day) through both the tubing and casing. Separator pressure was 10 psig (22 psia (151.7 kPaa)). Critical rate for the well (using the rigorous model because pressure was too low for the models based on Turner) was 343 MSCF/day (9.7 kSCm/day). With the 2-in (DN 50)

Figure 3.22 Tubing flow controller.

orifice meter with a 0.6 β-ratio and a 2 psid (13.8 kPad) differential pressure, the flow rate would be 362 MSCF/day (10.25 kSCm/day). The dead band on the dP controller was about 1 psid (6.9 kPad) so we set the controller to maintain 2–3 psid (13.8–20.68 kPad) which should have kept the tubing flow well above our calculated critical rate while allowing over 2 MMSCF/day (56.6 kSCm/day) to flow up the casing. This attempt failed in an epic manner and the well loaded up in less than a day. On post appraisal, we determined that while the pressure regulator was rated to operate at 10 psig (69.0 kPag), it could be expected to be sluggish and taking supply gas from the upstream meter tap as both sensor and actuator gas was poor measurement practice. This trial was scrapped, but we learned a lot:

- The dP of the orifice meter at low pressure was too high for repeatable measurement so we had introduced too much uncertainty.
- The large pressure regulator was too sluggish and too prone to overshoot its set point (which immediately dropped the tubing flow to zero).
- The dP required to actuate the dP controller was too large for accurate measurement.
- Pneumatic or hydraulic controls have a low probability of success, this application requires Program Logic Controller (PLC) control with a flexible program.

Other fields took these results and put together a system that works:

- They used PLC control (with a flexible program).
- They used V-Cone meters from McCrometer (see Chapter 5: Well-Site Equipment) instead of orifice meters.
- They used electric-drive chokes or (in bigger applications) a valve like a Fisher V-Ball with an electric actuator.

These refinements have created an environment that allows wells to remain unloaded for years after most operators are installing mechanical pumps.

3.5.2.4 Plungers

Early in a well's life, there tends to be enough flowing BHP that the well doesn't need assistance. Later in life the reservoir may not have enough pressure to provide the dP required to lift the plunger and the water load and the operator must look for other deliquification methods. Plungers are appropriate between these two extremes.

Gas is reasonably inefficient at moving liquid. As flow rates get too low for the gas to do an acceptable job, it is logical to place a mechanical barrier between the liquid phase and the gas phase to allow the gas to (efficiently) push the barrier and the barrier to (efficiently) push the liquid. This process is done using "plungers" as the mechanical barrier. The plunger acts like a pipeline pig (see Chapter 6: Gas Gathering Systems) to force any liquids or solids in the tubing to surface. The amount of mass that the plunger can move is limited by the available reservoir pressure. If flowing BHP is 10 psig (69 kPag) higher than flowing tubing pressure and tubing size is 2-3/8, 4.7 lbm/ft (1.996 in (50.7 mm) inside diameter), then the weight of the plunger and the weight of the water limits the carrying capacity to about 2.5 gallons/trip (9.5 L/trip), if you are able to cycle a plunger 6 times/h then 10 psid (69 kPag) can move 5 bbl/day (794 L/day) in 2-3/8 tubing. Bigger tubing increases the capacity some, but bigger plungers are heavier so the improvement is not linear.

A conventional plunger will not fall against flow, so some rudimentary controls are required. Fig. 3.23 shows a fairly typical setup (this well has a flow line from the casing to the separator which is not the norm, but is not all that uncommon either). The plunger controller in this case is a simple pneumatic timer that receives a signal from the arrival sensor that the plunger has arrived and starts a clock to allow after flow for a fixed period before shutting the motor valve to release the plunger for the next cycle. The motor valve will stay shut for a predefined period to allow the plunger to get to bottom, and then open to create the necessary differential pressure to cause the plunger to travel upward.

Figure 3.23 Plunger controls.

The controller in Fig. 3.23 happens to be a simple stop clock, the complexity of controllers go upward sharply from there. Several companies sell dedicated plunger controllers that look at various parameters such as the differential between casing pressure and line pressure, or other pressure differences. The next step-up is a generic PLC that can look at flow rates vs predefined parameters (such as critical flow rate) to facilitate more well-specific flow/no-flow decisions. The latest control schemes have "learning" algorithms to determine on their own what control parameters best suit a given well (e.g., one algorithm watches casing pressure from 1 second to the next and stops the after-flow when the rate of decrease in casing pressure levels off, indicating that the well has reached critical velocity and drops the plunger).

Plungers can be divided into "conventional" and "continuous" categories. Conventional plungers are much like pipeline pigs in that they are either facilitating the removal of liquids or blocking the line. To drop a conventional plunger you have to stop tubing flow.

Continuous plungers have a way to allow the plunger to bypass gas around themselves and can fall against tubing flow. It is common for these plungers to reach the surface, open the bypass port (either by forcing a ball off of a seat, or holding the sleeve and allowing the ball to fall, or with a valve-positioning tool on the surface), and then falling to the bottom to shut the port and make another trip without the well ever shutting in.

The operating sequence for a plunger in broad strokes is as follows:

- Drop the plunger (for conventional plungers this requires stopping tubing flow, for continuous they just fall).
- Allow enough time to let the plunger fall to the bottom (fall rate for conventional plungers tends to be around 160 ft/min (50 m/min), continuous plungers fall much faster, with some reporting 820 ft/min (250 m/min)).
- Open the tubing to sales and allow the plunger to rise to surface.
- When the plunger arrives (usually indicated by a "Magnetic Shut Off" or MSO) the well is allowed to flow to sales either for a set time or until some predefined condition is met. This period is called "after-flow" and is the time that the well is actually making money.
- At the end of the after-flow period for conventional plungers and immediately for continuous plungers, drop the plunger and start the process again.

An alternate technique that has worked well when it can be automated (and has failed miserably without automation or with stop-clock controllers) is as follows:

- Put the motor valve on the casing flow line instead of the tubing flow line.
- When the plunger arrives, open the casing flow as you drop the plunger.
- Let the plunger fall for a time (while flowing the casing).
- Shut the motor valve on the casing and let the plunger arrive.

This alternate technique allows the well to always be open to sales so it removes the risk of too much after flow (taking pressure too low to allow the plunger to turn around in the allotted time) or too little after flow (reducing the well's income).

3.5.2.5 Surfactants

When wells begin to lack the ability to reliably run a plunger, the next step is often to "drop soap" to either reduce the dP required to run a plunger or to improve the well's ability to lift liquids without mechanical assistance.

Soaps, foamers, and other surfactants are designed to foam and introduce voids that tend to lighten a liquid column and reduce the surface tension of the liquid droplets to minimize their size/weight. All soaps have to be activated by agitation, and you need to ensure that the environment where they are introduced is sufficiently turbulent to ensure activation. Liquid soap that gets back to surface without foaming will tend to foam in separators or flow lines.

Different formulations of surfactant are effective with different fluids. Each condensate mix requires a unique formulation, and each mixture of contaminants in water changes the effectiveness of a surfactant.

Surfactants are either introduced by dropping "soap sticks," injecting the chemical into the bottom-hole using a capillary string, or injecting it on the surface and hoping that it will dribble down the well-bore (rarely effective). Surface injection is ineffective because the adhesion between the soap and the pipe wall causes the soap to stick to the pipe more strongly than the downward gravitational force and it never gets to bottom. Since the reservoir fluids see a no-flow boundary at the pipe walls, the turbulence available to activate the soap is very low.

Soap sticks are surfactants enclosed in a degradable plastic that is designed to completely dissolve in produced water. They don't always completely dissolve and can plug up the end of a tubing string. Soap sticks can be effective, but they are often abused.

Capillary strings are run down the tubing through the tubing master valve. Shutting a master valve with a cap string in it will almost always cut the string and cause the whole string to wad up in the hole and can be very expensive to fish out. There have been reports of wells logging off from too much soap injected down a capillary string.

3.5.2.6 Intermitting

When a well is shut in, the pressure deep in the reservoir will tend to migrate toward the near-well-bore rock. We saw an example in Fig. 3.2 of a well that had been on intermittent cycles for some time. Stopping this process by installing a compressor will nearly always increase total production. About the only thing that can be said in favor of intermitting is that it makes slightly more revenue than plugging the well and throws away less gas than vent cycles. My recommendation to operators who chose to intermit wells is to stop and do the economic evaluation to install compression or a downhole pump.

3.5.2.7 Vent cycles

For the weakest wells, with marginal to poor economics, vent cycles are occasionally still used in spite of strong regulatory resistance to the practice (in many places the vented gas must go to a flare stack). Venting requires the same shut-in periods as intermitting, but at the end of the shut-in period instead of sending the well to sales you vent it to the atmosphere for a time and then send it to sales. Therein lies the rub—how do you know when the well is unloaded and you can shift to sales? You don't know and wells on vent cycles are constantly vacillating between too-short venting that doesn't unload the well and too-long venting that sends gas to the atmosphere after the last of the liquid has been lifted.

Fig. 3.24 shows why this occasionally works. If we have pipe with a 2.5 in (63.5 mm) ID (such as 3-1/2, 16.70 lbm/ft) with 100 psig (690 kPag) flowing tubing pressure you would be liquid loading. Opening the tubing to atmosphere drops flowing tubing pressure to 0 psig and puts you on the bottom curve in the "not loading" region. If the tubing were 3-1/2, 9.20 lbm/ft (2.992 in (76 mm) ID) instead, then opening the vent would take you from "loading" on the top curve to too

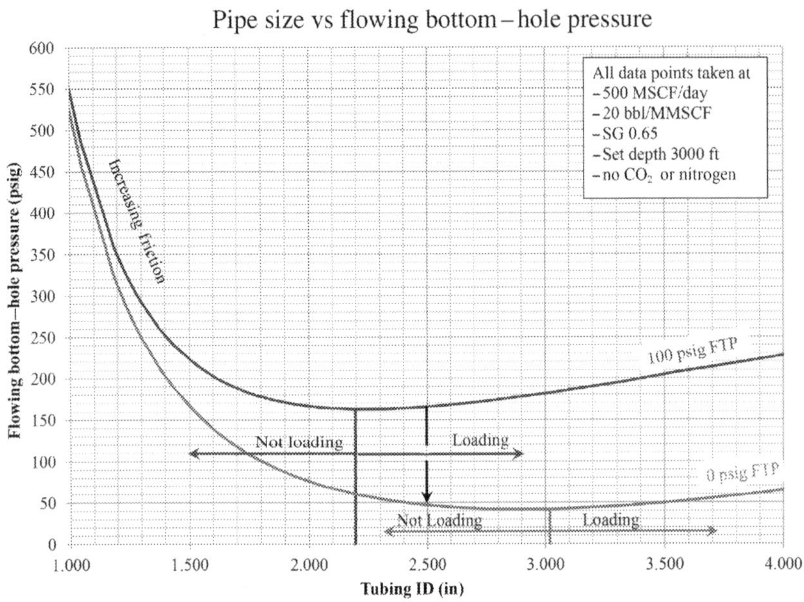

Figure 3.24 Pipe size vs flowing BHP.

close to "loading" on the bottom curve and would most likely not have the effect of unloading the well.

While in this example, the well-bore liquids might be lifted to surface, the gas would not be going to sales during the vent cycle so you have to wonder if the after-flow is worth the waste. The industry has recognized that this practice is very imprecise, not particularly effective, and discards significant saleable gas. Vent cycles are typically seen as an alternative to abandonment rather than as a viable deliquification method.

3.5.3 Deliquification with added energy

At some point in any well's life cycle, the operator has to make the economic decision to either add energy to the process or plug the well. Many of the added energy techniques that we'll discuss in this section are artificial lift techniques that have been adapted to deliquification by changing procedures and/or modifying equipment.

3.5.3.1 Pumping considerations

Not all techniques that add energy are actually "pumping liquid," it is common to think of them that way.

Emissions. None of the techniques in this section directly vent gas to the atmosphere, but they all require some external motive force. On-site engines are normally internal combustion engines that either burn well gas (most common for gas wells) or liquid fuel trucked to site. Electric motors result in about the same emissions as on-site engines, but those emissions take place elsewhere and someone else is accountable for them.

The major reporting concerns are oxides of sulfur (SO_x), oxides of nitrogen (NO_x), carbon monoxide (CO), and unburned fuel. Environmental regulators have added the so-called greenhouse gases to the list, often in direct contravention of local law (e.g., in the United States the "Clean Air Act" explicitly prohibits CO_2 and CH_4 from being classed as "pollutants," but in 2015 the US Environmental Protection Agency did in fact class these beneficial gases as pollutants and the regulation was later struck down by the courts). The actual pollutants of concern are all the result of burning fuel with inadequate oxygen. It is becoming common for operators to install fuel-air monitors to keep the combustion as close to stoichiometric as possible and to create a record of monitoring activities.

Environmental agencies have recently spent a lot of effort (and taxpayer money) on reducing "unloading emissions" which involve dumping raw "greenhouse gases" into the atmosphere unburned. Engine stack gas is not a factor in this category of emissions.

Net positive suction head (NPSH). In 1993, I was in a mixed engineering group made up of production engineers and facilities engineers. One of the production engineers proposed installing an electric submersible pump (ESP) in a gas well. Innocently, I asked "what is the net positive suction head required for that pump?" The production engineer got the look that all engineers get from time to time, the look that says "I have no earthly idea what you are talking about, but I refuse to look dumb so I'll make something up," and he said "it is zero." I looked over at our boss (another facilities engineer) and without another word he said to the production engineer "David will be helping you with specifying that pump." I've been very much involved in gas-well deliquification ever since. This seems to be an area with few facilities engineers involved which has always seemed odd to me since so much of facilities engineering is about shifting fluids from one place to another, vertical is not that much different.

Understanding NPSH is crucial to understanding how downhole pumps work (or don't). Let's start with three important definitions (Simpson 2):

- *NPSH*: the amount of external pressure at the inlet to the pump
- *Required NPSH (NPSH-r)*: The amount of external pressure required at the pump suction to ensure that the pump operates full of liquid without phase changes.
- *Available NPSH (NPSH-a)*: The amount of external pressure available at the pump suction.

It generally doesn't matter if NPSH comes from an actual column of liquid or from an imposed pressure (as long as the pump sees continuous-phase liquid at the pump suction). NPSH-r is a function of fluid properties, primarily boiling point and vapor pressure. Dissolved and entrained gases do not materially impact NPSH since they are not condensable.

Fig. 3.25 shows why NPSH in oil fields is less of a problem than in gas fields. In a static system, liquid will try to "seek its own level" meaning that in Fig. 3.25 the oil level will rise in the tubing to be equal to the highest point in the reservoir that is impacted by the well. Often several hundred feet of liquid will be above the pump in an oil well without impacting reservoir performance.

In a gas field, any liquid above the formation will add backpressure to the producing formation. Managing that total backpressure is an important part of production operations, but it is complicated by the competing requirements that downhole pumps need NPSH-r and the reservoir needs a certain flowing BHP. Trading these competing requirements against one another is far more complex than many operators realize.

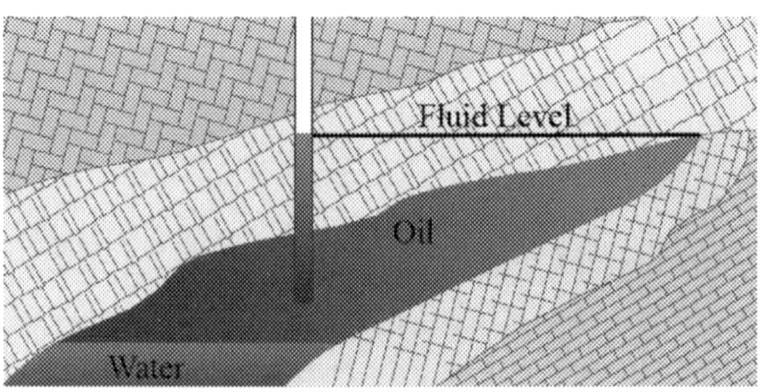

Figure 3.25 NPSH in oil fields.

If you need more NPSH-a than you have, you can:

- Change technology: an SRP requires less NPSH-r than a jet pump for example.
- Install or modify downhole equipment: you can look at changing tubing size, installing downhole separators, installing devices to trip traveling valves on SRPs, put vent holes in piping (be careful of this, any hole on the discharge side of the pump will reduce pump capacity and too big a hole can steal the entire pump capacity).
- Remove pressure drops: screens, tail pipes, standing valves all have some amount of pressure drop that reduces the NPSH-a, but you need to understand why these things are installed in the first place to be able to evaluate the impact of removing them.
- Change pump set depth: This will be discussed later.

Cavitation. When NPSH-a is less than NPSH-r in a dynamic pump (e.g., ESP or jet pump), then there is a significant risk of "cavitation." "Cavitation" is "the formation and subsequent collapse of vapor bubbles in a flow stream." The "subsequent collapse" phrase is the important part. When the vapor bubbles collapse, the surrounding liquid rushes into the void at sonic velocity and can tear metal from the surface of piping and fixtures. Fig. 3.26 shows a jet pump throat that was put into cavitation for less than 1 hour (the pump suction piping was clogged tight with coal fines and the pump could not find anything to pump so its "work" shifted to destroying itself). Cavitation is only an issue in dynamic pumps that rely on constant fluid-phase inside the pump. PD pumps can cavitate, but the outcome is rougher piping, not failed functionality.

Tubing set depth. The industry has observed that in oil fields, setting a pump deep in a well (often significantly below the perforations,

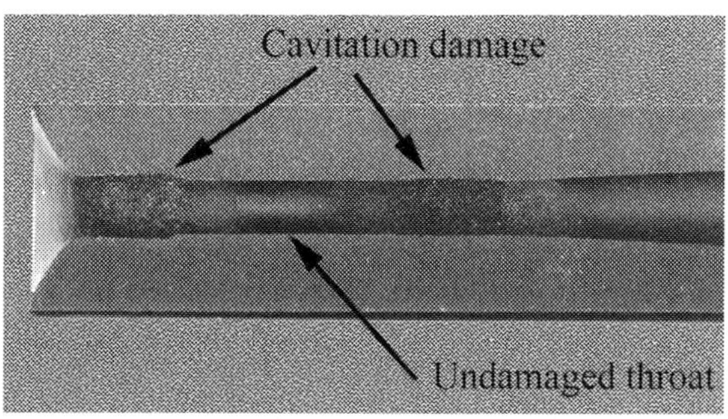

Figure 3.26 Cutaway of a damaged jet pump throat.

colloquially called "sumping the pump") will generally increase liquid production, in terms of both production rate and ultimate recovery. This seems to be due to the fact that gravity has a profound impact on liquids and the deeper you set the pump, the more pressure the pump will see at the suction added to the cohesive forces of a continuous-phase liquid. When pumping a commercial product, increasing both the production rate and ultimate recovery is a very good thing.

But is it good when pumping a waste product? If you'll remember back to Fig. 3.1, in a gas reservoir, the near-well-bore tends to "empty" with production and "refill" during shut-in periods. That process assumes that liquids are evenly disbursed as droplets or small pockets. Since gravity acts more strongly on liquids than on gases, putting a pump below the producing formation has a strong siphoning effect on the water (the same as it has on oil in an oil reservoir) allowing the isolated pockets and droplets of liquid to aggregate. The higher this differential is, the more of the reservoir will experience cohesive-liquid flow. As the cone of cohesive-liquid flow increases, the flow channels become liquid-full farther from the well-bore. Liquid-full flow channels increase the drainage area for liquids, at the expense of gas. The farther you get from the well-bore, the lower the differential pressure across any given volume. Lower dP means it is more difficult for gas to displace the liquid deep in the reservoir, and the only fluid that flows through that volume is liquid.

What is the impact of lowering tubing to sump the pump (Table 3.2) as many people do in gas fields around the world?

Data on sumping pumps in gas fields is quite difficult to come by. Companies that routinely set pumps below the producing formation never set pumps above the producing formation and vice versa. Getting comparative apples-to-apples data has proven impossible. The hardest

Table 3.2 Effect of sumping a pump

	Oil	Gas/Water
Provides NPSH-a for pump	Good	Good
Increases the rate you can remove liquid	Good	Only good if more water can correlate to more gas (generally the opposite is true)
Increases the liquid drainage area at the cost of gas production	Good	Very bad (you get water restricting flow paths into the well-bore)
Produces more liquid	Very good	Not good

thing to sort out is the question "does more water production correlate to more gas production and/or more ultimate recovery?" If not, does less water production (approximately equal to the native inflow rate) correlate to more or less production and/or ultimate recovery?" As an industry we do not have a universal answer to either of these questions.

In the POD project in the San Juan Basin mentioned earlier (Simpson, 1) we kept careful records on gas and water production along with everything else and we had the latitude to reposition the tubing as we saw the necessity. We found that wells with the tubing set below the lowest zone, we did make more liquid, but less gas. We further found that placing the tubing near the middle of the most productive zone maximized gas production while minimizing water production in both pumping and free-flowing wells. Our conclusion from this is that (at least in CBM) the water in the reservoir is holding the gas onto the coal and removing it faster than desorbed gas can flow to the well-bore causes that gas to migrate in other directions and become unrecoverable.

I can't say how well this concept translates to other formations, but I have observed in a very densely drilled tight gas field, sumping a pump did increase water production while decreasing both gas and condensate production on several wells when new engineers came into the field and demanded that tubing be lowered "because that is the way we did it in my last field." This was far from a scientific study and could have no relevance at all, but it is what I observed.

The area below the producing formation is typically called the "rat hole" because the original purpose of drilling past the target formation was to provide a location for fill and well-bore trash to accumulate. Keep that purpose in mind when deciding to set a pump in the rat hole. Some of the observed problems with sumping pumps in the rat hole have included the following:

- Concentration of well-bore trash (dropped tools, corrosion products, scale, sludge, and fill) in the pump suction
- Difficulty in removing heat from electric motors in a space with limited-to-no fluid outside of the production casing.

3.5.3.2 Surface compression

As we discussed earlier, critical flow rate can be altered by changing flowing tubing pressure. Lower surface pressure can also increase the differential pressure between the reservoir and the well-bore which can increase flow into the well. Well-site or gathering system compression is often a useful tool for reservoir management. It can also serve as a facilitating element for the deliquification techniques mentioned earlier (e.g., lower flowing tubing pressure lowers the critical flow rate for a given tubing size).

In oil fields, the addition of compression is almost always classified as "rate acceleration" in that its purpose is to remove interfering gas from liquid flow to get the reservoir flow closer to a continuous-phase liquid. In gas fields, rate acceleration economics are rarely robust enough to allow projects to be authorized. On the other hand, when we book gas reserves we assume an abandonment pressure. Compression can (and does) lower the economic abandonment pressure, adding reserves just as surely as drilling a new well does. The economics of projects that add reserves tend to be very robust.

3.5.3.3 Evaporation as deliquification

If your pressure is low enough, then it is sometimes possible to evaporate all of the liquid that flows into the well-bore. This technique works, but it requires you to be willing/able to work under vacuum conditions and generally requires you to remove production tubing from the well altogether. In a deep vacuum, none of the incompressible-flow equations will have any relationship to observed data. Friction factors based on the Colebrook equation (Chapter 4: Surface Engineering Concepts) will not be valid, and using them will understate the friction by at least an order of magnitude. Friction is still a strong function of velocity, but the relationship to Reynolds number becomes very tenuous. Providing all the flow area possible (by removing tubing) minimizes the velocity and therefore the friction.

A major concern is that evaporating the produced water will leave solids behind that can plug the formation. There is no theory that says that this is unlikely, but experience to date has not shown it to be a significant problem.

To get a feeling for the magnitude of inflow where this technique might be useful, remember from Fig. 3.4 that as pressure comes down (for a given temperature), the amount of water vapor that a gas can carry increases rapidly. At 106°F (41°C) and 50 psia (345 kPa) 100% RH is 1096 lbm/MMSCF (17.5 gm/SCm). If you drop the pressure to 3 psia (21 kPaa) the water content jumps to 17,980 lbm/MMSCF (288 gm/

SCm). The difference between those two data points is nearly 50 bbl/MMSCF (7.66 m^3/SCm). There are a very large number of gas wells that make less than 50 bbl/MMSCF.

The most significant impediment of widespread use of this technique is simple fear and superstition. This technique requires operating in vacuum conditions. As will be discussed at length in Chapter 4, Surface Engineering Concepts, some very bad science has led to widespread terror of oxygen-related corrosion modalities to compliment widespread uneasiness that introducing oxygen could cause compressor stations to explode. This uneasiness is unsupported in actual experience, but persists anyway. Many jurisdictions have responded to the bad science by placing significant regulatory impediments to operating gas wells below atmospheric pressure. Most regulations provide processes to allow engineers to successfully make the case to allow vacuum operations on a case-by-case basis, but the process is far from trivial or automatic.

Colorado has regulations requiring permission to operate at a vacuum. One of my clients elected to be the first company to attempt to acquire that permission on a group of 22 CBM wells in La Plata County, CO (Fig. 3.27). Since the client was both the producer and the gatherer, there

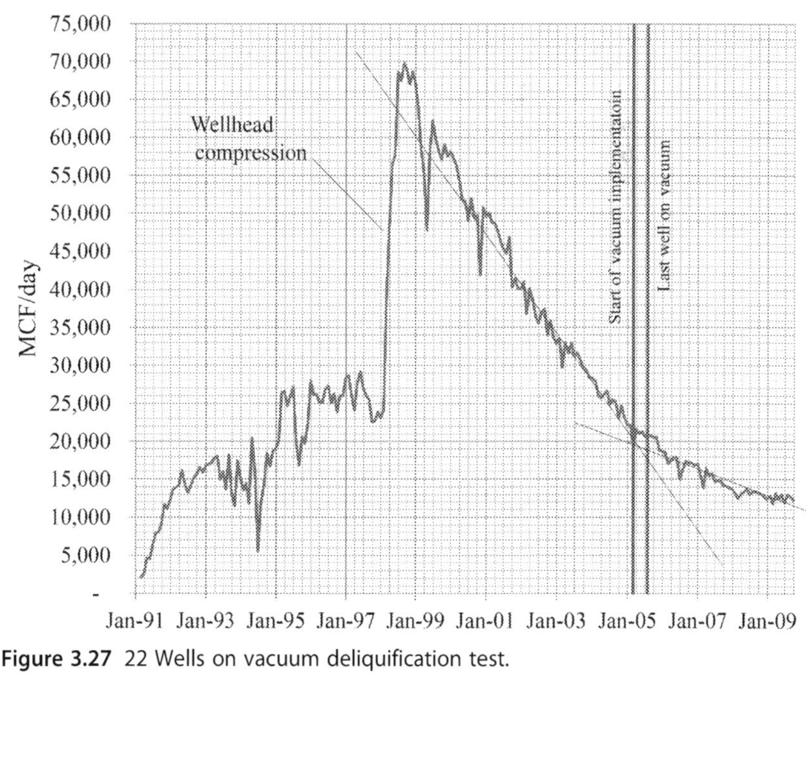

Figure 3.27 22 Wells on vacuum deliquification test.

were no gathering company objections to the application. We prepared the application with the following proactive provisions:

- Each well on vacuum would have an on-site oxygen sensor and a slam valve to shut the well in on high O_2 (without specifying what "high" meant which turned out to be important).
- At the aggregation points where the gas was delivered to a third party, there would be additional O_2 sensors connected to the station emergency shutdown system.

The application was approved and equipment was ordered. Fig. 3.27 shows the results of that test (in early 2010, both the company facilities engineer and company production engineer retired and their replacements decided that the field needed to go back to "tried and true" techniques and they installed tubing and pump jacks on all of the wells, the results were disappointing). During the implementation stage we learned:

- Setting the O_2 sensors at a spike to 10 ppm resulted in most wells being shut-in most of the time during the start-up phase of the project (we reset them to >25 ppm for 30 seconds which helped a lot).
- All pressure safety valves (PSV) are *designed* to leak when downstream pressure is greater than upstream pressure. We fixed this by putting rupture disks (best) or check valves (not as effective, but nearly) under the PSV.
- Sight glass packing nearly always leaks and the leaks can be very hard to find. We replaced sight glass packing, but eventually just isolated the sight glasses out of service.
- Finger-tight plugs in open-ended tubing are far worse than worthless (a visual inspection shows the line as plugged, but the plug doesn't really do anything).
- Threaded connections often leak, and leaks can be difficult to find.

We were able to solve most of the issues during the vacuum start-up, and had them all resolved within 6 months. I visited the field in early 2009 after 4 years of operation and talked to the field operator (who I didn't know) and asked him how often the O_2 sensors tripped. He looked at me funny and said "do you know how to reset them?" Meaning, of course, that in the year he had been operating the wells none of them had ever tripped and he lived in fear of one ever tripping. I showed him how to reset them.

3.5.3.4 Pump-off control

It is often important to match pump capacity to the reservoir inflow capacity. In oil wells it is very common to use start/stop controls quite effectively. The reason that this works can be traced back to Fig. 3.25, and the tendency of the fluids to seek their own level. It is common to pump the well-bore storage and then shut the pump off allow the well-bore to refill (basically intermitting the oil well).

Start/stop control is rarely effective in gas wells because the gas—oil ratio and/or the gas—water ratio (GWR) are high enough that accumulated liquid will be displaced by gas migrating through the liquid during the pump-off period, resulting in decreasing the NPSH-a instead of increasing it like happens in an oil well. In addition to gas-displacement problems, we often see cases where (on pumps installed without a standing valve, which is occasionally done to minimize suction dP and maximize NPSH-a) turning the pump off will cause the tubing to empty through the pump, usually turning the pump backward, which on electric-drive pumps can convert the motor into a generator running at a frequency incompatible with the grid. When there is a standing valve, solids will settle out of the fluid column and can plug off tubing or damage the pump on start-up.

In gas fields, we've looked for real-time indications that the well is approaching pumped-off that will alert the control system prior to the pump running dry. Some of the things that have been tried include:

- flow rate changes;
- tubing/casing pressure differential;
- rate of change in casing pressure (only effective when the casing is shut in, rarely useful);
- load indicators like dynamometer cards for SRP, torque for top-drive PCP, and power consumption for downhole electric motors.

Rather than using any criteria to start/stop a downhole pump, it is often more effective to install variable frequency drives (VFD) which have the added benefit of providing "soft start" to minimize the power consumption at start-up (which is the basis for a "demand charge" from the public utility and can be more expensive than the power-consumption portion of an industrial electric bill). VFD equipment significantly improves the performance of downhole pumps (both top drive and those with submersible motors).

A commercial system that has proven to be very effective in gas well deliquification is Weatherford's Wellpilot *Flow Control Technology (FCT) System* (Fig. 3.28). This deceptively simple device consists of two probes, one with both a heating element and a temperature-sensing element and the other probe with just a temperature-sensing element. As the flow

Figure 3.28 Weatherford Wellpilot FCT probe (Weatherford).

Figure 3.29 Installation suggestion for FCT.

stream gets less dense (because of more entrained gas), the heated probe gets hotter relative to the sensing probe and sends a signal to slow the pump down. As the flow stream gets more dense (because of less entrained gas), the heated probe temperature gets closer to the temperature of the sensing probe and sends a signal to speed the pump up. While these changes in density at the surface happen several minutes after an actual change, the device is sensitive enough to detect the smallest of changes and the lag tends to not be debilitating. This device along with a VFD has proven to be especially capable in keeping PCPs from filling up with gas and destroying themselves through the heat of compression (see Section 3.5.3.6). This is the only pump-off control device or process that I've seen to be consistently useful in gas-well deliquification. We found that if the flow line does not run full, then manufacturing a low place in the line (Fig. 3.29) can prevent the device from hunting.

3.5.3.5 Sucker rod pumps

Versions of the Sucker Rod Pump (SRP) technology have been used for over 6000 years. A simple chamber (the "barrel") with a nonreturn valve on the bottom (the "standing valve") and a moving plunger with a nonreturn valve (the "traveling valve") connected to the surface with "sucker rods" make up the system. As the plunger starts down (Fig. 3.30, left-hand image), the traveling valve is forced open and the plunger is allowed to "swallow" the liquid that is in the barrel. As the plunger starts back up (Fig. 3.30, right-hand image), the traveling valve is forced closed and the standing valve is pulled open. The ball portion of the standing valve controls the NPSH-r of the pump.

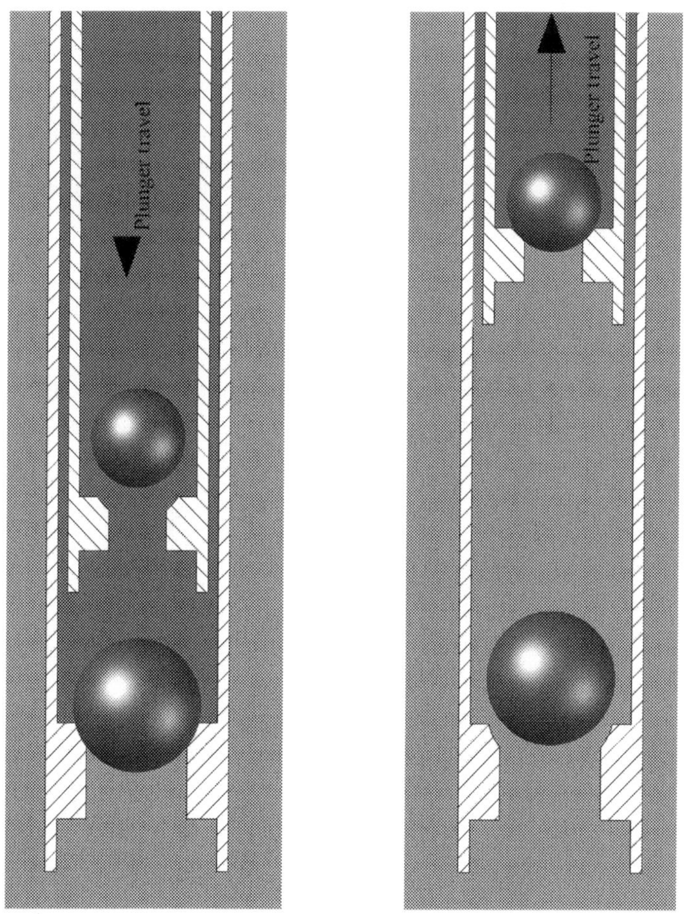

Figure 3.30 SRP.

Look at the left-hand image in Fig. 3.30 and note that the surface area exposed to the formation below the standing valve is much smaller than the surface are exposed to the tubing (otherwise the ball would fall through). In actual practice the effective area exposed to the reservoir is about 43% of total area under force. This causes the NPSH-r to open the standing valve to be about 14% higher than the pressure in the barrel when the plunger has moved some distance upward. For a barrel that is liquid-full, the incompressiblity of water makes this differential insignificant since the bulk modulus of water requires dropping pressure 319,000 psi (2.2 GPa) to expand liquid water by 1%. Rather than dropping pressure to an impossible value, the standing valve will lift. On the other hand, if there is any gas in the barrel it must expand until the pressure in the barrel is 14% lower than BHP. Since every pump stroke in a gas well invariably ingests some gas, this phenomena brings most pump manufacturers to the conclusion that the NPSH-r for SRP in gas wells is on the order of 100 ft (30.5 m) of water column.

SRP are "PD pumps." That means that for every complete cycle (full stroke in an SRP, a full 360-degree revolution on a PCP, etc.) the pump will displace the same physical volume, but *not* necessarily the same mass of fluid. If a PD pump is running at 100% efficiency (Fig. 3.31) there is no relationship between discharge pressure and flow (there is a very

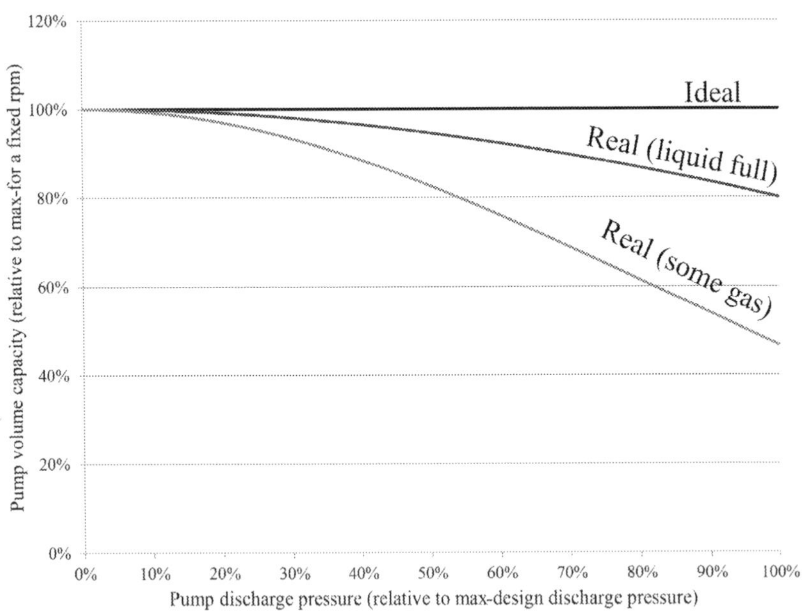

Figure 3.31 PD pump volumetric efficiency.

definite relationship between discharge pressure, mass flow rate, and power input required, but the discharge pressure and flow rate are independent of each other). We don't live in a world where 100% is even possible. When you push really hard on the fluid in the pump outlet, some will leak back through seals toward the suction. This slipped fluid is warmer than the bulk of the fluid so it can make micro changes in density and viscosity. The actual relationship between pump discharge pressure and volume flow rate is really quite complex in a PD pump, the details of this complexity are very much individual-pump specific so they don't lend themselves to a general discussion.

If the traveling valve does not open on every single stroke, then the pump is said to be "gas locked." When the barrel is full of gas, then the descending plunger must compress the gas in the barrel before the pressure in the barrel is high enough to lift the traveling valve. This compression is accompanied by the "adiabatic heat of compression" (Eq. (3.18)) which can raise the temperature of the gas high enough to boil any water that happens to enter the pump. This kind of gas lock will continue until enough liquid leaks between the plunger and the barrel to raise the mass in the barrel and increase pressure high enough to open the traveling valve and allow tubing liquid to rush in and quench the steam. At that point the pump will work for several strokes before beginning to accumulate gas again. Most SRP with conventional surface equipment are gas locked most of the time.

The risk of gas locking an SRP can be reduced by applying backpressure (Table 3.3) for the same reasons discussed in Section 3.4.3. The effect can be seen with an example. A CBM well has a conventional pump jack with the SRP set at 3000 ft (914 m) and designed for 20 bbl/day (3180 L/day). Flowing casing pressure is 0 psig.

In this real-life example, backpressure worked to return the pump to the zero-gas performance (which is not terribly good, but often acceptable). Downhole techniques like oversized pump barrels (or undersized

Table 3.3 SRP backpressure example

	0 MSCF/day		1 MSCF/day	
	0 psig tbg	**200 psig tbg**	**0 psig tbg**	**200 psig tbg**
Pump discharge at 3000 ft	1311 psig	1511 psig	41 psig	1248 psig
dP across plunger	1280 psi	1480 psi	38 psi	1245 psi
Slippage (gal/day)	7.9	9.1	0.004	7.3
Time to break gas lock	4 hours	3.5 hours	5 days	4 hours

Figure 3.32 Conventional pump jack.

plungers), installing a rod to trip traveling valve, tapered barrel, or downhole gas separators have had very limited success and while frequently tried, none have received widespread acceptance or long-term reliability.

Conventional pump jacks. The mainstay of the oil field for at least 100 years has been the "pump jack" called at various times in various places "beam unit" or "nodding donkey." The pump jack stand in Fig. 3.32 is a balance between requiring small lifting capacity for this well while having to clear the wellhead with the horse head. Pump jacks are often run by small natural-gas fired engines (single cylinder engines or "one lungers" are common) which are very difficult to adjust their speed remotely, so pumping rate is fixed in a macro sense (i.e., you can change it, but it requires a technician on site).

There are many online and slide-rule calculators to determine the expected flow rate and load requirements on a pump jack, and the underlying mathematics are reasonably complex and of questionable valve for conventional SRP in gas fields. Data from the analysis of hundreds of dynamometer cards in pump-jack gas wells indicate that it is rare for an SRP in gas wells to be less than 30% full of gas (Lea et al., 2008) which makes surface readings difficult to interpret (it is hard to tell the difference between gas in the barrel and other problems) and the computerized translation to a downhole card (which eliminates rod stretch) quite unreliable (since rod stretch is a function of the weight of the column of fluid and assumes no gas in the tubing).

The walking beam on a pump jack (the part that connects the rod string to the prime mover) translates rotational force from the prime mover to linear force lifting and lowering the SRP plunger. The prime mover is rotating at a constant speed, but depending on where the crank is in its travel, the pitman arm (and therefore the walking beam and polished rod) move at a faster or slower speed. When the crank is approaching the 3 o'clock or 9 o'clock positions, each degree of rotation causes the pitman arm to travel a much greater distance than when the crank is approaching the 12 o'clock or 6 o'clock positions. This results in very inconsistent plunger velocity that is completely unrelated to the needs of the pump or, more importantly, the reservoir. I have operated many pump jacks connected to SRP and have had consistently poor results. When a client asks for a recommendation on deliquification technology, nodding donkeys are not on the list.

There have been some recent reports of electric-drive pump jacks with highly capable VFD that can cause a pump to increase travel rate on the upstroke and decrease travel rate on the downstroke. Even though these pumps are not able to pause at the top of the stroke like the linear SRP (later) to facilitate pump leakage with minimum engagement of the plunger in the barrel, they are still able to significantly improve the performance of pump jacks.

Linear SRP. The observation that conventional pump jacks are generally a poor choice for deliquification should not be taken to mean that SRP is not a viable deliquification technology. In the late 1990s there was a field trial on two pneumatic linear rod pumps (Fig. 3.33). These pumps were completely pneumatic (with no electronic controls), operating on field gas at the very bottom of the acceptable pressure range. At the operating pressure we had, the pumps were very difficult to keep running reliably. One of the field techs was interested in making the technology work and spent considerable time dialing it in and keeping ahead of problems. The other field tech saw it as extra work and had no investment in its success or failure. The test failed, but we did learn:

- While the polished rod speed was not exactly constant (the pump paused at the bottom and top of travel while valves repositioned and pneumatic spaces repressurized), the speeds seemed very compatible with the operation of the SRP.
- Too much marketable gas was vented in pump operation (in our case the pump vented to the well-site compressor, but that is not the common configuration).
- The pump required far too much human intervention to maintain optimization, many of these activities would have been better done by a PLC.

Figure 3.33 Pneumatic SRP linear driver.

- The well with the optimized pump responded very well, and the pauses (especially at the top of the stroke with minimum plunger engagement) minimized gas locking, and possibly eliminated it.
- Success of this pump is very much dependent on the commitment of the lease operator, and without their buy-in this particular pump (like most technology) has to fail.

Over the last few years, several manufacturers have produced versions of linear rod pumps with electric or hydraulic drivers (one manufacturer has a hydraulic system with a pneumatic accumulator to recover a significant portion of the pumping energy, the pump is expensive and the manufacturer is not featuring it in their marketing).

As long as the system has flexibility in programming, either electric or hydraulic drivers provide an excellent deliquification solution. At a minimum the programming needs to:

- Allow programming a user-specified (and field changeable) pause at the top of the stroke. This seems to be the key element to allow an SRP to operate with zero NPSH-r. The pause at minimum engagement accelerates the exchange of gas from inside the chamber with liquid from the tubing so that the traveling valve will open immediately on the downstroke and (more importantly) immediately on the upstroke. With an adequate pause at the top of the stroke the pump should open both the standing valve and the traveling valve every stroke.

- Allow programming up-speed and down-speed independently and from the field.
- Allow programming a pause at the bottom of the stroke.
- Be compatible with a competent pump-off control mechanism such as Weatherford's Wellpilot FCT.

Linear drivers with SRP are near the top of my list of recommended devices for deliquification using external energy.

3.5.3.6 Progressing cavity pumps

My experience with PCP equipment has been mixed over the years. Many times I have been convinced that this is simply inappropriate technology for deliquification (Fig. 3.34), I've also seen them used effectively as headache racks on pickups and heard stories of a gentleman's club in an oilfield town using them in lieu of stripper poles). Other times I've seen encouraging results.

PCP are PD pumps and act much like the pump shown in Fig. 3.31. In top-drive, downhole applications, the pump housing and stator are screwed

Figure 3.34 PCP rotor as a truck bumper.

on to the bottom of the tubing and sucker rods are connected to the rotor. The rods are mostly similar to sucker rods used for an SRP (some of the nonmetallic rods used with SRP cannot be used in torsional applications, but the metallic rods and some nonmetallic rods work fine).

There are some electric submersible PCP on the market, but their market penetration has been very slow. One big problem is that the maximum rpm for a PCP is around 350 rpm which is not a comfortable speed for AC motors and speed reducers are generally required. Speed reducers are typically a significant maintenance item and when you have to pull the tubing to service the gearbox producers want to back away.

$$T_{disch} = T_{suct} \cdot \left(\frac{P_{disch}}{P_{suct}} \right)^{\frac{k-1}{k}} \tag{3.18}$$

PCP technology was developed to pump emulsions on the surface in food manufacturing. This is a challenging pumping application and seems ideal for a raw blood-guts-and-feathers downhole application, and it has proven very successful in oil wells. Fig. 3.35 shows the functional components of a PCP. The stator is made of an elastomer to allow the steel rotor

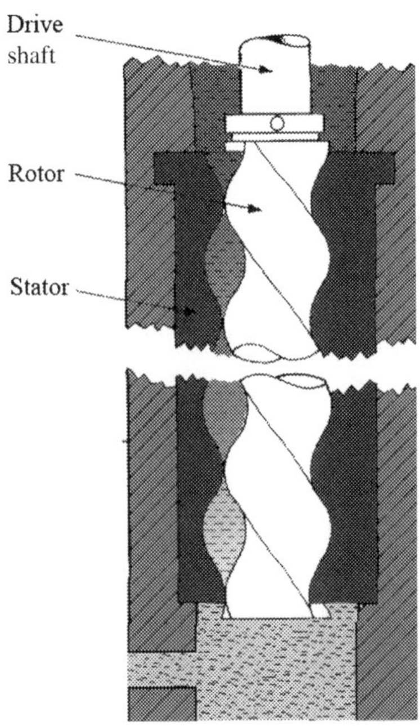

Figure 3.35 PCP cutaway.

to move inside the stator. It is a PD pump and every revolution of the rotor shifts the same volume from the suction toward the discharge. If that volume is filled with a gas, then the pump will still raise the pressure of the volume from suction conditions to discharge conditions. This creates heat of compression Eq. (3.18).

If a pump set at 3000 ft (914 m) that has been pumping a continuous stream of liquid with minimal gas with bottom-hole temperature of 100°F (38°C) and 15 psig (103 kPag) BHP gets hit with a slug of gas larger than the volume of the pump, the discharge temperature of the gas will be 1160°F (627°C)—much higher than the elastomer can tolerate. We frequently see pumps pulled out of gas wells with the top half of the stator cooked to (very nonresilient) glass. The pumps tolerate a small amount of gas, and as we saw in Section 3.4.3, the pump discharge can be significantly lower than the theoretical hydrostatic head, so we see many pumps survive gas slugs that should destroy them. We also see destroyed pumps in situations where logic would say they were not at risk.

The other issue with top-drive PCP with electric motors is that the motors put a significant cantilevered load on the wellhead. The motor and top-drive sheaves on the wellhead in Fig. 3.36 fell on the ground a few months before the picture was taken. It was very noisy and a lot of saleable gas was lost. After evaluating the techniques in general use, the client decided to have this motor stand designed. The stand has height adjustments to ensure that the stand is carrying the weight of the motor instead of it side-loading the pipe threads on the wellhead. It seemed to solve the problem for a few hundred dollars.

The difference between encouraging results and discouraging results is given in Table 3.4.

In certain conditions, PCP are high on a recommended technology list, but this definitely is not a "one-size-fits-all" or even a "silver bullet" solution to deliquification in general. Many operators have "standardized" on PCP and in those operations the failure rates tend to be unacceptable because replacing "analysis" with "policy" rarely ends well.

3.5.3.7 Electric submersible pump

ESP are multistage centrifugal pumps. A centrifugal pump is a dynamic device that brings fluid into the center of the "impeller" (the "eye"), the impeller slings it outward and forces it into a diverging section called a "volute." The note in Fig. 3.37 is important since the mass flow rate of

Figure 3.36 PCP support system.

Table 3.4 Success factors for PCP

	Good outcomes	**Bad outcomes**
Competent pump-off control	Most	Rarely
Flowing tubing pressure	>145 psia (10 bara)	<100 psia (6.9 bara)
Well configuration	Vertical top drive	Deviated
Drive	Top drive	Both top drive and ES drive
NPSH-a	>60 ft (18.3 m)	<40 ft (12.2 m)
Required capacity in gas well	<200 bbl/day (31.8 m^3/day)	>400 bbl/day (31.8 m^3/day)

the suction must equal the mass flow rate of the discharge (since the pump has no place to store extra fluid and no internal source of makeup fluid), if the pipes are the same size then (for an incompressible fluid) the velocity must be equal. Bernoulli's equation (Chapter 0: Introduction) explains the relationship between change in velocity and change in pressure for incompressible flow.

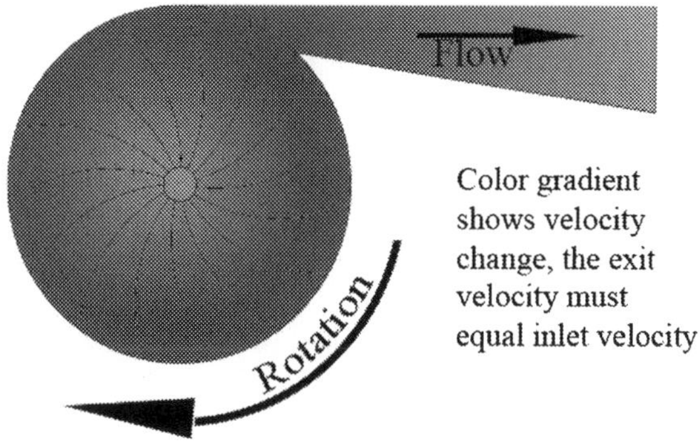

Figure 3.37 ESP operation.

In a multistage centrifugal pump, the first stage draws on the reservoir fluids and discharges into the next stage, and this process repeats until the last stage discharges into the tubing. Pump stages can be stacked as high as necessary. For example, if a pump can deliver 50 ft (15.2 m) (22 psi (162 kPa) of head per stage, and the BHP is 400 psig (2760 kPag) (924 ft (282 m)), then to overcome 3000 ft (914 m) of hydrostatic pressure and 200 psig (1380 kPag) flowing tubing pressure the pump probably needs something like 50 stages.

Fig. 3.38 shows the relationship between pump head and flow rate. The shaded area is the "operating envelope" which is a concept unique to dynamic pumps. While the pump can theoretically operate far outside of this region, in actual use at very low and very high flow rates, the pump simply stops operating. When we talk about centrifugal compressors in Chapter 8, Gas Compression, we'll see that for compressible fluids the upper and lower limits are defined as "choke line" and "surge line," respectively. While neither "choke" nor "surge" have physical meaning with flow of an incompressible fluid, analogous physical limitations prevent centrifugal pumps from operating outside of this envelope. In the Fig. 3.38 pump, slowing the motor down to reduce the capacity below 36% of theoretical maximum flow does not give you enough velocity at the entrance to the volute to achieve required discharge pressure within the volute. Speeding the motor up to increase the capacity above 83% of theoretical maximum flow rate risks phase change and cavitation in the volute even with adequate NPSH-a.

Centrifugal pumps have a narrow range of flow rates, but gas wells have a variable inflow rate that can change dramatically from moment to moment. The NPSH-r for these pumps is flow rate specific (larger flow

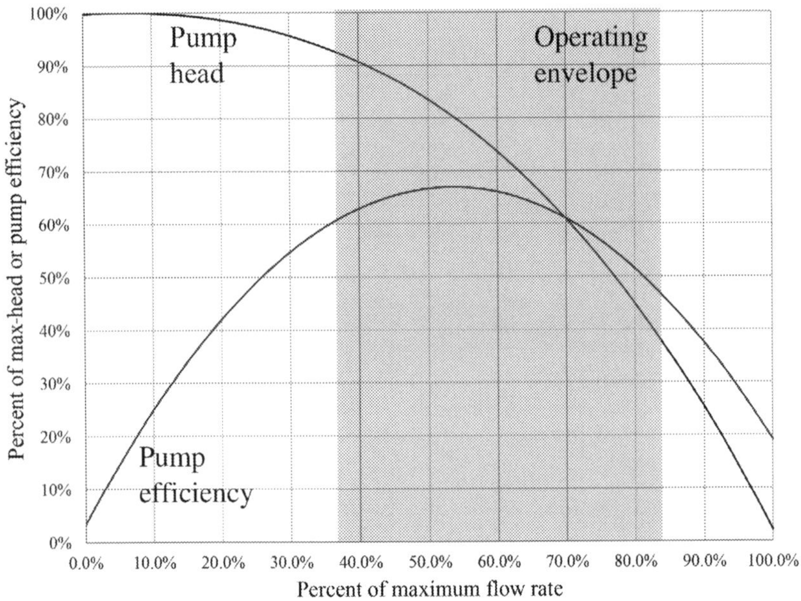

Figure 3.38 ESP pump head.

rate, more NPSH-r) but NPSH-r is never less than 150 ft (46 m) and very large (oil field) pumps can require 6000 ft (1829 m) NPSH-r. Some operators have put ESP on start/stop cycles to try to match inflow with the operating envelope, but there are risks:

- Stopping the pump will empty tubing back into the formation (and can turn the downhole motor into a generator).
- A standing valve on the pump discharge can prevent draining the tubing, but it allows solids to settle out and the solids can seize the pump.
- A hole in the tubing above the standing valve will allow the tubing to drain, but the hole will steal capacity and it is difficult to balance hole size with particle size. If you get it wrong it can steal the entire capacity of the pump.

There are operators of gas fields who only use ESP (a policy) and find the results good enough to continue the policy. Upon reviewing the performance of fields using ESP, I've found that they could only be considered a "success" if you have a very liberal interpretation of "success." Normal Oil & Gas definitions of success would call all of the installations I've reviewed to be "dismal failures" with failure rates, pulling jobs, and production rates all in the "unacceptable" category. Because of the limitations of ESP, operators that use them in gas wells try to always set the pump well below (sometimes as much as 400 ft (122 m) below the perforations) the producing formation which plays a role in marginal GWR. ESP are not on my list of recommended deliquification technologies (although for oil wells their very high flow capacity is often quite attractive).

3.5.3.8 Downhole jet pump

Jet pumps have been used downhole for nearly 75 years. The traditional oil field jet pump has the pump in a packer and you pump the power fluid down the tubing and the combined reservoir fluid and power fluid return up the tubing/casing annulus. This configuration is a problem for gas wells because there is really no space in the pump for the gas to exit the reservoir. In the 1980s the major manufacturer of downhole jet pumps developed a "tubing-free" pump. This pump starts with a seating nipple in the production tubing. The bottom-hole assembly (BHA) is attached to small-diameter coiled tubing (can be run with stick tubing, but the thread boxes can be a problem) and run into the production tubing. The pump is then dropped into the BHA and matches up with ports in the BHA that allow reservoir fluid to come into the bottom of the BHA, power fluid down the inner tubing, and spent power fluid along with reservoir fluid is exhausted into the tubing/tubing annulus. This configuration leaves the production tubing/casing annulus open for gas flow. There are pumps that use a dual tubing string instead of a concentric string, but they work the same way.

Fig. 3.39 shows a high-pressure wellhead for a tubing-free jet pump. With this configuration, by repositioning three valves, the power fluid can be used to push the downhole pump to the surface where the upper master valve can be shut to catch the pump when it arrives.

Jet pumps are in the family of "thermocompressors" and they are classed as "eductors" because they are designed for a liquid to pump a liquid (all flow is incompressible at all velocities). Fig. 3.40 shows the pressure and velocity traverse through the pump. In the prenozzle, the cross-sectional area is decreasing, which requires velocity to increase as a squared function while pressure drops as a linear function. Inside the nozzle, the velocity gets high enough to for dynamic pressure to be a significant portion of total pressure and the static pressure drops rapidly while the velocity increases rapidly. When the fluids enter the "throat," the power fluid is moving very quickly and the reservoir fluid is much

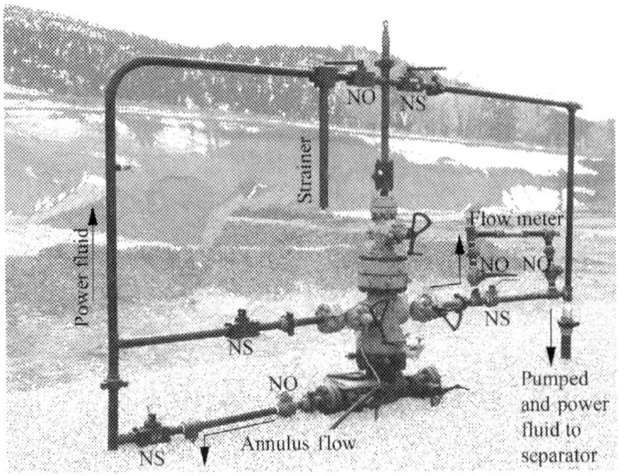

Figure 3.39 Jet pump wellhead.

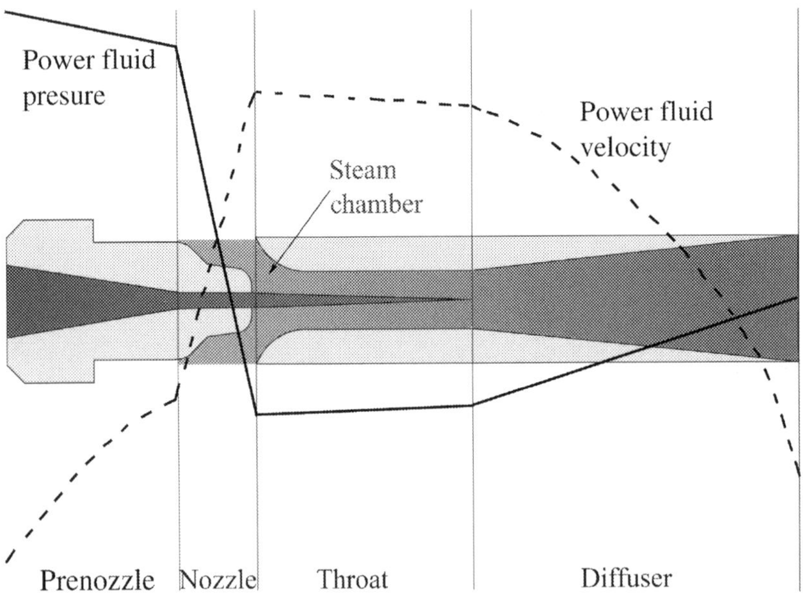

Figure 3.40 Jet pump pressure traverse.

slower. Where the two fluids touch, a "no-flow boundary" is created where the reservoir fluid that touches the jet stream must be accelerated to the speed of the jet (at the same time the inertia of the reservoir fluid tries to slow the jet stream, but there is considerably more power fluid than reservoir fluid). As the combined fluid enters the diffuser it is all moving at about the same speed. In the diffuser things have slowed and dynamic pressure is no longer a significant fraction of total pressure and the velocity vs pressure relationship is back to velocity change being a squared function.

If the density of the suction fluid decreases dramatically (e.g., the suction ports are blocked or the pump ingests a large slug of gas), the power fluid will not slow adequately in the steam chamber and throat which will increase the risk of cavitation in the throat (see Fig. 3.26 for a throat that saw no-flow for less than an hour) due to power fluid flashing in the steam chamber and condensing violently in the throat due to pressure increase as velocity bleeds off. NPSH-r for eductor-based downhole jet pumps are very dependent on nozzle and throat size, but it is rarely less than 460 ft (122 m). Eductor-based jet pumps are not on my list of deliquification technologies.

The previous section was careful to limit the discussion to "eductor-based" jet pumps. While many people have tried to drive an eductor-based jet pump with gas over the years, the results have been universally poor. Fig. 3.41 shows why. When gas exits an eductor nozzle the flow is choked and limited to sonic velocity (1.0 Mach). A small change in the exit

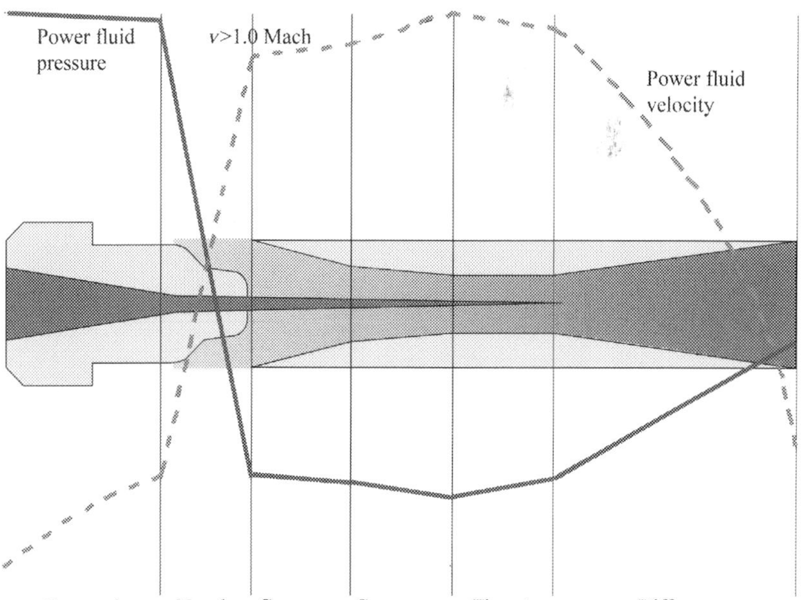

Figure 3.41 Ejector pressure traverse.

characteristics of the nozzle can turn the nozzle into a divergent nozzle that works far differently in compressible flow than we see with incompressible flow. In an ESP volute, increasing cross-sectional area decreases velocity. If you change the flow from "compressible" to "incompressible" increasing cross-sectional area actually increases velocity (the reasons are discussed in Chapter 4: Surface Engineering Concepts). If you can get velocity higher than about 1.3 Mach, then the density of the power fluid gets much closer to the density of liquid and the momentum (i.e., density times velocity) balance becomes much more favorable since the density of a gas increases rapidly as velocity increases above 0.6 Mach and the power fluid can dominate the interaction at the no-flow boundary. An ejector-based jet pump would have a 0 NPSH-r because there is no point in the process where the gas is going to change phases. There is a patent on this technique, but no commercial products at this time.

3.5.3.9 Gas lift

In producing oil, you hope that the tubing is completely full of oil. For an offshore platform in 1000 ft (305 m) of water with 12,000 ft (3700 m) of hole and 0.7 SG oil, that would be nearly 4000 psi (28 MPa) of hydrostatic head on the formation. There comes a time in reservoir development that the reservoir pressure is inadequate overcome that pressure. One technique to reduce the required pressure is to inject gas into the tubing to both add the potential energy of the gas and to replace part of the high-density liquid with low-density gas to lower the backpressure on the formation. Since gas dissolved in the liquid has a lower impact on the process than gas bubbles, you want to inject gas at a number of points in the tubing string to maximize the impact. This technique is called "gas lift" and in artificial lift it is typically done with a packer to isolate the tubing/casing annulus from the reservoir and "gas-lift mandrels" at carefully designed locations in the tubing. Gas-lift valves are put into the mandrels to allow gas-lift gas to move from the annulus into the tubing. An inventory of high-pressure gas is maintained in the annulus to supply the gas-lift valves.

Gas lift has been a staple of artificial lift offshore for decades. This is easy to understand since space is very limited on production platforms and with gas lift the bulky compression equipment can serve multiple wells minimizing the artificial lift footprint. Further, when gas production from the formation is low, flow interference between gas-lift gas and reservoir gas is minimal and you can get the benefit of lightening the column without too much fluid friction in the flow.

Gas lift has been so successful in artificial lift that it simply had to be brought to gas fields. The first issue is that energy requirements are about fivefold higher for gas lift than for a mechanical pump. Next, the minimum BHP achievable is very high. Finally, when you have to flow both reservoir gas and gas-lift gas in the tubing, the fluid friction can be very high, and small changes in reservoir flow can cause significant increases in BHP. Gas lift with a packer has shown that it can be effective in gas wells down to about 300 psig (2068 kPag) flowing BHP, so if a reservoir needs BHP about half of average reservoir pressure then gas lift is a reasonable tool to lower reservoir pressure to something like 600 psig (4137 kPag). Efforts to lower BHP further have been largely unsuccessful.

One variant that eventually gets tried in every gas field (always unsuccessfully, once the criteria for success is clearly laid out) is known colloquially as "Po' Boy Gas Lift." This process involves injecting gas (without packer or gas-lift valves) down the tubing/casing annulus and hoping that the gas will carry the liquids up the tubing instead of simply going into the formation. Wishful thinking is rarely a sound foundation for engineering decisions. I was once asked to review a Po' Boy gas-lift operation. The well was producing 800 MSCF/day (23 kSCm/day) of net gas over injection and 3 bbl/day (477 L/day) of water. I recommended turning the gas-lift supply off for 30 days and then turning it back on. The well immediately went to 1200 MSCF/day (34 kSCm/day) and 8 bbl/day (1270 L/day) of water, and was inclining. By the end of the 30 days the well was making 1350 MSCF/day (37 kSCM/day) and still making about 8 bbl/day (1270 L/day) of water. We turned the gas lift back on and the rate immediately dropped to 700 MSCF/day (19.8 kSCm/day) and the water rate dropped to zero. The advocate for Po' Boy Gas Lift said "See, it works." I have never known what his criteria for success was, but when he (voluntarily) left the company the following month I turned off the compressor and transferred it elsewhere. The rate immediately returned to the off-lift value and it was 3 years before natural decline took the well back to 800 MSCF/day (23 kSCm/day) again. I have had similar experiences with both production increasing after turning Po' Boy Gas Lift off and seemingly unshakeable support of the concept from its advocates in a half dozen fields around the world.

There is a technique generally called "continuous gas lift" which looks very much like Po' Boy Gas Lift, but it has one very important difference—it starts with an extended shut-in. When a gas well is shut-in the gas will bubble through the liquid column in both the tubing and the casing. Over time, gas will displace all of the liquid and the well-bore will be gas-filled. At that point (when tubing and casing pressure are equalized from below), injecting gas down the annulus with the tubing open can keep the liquid from overpowering the ability of the gas to lift it (i.e., an unloaded well will tend to stay unloaded as long as flow up the tubing is above the critical velocity, but a partially loaded well will tend to continue to load up at velocities near critical).

While gas lift is arguably the worst technique applied to deliquification of low-pressure wells (worse than intermitting and venting because you have to spend a lot of money to achieve the same results that those techniques provide for free), there is one variant that seems to have merit as a low-pressure deliquification technique. This subset is called "chamber lift." In that variant, a standing valve is set in the bottom of the tubing and a concentric tubing string is run inside the tubing. With chamber lift, periodically a burst of high-pressure gas is blown down the tubing/tubing annulus to shut the standing valve and blow any accumulated water up the inner tubing. The biggest problem with this technique is inducing the reservoir liquids to accumulate in the tubing instead of the casing. One technique that has aided the accumulation of liquid in the tubing is to install an ejector to pull on the tubing using the same gas source that is used to blow the liquid. This process is described in a case study under thermocompressors in Chapter 8, Gas Compression. Pulling the tubing to 6–8 psi (41–55 kPa) lower than flowing casing pressure seems to be adequate to remove up to about 20 bbl/MMSCF (0.11 m^3/kSCm).

3.5.4 Evolving requirements

Technology evolves in this industry like any other. When I started in this industry in 1980, CBM and shale gas/oil were known, but we didn't know how to economically produce them. As that knowledge has grown, the requirements for deliquification equipment have evolved.

3.5.4.1 Horizontal wells

All sorts of pumps perform best when presented with continuous-phase liquid. As the amount of gas that enters the pump increases, the effectiveness and longevity of a pump deteriorates. The lateral section of a horizontal well (Fig. 3.42) has much more in common with a pipeline

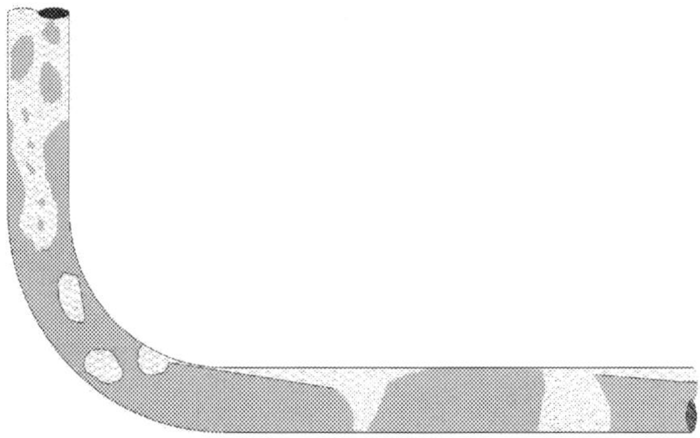

Figure 3.42 Multiphase flow in horizontal well.

or gathering system than it does with the concepts of flow in a vertical conduit presented in this chapter.

One important consideration that will be presented in Chapter 4, Surface Engineering Concepts (Section 4.2.2) is Eq. (4.6) (repeated here without development as Eq. (3.19)) which shows the flow rate required for a horizontal pipe to run full of liquid. For a 5-in (127 mm) ID liner in the lateral, this empirical equation shows that 20,000 bbl/day (3100 m^3/day) would be required to keep the liner full of liquid from wall to wall. Values less than this result in gaps in the liquid filled with gas. These gaps are generally called "slugs" (i.e., "gas slugs" and "liquid slugs") and represent the primary challenge to producing horizontal wells.

$$q_{\min\ gpm} = 10.2 \cdot \mathrm{ID}_{inches}^{2.5} \tag{3.19}$$

Several researchers have begun to address the problem of minimizing the impact of slugging on pump operation. One system that seems to have a high probability of successfully making the transition from horizontal flow to vertical flow is the patent pending HEAL system from Production Plus Energy Services, Inc. This system (Fig. 3.43) has four major sections:

1. The "HEAL Seal" isolates the lateral from the build section in the well and provides a conduit out of the isolated section.
2. The "Sized Regulating String" provides a conduit with a varying inside diameter that tends to homogenize the flow with minimal pressure drop (much like the "flow conditioners" used to homogenize the flow in front of square-edged orifice measurement, see Chapter 5: Well-Site Equipment).
3. The "HEAL Vortex Separator" employs centrifugal force to separate the gas from the solids and liquids. The gas exits the separator and

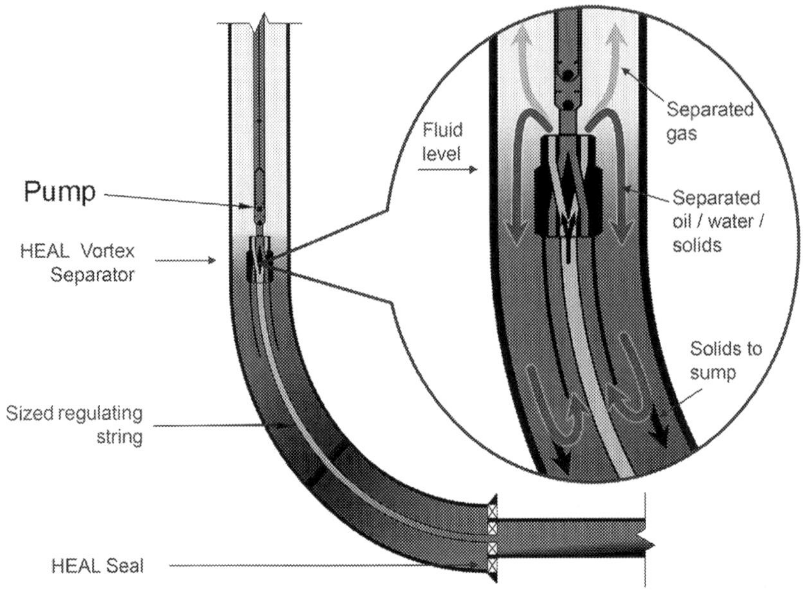

Figure 3.43 Heal system flow path (HEAL). *Source: http://pdnplus.com/heal-system/.*

rises in the tubing/casing annulus. The combined liquid/solid mixture is directed downward where another direction change tends to release the solids to fall to the Seal while sending the liquid upward.

4. Interface to vertical flow takes the liquid (now mostly degassed and solids-free) into some sort of deliquification equipment ranging from plungers to SRP to PCP to ESP to gas lift.

The HEAL Vortex Separator can be thought of as analogous to the perforations in a vertical well—it creates a transition from the very complex flow within the reservoir into the somewhat less complex wellbore flow. This system has been used in both high-pressure and low-pressure well-bores and in both cases, this "pseudo perforations" approach has converted complex multiphase flow into a more homogenized flow that is more conducive to deliquification approaches.

Another issue with horizontal wells is the transition from vertical to horizontal. That bend defines the longest tool that can be run to the toe of the casing. Pumps with electric motor drives tend to be too long to make the turn in most wells. Top-drive pumps will tend to wear on the tubing with the drive-rods and can break the rods and/or wear an actual hole in the tubing. So far options to pump from the horizontal section have met with very limited success both because of access to the lateral and ignoring the chaotic nature of the flow in the horizontal.

Many operators are putting pumps in vertical pipe above the dogleg and hoping for the best—wishful thinking is rarely completely effective. This decision makes pump selection even more critical than usual. If you want to install a jet pump 50 ft (15.24 m) above the lateral height, and the pump requires 600 ft (183 m) of NPSH-r, then the lowest pressure you can safely take the BHP to is 381 psig (2630 kPag) with zero pressure on the wellhead. If the reservoir wants BHP to be half of average reservoir pressure, then the minimum abandonment pressure is 760 psig (5240 kPag). In a CBM well this abandonment pressure could easily leave 75% of the OGIP behind at abandonment. Changing technology to a linear SRP changes the pump NPSH-r to zero, so for the same pump set depth the minimum abandonment pressure drops to 43 psig (297 kPag) and abandoned OGIP can go as low as 15%.

Single lateral. A considerable amount of work is being done to deliquify single lateral horizontal wells with marginal success. The HEAL system has been able to improve that success by effectively moving the reservoir access to the vertical section (via the "pseudo perforations" mentioned earlier). With that innovation, Fig. 3.44 shows that achievable flowing BHP for various deliquification technologies. Without the HEAL system determining the achievable flowing BHP has been largely hit and miss.

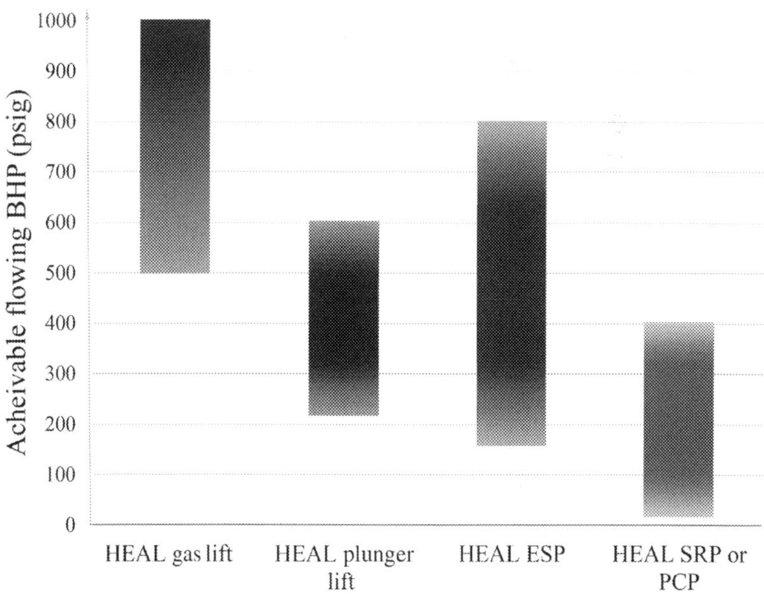

Figure 3.44 Achievable pump suction pressure with HEAL (HEAL). *Source: http://pdnplus.com/heal-system/.*

Multilateral. As drilling technology continues to improve, wells are starting to be designed with multiple laterals in a single zone and even multiple zones in the same well. These highly complex drilling wells introduce a whole new level of complexity for operators. A "simple" multilateral well would be exceedingly difficult to ever pump and today's technology is simply not up to the task. The looks I've had at emerging technologies don't hold out much hope that multilateral deliquification will ever be viable (beyond using evaporation where the water volume is appropriate).

3.5.4.2 Interconnected series of wells

One technique that seems to be getting very popular is to drill a single vertical well and then drill multiple horizontal wells (all with down-dipping lateral to drain liquids toward the vertical well) that intersect the vertical well. The two major approaches so far are: (1) gas production from the horizontal well-bores with liquid production from the vertical well; and (2) all production from the vertical well (to minimize the number of sites with equipment and operator traffic). This approach seems to be the flavor of the day in 2016, but there is not much data on overall effectiveness.

3.5.4.3 Slim-hole wells

An idea from the 1980s that simply will not die is "slim-hole wells." In this process, you attach the drill bit to pipe that will become your casing and drill to total depth (TD), then abandon the bit in place and cement the "drill pipe" in place. This has been done with 2-3/8 (up to about 3-1/2) tubing-as-casing. The thinking is that these wells are typically drilled without ever tripping pipe, and you save so much money on drilling that you can afford to drill a lot of them.

If (when) the well begins liquid loading you plug the slim hole and if late-life economics justify drilling a well that can be pumped (rare) then you redrill the well. The only technique that has any potential for deliquifying a slim-hole is compression and (possibly) evaporation (see Fig. 3.2), but even plungers have a problem since there is no well-bore gas storage to help drive the plunger.

A potential client asked me a few years ago to provide a deliquification solution for 6000 slim-hole wells with 2-7/8 casing that were all liquid loading. After a month of talking to manufacturers and various inventors of my acquaintance I had to tell the potential client that I couldn't take their work because there is no answer and they would have to run the economics to decide if they could afford field-wide compression (they couldn't), redrill the wells, or plug them. Four years later the field is largely shut in and they are still shopping for a solution.

3.5.4.4 Multiwell pads

Rather than creating problems like most of the evolving requirements, multiwell pads create some unique opportunities. The biggest advantage to the facilities engineer is economies of scale in power generation. Where it is rarely economic to install a genset for a single well, when one unit can provide power for 10 wells, the economics can change dramatically. To a large extent this is due to economies of scale. For example, the cost of the structural portion of the genset skid costs about the same to engineer and build for 25 kW as for 75 kW, and not a lot less than the skid for a 300 kW unit. If for example there are 10 wells, each with a need for 25 kW, then a pair of 300 kW skids can supply all 10 (with 100% backup) for about the cost of 8 of the 25 kW units (that would not have any backup at all). The same discussion holds true for compression, vapor recovery, and produced water handling/disposal.

One risk is that people will start thinking of a multiwell pad like an offshore platform, and implement gas lift because "you only need one compressor for all 10 wells." Fig. 3.45 is an example of this thinking. Gas lift (with packers and mandrels) was installed on all the wells during the initial completion. The gas-lift system had to buy gas from the gathering system to supply the wells, and sales back to the gathering system were significantly less than purchased gas (i.e., pipeline gas not used for

Figure 3.45 Multiwell pad.

compressor fuel was going into "storage" in the gas-lift wells). This system operated for 4 months before sense broke out and the compressor was converted to suction (the wells did not need deliquification at all) and the wells started selling gas.

Multiwell pads are a facilities-opportunity, but you cannot forget that every well is going to have a personality, and most of them are unpleasant. Try to fight the urge of trying to find a single solution that is required to fit all wells. The best multiwell operation I ever saw had two linear rod pumps, two PCP, and three flowing wells that were all making pretty much all the gas those wells were going to make.

3.5.4.5 Emerging technologies

Research directions seem to be toward adapting gas-compression equipment to moving liquid. This requires liquid quantities to be fairly low (5–200 bbl/day (0.8–32 m^3/day). Different technologies have different ideas on how to manage the pump discharge pressure. So far most of the proposed technologies are limited in depth to less than 2500–5300 ft (760–1600 m), but everyone is working to extend this.

There is some work going on to adapt jet pumps to gaseous power fluid, but there is considerable inertia in the industry saying "we tried that and it didn't work." That phrase has been death to more good ideas than any other.

Hydraulic fluid used to drive a downhole hydraulic motor connected to a downhole pump has been tried many times with varying success. Most of these products have had limited success at shallow depths, but quickly reached saturation of the market where the technology would work and tried to push their limits without redesign, almost always unsuccessfully.

Two problems that have showed up repeatedly in field trials of hydraulic pumps have been the compressibility of the hydraulic fluid and the reliability of the switching valve. When you use dual hydraulic strings, one string is pressurized while the other is depressurized. The compressibility of hydraulic fluid is low, but not zero and it can take 50 gallons (190 L) of fluid to compress the line and start building pressure. Keeping enough inventory of fluid on location to allow this quantity of makeup starts being a liability for some of the toxic (and expensive) hydraulic oils. The other issue is the switching valve. This valve is the highest maintenance item on any hydraulic system. When the valve is downhole, you have to pull the pump to fix it and it can take months to get on a rig

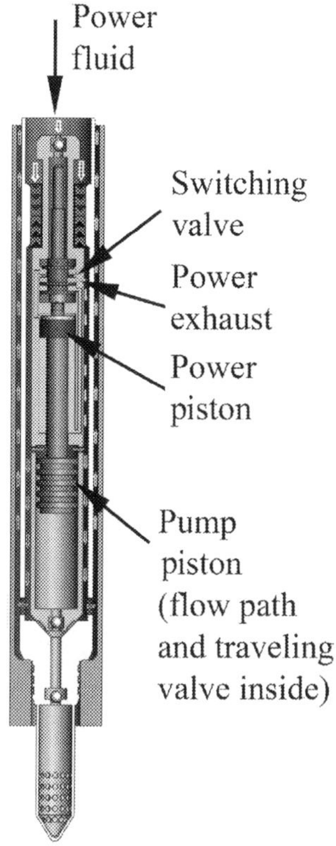

Figure 3.46 Self-reciprocating pump (Cormorant Engineering).

schedule in some fields. When the valve is on the surface, maintenance is easier, but the makeup volume can be higher.

Cormorant Engineering from New Waverly, TX, has started marketing a downhole hydraulic pump (Fig. 3.46) that seems to address the worst of these problems. The pump uses produced water for hydraulic fluid, so they don't have to depressurize the system to reverse the pump stroke (since the LP side can be exhausted to the produced water stream). This pump has a single power fluid line (inner tubing on a concentric tubing–tubing string), and the low-pressure side of the pump exhausts into the tubing/tubing annulus along with the new produced water. The switching valve is operated by a mechanical rod in the pump at either end of travel. If the operator needs to check the condition of the pump it can be floated to the surface using wellhead valving and the power fluid pump without the need for a rig or even a slick-line unit. The manufacturer claims the pump will produce up to 100 bbl/day (16 m^3/day) from

Table 3.5 Deliquification technology summary

	Typical capacity (bbl/day)	NPSH-r (ft)	Failure Method
Velocity strings	<100	0	Well capacity falls below critical
Tubing flow controller	<100	0	Well capacity falls below critical
Plungers	<10	0	Reservoir pressure falls below required
Surfactants	<10	0	Soap fails to activate
Intermitting	<5	0	Time required for the shut-in gets too long
Vent cycles	<1	0	Time required for the shut-in gets too long
Surface compression	<3	0	Water rate too high
Evaporation	<20	0	Formation scale
Pump-jack SRP	20–10,000	75–100	Gas lock
Linear SRP	0–300	0	Rod/tubing wear
PCP	4–2000	60–100	Heat of compression
ESP	70–20,000	150–2000	Cavitation (NPSH-r is proportional to flow rate)
Downhole jet pump	10–200	450–1000	Cavitation (NPSH-r proportional to the combination of suction and power liquid flow rate)
Gas lift	1000+	200–500	Fall below critical rate

depths up to 14,000 ft (4270 m). At the speed that this pump runs, it should be able to handle considerable ingested gas and the NPSH-r should be similar to a linear SRP. The engineering looks sound, and the concept is intriguing, but they hit the market during a downturn and haven't installed very many of the pumps yet. An interesting idea, with a lot of potential, but many interesting ideas with high potential have failed to live up to their potential in real use, the usefulness of this pump, like any other technology, will reveal itself over time.

3.5.5 Deliquification conclusion

Deliquification is different from artificial lift. It requires different:

- Tools (gas wells want more attention)
- Mind set (e.g., pipeline operation is a tool of production, and pigging is not a "necessary evil," it is critical)
- Staffing levels (more stuff to do takes more folks).

No technology is set-and-forget. Be prepared for any given technology to work or fail to work in any given well regardless of its performance in the last (or next) well. Expect to spend considerable field and engineering effort to "get it right" only to find that at next week's pressure regime it no longer works. The technologies and processes we've discussed in this section have been given in Table 3.5.

The only "silver bullet" for deliquification is great data, appropriate staffing, and a flexible approach.

REFERENCES

Atkinson, C.H., Ward, D.C., & Lemon, R.F. (1969). *Gasbuggy Reservoir Evaluation—1969 Report.* https://inis.iaea.org/search/search.aspx?orig_q = RN:36010815

Bradley, H. B., & Gipson, F. W. (1987). Petroleum Engineering Handbook. Richardson, TX: Society of Petroleum Engineers.

Coleman, S.B., Clay, H.B., McCurdy, D.G., & Norris, III, H.L. (1991). A New Look at Predicting Gas-Well Load-Up. Journal of Petroleum Technology (also published as SPE 20280).

Cormorant Engineering. http://www.cormorant.us.com/srp.htm (accessed September, 21, 2016).

EIA. "Annual Energy Outlook 2008", DOE/EIA-0383 (2008). https://www.eia.gov/outlooks/archive/aeo08/index.html.

Fetkovich, M.J. "The Isochronal Testing of Oil Wells," paper SPE 4529 presented at the 1973 SPE Annual Meeting, Las Vegas, Nevada, 30 September–3 October.

GPSA. Engineering Data Book, twelfth edition, Gas Processors Association, 2004.

HEAL. Production Plus Energy Services, Inc. Retrieved from http://pdnplus.com/heal-system/. February 23, 2017.

Lea, J. F., Nickens, H. V., & Wells, M. R. (2008). Gas Well Deliquification. Amsterdam: Gulf Professional Pub./Elsevier.

Li, M., Li, S.L., & Sun, L.T. "New View on Continuous-Removal Liquids From Gas Wells", SPE75455, 2001 SPE Permian Basin Oil and Gas Recovery Conference, Midland, TX, 15–16 May 2001, Revised paper submitted to SPE for peer review, re-published 3 December 2001.

McKetta, J.J. & Wehe, A.H. Hydrocarbon Processing, August, 1958.

Simpson 1. "Producing Coalbed Methane at High Rates at Low Pressures", SPE 84509, David Simpson, Mike Kutas, SPE Annual Technical Conference and Exhibition, Denver, CO, 5–8 October 2003, Journal of Petroleum Technology.

Simpson 2. "Low-Pressure Gas Well Deliquification Requires Different Approaches", February 27, 2006, Oil & Gas Journal.

Turner, R.G., Hubbard, M.G. &, & Dukler, A.E. (1969). Analysis and Prediction of Minimum Flow Rate for the Continuous Removal of Liquids from Gas Wells. Journal of Petroleum Technology.

Vogel, J. V. (1968). Inflow Performance Relationships for Solution-Gas Drive Wells. Journal of Petroleum Technology 20 (1): 83–92. SPE 1476-PA. http://dx.doi.org/10.2118/1476-PA.

Weatherford Wellpilot® FCT System. http://www.weatherford.com/doc/wft255784 (accessed September 12, 2016).

Webster (n.d.). Retrieved April 21, 2016, from http://www.merriam-webster.com/dictionary/biogenic (other dictionary references replace "biogenic" with the referenced term).

NOMENCLATURE

Symbol	Name	fps Units	SI Units
b	Langmuir shape factor	1/psi	1/kPa
c	Cohesion	psi	kPa
c/b	Matrix shrinkage	psi^2	kPa^2
c_p	Flow constant (well specific)	MSCF/day	kSCm/day
d_{eff}	Effective diameter. For flow inside a pipe it is inside diameter, for annular flow it is effective diameter	in	mm
f_m	Moody friction factor	decimal	decimal
g	Gravitational constant	32.174 ft/s^2	9.81 m/s^2
g_c	Unit converter	32.174 ft-lbm/s^2-lbf	N/A (for now)
h	Height	ft	m
h_{liq}	Height of a liquid column	ft	m
h_{td}	Height (length) of a tubing string	ft	m
k	Permeability	mD	mD
k	Ratio of specific heats	decimal	decimal
ID	Inside diameter	in	mm
n	Nonlinearity term used to account for deviations from semilog linear flow (1.0 for everyone except reservoir engineers)		
OD	Outside diameter	in	mm
P	Pressure	psia	kPaa
$\overline{P}$	Average reservoir pressure	psia	kPaa
q	Gas flow rate	MSCF/day	kSCm/day
$q_{0(press)}$	Gas flow rate at a fixed time as a function of pressure	MSCF/day	kSCm/day
$q_{0(max)}$	Gas flow rate at a fixed time at maximum differential pressure	MSCF/day	kSCm/day
r_e	Drainage radius of well	ft	m
r_w	Well-bore radius	ft	m
SG	Specific gravity	none	none
Skin	Resistance to flow in a reservoir	none	none
T	Temperature	Rankine	Kelvin
v	Velocity	ft/s	m/s
Z	Compressibility	decimal	decimal

(*Continued*)

(Continued)

Symbol	Name	fps Units	SI Units
β-ratio	Ratio of orifice ID to Pipe ID	decimal	decimal
Δ	Change	none	none
ε	Efficiency	decimal	decimal
ϕ_0	Original porosity	decimal	decimal
μ	Viscosity	lbm/ft/s	mPa/s
ρ	Density	lbm/ft^3	kg/m^3
σ	Interfacial tension	dyne/cm	dyne/cm
Subscripts			
avg	Average		
bh	Bottom-hole		
csg	Casing (or tubing/casing annulus)		
disch	Condition at the discharge of a pump or compressor		
gas	Gas		
liq	Liquid		
mud	Drilling fluid		
std	Standard condition		
suct	Condition at the suction of a pump or compressor		
tbg	Tubing		

EXERCISES

1. Using the CBM gas analysis from Chapter 0, Introduction, a gas is at 30 psig (206.8 kPag) (assume atmospheric pressure is 12 psia (82.7 kPaa)) and 60°F (15.6°C) and in contact with 10,000 mg/L TDS produced water. If you raise the pressure to 250 psig (1724 kPag) and 316°F (157.8°C) find:
   - The water content of the gas before and after the pressurization
   - The amount of solids that will drop off with this change in water-vapor content
   - Adjust the gas analysis for water in both cases
2. The lease operator opens both the tubing and casing on a CBM well that is flowing 300 MSCF/day (8.5 kSCm/day) of gas and 10 bbl/MMSCF (0.0561 L/SCm) of water into 75 psig (517 kPag) and 125°F (51.7°C) on the surface. The tubing line has a check valve that has a

Table 3.6 Data for exercise 2

	Casing	Tubing
OD	5.5 in (139.7 mm)	2.375 in (60.3 mm)
ID	4.670 in (118.6 m)	1.995 in (50.7 mm)
Length	6.500 ft (1981 m)	6,500 ft (1981 m)
Friction factor (f_{m})	0.02	0.02

cracking pressure of 0.5 psid (3.4 kPad). Using the data from Table 3.6 estimate how much gas and water is flowing up the tubing.

3. Which of the following deliquification techniques require the most energy input per bbl of liquid lifted
   a. Jet pump
   b. Gas lift
   c. Electric submersible pump
   d. Plunger
4. Turner's critical flow equation does not include a liquid production rate. Why not?
5. On a well that is flowing up the tubing with the casing open to the reservoir, but not flowing, what are the two major components of the difference between tubing pressure and casing pressure?
6. Which of the following conditions would maximize evaporation:
   a. 30 psia (207 kPaa) and 80°F (26.7°C)
   b. 45 psia (310 kPaa) and 105°F (40.6°C)
   c. 100 psia (689 kPaa) and 120°F (48.9°C)
   d. 200 psia (1379 kPaa) and 140°F (60.0°C)

CHAPTER FOUR

Surface Engineering Concepts

Contents

Practical Onshore Gas Field Engineering.
DOI: http://dx.doi.org/10.1016/B978-0-12-813022-3.00004-3

When we think of "surface engineering" we tend to include facilities at the well-site, gas gathering, compressor stations, midstream pipelines, and plants. Every one of these has fluid flow, fluid friction, corrosion, and a need to purge air from lines and vessels. Pulling these concepts out of each individual chapter allows a focus on what is important specifically to those individual topics.

4.1 FLUID FRICTION

As a fluid moves from one place to another, it interacts with its environment and with itself. These interactions are characterized by collisions and shear forces. Each collision and each shear event convert energy from kinetic to thermal. This nonreversible energy conversion is fundamentally an "energy loss" to the flowing stream and is called "fluid friction." Fluid flow with friction is called "viscous flow," meaning that the viscosity of the fluid is nonzero (in some fields the term "viscous flow" is limited "creep flow," "laminar flow," and other low Reynolds number flows, but these limitations are far from universal and in general terms "viscous flow" meaning is as given earlier).

4.1.1 Viscosity

Viscosity is a measure of the ability of a fluid to resist shear forces. Dynamic (or absolute) viscosity is an expression of a fluid's ability to resist shear flows. Kinematic viscosity can be thought of as resistance to fluid momentum.

4.1.1.1 Dynamic viscosity (μ)

Dynamic viscosity can be determined via an apparatus that has concentric tubes whose annular space can be filled with the fluid under test. One of the tubes is rotated and the amount of force that it takes to rotate at a fixed angular velocity is proportional to the dynamic viscosity. It can also be predicted through an equation of state or one of many empirical relationships. The various methods can vary by more than 20% one from the next for the same fluid. None of the empirical methods provide results that match field conditions when there is any acid gas in a gas stream, and the equations of state methods are not a lot better for acid gases. With the wide range of answers that you get from repeated lab measurement and/or using various equations, it is generally best to pick a single

Table 4.1 Relationship between viscosity units

	Poise	cP	lbm/ft/s	Pa × s
poise	1	0.001	0.0672	0.1
cP	100	1	0.000672	0.001
lbm/ft/s	14.88	1488	1	1.488
Pa × s	10	1000	0.672	1

method and pretend that you believe it—repeatable results will not be available for viscosity measurements.

The basic units of dynamic viscosity are Poise (poise), Centipoise (cP), lbm/ft/s, or Pa × s. The relationship between various units is given in Table 4.1.

Gas viscosity tends to be in the range of 0.01–0.02 cP (10×10^{-6} to 20×10^{-6} Pa × s) (air at 60°F (15.6°C) is usually reported as 0.0179 cP (17.9×10^{-6} Pa × s)). Liquid viscosities tend to be much higher (water at 60°F (15.6°C) is usually reported as 1 cP (0.001 Pa × s)). Dynamic viscosity is a strong function of temperature and a weak function of pressure.

4.1.1.2 *Kinematic viscosity ($\nu = \mu/\rho$)*

Kinematic viscosity is the ratio of dynamic viscosity and density is used in some functions. The units are Stokes (St), centistokes (cSt), or in^2/s (m^2/s). One St is equal to 1 cm^2/s. One cSt is equal to 1 mm^2/s.

4.1.2 Reynolds number

We can use similitude (Chapter 0: Introduction) to develop scalable models that are a very reasonable place to begin the process of developing a general description of the impact of friction on a flowing stream. Starting with Reynolds number (Eq. (4.1)):

$$Re = \frac{\rho \cdot \vec{v} \cdot \mathrm{ID}}{\mu} \tag{4.1}$$

Observing that there are ranges of Reynolds numbers in which the fluid behaves very differently from the other ranges provides some insights:

- $Re < 2000$—flow is laminar (i.e., all flow is mostly in the same direction with very few perturbations).
- $Re > 6000$—Flow is turbulent (i.e., random, three-dimensional motion is superimposed on the bulk flow direction).
- $2000 < Re < 6000$—Flow is in transition and no conclusions can be drawn.

With Reynolds numbers providing a clear understanding of gross-level flow regime, it is only required to find another parameter that clarifies how fluid acts in each flow regime to begin to affix numerical values for fluid friction.

4.1.3 Absolute pipe roughness (ε)

Boundary layer thickness (Chapter 0: Introduction) has a definable relationship to the roughness of the pipe. Pipe roughness is reported in (length-of-travel minus unit-length-of-pipe) ÷ unit-length-of-pipe, meaning that if you had a molecular-scale ant crawling one foot of the length of a pipe, including the dips and valleys how far would he crawl? The answer to that is a published factor determined by each pipe manufacturer (typically experimentally by backing into a roughness value that matches the calculated pressure drop in a known flow rate to measured values—a much superior approach to trying to actually measure the dips and peaks). In general terms you would expect values to be in the neighborhood of the values in Table 4.2, but these values should be confirmed with manufacturer's data.

Any given sample of pipe will have an absolute roughness that varies by ±20%–30% so the values in Table 4.2 need to be taken with a bit of skepticism.

Absolute roughness has units of "length," and that will not do for similitude. To get to a nondimensional value called "relative roughness" divide absolute roughness by pipe inside diameter (Eq. (4.2)). One of the most common mistakes that people make in doing this simple calculation is not using the same units for absolute roughness and pipe ID. If your pipe diameter is in "mm," then you have to convert it to "m" before

Table 4.2 Pipe roughness table

Material	Absolute pipe roughness (fps)	Absolute pipe roughness (SI)
Concrete pipe	4000×10^{-6} ft	1220×10^{-6} m
Commercial steel pipe	150×10^{-6} ft	46×10^{-6} m
Glass reinforced plastic (GRP or fiberglass)	100×10^{-6} ft	31×10^{-6} m
HDPE	50×10^{-6} ft	15×10^{-6} m
Fiber reinforced plastic (FRP or spoolable composite)	50×10^{-6} ft	15×10^{-6} m

comparing it to the absolute roughness. This may be obvious, but there are too many examples where this was done improperly on systems that did not work as designed.

$$\varepsilon\_D = \frac{\varepsilon}{\mathrm{ID}_{pipe}}$$
$$\text{If } \varepsilon\_\mathrm{D} \leq 50 \times 10^{-6}\ \text{ft then you are outside the useful range and should fix } \varepsilon\_D \text{ at } 50 \times 10^{-6}\ \text{ft} \tag{4.2}$$

4.1.4 Friction factor

Empirical equations (such as Eq. (3.10)) for flow will have a parameter called a "friction factor." These factors are nondimensional and a given flow will have the same factor in either fps or SI units. The most widely accepted relationship between Reynolds number and relative pipe roughness in the turbulent flow regime is called the "Colebrook equation" (Eq. (4.3)). This equation obviously requires an iterative solution since the Moody friction factor is on both sides of the equation. Eq. (4.4) is Poiseuille's law or the "Hagen–Poiseuille equation" and applies to smooth pipe flow (i.e., at low Reynolds numbers the pipe roughness becomes immaterial). Eq. (4.5) is Churchill's 1973 noniterative approximation of the Colebrook equation (Churchill revised this equation in 1977 with considerable added complexity which results in values that are very close to the 1973 equation) which provides very reasonable values as long as you honor the limits shown in the equation statement.

$$\frac{1}{\sqrt{f_m}} = -2 \cdot \log_{10}\left(\frac{\varepsilon\_D}{3.7} + \frac{2.51}{Re \cdot \sqrt{f_m}}\right) \tag{4.3}$$

$$f_m = \frac{64}{Re} \quad \text{for Reynolds numbers} \leq 2000 \tag{4.4}$$

$$f_m = \frac{1.325}{\left(\ln\left(\frac{\varepsilon\_D}{3.7} + \frac{5.74}{Re^{0.9}}\right)\right)^2} \quad \text{for } 6000 \leq Re \leq 10^8 \text{ and } 50 \times 10^{-6} \leq \varepsilon\_D \leq 10^{-2} \tag{4.5}$$

If you have "smooth pipe flow" by the description in Eq. (4.2) then you set ε_D to 50×10^{-6} which corresponds to the last ε_D line that crosses the fully turbulent line on Fig. 4.1.

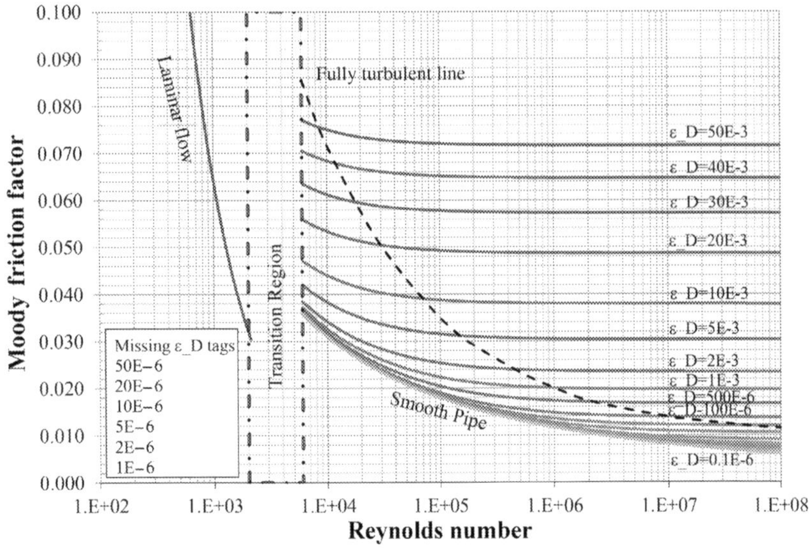

Figure 4.1 Moody friction factors.

Fig. 4.1 is usually presented as log–log scales, but I find that a logarithmic y-axis creates more confusion with interpolation than it provides in improved visibility.

Friction factors from Fig. 4.1 are used in various equations that use assumed flow rates to estimate a pressure drop, estimate a flow rate from measured pressure drop, or predict a flow rate from assumed pressure drop. For the last two uses, velocity is not known at the start of the calculation so the flow equation must be iterated. You guess a Reynolds number, solve for a friction factor, calculate a flow rate, calculate a new Reynolds number with the new velocity, use that value to calculate a new friction factor, and then repeat until the change from one step to the next is acceptably small.

4.1.4.1 Moody (D'Arcy) friction factor

The friction factor is a dimensionless parameter that can either be calculated using something like Eqs. (4.3)–(4.5), or extracted from a version of Fig. 4.1.

4.1.4.2 Fanning friction factor

As researchers started moving the study of fluid flow in pipes from strictly liquid flows to gas flows, the Moody friction factors appeared to be a bit coarse. John Thomas Fanning (1896) looked at the stresses in gas flows and found that:

$$f_f = \frac{f_m}{4} \tag{4.6}$$

Many equations developed for gas flow are designed to use the Fanning friction factor. Authors are not always clear as to which friction factor to use. A rule of thumb that might be useful is:

- $f_m \rightarrow$ Moody friction factor
- $f_d \rightarrow$ Moody friction factor (the subscript "*d*" refers to D'Arcy's work which predated Moody's work)
- $f_f \rightarrow$ Fanning friction factor
- $f \rightarrow$ *Assume* Moody friction factor for liquids and Fanning friction factor for gases. This can be very problematical since some researchers are too clever for their own good and roll the ¼ factor into a constant and use f without a subscript, but if you use Fanning you'll fail to match field conditions by a factor of 4. My approach in this case is that if you find an author that is too lazy to differentiate then you should look for another source.

4.1.4.3 Average pressure

In Eq. (4.1) we need to calculate density, viscosity, and velocity. For liquids, viscosity is a function of pressure and temperature. For gases, density, viscosity, and velocity are all strong functions of pressure. While we can assume that a flowing pipeline is isothermal, it is anything but isobaric; so what pressure do you use to calculate these important parameters? Choices seem to be upstream, downstream, or average. Average seems to make the most sense, but is it a simple average or is it somehow weighted? Analysis of pressure drop in a pipeline is an important field of study, and researchers have shown that the pressure drop per unit length is higher early in a pipeline than later in the line, so an average pressure should be front-end weighted. We do this by:

$$P_{avg} = \frac{2}{3} \cdot \left(P_1 + P_2 - \frac{P_1 \cdot P_2}{P_1 + P_2} \right) \tag{4.7}$$

If we compare this to a simple average (i.e., $(P_1 + P_2)/2$) you can see (Fig. 4.2) that for short lengths of pipe it simply doesn't matter. In this example, you don't have a 5% error until over 50 miles (80.5 km). Eq. (4.7) is simple enough that my approach is to always use it and eliminate that (admittedly small) error from the body of the calculation. In the

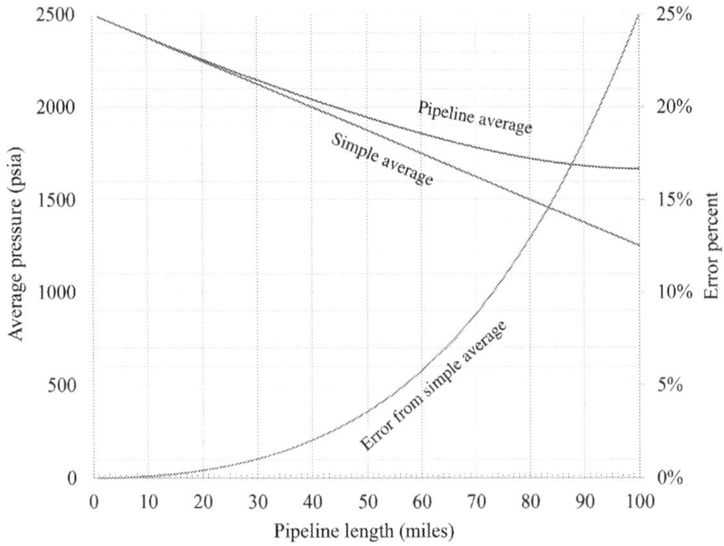

Figure 4.2 Comparison of average pressure methods (25 psi/mi pressure gradient).

big picture it rarely makes a material difference, but I always wonder "is this the scenario where I would reach a different decision if I used the right value?" You can avoid that particular nagging doubt by always using the front-end loaded average. People who think "every line in this project is under 10 miles (16 km) so why bother?" are not wrong, and usually reach the same decision that they would reach using the other equation. There is nothing "wrong" with that thought process, as long as the thought process actually takes place—many engineers have decided to use a simple average and have erased the possibility of an alternative method from their minds, and approach a long line with the simple average which results in density being low, velocity being high, and viscosity being low; all by different amounts. For a gas, the end result is complex, but often results in Reynolds number being low, which for modern pipe tends to put you in a steeper portion of Fig. 4.1 and small changes in pressure can give you a sharp increase in friction factor.

4.2 LIQUID FLOW

You might wonder why a book titled "*Practical Onshore Gas Field Engineering*" would be interested in liquid flow. There really is no way to get gas out of the ground without some amount of liquid. That liquid has to be pumped, piped, trucked, stored, sold, and/or disposed of, and facilities engineers are accountable for all of these things.

4.2.1 D'Arcy–Weisbach equation

Julius Weisbach, working with Henry D'Arcy's equations for flow through a porous medium developed the equation generally known as the Darcy–Weisbach equation (Henry D'Arcy's name is regularly anglicized to "Darcy" today, publications during his life spelled it "D'Arcy") to describe flow through a closed conduit.

$$\begin{aligned} dP_{Darcy} &= f_m \cdot \left(\frac{L_{Darcy}}{\mathrm{ID}_{Darcy}}\right) \cdot \left(\frac{\rho_{liquid}}{g_c}\right) \cdot \left(\frac{\vec{v}\ 2}{2}\right) \\ &= f_m \cdot \left(\frac{L_{Darcy}}{\mathrm{ID}_{Darcy}}\right) \cdot \left(\frac{\rho_{liquid}}{g_c}\right) \cdot \left(\frac{{q_{Darcy}}^2}{2 \cdot A_{pipe}^2}\right) \\ q_{Darcy} &= \sqrt{\frac{2 \cdot g_c \cdot dP_{Darcy} \cdot \mathrm{ID}_{darcy} \cdot A_{pipe}^2}{L_{Darcy} \cdot \rho_{liquid} \cdot f_m}} \end{aligned} \tag{4.8}$$

Eq. (4.8) is an empirical equation (even though it has none of the usual telltales) (Wikipedia 1) and extreme care needs to be taken with units. It is common in gas flow equations to specify the length in miles (or km) and ID in inches (or mm), but the units in the L/ID term must cancel so this equation uses ft (or m) for both. The volume flow rate is in actual volume/second (rather than gallons, barrels, or SCF). Eq. (4.8) can be used for gas flows where the total pressure drop is very small (on the order of 1%–2% of upstream pressure), but at the end of the calculation you must remember to convert to flow rate at standard conditions.

As we will see in gas flow equations, before computers there was a real need to replace the friction factor in Eq. (4.8) with something that could be calculated in a single step and several closed-form equations were developed for liquid flow taking advantage of specific conditions in a flow stream. None of these equations will provide results as good as Darcy–Weisbach.

4.2.2 Full-pipe determination

In a fluid-statics problem, intermediate hills and valleys are irrelevant to liquid pressure—the only thing that really matters is the height difference between the "inlet" and the "outlet." This is the classic manometer problem discussed elsewhere in this book. A disruption in the continuity of the liquid changes the relative heights of the two ends of the manometer.

Flowing liquid has a component represented by this fluid-static problem. If a flowing line is completely full of liquid, then the discharge pressure is a function of the "relative head" (i.e., the net height difference between the position of the inlet and the position of the outlet plus the pump discharge pressure minus friction losses) and intermediate heights are irrelevant. The reason that this works is that the downhill side of a hill will syphon the fluid moving uphill and the hydrostatic portion of the total pressure at any elevation on the uphill part of the flow will equal the pressure at the same elevation on the downhill part of the flow. This works very much like the description of the plunger movement in an SRP in Chapter 3, Well Dynamics—in a continuous-liquid pulling 1% of a control volume from the flow will try to lower the pressure by the bulk modulus (319,000 psia (2200 MPaa) for pure water) which of course is impossible in a real flow so the uphill liquid is dragged by the downhill liquid falling.

It is rare for a liquid line to see terrain features as irrelevant. I can point to a system in hilly country where the pipe outlet is 256 ft (78 m) lower than the pump location. In a full system with minimal friction losses, the line should be on a vacuum at the pump location. In real life, the pump is having to supply 90 psig (621 kPag). Fig. 4.3 shows why this is. As the inflow approaches the top of a hill, it hits a void space at the top (the void is full of gas, either entrained gases that evolved out of the liquid or phase-change gas that "boiled" in the low pressure created by

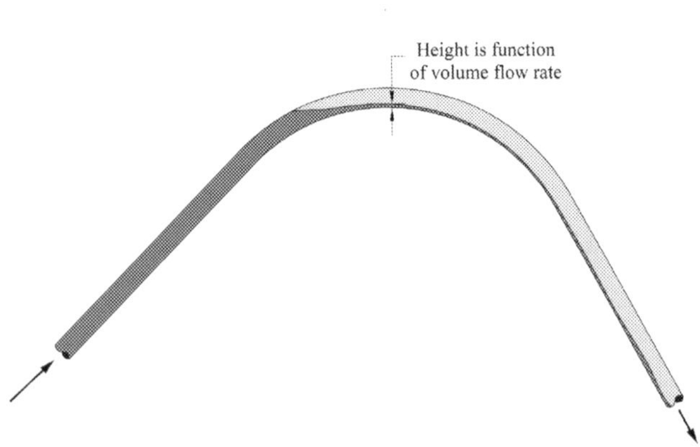

Figure 4.3 Liquid flow in a pipe.

the liquid falling away). This gas pocket has a very different bulk modulus than the liquid and is quite able to expand and contract at will. This means that the falling liquid, rather than dragging the rising liquid down the hill, simply flows under the gas pocket without impacting the rising liquid at all.

If a line is not running full, then you have to design your system with adequate pumping horsepower to overcome fluid friction and to overtop the sum of uphill runs in the system (not necessarily the highest elevation, you need to be concerned about every continuous up-slope from the start of rise to the end of rise, which may have a peak that is physically at a lower height above sea level than the pump).

Engineers have observed that if the flow rate (in US gallons per minute, sorry) is higher than Eq. (4.9) then the line will tend to run full with any topography.

$$q_{\min\_gpm} = 10.2 \cdot \mathrm{ID}_{inches}^{2.5} \tag{4.9}$$

This results in some fairly high fluid velocities in larger pipe. In lines larger than 8-in (200 DN), the velocity gets high enough to be concerned about erosion (see Section 4.4.1). The 8-in and smaller lines end up with velocities that are workable (e.g., 2 in results in a minimum velocity of 5.8 ft/s (1.80 m/s) while the erosional velocity of pure water is 12.7 ft/s (3.9 m/s)) which is still higher than we normally see for liquid flows in upstream Oil & Gas.

4.2.3 Pumping HP

The energy required to pump liquid is a function of how much stuff you are pumping (mass flow rate) and how high you need it to go (differential pressure across the pump).

$$\mathrm{BHP} = \frac{dP \cdot q_{liquid}}{C_{pumping}} \cdot \left(\frac{1}{\eta}\right) \tag{4.10}$$

Efficiency for a centrifugal pump tends to be around 0.60. Absent specific pump data from the manufacturer, 0.6 is a reasonable approximation. Since motors for pumps tend to come in discrete sizes, most engineers will take this calculation and round it up to the next available motor size and then jump to next size above that. The price differential between adjacent sizes of motor tends to be reasonably small and no one ever wants their pump design to be the limiting factor in any operation.

4.3 GAS FLOW

The way that engineers and scientists deal with gas displays a distinctly split personality. On the one hand, density, viscosity, and velocity all change with changes in pressure (a characteristic of a *compressible* fluid). On the other hand, over short(ish) lengths of pipe the density, viscosity, and velocity changes are reasonably small and if you treat them as constant then you can discard many of the terms in Navier–Stokes equation (Eq. (0.24)) and develop closed-form solutions to real-world problems that allow you to reach defensible decisions (a characteristic of an *incompressible* fluid).

We address this split personality by being especially careful to state assumptions and to honor them. For example if an equation is only valid above and to the right of the dashed fully turbulent line on Fig. 4.1, then you need to check the Reynolds number and the relative roughness to see if the data fits the equation you want to use.

Not all gas flow meets the definitions of incompressible flow (e.g., at very high velocity, the dynamic pressure component of total pressure increases far more than the static-pressure component decreases and total pressure is no longer constant) and you have to use the less robust and more limited compressible-flow equations.

4.3.1 Compressible flow

Compressible gas flow generally represents a flow that the downstream density is less than 90% of the upstream density. For long lines, you can break up the line into segments that each have a pressure drop that keeps the change in density to under 10%, but you have to be serious about it. An equation like the *AGA fully turbulent equation* (Section 4.3.3.1) does not have any parameters that change much from segment to segment, so breaking the line up into small segments results in the same answer as not breaking the line into segments (i.e., the only component that changes from one step to the next is the square root of a difference in squared pressure, if one step is going from 1000 to 900 psia and the next step is going from 900 to 810 psia, the impact is exactly 10% and the "calculated" flow rate over the two segments is the same as if you hadn't segmented the line). An equation like the *Isothermal single-phase gas flow equation* (Section 4.3.1.4) requires you to recalculate the Reynolds number and friction factor for each step and provides valid results on a segmented line.

Compressible flow is fairly rare in Oil & Gas, but it happens in several specific situations. We see compressible flow for most of a pipeline or site blowdown, and flow through regulators can get fast enough to provide compressible flow.

4.3.1.1 Sonic velocity and choked flow

"Choked flow" is caused by a standing shock wave in the flow that inhibits communication from downstream back to upstream that the downstream pressure has changed. High-velocity flow upstream of the shock wave is referred to as "Fanno flow" (i.e., flow is adiabatic and isothermal with viscous effects causing irreversible energy loss from the flowing stream, Fig. 4.4). You can calculate a "Fanno friction factor," but the field data resists matching differential pressure calculated using it.

Real compressible flows tend to have a change in temperature with flow. In some velocity ranges the flow will tend to heat up and in other velocity ranges the flow will tend to cool. This phenomenon is evident in the Joule–Thomson effect and is described by Rayleigh flow. Before the shock

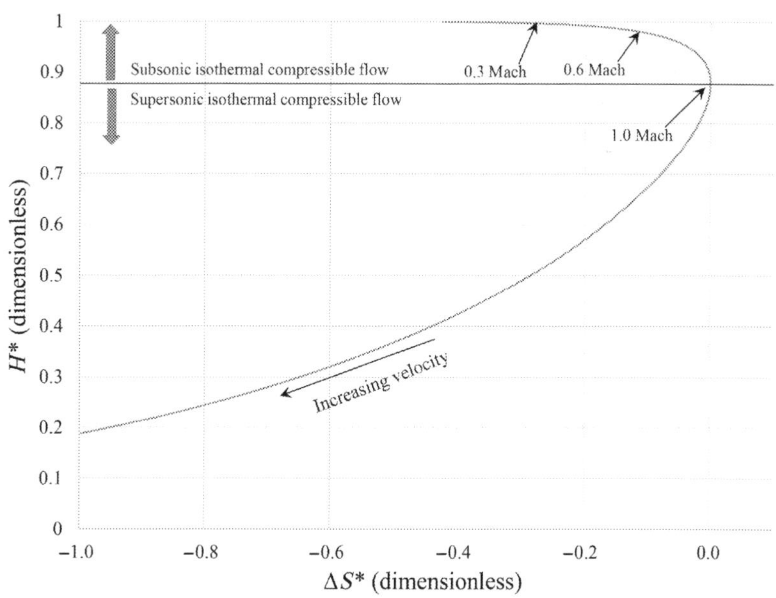

Figure 4.4 Fanno flow (methane, $k = 1.28$).

wave the flow decently matches the Fanno flow model. After the shock wave the flow more closely follows the frictionless Rayleigh flow model.

The shock wave that represents the transition from isothermal, viscous flow to varying temperature frictionless flow happens whenever downstream pressure satisfies Eq. (4.11).

$$P_2 \leq P_1 \cdot \left(\frac{2}{k+1}\right)^{\frac{k}{k-1}} \tag{4.11}$$

This says that downstream pressure becomes irrelevant when it is less than a multiple (which is only a function of gas composition) of upstream pressure. Whenever Eq. (4.11) is satisfied, flow will be "choked" to a velocity of 1.0 Mach (i.e., the speed of sound). Sonic velocity is:

$$v_{sonic} = \sqrt{k \cdot R_{gas} \cdot T_1} = \sqrt{k \cdot \frac{R_{air}}{\text{SG}} \cdot T_1} = 1.0 \text{ Mach} \tag{4.12}$$

"Mach number" is a dimensionless parameter that is actual velocity divided by sonic velocity. Note in Eq. (4.12) that sonic velocity of (or in) a gas is only a function of gas composition and temperature, not pressure or density (the adiabatic constant actually does change measurably with changes in pressure, but as a general rule the same value is used at all pressures and a rigorous calculation will lead you to the same decision as you get from this simplification).

Many textbooks classify choked flow as "constant mass flow rate" flow which is anything but accurate, and has led to some very significant errors in actual calculations. In choked flow "velocity" is constant, but mass flow rate is made up of more than velocity (Eq. (4.13)).

$$\dot{m} = q \cdot \rho = (v \cdot A) \cdot \rho \tag{4.13}$$

Velocity and flow area are constant as long as the flow is choked, but in many situations the density can change dramatically. If we are talking about an airplane flying through the air, the bulk density of the atmosphere is constant, and the mass flow rate over the wings can be thought of as constant, but for something like a pipeline blowdown, we are actively working to reduce upstream pressure and mass flow rate can decrease several percentage points over a very short time.

4.3.1.2 Pipeline blowdown example

For a pipeline blowdown, there is not an easy way to determine how long the event will take. The most effective method is to consider mass

Table 4.3 Pipeline blowdown example

Elapsed Minute	P_{start} (psia)	P_{crit} (psia)	rho (lbm/ft^3)	m_{dot} (lbm/h)	$m_{remaining}$ (lbm)
0.5	1987.2	968.9	3.566	396,000	394,700
1.0	1975.1	909.1	3.325	364,200	391,400
1.5	1963.0	907.6	3.319	363,500	388,400
⋮					
160.0	277.4	151.3	0.949	46,127	54,652
160.5	275.5	150.3	0.942	45,809	54,267
161.0	273.7	149.3	0.935	44,494	53,886
⋮					
326.0	26.8	14.64	0.048	4,378	5,088
326.5	26.6	14.53	0.048	4,347	5,051
327.0	26.5	14.50	0.048	4,315	5,015

flow rate constant for a time interval and calculate the mass removed from the line during the interval, calculate a new remaining mass, new starting pressure, new critical pressure, new density (which is based on critical pressure, not upstream pressure), and new mass flow rate for the next interval (Table 4.3).

Example conditions (flow rates used Eq. (4.32)):

- Gas:
  - Species: coalbed methane (CBM) gas
  - SG: 0.6315
  - *k*: 1.3016
- Flowing conditions
  - Initial pressure: 2000 psia (13.8 MPaa)
  - Temperature: 60°F (520 R (289K)) and constant
- Pipe parameters
  - Length: 5 miles (8.05 km)
  - ID: 20 in (500 DN)
  - Volume of pipe: 57.6×10^3 ft^3 (1630 m^3)
  - Mass of gas in line: 476,500 lbm (216,100 kg)
  - Volume of gas in line: 10.4 MMSCF (295 kSCm)
- Blowdown valve ID: 1.996 in (50.7 mm)
- Sonic velocity: 1357 ft/s (414 m/s)
- Time increment: 30 seconds

Note that the fluid velocity through the whole calculation was constant, but mass flow rate decreased 98.7%. The line initially held 10.4 MMSCF (295 kSCm), and at the end of choked flow it still holds

0.105 MMSCF (2.96 kSCm). As we continue to blow this line down, the flow rates become progressively less predictable, and even at the point where no gas is flowing out the vent, there is still gas in the line.

Increment size is an important decision. It is obvious that the mass flow rate at the end of an increment will be less than the mass flow rate at the beginning, but this analysis ignores that (hopefully slight) change. The longer we apply an overstated flow rate, the worse the analysis will be. For example:

- 30-minute increment → 270-minute blowdown time
- 15-minute increment → 300-minute blowdown time
- 10-minute increment → 310-minute blowdown time
- 1-minute increment → 326-minute blowdown time
- 30-second increment → 327.0-minute blowdown time
- 10-second increment → 327.3-minute blowdown time
- 1-second increment → 327.6-minute blowdown time
- 0.1-second increment → 327.7-minute blowdown time

The blowdown will run out of gas during the last increment, but not necessarily at the beginning or end of the increment, using a small enough step size that this is immaterial to the calculation is about as small as you need to go. In this example, a 30-minute step size estimates the blowdown time to the end of choked flow to be 4.5 hours, obviously too short. One minute increments make it 5.43 hours which is certainly close enough to provide operator guidance during a blowdown procedure. Using 0.1 second requires 166,920 steps and refines the number to 5.46 hours. Experience led me to pick 30 seconds as an increment that nearly always leads to acceptable results (but I always run the MathCad program one more time with 1 second to see if it suggests a different decision with finer data). If you are working in Excel, then changing the increment is not very easy and is rarely done, when I have to work in Excel I pick a 1-minute time interval and pretend that it doesn't matter much.

4.3.1.3 Dynamic pressure during compressible flow

Remember that for incompressible flow inside of a control volume with no additions or removals of mass or heat:

$$P_{total} = P_{static} + \frac{\rho \cdot v^2}{2} + \rho \cdot g \cdot h = \text{constant} \tag{4.14}$$

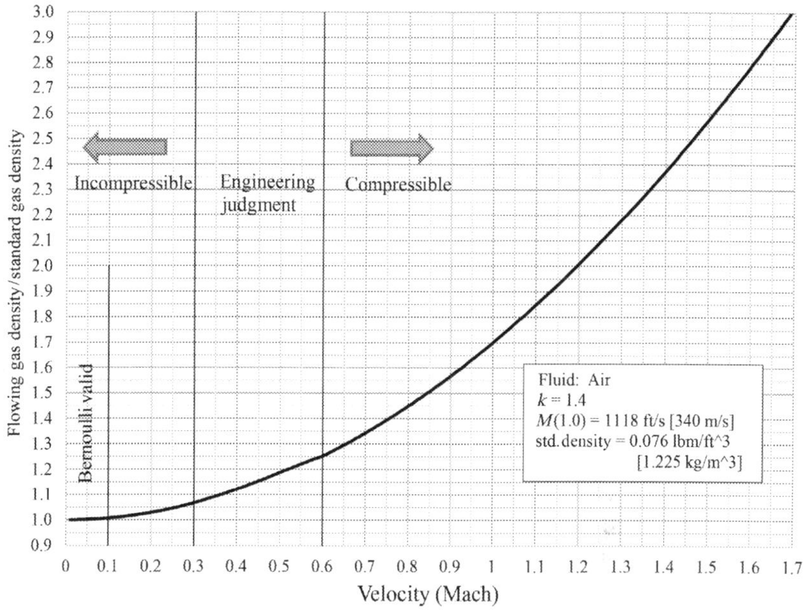

Figure 4.5 Velocity vs total pressure.

Once you can no longer consider total pressure as constant, you need to shift to the compressible-flow calculations:

$$P_{total} = P_{static} \cdot \left(1 + \frac{k \cdot M^2}{2}\right) + \rho \cdot g \cdot h \neq \text{constant} \tag{4.15}$$

Combining Eq. (4.14) for flow less than 0.3 Mach, Eq. (4.15) for flow greater than 0.6 Mach, and a linear interpolation to shift from one curve to the other between 0.3 and 0.6 Mach, you get Fig. 4.5. You can see from this graph that below 0.1 Mach, the flow is incompressible for any purpose, and below 0.3 Mach it is incompressible for virtually any practical purpose. You can also see that above 0.6 Mach the deviation from constant density becomes too great to ignore. Where you draw the compressible/incompressible line is a matter for engineering judgment.

4.3.1.4 Compressible versus incompressible flow

The flow behavior differences between compressible and incompressible flow are somewhat counterintuitive. We can all accept that in a divergent nozzle during incompressible flow, as the cross-sectional area increases the velocity has to decrease, it is just logical. Accepting that in compressible

Table 4.4 Compressible vs incompressible flow

	Incompressible	Compressible
Density	Dominated by static pressure Nearly constant with time and position, step changes don't happen	Dominated by dynamic pressure Changes rapidly and large step changes are common
Viscosity	Use static viscosity	Viscosity nearly impossible to determine
Inside pipe fluid friction	Use Moody diagram	No known reliable general method to quantify
Divergent nozzle	Pressure increases Temperature constant Velocity *decreases*	Pressure nearly constant Temperature nearly constant Velocity *increases*
Convergent nozzle	Pressure decreases Temperature constant Velocity *increases*	Pressure nearly constant Temperature nearly constant Velocity *decreases*

flow the velocity is controlled by shock waves and a divergent nozzle loosens the physical restrictions on the flow, actually increasing velocity at constant pressure is a bit more difficult. Some of the differences are highlighted in Table 4.4.

4.3.2 Isothermal single-phase incompressible gas flow

Virtually all real gas flows in Oil & Gas facilities can safely be approximated as incompressible. People have real difficulty internalizing the concept of "incompressible flow of a highly compressible fluid." It just doesn't make logical sense to most of us. Hopefully the discussion in the last section clears up that logical difficulty, if not then it is still required to memorize that "virtually all real gas flows in Oil & Gas facilities can safely be approximated as incompressible" if you are going to work in this industry.

That being said, it is rare for Eq. (4.8) to provide an adequate representation of the flow. Repeating that equation here for convenience:

$$q_{Darcy} = \sqrt{\frac{2 \cdot g_c \cdot dP_{Darcy} \cdot \mathrm{ID}_{Darcy} \cdot A_{pipe}^2}{L_{Darcy} \cdot \rho_{liquid} \cdot f_m}} \tag{4.16}$$

Because of the derivation of Eq. (4.16) and its resulting actual volume (rather than converting to some "standard") it is only valid for gas in the portion of Fig. 4.5 marked "Bernoulli valid." For the rest of the time the relationship is a bit more complicated (Eq. (4.17))

$$q_{MSCFd} = \frac{38.77}{1000}\left(\frac{T_{std}}{P_{std}}\right)\eta\left(\sqrt{\frac{1}{f_f}}\right)\left(\frac{\left(P_{up}^2 - P_{dwn}^2\right)\cdot \mathrm{ID}^5}{\mathrm{SG}\cdot L_{miles}\cdot T_{avg}\cdot Z_{avg}}\right)^{0.5} \tag{4.17}$$

The first term in Eq. (4.17) is intended to adjust units. This is an empirical equation and I have been unable to find or develop a set of units and conversion factors that allow it to be presented in SI.

Volume flow rate (q) is velocity times flow area. Velocity is a first-order component of Reynolds number which is the determinant of Fanning friction factor (f_f). The term $\sqrt{1/f_f}$ is called the "transmission factor" and it is solved iteratively (i.e., guess a Reynolds number, calculate a transmission factor, calculate a flow rate, repeat until the change in Reynolds number is acceptably small from step to step).

The "efficiency (η)" term is a plug value. Clean, single-phase flow in new steel pipe should get a value of 0.95. Clean, new plastic pipe should get a value of 1.0. For evaluating lines for water, scale, or slime it is usual to adjust efficiency until the measured dP matches the measured flow.

4.3.2.1 Assumptions in the derivation

This equation is only valid when all of the following are satisfied:

- Flow is steady state ($dq/dt \approx 0$)
- Flow is isothermal ($T_1 \approx T_2$)
- Pipe size doesn't change
- All flow is gas (no liquid carryover or condensation)
- No change in Kinetic energy ($\mathrm{KE} = 1/2\cdot m\cdot \vec{v}^2$)
- Change in density is not material (e.g., <10%)
- No fluid is added or removed between the upstream and downstream pressure measurement points
- No work is done on or by the flowing gas

Since it is quite rare in Oil & Gas facilities to have a line that is perfectly free of liquid accumulation, people generally fudge "no liquid" assumption with reasonable results. The problem with liquid in the lines is that the gas will do work on the liquid and the pressure loss due to that work can only be represented as "friction" or "efficiency." Since multiphase friction factors are notoriously inaccurate, we put the loss into efficiency and this number must be dropped precipitously (e.g., on one line, before pigging the efficiency was 0.03, and after pigging it went up to 0.85).

4.3.2.2 Useful restructures of isothermal gas flow equation

The basic equation is useful, but often you have a flow rate and either upstream or downstream pressure and want to calculate the other pressure. You can approximate compressibility by using the available pressure and setting the other pressure to $P_{unknown} = P_{known} \pm (15 \text{ psi/mi} \times \text{length})$

$$P_1 = \left[P_2^2 + \left(\frac{25.793 \cdot q_{MSCFd}}{\text{ID}^{2.5}\eta}\right)^2 \left(\frac{P_{std}}{T_{std}}\right)^2 f_f \cdot \text{SG} \cdot L_{mile} \cdot T_{flowing} \cdot Z_{avg}\right]^{0.5} \quad (4.18)$$

$$P_2 = \left[P_1^2 - \left(\frac{25.793 \cdot q_{MSCFd}}{\text{ID}^{2.5} \cdot \eta}\right)^2 \left(\frac{P_{std}}{T_{std}}\right)^2 f_f \cdot \text{SG} \cdot L_{mile} \cdot T_{flowing} \cdot Z_{avg}\right]^{0.5} \quad (4.19)$$

4.3.2.3 Example of isothermal gas flow

An early life CBM well has a flow line running 8 miles (12.8 km) to a compressor station (called a "home run" flow line, very common in oil fields, quite rare in gas fields). Conditions:

- $P_{atm} \rightarrow$ 12 psia (83 kPag)
- $P_{well} \rightarrow$ 533 psig (3.67 MPag)
- $P_{booster} \rightarrow$ 493 psig (3.40 MPag)
- $T_{up} \rightarrow$ 60°F (15.6°C)
- Pipe $\rightarrow$ 2-in schedule 40 (ID = 2.067 in (52.5 mm))

To determine the flow rate in these conditions:

$$P_{avg} = \frac{2}{3}\left(P_1 + P_1 - \frac{P_1 P_2}{P_1 + P_2}\right)$$

$$= \frac{2}{3}\left(545 \text{ psia} + 505 \text{ psia} - \frac{(545 \text{ psia})(505 \text{ psia})}{545 \text{ psia} + 505 \text{ psia}}\right) = 525.25 \text{ psia}$$

$$\mu_{avg} = 7.75 \times 10^{-6} \frac{\text{lbm}}{\text{ft} \cdot s}$$

$$\rho_{avg} = \frac{P_{avg}\text{SG}}{R_{air} T_{avg} Z_{avg}} = \frac{525 \frac{\text{lbf}}{\text{in}^2}\left(\frac{144 \text{ in}^2}{\text{ft}^2}\right)(0.6312)}{53.355 \frac{\text{ft} \cdot \text{lbf}}{\text{R} \cdot \text{lbm}}(520R)(0.924)} = 1.869 \frac{\text{lbm}}{\text{ft}^3}$$

$$\rho_{std} = \frac{14.73 \frac{\text{lbf}}{\text{in}^2}\left(\frac{144 \text{ in}^2}{\text{ft}^2}\right)(0.6312)}{53.355 \frac{\text{ft} \cdot \text{lbf}}{\text{R} \cdot \text{lbm}}(520R)(0.998)} = 0.0484 \frac{\text{lbm}}{\text{ft}^3}$$

$$\rho_1 = \frac{505\,\frac{\text{lbf}}{\text{in}^2}\left(\frac{144\ \text{in}^2}{\text{ft}^2}\right)(0.6312)}{53.355\,\frac{\text{ft}\cdot\text{lbf}}{\text{R}\cdot\text{lbm}}(520R)(0.921)} = 1.939\,\frac{\text{lbm}}{\text{ft}^3}$$

$$\rho_2 = \frac{453\,\frac{\text{lbf}}{\text{in}^2}\left(\frac{144\ \text{in}^2}{\text{ft}^2}\right)(0.6312)}{53.355\,\frac{\text{ft}\cdot\text{lbf}}{\text{R}\cdot\text{lbm}}(520R)(0.926)} = 1.786\,\frac{\text{lbm}}{\text{ft}^3}$$

$$\frac{\rho_2}{\rho_1} = \frac{1.786}{1.939} = 92.1\% \quad \text{Must be} > 90\%$$

From Table 4.2 for commercial steel pipe ($\varepsilon = 150 \times 10^{-6}$ ft), and guess that Reynolds number is 1×10^6 (an arbitrary value to the right of the fully turbulent line):

$$\varepsilon\_D = \frac{150 \times 10^{-6}\text{ft}}{2.067\ \text{in}\cdot\left(\frac{\text{ft}}{12\cdot\text{in}}\right)} = 8.71 \times 10^{-4}$$

$$f_m = \frac{1.325}{\left(\ln\left(\frac{\varepsilon\_D}{3.7} + \frac{5.74}{R_e^{0.9}}\right)\right)^2} = \frac{1.325}{\left(\ln\left(\frac{8.71 \times 10^{-4}}{3.7} + \frac{5.74}{\left(10^6\right)^{0.9}}\right)\right)^2}$$

$$= 0.0193 \text{ (identical to Colebrook to 5 decimal places)}$$

$$f_f = \frac{f_m}{4} = \frac{0.0193}{4} = 0.00483$$

Now we can calculate a first-guess flow rate:

$$q_{MSCFd} = \frac{38.77}{1000}\left(\frac{520}{14.73}\right)0.95\left(\sqrt{\frac{1}{0.00483}}\right)\left(\frac{(545^2 - 505^2)\cdot(2.067)^5}{0.6312(8)(520)(0.924)}\right)^{0.5}$$

$$= 478\,\frac{\text{MSCF}}{\text{day}} \Leftarrow \text{Calculated flow rate, first guess}$$

$$q_{acfd} = q_{MSCFd}\frac{\rho_{std}}{\rho_{avg}} = 478\,\frac{\text{MSCF}}{\text{day}}\left(\frac{0.0484}{1.863}\right)\left(\frac{1000\text{ft}^3}{\text{MCF}}\right)\left(\frac{\text{day}}{86400\text{s}}\right) = 0.1432\,\frac{\text{ft}^3}{\text{s}}$$

$$A_{pipe} = \left(\frac{\pi}{4}\right)\left(\frac{\text{ID}}{12}\right)^2 = \left(\frac{\pi}{4}\right)\left(\frac{2.067}{12}\right)^2 = 0.0233\text{ft}^2$$

$$\vec{v} = \frac{q_{acfd}}{A_{pipe}} = \frac{0.1499\,\frac{\text{ft}^3}{s}}{0.0233\text{ft}^2} = 6.146\,\frac{\text{ft}}{\text{s}}$$

$$Re = \frac{\rho_{avg} \cdot \vec{v} \cdot ID_{pipe}}{\mu_{avg}} = \frac{\left(1.869 \frac{lbm}{ft^3}\right)\left(6.146 \frac{ft}{s}\right)\left(\frac{2.067}{12} ft\right)}{7.75 \times 10^{-6} \frac{lbm}{ft*s}} = 2.55 \times 10^5$$

Now you have to verify the friction factor. This result can have one of three outcomes:

- Both Reynolds number guess and calculated value are to the right of the fully turbulent line in Fig. 4.1 → accept calculation
- Guess and calculation on opposite sides of fully turbulent line → iterate
- Guess and calculation both to the left of fully turbulent line → iterate until Re_{before} is more than 90% of Re_{after}.

In this case, the calculated value is on the wrong side of the fully turbulent line, so you use the calculated Reynolds number in the next iteration.

$$f_f = \frac{1.325}{\left(\ln\left(\frac{1.475 \times 10^{-4}}{3.7} + \frac{5.74}{(2.55 \times 10^5)^{0.9}}\right)\right)^2} \cdot \frac{1}{4} = 0.00505$$

$$q_{MSCFd} = \frac{38.77}{1000}\left(\frac{520}{14.73}\right)0.95\left(\sqrt{\frac{1}{0.00505}}\right)\left(\frac{545^2 - 505^2}{0.6312(80)(520)(0.924)}\right)^{0.5} \times (2.067)^{2.5} = 467 \frac{MSCF}{day}$$

$$q_{acfd} = 467 \frac{MSCF}{day}\left(\frac{0.0484}{1.863}\right)\left(\frac{1000ft^3}{MCF}\right)\left(\frac{day}{86400s}\right) = 0.1400 \frac{ft^3}{s}$$

$$\vec{v} = \frac{0.1400 \frac{ft^3}{s}}{0.023ft^2} = 6.01 \frac{ft}{s}$$

$$Re = \frac{\left(1.869 \frac{lbm}{ft^3}\right)\left(6.01 \frac{ft}{s}\right)\left(\frac{2.067}{12} ft\right)}{7.75 \times 10^{-6} \frac{lbm}{ft*s}} = 2.496 \times 10^5$$

Same side of fully turbulent line

$$\Delta Re = \frac{2.496 \times 10^5 - 2.55 \times 10^5}{2.496 \times 10^5} = 2.22\% \text{ Flow rate acceptable}$$

With this very small pipe, the difference between an outright guess of friction factor and a friction factor calculated from field conditions was a difference in flow rate of 2.2%. For larger pipes and/or larger flow rates the difference can be considerably greater (Fig. 4.1).

4.3.3 Closed-form equations

Eq. (4.17) was developed far before the era of digital computers and hand-held calculators. Engineers using this equation were running the numbers on slide rules. As someone who started doing engineering calculations prior to the end of the slide-rule era, I can attest to doing iterative calculations on a slide rule being a serious pain in the posterior. Consequently, researchers searched for relationships to give them a meaningful, closed-form representation of friction that did not require iterating the transmission factor. Several were developed, each with its own limitations.

4.3.3.1 AGA fully turbulent

If you take the section of Fig. 4.1 to the right of the fully turbulent line, the relative roughness values tend to be acceptably close to horizontal. Using this observation:

$$\sqrt{\frac{1}{f_f}} = 4\,\log_{10}\left(\frac{3.7}{\varepsilon\_D}\right)$$

$$q_{MSCFd} = \frac{38.77}{1000}\left(\frac{T_{std}}{P_{std}}\right)\eta\left(4\cdot\log_{10}\left(\frac{3.7}{\varepsilon\_D}\right)\right)\left(\frac{P_1^2 - P_2^2}{\mathrm{SG}\cdot L_{miles}\cdot T_{avg}\cdot Z_{avg}}\right)^{0.5}\mathrm{ID}^{2.5} \tag{4.20}$$

You need to verify that the resulting velocity is in the fully turbulent region of Fig. 4.1, and that all of the assumptions of Eq. (4.17) are satisfied. Modern steel pipe has much lower absolute roughness than historical pipe, and modern plastic and fiberglass pipes are several orders of magnitudes smoother than the steel pipe used in the development of Eq. (4.20). An honest look at the results of the AGA fully turbulent equation shows that it is almost never applicable to flows in Oil & Gas.

AGA also has a Partially Turbulent equation, but it contains f_f so it doesn't have much to recommend it.

4.3.3.2 Weymouth

In 1912 it was postulated that:

$$\sqrt{\frac{1}{f_f}} = \sqrt{\frac{1}{\left(\frac{0.008}{\mathrm{ID}^{1/3}}\right)}} = 11.18\cdot\mathrm{ID}^{1/6}$$

which makes Eq. (4.17):

$$q_{MSCFd} = \frac{433.5}{1000}\left(\frac{T_{std}}{P_{std}}\right)\eta\left(\frac{P_1^2 - P_2^2}{\mathrm{SG}\cdot L_{miles}\cdot T_{avg}\cdot Z_{avg}}\right)^{0.5}\mathrm{ID}^{2.667} \tag{4.21}$$

This equation adds two very important assumptions to the list from Eq. (4.17):

- Pressure less than 130 psig (896 kPag)
- No CO_2 or H_2S.

The limitation on acid gas stems from the way that they calculated viscosity (which was invalid for acid gases) to develop their transmission factor.

4.3.3.3 Panhandle A

In the 1940s the Panhandle Eastern Pipe Line Company postulated that the transmission factor was a function of the flow rate over a narrow range of flow rates:

$$\sqrt{\frac{1}{f_f}} = 7.211\left(\frac{\mathrm{Q}\cdot\mathrm{SG}}{\mathrm{ID}}\right)^{0.0736} \quad \text{where “}Q\text{” is in SCF/day}$$

This results in Eq. (4.22):

$$q_{MSCFd} = \frac{435.87}{1000}\left(\frac{T_{std}}{P_{std}}\right)^{1.0768}\eta\left(\frac{P_1^2 - P_2^2}{\mathrm{SG}^{0.853}\cdot L_{miles}\cdot T_{avg}\cdot Z_{avg}}\right)^{0.5392}\mathrm{ID}^{2.6182} \tag{4.22}$$

This equation adds the assumptions:

- $5\times10^6 < Re < 1\times10^7$
- Pipe roughness is constant and high enough for the friction assumptions (cannot be used with plastic or fiberglass pipe).

This equation gives the best results when the efficiency is 0.9–0.92.

In 1956, the company published Panhandle B, but it has the same limitations as Panhandle A and never really caught on.

4.3.3.4 Oliphant

As pressures drop below 45 psig (310 kPag) toward atmospheric pressure, the incompressible assumption becomes more difficult to justify. One solution to the horrid complexities of the transition from incompressible to compressible-flow calculations is to take field values back to reference values to remove some of the complexity and narrow the field of view down to deviations from known conditions.

$$q_{MSCFd} = \frac{1008}{1000}\left(\mathrm{ID}^{2.5} + \frac{\mathrm{ID}^3}{30}\right)\left(\frac{14.4}{P_{std}}\right)\left(\frac{T_{std}}{520}\right) \times \left[\left(\frac{0.6}{\mathrm{SG}}\right)\left(\frac{520}{T_{avg}}\right)\left(\frac{P_1^2 - P_2^2}{L_{miles}}\right)\right]^{0.5} \cdot \frac{\mathrm{MSCF}}{\mathrm{day}} \quad (4.23)$$

Note the "30" in the "$\mathrm{ID}^3/30$" term is approximately 3.068^3 which is the ID of 3-in (75 DN) schedule 40 pipe. That number is quite telling and the farther you get from 3-in pipe, the less reliable this equation is going to be. Eq. (4.23) uses all of the assumptions from Eq. (4.17) plus a maximum pressure of 45 psia (300 kPaa).

4.3.3.5 Spitzglass

As the flow approaches a vacuum, the incompressible-flow assumption becomes untenable and you have to use compressible-flow calculations. The Spitzglass formula can be used to approximate flow in vacuum conditions, it doesn't do a great job, but engineers seem to have an insatiable need to have an equation for everything. This formula is in Eq. (4.24):

$$q_{MSCFd} = \frac{85,200}{1000}\left[\frac{27.69 \cdot \frac{(P_1 - P_2)}{\mathrm{psi}} \cdot \mathrm{ID}_{inches}^5}{\mathrm{SG} \cdot L_{pipeFt} \cdot \left(1 + \frac{3.6}{\mathrm{ID}_{inches}} + 0.03 \cdot \mathrm{ID}_{inches}\right)}\right]^{0.5} \frac{\mathrm{MSCF}}{\mathrm{day}} \quad (4.24)$$

This equation is limited to flows where at least one end of the pipe is under vacuum.

4.3.3.6 Comparison of closed-form solutions

If I had to solve a real-world flow problem with a slide rule today, I would likely evaluate the problem and try to find the closest to its assumptions matching field conditions, failing this (since it is very rare to find modern pipe that is anywhere near the fully turbulent region of

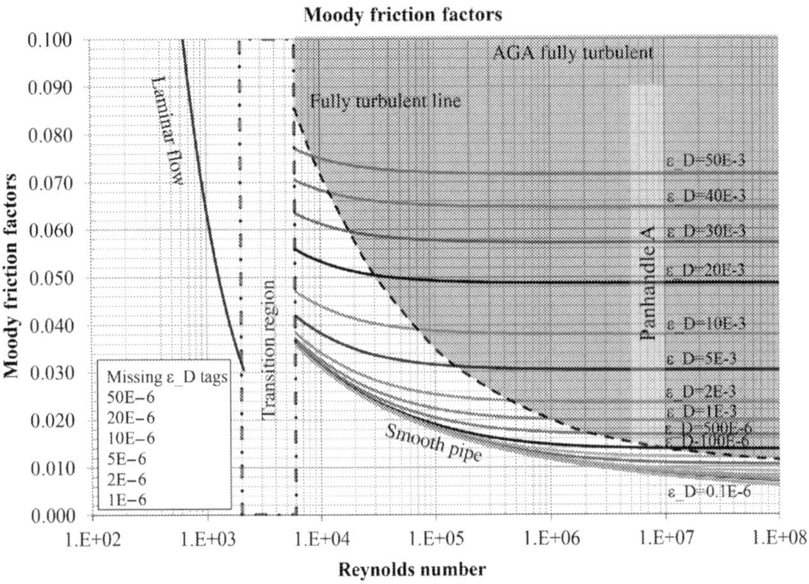

Figure 4.6 Valid regions of Moody diagram for closed-form equations.

Fig. 4.1) I would likely use the AGA fully turbulent equation. With access to a computer I never use any of the closed-form equations.

Fig. 4.6 shows where the AGA fully turbulent and Panhandle A equations are valid. To compare one equation to another, look at Fig. 4.7. This figure is a single pipe size and a single *dP* scenario, regenerating the data with different pipe size or *dP* changes the differences between the lines.

4.3.4 Multiphase flow

In most of this chapter we have pretended that we can treat real flows as "single phase." That is because we have reasonable arithmetic for single-phase gas or liquid flow that does a pretty good job at representing the physical world when we are able to force the flow into a single phase for short periods of time (e.g., right after a pig run, a gathering line will be single-phase for dozens of seconds). Multiphase flow is a totally different kettle of fish. First off, fluids are unable to withstand applied stresses, so when a line fills up with liquid with gas pressure increasing behind it, the liquid inertia will only be able to withstand that build-up in energy for a few milliseconds before it breaks free. This leads us to the observation

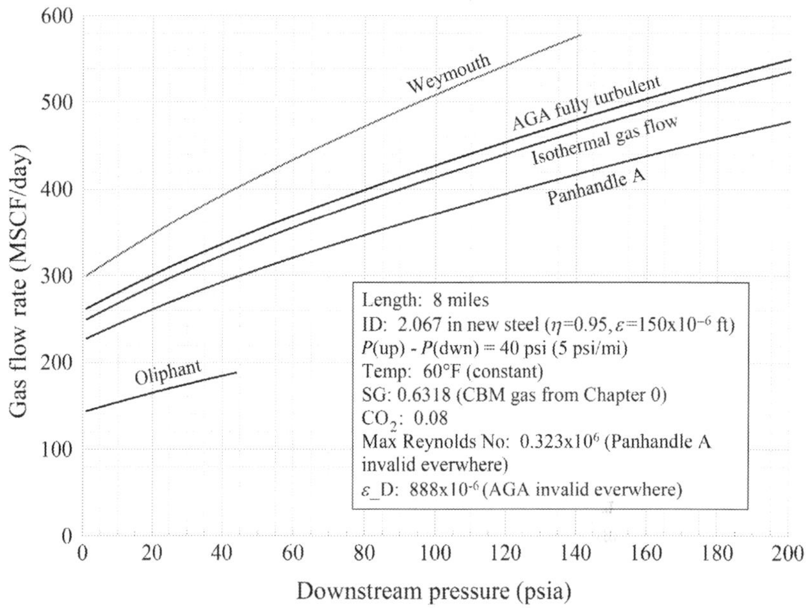

Figure 4.7 Comparison of flow equations.

that multiphase flow is inherently unstable relative to time, and no condition can exist for more than a very brief instant.

We talked about multiphase vertical flow in Chapter 3, Well Dynamics, without elaborating on transitions and types of flows. The representations in Figs. 3.14, 3.17, and 3.18 would have been more accurate if shown as in Fig. 4.8 where the "flow regime" changes from inch to inch and from second to second as you move up the tubing. Envisioning vertical multiphase flow is considerably easier than horizontal multiphase flow since gravity is operating counter to the bulk flow and tends to make liquid flow require higher energy than gas flow, and the liquid will exhibit a tendency to "run out of petrol" and fall back down.

"Flow regime" has historically been demonstrated on a map like Fig. 4.9. The axes on these maps have been "superficial gas velocity" and "superficial liquid velocity" (i.e., the velocity that the given volume flow rate of the particular fluid species would exhibit if that species were the only thing in the pipe). Modern computer models have demonstrated that flow regime has less to do with superficial velocity than

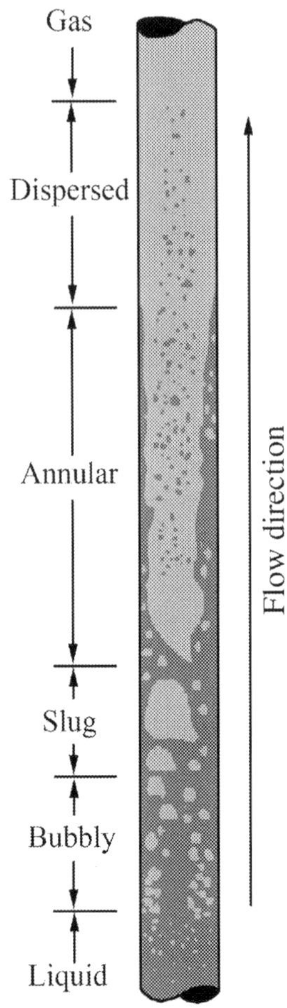

Figure 4.8 Vertical multiphase flow.

we previously thought and have shifted the axis to a "mass flux" (i.e., mass flow rate divided by cross-sectional area of the pipe), or "superficial momentum" whatever that means. Researchers in this area have also moved away from a discrete boundary between the various flow regimes in favor of a transition region. Even with these refinements, the maps are far more useful as visualization aides rather than computation aides—in other words the numbers on the *x*- and *y*-axis are mostly useless and can be misleading. Looking at Fig. 4.9 if we start with a fairly energetic liquid and a moderately energetic gas, say in the dead center of the "dispersal" area and flow the fluid down the line.

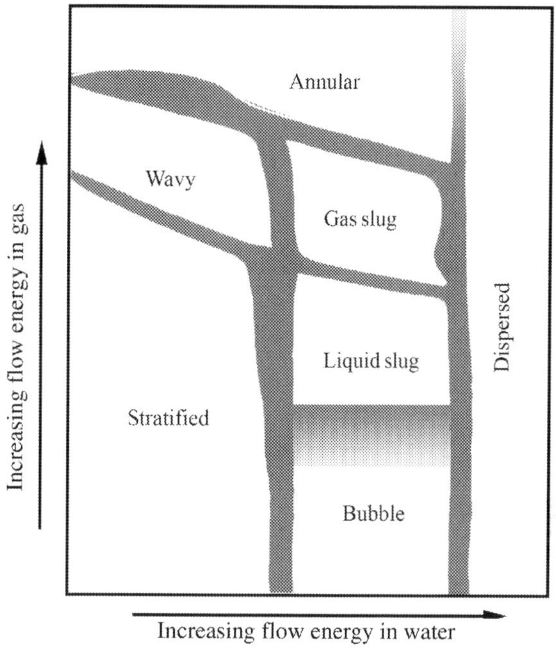

Figure 4.9 Flow regime map.

The tendency is for interfacial effects and friction to lower the liquid energy first, which causes the dispersed liquid droplets to fall out and accumulate in the bottom of the pipe. Shortly, the accumulated liquid builds up and momentarily increases the energy in the gas behind the accumulated liquid and it "launches" the slug down the line at high velocity while dropping the energy in the gas, and the flow moves to either the higher-energy portion of the "stratified" region or into the Wavy region. Annular flow only occurs with very high gas flow energy with low to moderate liquid energy, and is generally very short-lived since it maximizes both fluid friction and interfacial energy transfer.

4.3.4.1 Calculations with horizontal multiphase flow

Computational fluid dynamics (CFD) is the darling of early 21st century researchers. I can't remember the last time I saw a doctoral thesis that did not have a CFD model as a center point. When I hired scientists to analyze a device I was trying to patent, the outcome was a CFD model. The problem is that the underlying arithmetic is far from robust. The models are generally unable to change state as often or as dramatically as happens within the pipe. I reviewed one doctoral thesis that went on for hundreds of pages about a multiphase CFD model that the author purported to represent a breakthrough in fluid measurement. Somewhere around page 250, there was a statement that the ability of the model to match measured data was consistently $\pm 60\%$ normalized for measured data. This means that the CFD was not quite as useful as a random number generator. There are some very expensive CFD programs on the market and they are actually useful in helping understand the strengths and weaknesses of a design in a qualitative manner, but if you use the quantitative data from one of them you will reach the wrong conclusion more often than the right conclusion.

The *GPSA Engineering Data Book* (GPSA) has a section on closed-form multiphase flow equations that ends up doing about the same quality of calculation as a very expensive CFD model for a lot less money. One process to put numbers on multiphase flow is as follows:

- Calculate the pressure drop due to gas–liquid interactions using the *Duckler method* (Eq. (4.29)).
- Calculate the pressure drop due to elevation changes using the *Flannigan method* (Eq. (4.30)).
- Calculate the pressure drop due to gas flow using the *isothermal gas flow equation* (Eq. (4.18) or Eq. (4.19) depending on what data is available) assuming the liquid is absent.
- Add the three pressure drops together.

Using this method you should expect to match field data $\pm 20\%$, about 75% of the time. That is considerably better than people have done with CFD.

Liquid holdup. Liquid holdup is the traditional method for assessing a total stream effective density to a multiphase mixture:

$$q_{gasACFd} = q_{mscfd} \cdot \frac{1000 \cdot \mathrm{SCF}}{\mathrm{MSCF}} \cdot \frac{\rho_{gasSTD}}{\rho_{gasAct}} \cdot \frac{\mathrm{ft}^3}{\mathrm{SCF}}$$

$$\rho_{liquid} = \frac{\rho_{pureWater} \cdot \sum \mathrm{SG} \cdot q_{liquid}}{\sum q_{liquid}}$$

$$\lambda = \frac{q_{liquid}}{q_{liquid} + q_{gasACFd}} = \frac{q_{oil} + q_{water} + q_{condensedWater}}{q_{oil} + q_{water} + q_{condensedWater} + q_{gasACFd}} \tag{4.25}$$

Using the liquid holdup (λ, Eq. (4.25)) you can get a first estimate of full stream density:

$$\rho_0 = \rho_{liquid} \cdot \lambda + \rho_{gas} \cdot (1 - \lambda) \tag{4.26}$$

4.3.4.1.1 Duckler Method

The Duckler method of determining fluid velocity starts with an odd combination of parameters:

$$\begin{aligned}
v_{gasSuperficial} &= \frac{q_{gasACFd}}{A_{pipe}} \\
\mu_0 &= \mu_{liquid} \cdot \lambda + \mu_{gas} \cdot (1 - \lambda) \\
Re &= \frac{\rho_0 \cdot v_{gasSuperficial} \cdot \mathrm{ID}_{pipe}}{\mu_0} \\
H_{liqD} &= \lambda \cdot \left(1 + \frac{Re}{250,000}\right) \\
\rho_k &= \frac{\rho_{liq} \cdot \lambda^2}{H_{liqD}} + \frac{\rho_{gas} \cdot (1 - \lambda)^2}{1 - H_{liqD}}
\end{aligned} \tag{4.27}$$

With ρ_k you can calculate a full stream velocity:

$$\begin{aligned}
\dot{m}_{gas} &= q_{mscfd} \cdot \rho_{gasSTD} \\
\dot{m}_{liq} &= \rho_{pureWater} \cdot \sum (\mathrm{SG}_{liquid} \cdot q_{liquid}) \\
\dot{m}_{fullStream} &= \dot{m}_{gas} + \dot{m}_{gas} \\
q_{fullStreamACF} &= \frac{\dot{m}_{fullStream}}{\rho_k} \\
v_{fullStream} &= \frac{q_{fullStreamACF}}{A_{pipe}}
\end{aligned} \tag{4.28}$$

Now you can get a pressure drop using Duckler's empirical equations:

$$\begin{aligned}
Re_y &= \frac{\rho_k \cdot v_{fullStream} \cdot \mathrm{ID}_{pipe}}{\mu_0} \\
f_{Duckler} &= 0.0056 + 0.5 \cdot Re_y^{-0.32} \\
y &= -\ln(\lambda) \\
f_{tpr} &= 1 + \frac{y}{1.281 - 0.478 \cdot y + 0.444 \cdot y^2 - 0.094 \cdot y^3 + 0.00843 \cdot y^4} \\
dP_{Duckler} &= \frac{f_{duckler} \cdot f_{tpr} \cdot \rho_k \cdot v_{fullStream}{}^2 \cdot L_{pipeFt}}{\mathrm{ID}_{pipeFt}}
\end{aligned} \tag{4.29}$$

4.3.4.1.2 Flannigan method

The Flannigan method is based on the lines not being liquid-full (i.e., there is no pressure recovery on the downhill slopes):

$$\begin{aligned} h_e &= \sum \Delta h_{upHillSlope} \\ H_{Flannigan} &= \frac{1}{1 + 0.3264 \cdot v_{superficialGas}^{1.006}} \\ dP_{Flannigan} &= \rho_{liquid} \cdot h_e \cdot g \cdot H_{Flannigan} \end{aligned} \tag{4.30}$$

4.4 CORROSION

Corrosion is the wastage of material by the chemical action of the environment. It does not include mechanisms such as erosion or wear, which are both mechanical. I have been a member of the National Association of Corrosion Engineers (NACE) for many years and I study their publications carefully to find out if the state of corrosion research has improved, it hasn't. When original research is done (fairly rarely), it tends to be structured to maximize corrosion rates (e.g., very high temperatures and the corrosion element under consideration pumped up to levels not seen in industry) instead of developing tools to allow users to assess the corrosion risk of real flows at real field conditions.

4.4.1 Erosion

While erosion is not the same as corrosion, the results of erosion are similar—the pipe integrity can be compromised. For liquid flow, the API has published a maximum velocity to prevent erosion in API RP 14E (*Recommended Practice for Design and Installation of Offshore Production Platform Piping Systems*) and while the scope of this document is limited to offshore, the pipe we have onshore is similar even though the impact of a failure is assumed by the industry to be lower onshore we can still use the equation (ρ_0 comes from Eq. (4.26) and must be in lbm/ft^3).

$$v_e = \frac{c_e}{\sqrt{\rho_0}} \tag{4.31}$$

The constant c_e is application specific:

- Solids-free fluids in continuous service → 100 ft/s
- Solids-free fluids in intermittent service → 125 ft/s

- Solids-free fluids with corrosion controlled in continuous service → 150 ft/s
- Solids-free fluids with corrosion controlled in intermittent service → 200 ft/s.

If solids are expected, then lower values should be used based on experience. Eq. (4.31) blows up for densities less than about 20 lbm/ft^3 and should never be used for flow that is primarily gas.

Erosion in gas flows requires velocities approaching 1.0 Mach, and jets from holes in pipelines can do considerable damage, but if you have a low probability of exposing piping, vessels, and fixtures to a choked jet then it is not a consideration and do not use Eq. (4.31) to justify unreasonably restrictive velocities.

4.4.2 Common corrosion modalities

The corrosion that concerns us is an anode/cathode reaction where an anode (something like steel pipe) has a lower electrical potential than a cathode and gives up electrons to the cathode. This reaction requires:

- anode,
- cathode,
- electrolyte.

In most cases the anode is the steel pipe, the cathode is a substance within standing water, and the standing water is the electrolyte.

4.4.2.1 Microbiologically influenced corrosion

MIC failures in piping occur more often in Oil & Gas operations than all other corrosion modalities combined. Every single water sample that I've ever had tested for microbial activity has swarmed with both aerobic and anaerobic bacteria. The primary culprits in MIC are:

- sulfur-reducing bacteria (SRB),
- acid-producing bacteria (APB).

In both cases the bacteria consume something in their environment and the biological process results in waste that is a strong cathode. It is very common in the world to see co-colonies of APB and SRB. In addition to a very aggressive cathode, the APB secrete a biofilm that tends to protect the colonies from external chemical attack, but the APB themselves can't live in the environment that they create and they generally die off fairly rapidly, creating an ideal environment for SRB to set up shop. Fig. 4.10 shows a distressingly common result of these co-colonies. All over that picture you see the straight-walled pits that are the characteristic

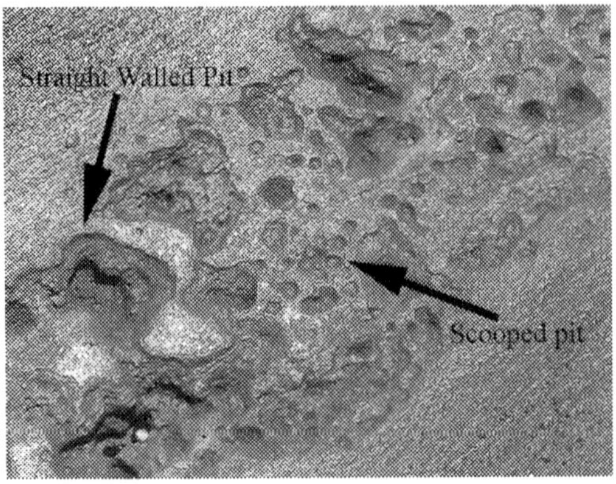

Figure 4.10 MIC corrosion.

look of APB attack. You also see the "ice cream scoop" pits that are the characteristic look of SRB attack. This failure was interesting in that it was in the middle of an 8-in (200 DN) ERW ("electric resistance weld" pipe with a longitudinal weld seam) flow line. When we cut out the leak we inspected the pipe on either side and it was also damaged. We cut out pipe until we found clean pipe, and both upstream and downstream pipe joints were pristine. We sent a sample to a testing laboratory and the failure happened on the longitudinal weld seam which happened to be at the 6 o'clock position. The other joints had the weld seam off the bottom. After that experience, my design documents have specified that the longitudinal weld seam on ERW pipe must be located above the plane of the centerline (i.e., located somewhere before 3 o'clock or after 9 o'clock to use the clock analogy).

4.4.2.2 CO_2 corrosion

More dumb decisions are made in upstream Oil & Gas operations through an unreasoning fear of CO_2 corrosion than any other single concept. You can see stainless steel wellheads because "there is 5% CO_2 somewhere in the field" and high chrome tubing and separators to prevent corrosion. CO_2 is a factor in two different kinds of corrosion (Fig. 4.11):

- Mesa attack: Areas of the affected pipe are intact, surrounded by steep-walled pits over a wide area. This typically happens at pH from 5.0 to 6.5.

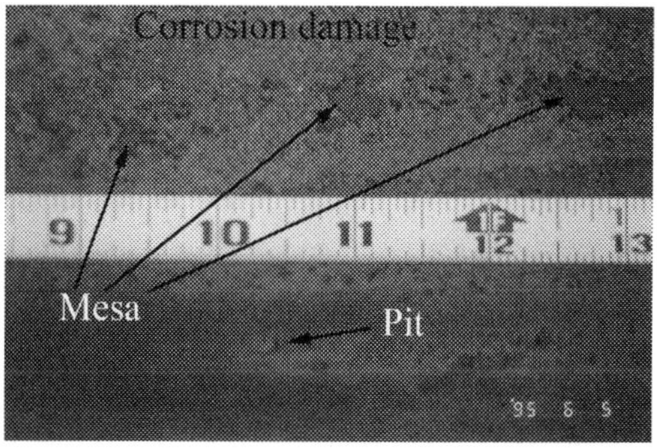

Figure 4.11 CO_2 corrosion.

- Acid attack: Deep localized pitting without evident general corrosion. This typically happens at pH less than 5.5.

Both of these corrosion modalities are aggressive and can cause pipe to fail fairly rapidly, but they only happen in acidic pH. The CO_2 has been in the formation for millions of years. All of the CO_2 that is going to dissolve in water and become carbonic acid had already reacted to its environment before the dinosaurs died off. What we see in unconventional wells is that the produced water is decidedly basic (the spent carbonic acid has become carbonate and bicarbonate) and is as benign as a fluid can be. The example in Fig. 4.11 is the only one I have ever seen in spite of the fact that I've been asked to investigate dozens of failures purported to have been caused by CO_2, all of them were MIC (the corrosion in Fig. 4.11 was discovered when a piece of a trunk line was removed to tie in another facility and the corrosion was visible in the cutout pup, it was not a failure).

4.4.2.3 H_2S

Hydrogen sulfide has a tendency at pH above 2.0 to form scale in contact with steel instead of creating pits in steel pipe. This scale can be a huge problem in processing facilities and can actually clog lines. Most of the serious corrosion failures we see in upstream due to H_2S have some component of the line with a copper or brass component (i.e., the so-called yellow metals) which pit at higher pH than steel.

4.4.2.4 Oxygen corrosion

In general usage, the terms "corrosion" and "oxidation" are used interchangeably to highlight the role of oxygen in the wastage of metal. Oxygen participates in three flavors of corrosion:

Figure 4.12 Oxygen corrosion.

- Uniform corrosion (rust). This ubiquitous corrosion modality poses approximately zero risk to the integrity of pressure-containing pipes and vessels. The piping in Fig. 4.12 has been exposed to the elements in a semiarid region for nearly 100 years (many of the components have been upgraded and replaced over decades, but the major structures are the material that was installed before World War I) and the wall thickness continues to be adequate for the design task.
- Stress corrosion cracking (SCC). Oxygen can accelerate cracking in carbon steel that is under high stress and in stainless steel. In normal upstream usage, steel piping tends to be limited to fairly mild carbon steel with very low native stresses and operated at pressures distinctly toward the low-stress end of the pressure range. SCC happens, but it is not common.
- Galvanic pitting. Dissolved oxygen can create cathodic cells that can cause aggressive pitting. This galvanic action requires at least 450 ppm (0.045%) oxygen in standing water. To sustain 450 ppm dissolved in water requires something on the order of 2000 ppm (0.2%) O_2 in the gaseous stream at 145 psig (1000 kPag). At 50 psig (345 kPag) the gaseous O_2 would need to be over 14,000 ppm (1.4%) to reach 450 ppm dissolved.

Historically, third-party gas gathering agreements have set the limit for free oxygen above 2000 ppm (0.2%) in the gas stream. A widely published document from a well-known testing laboratory (I will not reference the specific paper because I don't want to imply that anyone should read it) did a series of "experiments" including one that had a 50/50 mix of CO_2 and O_2 bubbling through water with a pH of 6.0 and they saw accelerated corrosion in very low carbon-content steel. This spurred them to recommend a limit for oxygen content in natural gas streams of 10 ppm

(0.001%). There is no attempt to link the 500,000 ppm in their test to the 10 ppm in their recommendation within the report.

At about the time of the publication of the report, many third-party gas gathering companies were decommissioning well-site dehydrators as "inefficient" (discussed at length in Chapter 9: Interface to Plants) and saw the potential of liquid water accumulation in their lines from the condensation of water vapor and they were not prepared to deal with it properly, so they cynically began requiring contract revisions to the new 10 ppm limitation in the hopes that keeping oxygen out of their system would inhibit corrosion. They knew full well that it wouldn't, but there was a need for a scapegoat when pipes standing full of water started to fail (likely due to the action of anaerobic bacteria which would die off if oxygen were present). As a facilities engineer, I have been involved in six of these contract renegotiations and been an expert witness in two lawsuits and the same report gets pulled out as "proof" every time. In the end, most companies just accept the new limits and the requisite monitoring and control equipment expense without challenge, those that challenge the arbitrary and unilateral change always win because the gathering companies know that they don't have a leg to stand on.

4.4.2.5 External galvanic corrosion

If there is a single concept that the industry and regulators have embraced with vigor it is protection against external galvanic corrosion. Pipeline regulations around the world all have extensive and detailed laws intended to prevent external galvanic corrosion. These laws were both important and effective when they were devised in the 1950s, but this is an area where technology has far surpassed regulations. External galvanic corrosion implies that groundwater can serve as an electrolyte, the earth itself can serve as an (infinite) cathode, and unprotected steel pipe serves as the (very much finite) anode. The result can be quite rapid wastage of metal and early failure of the pipe (Fig. 4.13). Modern coatings make this type of corrosion rare (see later).

4.4.3 Corrosion control

By far the most effective corrosion-control method in gas flow lines is to remove access to steel by any electrolyte. Everything else that we do to maintain pipe integrity is a Band-aide at best and many of the things we do are actively harmful.

Figure 4.13 External galvanic corrosion.

4.4.3.1 External

The two main methods of combating external corrosion are:

1. Proper pipe coating, properly applied, and properly installed
2. Providing an alternate anode.

Alternate anodes (called "sacrificial anodes") are metals with a lower Gibbs free energy level than carbon steel that will corrode before the steel will. One example of sacrificial anodes is the zinc rods that are installed in residential hot-water heaters. A similar arrangement is common to install zinc components in the vicinity of the propeller on pleasure boats to protect the metal of the propeller.

Providing alternate anodes can be done more effectively by impressing a current on the pipe to make its electrical potential more positive than the environment. This is called "impressed current cathodic protection" and requirements for its installation, monitoring, and maintenance are common in the laws of many countries. Whole industries have grown up around the requirement to monitor and maintain cathodic protection systems.

When the regulations were being developed to require cathodic protection, the quality of external pipe coatings was quite poor and the cathodic protection system required considerable current flow to protect the pipe at the points where failed coating left the steel in direct contact with damp soil. As coatings have improved year-on-year, the amperage requirements have dropped dramatically to where it is common today to have current flow too low to measure on historical instruments. With 21st century coatings, disbondment of the coating from the pipe has become rare, and the new material is so tough that field handling rarely creates "holidays" in the coating field (i.e., points where the pipe is uncoated due to damage). Coatings have gotten so good that many facilities engineers have never seen external corrosion on buried piping. High-quality pipe coating, properly applied, and properly installed is a far better protection against external galvanic corrosion than impressed cathodic protection system, but the laws will likely never accept that fact.

Before you get the idea that all impressed cathodic protection systems are junk, let me say that downhole tubular goods are completely uncoated, and any groundwater will create corrosion cells. Not providing, monitoring, and maintaining well-bore cathodic protection is irresponsible.

4.4.3.2 Internal

The first time I assumed responsibility for operating a gas gathering system, our budget for corrosion chemicals was over $20k/month. I canceled that contract and stopped injecting any corrosion-control chemicals. Five years later, that system had lower incidence of corrosion than any similar sized gathering system in that field. Injecting corrosion chemicals into a fluid requires that the fluid disperse the chemicals to the locations where corrosion cells might form. With a liquid system, the process fluid has adequate density to transport the corrosion chemicals throughout the system. Not so with gas. When you inject a liquid into a gas (even as an aerosol), the droplets start bashing into each other and/or the pipe walls immediately. Every collision causes the droplets to get bigger, and the effect of gravity on them will increase their size and mass until they fall to the bottom of the pipe. Shortly after I took over the gathering system we had to cut open a line to tie in another well. The line being cut had been treated with biocide. The job was within sight of the biocide injection point, and when we cut the pipe we filled the bottom of the ditch with pure biological poison and had to evacuate for several hours while we sucked the poison out of the ditch and dug up considerable dirt for disposal away from people. This is common, injected chemicals are rarely seen more than a dozen pipe joints downstream of the injection point.

Corrosion-control chemicals in gas lines are ineffective (no transport mechanism), dangerous (many are active poisons and all can create chronic health problems), and expensive. When I see them in my client's lines in my consulting business I always recommend ceasing the chemical injection immediately.

If injecting chemicals is a bad idea, what can you do? First, every steel line in a system needs to be piggable. There are inexpensive ways to accomplish this (such as "pigging valves" discussed in Chapter 6: Gas Gathering Systems), but even stick-built launchers and receivers cost less than injecting chemicals and fixing leaks. Not all lines need frequent pigging, but all lines need the ability to be pigged to eliminate standing liquids (and the electrolyte they represent). The pigging schedule needs to be adapted to the requirements of the particular gathering system. On one system, a pipeline model was used to adjust pipe efficiency on each line independently. A complete set of pressure data was collected once a quarter and fed the new flow rates and pressures into the model and tweaked the efficiency values until the model matched the field data within very close tolerances. For each line there was one of three possible outcomes:

- The efficiency did not change from the previous quarter → leave the pigging schedule on that line the same for the next quarter.
- The efficiency got higher than the previous quarter → if it remains the same or improves again the next quarter then reduce the pigging frequency.
- The efficiency got worse than the previous quarter → increase the pigging frequency.

This schedule was in place for 7 years and the next engineer left the schedule the way that was calculated the last quarter. They had their first corrosion failure 3 years after the schedule became stagnant and corrosion failures became a quarterly event after that.

4.4.4 Flow through a hole

Often the first indication of a line failure is a very noisy "ant hill" structure above the line (Fig. 4.14). The gathering line failure in Fig. 4.14 was a result of MIC, and the sum of the 15 holes through the pipe had a cross-sectional area of less than the flow area of a 1-in (25 DN) pipe. There is no way to know how long the line was leaking before it acquired enough energy to blow a path from the bottom of the line 4 ft (1.22 m) below the surface, but the amount of natural gas stored in the surrounding soil

Figure 4.14 Indication of a line failure.

indicates it was probably weeks. Once the accumulated gas was at a pressure greater than the limiting pressure for choked flow, the rate of failure of the soil accelerated.

Part of the corrective action for any line leak is to assess how much gas leaked at that site. The primary tool is comparing well-site gas measurement to off-system gas measurement, but that number contains all of the sins of the line and often significantly overstates the leak.

Calculating the flow is often more reasonable. The flow can be expected to have a period of compressible flow followed by a period of incompressible flow. The compressible-flow portion can be assessed by:

$$q_{comprMSCFd} = \frac{C_d \cdot A_{open} \cdot P_1 \cdot C_{unit}}{\rho_{std}} \cdot \sqrt{\left(\frac{SG \cdot k}{R_{air} \cdot T_1 \cdot Z_1}\right) \cdot \left(\frac{2}{k+1}\right)^{\frac{k+1}{k-1}}} \tag{4.32}$$

When velocity drops to below 0.6 Mach, you have to go to an incompressible-flow model:

$$q_{incomprMSCFd} = \frac{C_d \cdot A_{open} \cdot P_1 \cdot C_{unit}}{\rho_{std}} \cdot \sqrt{\left(\frac{SG \cdot k}{R_{air} \cdot T_1 \cdot Z_1}\right) \cdot \left(\frac{2}{k+1}\right)^{\frac{k+1}{k-1}} \cdot \left[\left(\frac{P_2}{P_1}\right)^{2/k} - \left(\frac{P_2}{P_1}\right)^{(k+1)/k}\right]} \tag{4.33}$$

4.4.5 Corrosion prediction

The ultimate quest for corrosion engineers is to be able to predict corrosion in real flows in varying conditions. Researchers are not providing much help in that quest. Virtually every research study contains pure liquids and limited gas species to try to get an understanding of how a specific corrosion modality is impacted by a specific environment. If you get enough of these one-off relationships, the researchers feel that they can be added together to reach a real-world prediction. This approach disregards species interactions that may accelerate or inhibit corrosion reactions, but researchers don't have an alternative so they use this one.

Prediction methods like the DeWaard–Milliams nomograph for CO_2 have not done a very good job of predicting corrosion rates. The newer computer programs based on recent research don't look to be doing much better. Neither one is actually as good as a bald faced guess.

For many years people have used corrosion coupons to assess the current rate of corrosion in their actual systems. These coupons are slabs of steel that are designed to go in and out of a lubricator built into the pipeline. The slabs are carefully weighed and measured before installation and again at postinstallation. Loss of mass for a known volume of metal yields the volume loss or the rate of wastage. Sounds good, except we always place the coupons in locations that are accessible from the road, not necessarily at locations with standing water. I have never seen coupon results that led to a decision that avoided future corrosion failures.

My approach is to pig the water from the flow lines which lowers the corrosion failure rate to approximately zero.

4.4.6 Corrosion summary

For upstream gas piping corrosion evaluation and mitigation is fairly straight forward—install the pipe with good coatings and pig it regularly. Table 4.5 summarizes my views on managing it.

Table 4.5 Corrosion summary

	Requirements	Frequency	Consequence	Mitigation
SCC	Catalyst, high stress	Rare	High	Avoid high-strength or high chrome steel
Uniform	Catalyst, electrolyte	Common	Low	Paint, coatings
MIC	Electrolyte, microbes	Common	High	Pigging
Galvanic (external)	Cathode, electrolyte	Moderate	High	Coatings, cathodic protection
Other galvanic (internal)	Cathode, electrolyte	Rare	High	Pigging

4.5 PURGING AIR FROM GAS LINES

Improper purging of flow lines is the number 2 cause of fatalities in Oil & Gas around the world (of course the number 1 is driving vehicles on public roads). Nearly everyone who has been in the industry for a few years has seen a case study from their own company of a purge-related fatality, and far too many of us knew people killed in purging incidents. The reaction of the safety community to purging is very odd. Few safety manuals cover it. No regulations cover it. Few engineers consider it in their designs. Every field tech thinks that she understands it, and she is generally wrong.

The danger in bringing a line back into service after opening it for work is that somewhere in the system you will form an explosive mixture of air and natural gas—two legs of the fire triangle waiting for the ignition source to blow up. As you raise the pressure in the line, the explosive range of the mixture also increases (Figs. 2.7 and 4.15) so relying on the mixture being "too rich to burn" becomes progressively less effective as pressures increase.

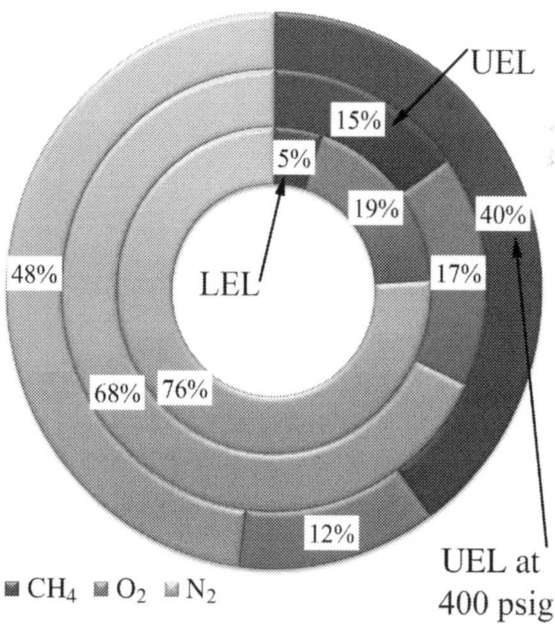

Figure 4.15 LEL for methane.

Nearly everyone has inserted a flowing garden hose into a swimming pool and noticed that the high-velocity stream does not mix with the pool water until it has lost enough energy pushing the pool water out of the way to approximate the same velocity (zero) as the pool water. This is true (although more difficult to see) for any two fluids.

- A stream flowing at a laminar Reynolds number will mix with a static volume or another laminar stream.
- A stream flowing at a turbulent Reynolds number will not mix with a static volume, laminar stream, or another turbulent stream that has a significantly different bulk velocity (something like 20% difference or more than about 30-degree difference in angle will prevent mixing).
- A sonic stream will not mix with anything (not even another sonic stream going in the same direction). A sonic stream will act like a piston in a pipe, compressing the gas in front of it until the pressure downstream of the shock wave increases to above the critical pressure for choked flow.

Just like the internal process within a diesel engine, blowing a sonic stream into static gas volume will tend to heat the gas compressed ahead of it. The heat of compression is:

$$T_D = T_S \cdot \left(\frac{P_D}{P_s}\right)^{\frac{k-1}{k}} \tag{4.34}$$

If you have a gas source (like a well-bore, for example) pressurized with methane to 650 psig (4.5 MPag) and 80°F (26.7°C), and an isolated piping system full of air ($k = 1.4$) at atmospheric pressure and you rapidly open a wellhead wing valve to pressurize the piping system, the results can be dramatic. Initial flow is 1.0 Mach, and critical pressure for choked flow is 364 psia (2.507 MPaa) (assume atmospheric pressure at this location is 12 psia (82 kPaa)) so when the well-bore stream slows enough to begin mixing, its temperature would be 971°F (521°C). Slightly below the autoignition temperature of methane (which is 999°F (537°C)) but if mixing were to be delayed to 0.6 Mach (which it will be), the additional compression would certainly exceed the autoignition temperature.

The first time I did this analysis, I thought "we open meter runs all the time and never purge them or slowly pressurize the lines, why have we never blown up a meter run?" I went to one of the wells I was responsible for and looked at the 6-in (150 DN) meter run, and there was clear evidence on the last flange before the block valve that there had been (at least one) fire in the tube, but apparently there was not enough air in the line to support an explosion. Then I visited 20 other 6-in meter runs and all of them showed evidence of the paint having been burned off the pipe. It was very eye opening and I started writing purge procedures for well-sites and meter runs (and taking more care in writing purge procedures for pipeline construction/repair).

The goal of purging into service is to ensure that:

- no mixture is exposed to an ignition source,
- you never form an explosive mixture.

Succeeding at either goal promises a safe start-up, but working toward both helps provide confidence that no error will be fatal.

There are basically three types of purge:

1. Dilution. Introduce enough inert gas to a closed system to ensure that an explosive mixture cannot form, then purge the inert system with process gas at a high enough pressure to ensure that flow out the vent is choked.
2. Displacement. In an open-ended system, introducing inert gas at a turbulent injection rate will tend to push the air out of the system rather than diluting the air. Done at high enough pressure to ensure that the flow out the vent is choked, followed by a clearing purge with process gas.
3. Clearing. Displacing the air with process gas (natural gas) so rapidly that that there is minimal mixing. Clearing purges are done at a high enough pressure to ensure that the flow out the vent is choked.

4.5.1 Dilution purges

For systems that do not have a clear path from an injection point to a vent point, the only option available is a dilution purge.

By the nature of the methane equation of state, slowly raising the pressure in an air-filled line to 150% of local atmospheric pressure will be sufficient to guarantee that you cannot burn or explode the contents of the line. Once the system has been rendered inert, introduce process gas at a slow rate to raise the line pressure high enough to ensure choked flow out the vent and purge three pipe volumes. See later for the technique to determine purge times.

4.5.2 Displacement purges

On the face of it, displacement purges sound like the clear winner until you consider a purge in the Texas City, TX, area in the 1940s.

The operator was purging a 10 mile (16 km) long 12-in (300 DN) diameter ethylene line. Because of the extremely explosive nature of this product, the operator decided to put a pig in the line, follow that with potable water (pumped until water appeared at the pig receiver) and another pig followed up with air. When the second pig passed the electrical isolation kit on the receiver, it created a shunt around the isolation kit and allowed the accumulated static electricity from the pig run to spark hot enough to ignite the fuel–air mixture behind the second pig. The operator never found any of the pipe, it was vaporized over its entire length. While this was an extreme example (typical of the Texas City area, all the best war stories come from there), it made me think that displacement purges require more control than is actually possible, and I don't recommend displacement purges.

If you must do a dilution purge inject the purge gas fast enough to raise the pressure in the system above 14.5 psig (1 barg) and purge three pipe volumes. When the inert-gas purge is complete, use process gas to purge three more pipe volumes at 14.5 psig (1 barg).

4.5.3 Clearing purge

Initially, clearing purges look very risky because you were purposely introducing an explosive gas to an air-filled system. This is the only kind of purges that field techs have access to, and it is the kind of purge done most often by a large margin. At any given instant in time, you can be certain that somewhere in the world someone is doing a clearing purge. Mostly they are still alive at the end of the purge.

To safely perform a clearing purge you need to verify that your source gas is oxygen free, then you need to open a single vent, raise the system pressure above 14.5 psig (1 barg), purge three pipe volumes, shut the vent, and raise the system pressure to nominal.

4.5.4 Determining purge pressure and required time

I have presented this information many times to field techs, engineering societies, and pipeline companies. At one pipeline company, the local manager provided me with their purge procedure and the procedure started with "raise the pressure in the line being purged to 1000 psig." I gasped and used their procedure in my presentation. In my estimation there are three possible pressures to use for a purge with process gas:

- Less than the pressure required to ensure choked flow out the vent
- Approximately the same pressure required to ensure choked flow out the vent
- Too much pressure.

A study in the 1950s showed that it is statistically sound to assume that when you have flowed 2.5 times the pipe volume through the pipe you are assured that the gas in the line has been replaced completely by the purge gas. Typically we round that up to three times, and any published purge procedure will have "three pipe volumes" somewhere in it.

How do you know that you've flowed three pipe volumes? If the pressure in the line is less than the pressure required for choked flow, you don't know, and there is really no reliable way to assess the flow rate. If the pressure is high enough to ensure choked flow then you can use Eq. (4.32) to determine the flow rate quite accurately. Convert the result of Eq. (4.32) to actual volume (at the pressure upstream of the choke) instead of standard volume and continue the purge until you've flowed three times the volume of the pipe (i.e., divide pipe volume times 3 by flow rate at actual conditions). As we've seen, the higher the pressure, the higher the mass flow rate (and the lower the volume flow rate at actual conditions for a given velocity) so raising the pressure increases the required purge time without improving purge efficiency at all. Purging at 1000 psig requires increasing the purge time by a factor of 35 over the required purge time using the minimum pressure required for choked flow.

The minimum line pressure that will ensure choked flow is:

$$P_{purgeMin} = P_{atm} \cdot \left(\frac{2}{k+1}\right)^{\frac{k}{1-k}} \tag{4.35}$$

At most elevations this number works out to around 12 psig (82 kPag) (10 psig (67.9 kPag) at the elevation of Denver, CO, or Windhoek, Namibia). Procedures I write use 15 psig or 100 kPag to get to values that people can actually read on a field gauge.

4.5.5 Purge conclusion

I've reviewed many purge procedures and too many of them show a remarkable lack of knowledge while some of them are downright dangerous. One technique in frequent use is to sneak up on a purge by pressuring the line to some lower pressure, blow it down, pressure it to a slightly higher pressure, blow it down, and repeat three to five times. This process is very wasteful both of purge gas and people's time with no real benefit.

Another process that is in common use (primarily in plants) is to require an operator to hold an LEL meter in the exhaust stream until it registers maximum. There are so many problems with this. First, the vent is usually not very accessible to the operator and it is not uncommon to see them standing on the roof of a pickup truck to access it. Second, LEL meters are calibrated in percent of the lower explosive limit—when your LEL meter says "100%," it means "100% of 5%" and at its maximum reading the flow stream is very explosive. Finally, the LEL meter is intended for personnel protection, not process evaluation, and is designed to work in a static gas volume; sticking it in a choked stream simply floods the sensors and turns it into a random number generator, some plants get around this by requiring a bag to be placed over the vent and testing the exhaust of the bag, generally resulting the bag being shredded or launched into outer space. There is no instrumentation that can tell you that a purge should be complete.

Steps to ensure a safe and effective purge are as follows:

- Ensure that the source gas is oxygen free.
- Make sure that every valve in the purge path can be operated.
- Ensure that automated valves are disabled (you don't want an automated valve to slam shut in the middle of a purge because it now has pneumatic pressure to satisfy the last command to close).
- Raise pressure above pressure required to provide choked flow, typically 15 psig or 100 kPag.
- Open only one vent (multiple vents can short cycle the purge and suck in air).
- Purge three pipe volumes (if the pressure creeps up during the purge, increase the purge time 1 minute for every psi (10 kPa) increase).

Using these steps will ensure that you don't cause the fatality statistics to increase.

REFERENCES

Fanning, J.T. (1896). A practical treatise on hydraulic and water-supply engineering. D. Van Nostrand, ISBN 978-5-87581-042-8.

GPSA. Engineering Data Book, twelfth edition, Gas Processors Association, 2004.

Wikipedia, 1. Darcy–Weisbach equation. Retrieved Octoby 4, 2016, from https://en.wikipedia.org/wiki/Darcy%E2%80%93Weisbach_equation.

FURTHER READING

Simpson 1. "Producing Coalbed Methane at High Rates at Low Pressures", SPE 84509, David Simpson, Mike Kutas, SPE Annual Technical Conference and Exhibition, Denver, CO, 5–8 October 2003, Journal of Petroleum Technology.

Webster. (n.d.). Retrieved April 21, 2016, from http://www.merriam-webster.com/dictionary/biogenic (other dictionary references replace "biogenic" with the referenced term).

NOMENCLATURE

Symbol	Name	fps units	SI units
A	Flow area	ft^2	m^2
BHP	Energy requirement	hp	W
C_d	Discharge coefficient (square edged hole = 0.8, ragged hole = 0.72, ductile failure = 1.0)	decimal	decimal
c_e	Erosion constant	ft/s	n/a
c_p	Flow constant (well specific)	MSCF/day	kSCm/day
$C_{pumping}$	Unit conversion factor	1714	1
C_{unit}	Unit conversion	490	86,400
d_{eff}	Effective diameter. For flow inside a pipe it is inside diameter, for annular flow it is effective diameter	in	mm
dP	Differential pressure	psi	kPa
f_f	Fanning friction factor (f_m/4)	decimal	decimal
f_m	Moody friction factor	decimal	decimal
$f_{vortexShedding}$	The frequency that von Karman Streets are shed around a bluff body in flow	decimal	decimal
g	Gravitational constant	32.174 ft/s^2	9.81 m/s^2
g_c	Unit converter	32.174 ft-lbm/s^2-lbf	N/A (for now)
h	Height	ft	m
$H^\star$	Enthalpy divided by reference enthalpy	decimal	decimal
k	Ratio of specific heats	decimal	decimal
ID	Inside diameter	in	mm
ID_{Darcy}	ID used in Darcy–Weisbach	ft	m

(Continued)

(Continued)

Symbol	Name	fps units	SI units
L_{miles}	Length of pipe for gas flow	mi	N/A
L_{Darcy}	Length of pipe for Darcy–Weisbach	ft	m
$L_{characteristic}$	Some length related to an analysis, can be OD, length, hydraulic diameter, etc.	ft	m
m	Mass (not used without subscript in any equation with explicit units)	lbm	kg
$\dot{m}$	Mass flow rate	lbm/h	kg/h
MW_{air}	Molecular weight of air	28.962 lb/lb-mole	28.962 gm/gm-mole
P	Pressure	psia	kPaa
P_{atm}	Atmospheric pressure	psia	kPaa
$P_{imposed}$	Imposed pressure	psig	kPag
OD	Outside diameter	in	mm
q	Gas flow rate	MSCF/day	kSCm/day
$q_{comprMSCFd}$	Gas flow rate during compressible flow	MSCF/day	kSCm/day
q_{Darcy}	Fluid flow rate	ft^3/s	m^3/s
$q_{incomprMSCFd}$	Fluid flow rate	MSCF/day	kSCm/day
q_{liquid}	Liquid flow rate	gpm	L/s
$\overline{R}$	Universal gas constant	1545.3 ft*lbf/R/lb-mole	8.314 J/K/mole
R_{air}	Specific gas constant for air ($\overline{R}/MW_{air}$)	53.355 ft × lbf/R/lbm	287.068 J/K/kg
R_{gas}	Specific gas constant	R_{air}/SG_{gas}	R_{air}/SG_{gas}
Re	Reynolds number	decimal	decimal
ΔS^*	Change in dimensionless entropy compared to reference entropy	decimal	decimal
SG	Specific gravity	none	none
T	Temperature	Rankine	Kelvin
v	Velocity	ft/s	m/s
V	Volume	ft^3	m^3
W	Weight	lbm	kg
Z	Compressibility	decimal	decimal
Δ	Change	none	none
ε	Efficiency	decimal	decimal
ε	Absolute pipe roughness	ft	m
ε_D	Relative pipe roughness	decimal	decimal

(Continued)

(Continued)

Symbol	Name	fps units	SI units
μ	Viscosity	lbm/ft/s	mPa/s
ρ	Density	lbm/ft^3	kg/m^3
η	Efficiency	decimal	decimal
λ	Liquid Hold Up	decimal	decimal
Subscripts			
1	Upstream value		
2	Downstream value		
avg	Average		
disch	Condition at the discharge of a pump or compressor		
gas	Gas		
liq	Liquid		
std	Standard condition		
suct	Condition at the suction of a pump or compressor		
tbg	Tubing		

EXERCISES

1. Determine the amount of work required to pump the required 1.01 SG water volume from Table 4.6 for each of the pipe sizes and materials in the table.
2. For a 3 mile long RTP pipeline with a 2.0 in (50.8 mm) ID and an absolute roughness of 50×10^{-6} ft (15×10^{-6} m) flowing 480 MSCF/day (13.6 kSCm) of natural gas with an upstream pressure of 533 psig (3.67 MPag) and a downstream pressure of 493 psig (3.40 MPag) at 60°F (15.6°C), which closed-form equation would be

Table 4.6 Exercise 1 data

	fps	Both	SI
Elevation ASL	5588 ft		1703 m
q	550 bbl/day		87.4 m^3/day
Length	7 miles		11.27 km
h_{max} (highest point on system)	+130 ft		+39.62 m
Δh (start vs end)	−256 ft		−78.03 m
Σh_{up} (sum uphill runs)	446 ft		136 m
Pipe 1 ID and material	1.049 in	Steel	26.64 mm
Pipe 2 ID and material	2.96 in	FiberSpar (FRP)	75.18 mm

most appropriate for these conditions? What flow rate does that equation predict?

3. Determine the Reynolds number for the following (state any required assumptions):
   - Gas → natural gas
   - Flow rate → 480 MSCF/day (65.1 SCm)
   - Pipe ID → 2.0 in (50.8 mm)
   - Pressure → 525 psia (3620 kPaa)
   - Temperature → 60°F (15.6°C)
4. For the pipe in #2 and the Reynolds number from #3, determine a friction factor.
5. Would the flow in #2 be appropriate for a calculation based on Bernoulli's equation? Why or why not?
6. A system purge that uses only process gas is called a ________ purge.

CHAPTER FIVE

Well-Site Equipment

Simpson's Third Postulate: Any process or procedure that inhibits achieving meaningful assessment of well site conditions will reduce the profitability of the reservoir

Contents

Practical Onshore Gas Field Engineering.
DOI: http://dx.doi.org/10.1016/B978-0-12-813022-3.00005-5

5.1 INTRODUCTION

Onshore gas well equipment and piping selection and configuration is a subject that our industry has always treated as "easy" and it has historically been rare for engineers to be involved in these activities at all. Unconventional gas has made in-roads into that mindset, but it still seems to be difficult for the industry to accept that applying engineering expertise to surface equipment can improve the performance of a reservoir.

For example, there was a coalbed methane (CBM) gathering system in a field where wells had been (prior to CBM development) expected to make less than 500 MSCF/day (14 kSCm/day) and all well-site facilities were specified by the field foreman without engineering input. The CBM gathering system had been built for the same rates, but CBM production rates were far higher.

One well was especially telling:

- Production rate: 28 MMSCF/day (793 kSCm/day) of gas and 340 bbl/day (54 m^3/day) of water
- Compression: Two 500 hp (373 kW) oil-flooded screw compressors with 4 psig (24.6 kPag) suction pressure
- Pipe from wellhead to separator: 90 ft (27 m) of 12 in (300 DN) pipe
- Separator: 90 in × 25 ft (2.28 m × 7.62 m) vertical two phase
- Pipe from separator to compressor: 110 ft (34 m) of 12 in (300 DN) pipe

This volume of fluids at this compressor suction pressure should result in a wellhead pressure about 6 psig (41.4 kPag), but field measurements showed that separator pressure was 47 psig (324 kPag) and wellhead pressure of 86 psig (593 kPag). The lease operator was at a loss for how to fix this problem. Looking at the flows it is clear that the observed pressure drops could be caused by the velocities indicated in Fig. 5.1 along with the line being nearly full at the riser into the separator (Fig. 5.2), but that when the operator tried to blow the water out of the line he was only able to lower the liquid inventory a small amount before the velocity dropped too low to effectively shift the liquid.

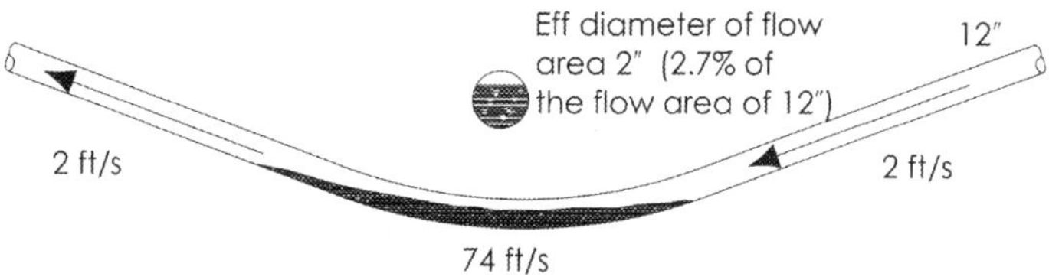

Figure 5.1 Standing water in sag.

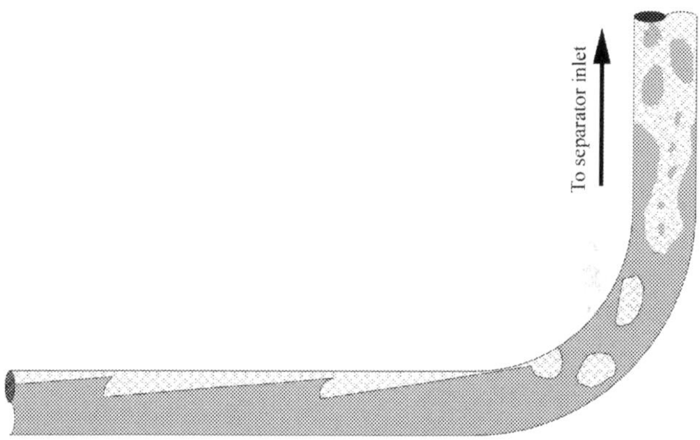

Figure 5.2 Full line.

This problem was completely corrected by replacing the 12 in (300 DN) line from the wellhead with a 6-in (150 DN) line from one casing wing valve to a 30 in × 10 ft (0.76 m × 3 m) separator; and a 4-in (100 DN) line from the other wing valve with a 3 in (75 DN) line from the tubing wing valve both going to a second 30 in × 10 ft (0.76 m × 3 m) separator; and a 6 in (150 DN) line from each separator to the compressor suction header. This design reduced the flow area from the wellhead to the separator by 58%, separation area by 78%, and the flow area from the separator to the compressor by 50%. Huge reductions in flow area. The end result was that wellhead pressure dropped from 86 psig (593 kPag) to 9 psig (62 kPag), but the well would only yield 23 MMSCF/day (651 kSCm/day).

This design allowed the operator to send the entire well down a single line (6, 4, or 3 in) to blow liquids from the line, but reduced gas production. Further analysis suggested that the lower wellhead pressure had impacted the flow efficiency of the wellbore tubulars and the near-wellbore portion of the reservoir. Compressor suction was raised to 50 psig (345 kPag), resulting in wellhead pressure increasing to 54 psig (372 kPag) and the flow rate steadied out at 29.5 MMSCF/day (835 kSCm/day).

Certainly, not every onshore gas well-site design *always* needs detailed engineering analysis, but it is quite certain that *never* having this engineering analysis has resulted in many suboptimum outcomes.

5.2 PIPING DESIGN CODE

The difference that engineering practices bring to well-site equipment selection and construction is "an objective standard" to measure the design against. Without this objective standard, there really is not any way to determine if a design is fit for purpose. There is no way to evaluate the design to see if it is strong enough or if it is too strong (and therefore more expensive than it needs to be), to assess whether it is adequate for the expected fluids or whether it is simply too large for the expected fluids creating operational difficulties like the example given earlier.

The well-site equipment ties into a gas gathering system that was likely designed to the specifications of ASME B31.8 (Gas Transmission and Distribution Piping Systems) (ASME B31.8), but that code explicitly excludes: "802.12.f. wellhead assemblies, including control valves, *flow lines between wellhead and trap or separator*, offshore platform production facility piping, or casing and tubing in gas or oil wells (emphasis added)." We could quibble and say that this exclusion does not include piping from the separator to the gathering system, but then we still need a basis for design of the piping from the wellhead to the separator. The only one of the ASME B31 series that does not include this wellhead-to-separator exclusion is ASME B31.3 (Gas Processing Plants) (ASME B31.3) which is also labeled ISO 15649:2001 (petroleum and natural gas industries—piping), many federal regulations around the world, many state/provincial regulations, and many company standards include ASME B31.3 by reference.

It is common for company standards to have stricter requirements than the code. For example, one company specifies that ASME B31.3 will be used for well-site construction, and that API 1104 (Standard for Welding Pipelines and Related Facilities) will be used for welder qualification—a fairly common combination. API 1104 specifies that each qualification weld will be evaluated by nondestructive testing (NDT) *or* a break test. This same company specifies that each weld will be evaluated by NDT *and* a break test, it still meets or exceeds the standard.

The hierarchy of code compliance is as follows:

- Apply the requirements of the standard.
- Add any additional requirements of regulations.
- Eliminate any standards requirements that are prohibited by regulations.
- Add any additional requirements of company standards (unless prohibited by regulation).
- Eliminate any standards requirements that are prohibited by company standards (unless required by regulations).

The area that this hierarchy is becoming increasingly important is threaded connections. The code allows threaded connections (with some very minor restrictions on their use). Regulations are typically silent on threaded connections. Company standards have evolved over the last few years to either prohibit threaded connections altogether or to limit the size of acceptable fittings. Company standards do not undergo the extensive vetting process that industry standards see and this prohibition is often ambiguous. One company standard that had the statement "threaded connections smaller than 2 in (50 DN) are allowed in nonvibrating service." The intent of the authors (I asked) was that 2 in and smaller threaded connections would be allowed except on compressor skids. The actual result was a significant increase in 1-1/2 in piping in applications where 2 in would be more appropriate because of the significantly increased availability and scope of 2 in fittings over 1-1/2 in fittings.

5.2.1 Pipe wall thickness

To calculate the minimum wall thickness, you need to calculate it twice:

$$t_{3a} = \frac{P_{mawp} \cdot \mathrm{OD}_{pipe}}{2 \cdot (S_{tableA1} \cdot E_{tableA1B} \cdot W + P_{mawp} \cdot Y)}$$

$$t_{3b} = \frac{P_{mawp} \cdot (\mathrm{ID}_{pipe} + 2 \cdot t_{corr})}{2 \cdot (S_{tableA1} \cdot E_{tableA1B} \cdot W - P_{mawp} \cdot (1 - Y))} \tag{5.1}$$

Table A1 is in Appendix A and the title is "Basic Allowable Stresses for Metals in Tension." It has 48 pages of metal choices, but mostly we use API 5L Grade B, X42, X60, or X70, and these grades are in the front of the appendix.

Table A1B is also in Appendix A and the title is "Basic Quality Factors for Longitudinal Weld Joints in Pipes, Tubes, and Fittings." It is a derate for pipe developed from plate and longitudinally welded. This derate is larger than the other B31 series codes apply to longitudinally welded pipe.

"*Y*" is a derate factor for pipe metallurgy. For pipe with a wall thickness less than OD/6, use the temperature-specific values from ASME B31.3 Table 304.1.1. For thick walled pipe, "*Y*" is:

$$Y = \frac{\mathrm{ID}_{pipe} + 2 \cdot t_{corr}}{\mathrm{OD}_{pipe} + \mathrm{ID}_{pipe} + 2 \cdot t_{corr}} \tag{5.2}$$

"*W*" in equation 5.1 is the "weld joint strength reduction factor." This table starts at 800°F (427°C) and goes up. Since temperatures that high are largely unheard of on a well-site, the notes for the table point you to $W = 1.0$.

The corrosion allowance (t_{corr}) is intended to be annual wastage of the pipe times the number of years of expected service. It is almost never that scientific. ASME B31.3 requires at least 3/64 in (1.2 mm), but many companies require greater corrosion allowances for well-site piping. If you pick a pipe size/schedule/grade you have a nominal wall thickness, subtract the corrosion allowance from that nominal value and you can calculate hoop stresses using Barlow's formula (Eq. (5.3)):

$$S = \frac{P_{gauge} \cdot \mathrm{OD}_{pipe}}{2 \cdot t_{actual}} \tag{5.3}$$

Pressure in Eq. (5.3) can be normal operating pressure, static test pressure, or maximum allowable working pressure (MAWP also known as maximum allowable operating pressure or MAOP); it is best to calculate all three and ensure that they are within company limits. If company standards are silent, then compare Eq. (5.3) to the appropriate temperature column in ASME B31.3 Table A.1. Note that there is a "Specified Min. Strength" column in this table that you cannot use for anything in B31.3. For example, API 5L Grade X42 has a specified minimum yield strength (SMYS) of 42 ksi, but at any temperature below 300°C (672°F) the maximum design stress is 20 ksi. Company standards often specify a maximum percent of minimum yield strength, and in this example you multiply the company maximum times 42 ksi, not 20 ksi. If your company only allows 20% of minimum yield strength at normal operating pressure, then any value from Eq. (5.3) less than 8.4 ksi would satisfy company standards while any value less than 20 ksi would satisfy B31.3.

ASME B31.3 is a plant code, and every statement in the 386 page document has several qualifiers and much branching logic. It only makes sense to enter this document with a specific project in mind and follow all of the paths for that specific project—trying to read the document from cover to cover is a very confusing task (and it is easy to get bogged down in a once-in-a-millennium side branch).

5.2.2 Pipe wall thickness example

A well needs a line from the wellhead to the separator (Table 5.1). An analysis of other wells in the field and company standards developed the conditions given in Table 5.2. Selected extracts from the various ASME B31.3 tables are given in Table 5.2, Table 5.3, and Table 5.4.

If you assume Schedule 80 Grade B pipe, then:

$$t_{3a} = \frac{P_{mawp} \cdot OD_{pipe}}{2 \cdot (S_{tableA1} \cdot E_{tableA1B} \cdot W + P_{mawp} \cdot Y)}$$

$$= \frac{600\ \text{psi} \cdot 10.75\ \text{in}}{2 \cdot (20000\ \text{psi} \cdot 0.85 \cdot 1.0 + 600\ \text{psi} \cdot 0.4)} = 0.187\ \text{in}$$

$$t_{3b} = \frac{P_{mawp} \cdot (ID_{pipe} + 2 \cdot t_{corr})}{2 \cdot (S_{tableA1} \cdot E_{tableA1B} \cdot W - P_{mawp} \cdot (1 - Y))}$$

$$= \frac{600\ \text{psi} \cdot (9.562\ \text{in} + 2 \cdot 0.187\ \text{in})}{2 \cdot (20000\ \text{psi} \cdot 0.85 \cdot 1.0 - 600\ \text{psi} \cdot (1 - 0.4))} = 0.179\ \text{in}$$

Use the larger value (0.187 in (4.751 mm)) for the minimum wall thickness. This is considerably less than the nominal wall thickness of

Table 5.1 Pipe thickness example data

	fps	Both	SI
MAWP	600 psig		4140 kPag
Design temperature	100°F		37.8°C
Nominal pipe	10 in		250 DN
Type test		Pneumatic	
Test pressure		150% of MAWP	
Corrosion allowance	3/16 in		4.8 mm
Normal operating pressure	150 psig		1034 kPag

Table 5.2 Pipe schedules for example

	OD	ID	*t*
Sched 20	10.750 in (273 mm)	10.250 in (260 mm)	0.250 in (6.35 mm)
Sched 40 (std)	10.750 in (273 mm)	10.020 in (255 mm)	0.365 in (9.27 mm)
Sched 60 (X)	10.750 in (273 mm)	9.750 in (248 mm)	0.500 in (12.70 mm)
Sched 80	10.750 in (273 mm)	9.562 in (243 mm)	0.594 in (15.09 mm)

Table 5.3 Extracts from table A1

Table A1	SMYS (psig (MPag))	$S_{tableA1}$ (psig (MPag))	Max. temp before derate (°F (°C))
API 5L Gr B	35,000 (241)	20,000 (138)	400 (204)
API 5L X42	42,000 (290)	20,000 (138)	400 (204)
API 5L X60	60,000 (414)	25,000 (172)	400 (204)
API 5L X70	70,000 (483)	27,300 (188)	400 (204)

Table 5.4 Extracts from table A1B

Table A1B	Description	$E_{tableA1B}$
API 5L	Seamless pipe	1.0
API 5L	Electric resistance welded	0.85
API 5L	Electric fusion welded pipe	0.95
API 5L	Furnace butt welded pipe	0.95

Schedule 80 pipe (0.594 in (15.09 mm), but when you run the calculations with thinner walled pipe with our (very large) corrosion allowance, the nominal pipe thickness is less than the required thickness.

Now we need to calculate stresses using Eq. (5.3) we see that at MAWP the hoop stress is 22.7% of SMYS, at test conditions it is 34.0% of SMYS, and normal operating pressure it is 5.7% of SMYS.

5.3 PIPING SELECTION

The equations in the last section result in very safe piping, but company standards often go considerably further to specify that all well-site piping must be at least Schedule 80, and pipe that is 2 in and smaller must be Schedule 160. The source of these arbitrary limits seems to have been lost in the murky past, but they are very common. It can be disconcerting to calculate that the right pipe for a job is 2 in, but the inside diameter of Schedule 160 is too small for the task and you need to step up to 3 in Schedule 80 which is too big for the job (each step-up in standard pipe size has about twice the flow area of the next smaller standard size). The only solution that I've found to this (seemingly) irrational requirement is to use the 3 in and put in pigging valves to allow pigs to be run from the wellhead to the separator or the separator to the outlet meter.

In addition to strength and safety considerations, when designing well-site piping you need to think about:

- When liquids accumulate in the lines, how will you get them out?
- If you decide to flow the casing, do you have enough pipe?
- How long do you expect the piping to last?

To address these issues it is useful to run multiple lines from the wellhead to the separator, and generally cross connect them so that you can send either or both the tubing and/or the annulus down either line. Address longevity issues by choosing reinforced thermoplastic (see Chapter 6: Gas Gathering Systems) piping for the well-site instead of steel when company standards allow it. Size piping to provide expected flow at expected velocity of 11 ft/s (3.4 m/s) to 120 ft/s (36.6 m/s) using the Duckler method from Chapter 4, Surface Engineering Concepts.

5.4 PRODUCTION VESSELS

Production units on well-sites are a mixture of "separators" or "scrubbers"; oriented horizontally or vertically; able to separate gas from liquid (two phase) or able to separate the stream into gas, oil, and water (three phase). All have their proponents and detractors for well-site use.

Both separators and scrubbers employ one or more of gravity settling, centrifugal force, impingement, electrostatic precipitation, sonic precipitation, filtration, absorption, adsorption, and thermal effects. For well-sites, we tend to limit that list to gravity settling, centrifugal force, and impingement.

Liquid-storage vessels are common on many well-sites and they are called "tanks" or "pits." All are designed for atmospheric storage of liquids and the only significant difference is that tanks are above grade and pits are below grade. Production pits should not be confused with the lined holes that are used during the drilling process, those earthen pits are required to be cleaned up before the drilling process is completed.

5.4.1 Vessel design code

A rash of fatalities related to boiler explosions in the early days of mankind harvesting the power of steam led to the development in 1914 of the ASME Boiler and Pressure Vessel Code (BPVC). This code defined what did and did not constitute a "pressure vessel," what components a pressure vessel had to include, how to determine acceptable pressure (at what temperature) a vessel could contain, and fabrication and inspection standards for building pressure vessels. Pressure vessels are covered in Section VIII Division 1 (Rules for Construction of Pressure Vessels) of this 28-volume document (Wikipedia, 1). The BPVC scope for Section VIII Division 1 "provides requirements applicable to the design, fabrication, inspection, testing, and certification of pressure vessels operating at either internal or external pressures exceeding 15 psig. Such pressure vessels may be fired or unfired" (BPVC), but it does not include piping not immediately associated with the pressure vessel.

The BPVC has been adopted in whole or in part by over 100 countries, all US states and territories, all Canadian provinces, and states, provinces, and municipalities of countries all over the world. There is some confusion in our industry about the distinction between "Code" jurisdictions and "Non-Code" jurisdictions. A "Non-Code State" does not mean that industry is allowed to ignore the BPVC, far from it. "Non-Code" status simply means that the state does not have an appointed official called "Boiler Inspector" (or similar) and that the state (or municipality) does not issue "inspection certificates" (or other de facto licenses) on operating or new-construction pressure vessels. It should never be taken as a license to ignore the BPVC. Even in those few jurisdictions where the BPVC is not made part of regulations, failure to follow the design requirements of the BPVC generally meets the standards of "gross negligence" in civil proceedings which usually allow courts of law to impose multiples of the actual damages a plaintiff incurs.

The BPVC definition of a "pressure vessel" is "a pressure-containing device that is exposed to internal or external pressure in excess of 15 psig and has an inside diameter, width, height, or cross-sectional diagonal exceeding 6 in" (BPVC). This definition is quite broad and on its face would include several million miles of piping. That is not the intention of the BPVC, and they have published a large number of code cases that together exclude a class of pressure-containing devices that can be classed as "pipeline and pipeline accessories." Something that satisfies all of the following is excluded from the BPVC:

1. Can be described by a piping sketch or isometric drawing *and referenced piping specs.*
2. Is not intended for storing or processing fluids. The exception is items such as mixers, tees, headers, metering devices, or other items that are typically recognized as piping components.

3. Its primary function is to transport liquids and gases from one location to another within a piping system of which it is an integral part.
4. It is not intended to act as an air receiver.
5. The item will not be subjected to more frequent test and inspection intervals than the remainder of the attached system.

ASME B31.8 (Section 831.35(d), Special Components Fabricated by Welding) formalizes this some by saying [emphasis added] "Prefabricated units, *other than regularly manufactured butt welding fittings*, that employ plate and longitudinal seams *as contrasted with pipe that has been produced and tested under one of the specifications listed in this Code*, shall be designed, constructed, and tested under the requirements of the BPV Code." B31.8 clearly draws the line at rolled plate with longitudinal welds as opposed to "pipe" (including electric resistance welded and other piping fabricated by the manufacturer from plate to an included specification) and butt weld fittings.

A BPVC "pressure vessel" must have a "U-Stamp" with very definite information permanently affixed to a nameplate so that an auditor can go back to a set of documents that describe the design limits, metallurgy, welding procedures, types of NDT performed and their results, and inspector's certification of the required documents. If you modify a pressure vessel, then you have another set of requirements that result in an "R-Stamp" that must also be permanently affixed to the vessel.

Well-site equipment will either be "pipe" or "vessels." Silly games like making a fuel-gas scrubber out of 6 in (DN 150) Schedule 80 with 5.761 in (146.3 mm) ID to avoid having to affix a U-Stamp to the scrubber are within the letter of the law, but in the end it is better to design the vessel to meet the design conditions and if the ID is bigger than 6 in (152.4 mm) build it in a code shop to the BPVC, it is better to have a vessel that meets design conditions than one that skirts the BPVC.

5.4.2 Separator selection

A "separator" must have all of:

1. the ability to separate gases from liquids,
2. sufficient liquid capacity to handle expected surges in liquid,
3. sufficient height to allow droplets to settle out via gravity,
4. a means of reducing turbulence in the main body of the separator so that proper settling can occur,
5. a mist extractor to capture droplets too small to settle via gravity,
6. proper liquid-level controls.

A "scrubber" is a separator that is missing one or more of the list above. It is common for scrubbers to lack (#2) adequate liquid holding capacity; (#5) a mist extractor; and/or (#6) liquid-level controls. There are people who prefer scrubbers for well-sites, but I have been unable to follow their arguments which seem to me to be based solely on costs, not on the vessel's ability to facilitate reservoir performance.

There are large numbers of both vertical and horizontal separators around the world. Horizontal vessels have historically been the clear winner because they can tolerate higher maximum velocity (see later) and their ability to further separate oil or condensate from water is far superior to vertical vessels. If you can ignore evaporation, you would nearly always select horizontal vessels, but can you ignore evaporation?

At the gas/liquid interface point, liquid will evaporate until the gas in contact with the surface is at 100% relative humidity (RH) for that pressure and temperature. A horizontal vessel has a very large coherent gas/liquid interface.

The pressure traverse in Table 5.5 shows that because of the large surface area of liquid, the outlet of the unit will be 100% RH and about 37% more water (by mass) will pass out of the separator. If the separator outlet pressure were 150 psig (1034 kPag), the gain would be a more tolerable 13%. The contact area of a vertical vessel is much smaller and a much smaller proportion of the gas contacts the liquid surface so it is rare for the gas stream out of a vertical vessel to be at 100% RH or even at a measurably higher water vapor content than the inlet.

Because of water vapor inventory, my preference on dry-gas well-sites (i.e., wells without the expectation of condensate) is vertical vessels. If there is a significant potential for economic quantities of liquid hydrocarbons, a horizontal separator will recover more of that condensate and oil than a vertical vessel will and is preferred.

Table 5.5 Horizontal separator re-saturation

	Bottom-hole	**Separator inlet**	**Separator outlet**
Pressure	20 psig (138 kPag)	5 psig (34.4 kPag)	2 psig (13.8 kPag)
Temperature	180°F (82°C)	160°F (82°C)	160°F (82°C)
Water content	11,728 lbm/MMSCF (187,900 mg/SCm)	88% RH (same water content as bottom-hole)	16,123 lbm/MMSCF (258,300 mg/SCm)

It is starting to be common on wet shale wells to send the well to a vertical, two-phase scrubber and dump the liquid into the low side of a high-low producer (HLP) (see later). This configuration allows the high side of the HLP to function normally without being flooded.

5.4.3 Separator sizing

The sizing calculations are the same for either horizontal or vertical vessels.

5.4.3.1 Shell sizing

The vessel minimum ID is a function of maximum velocity:

$$v_{\max} = K_s \cdot \left(\frac{\rho_{liq} - \rho_{gas}}{\rho_{gas}} \right)^{0.5} \tag{5.4}$$

The term K_s is an empirical constant that is generally taken to be:

- Vertical vessels with mist extractor
  - <10 ft seam-to-seam → 0.18 ft/s
  - >10 ft seam-to-seam → 0.21 ft/s
  - ≫10 ft seam-to-seam → 0.35 ft/s
- Horizontal vessel with mist extractor → 0.45 ft/s

For pressures above 150 psia K_s is reduced (use a linear interpolation from 150 psia $= 0.9 \times K_s$ to 1150 psi $= 0.75 \times K_s$).

Using the Duckler method from Chapter 4, Surface Engineering Concepts, you can calculate ρ_k (Eq. (4.24)) and $q_{fullstream}$ (Eq. (4.25)) and the minimum ID of the vessel becomes:

$$\text{ID}_{\min} = \left(\left(\frac{4}{\pi} \right) \cdot \left(\frac{q_{fullstream}}{v_{\max}} \right) \right) \tag{5.5}$$

It is normal for Eq. (5.5) to result in a messy number, this number must be rounded up to a nominal pipe size (even though virtually all pressure vessels are made from postmanufacturer rolled plate and can be any size, the end caps are forged in standard sizes). Now you need to check the sizing:

$$v_{shell} = \frac{q_{fullstream}}{A_{shell}}$$

$$q_{designMSCF} = v_{shell} \cdot A_{shell} \cdot \left(\frac{\rho_k}{\rho_{std}} \right) \tag{5.6}$$

If the $q_{designMSCF}$ is greater than expected gas flow rate, then verify that the Reynolds number is greater than 6000 (if it isn't then drop the vessel ID one pipe size and recalculate, the flow needs to be turbulent).

The shell design length is measured from the centerline of the float connection to the bottom of the mist pad. This value should be four to six times the ID of the shell. If the mist pad is outside the vessel and/or the liquid dump is outside of the vessel (common with vertical separators equipped with blow cases, see later) use seam-to-seam length.

5.4.3.2 Nozzle sizing

Nozzle sizing is more art than science and I've gotten more push-back from vessel designers over specifying nozzle size than any other single argument. This intractability seems to stem from their having learned this skill by rote rather than understanding the underlying physics like they tend to be able to do with shell sizing and mist extractor sizing. Many vessel manufacturers have proprietary nozzle sizing algorithms, but I have never been able to differentiate one manufacturer from another by the performance of their nozzles. All of the designs eventually come down to wanting $\rho \cdot v^2 \approx 225 \cdot \mathrm{lbm}/(\mathrm{ft} \cdot \mathrm{s}^2)$ which is a kind of funny choice of units since that number works out to 0.049 psi. Minimum nozzle ID is:

$$\mathrm{ID}_{nozzlemin} = \left(\left(\frac{4}{\pi}\right) \cdot q_{fullstream} \cdot \left(\frac{\rho_k}{225 \cdot \frac{\mathrm{lbm}}{\mathrm{ft} \cdot \mathrm{s}^2}}\right)^{0.5}\right)^{0.5} \tag{5.7}$$

Round $\mathrm{ID}_{nozzlemin}$ up to the next nominal pipe size.

5.4.3.3 Mist extractor

Droplets smaller than about 200 μm (0.2 mm (0.00078 in)) will not fall at separator velocities, they must be coalesced into larger droplets that will fall at separator velocity. We do this in a "mist pad" (Fig. 5.3). Mist pads are generally either vane packs or wire mesh. There is not a lot of difference in the performance of the two options, vane packs come in discrete sizes and wire mesh pads can be adapted for any size outlet plenum.

The outlet plenum is an important part of the mist pad design. Small droplets need enough velocity to make changing directions difficult, but not so much velocity that they can traverse the mist pad so fast that they avoid most collisions. Optimum velocity is 5 ft/s (1.5 m/s). If you slow to below 2 ft/s (0.91 m/s) the effectiveness drops by over 10%. Above 20 ft/s (6 m/s) the effectiveness drops by about 5%. The target velocity that I use for design is 10 ft/s (3 m/s) since too fast has a smaller penalty than too slow.

Fig. 5.4 shows the impact of getting the diameter of the mist pad wrong. The vessel upstream of this reciprocating compressor had a mist pad diameter equal to the vessel ID, velocity through the mist pad was under 1 ft/s (0.305 m/s), and based on the mass of salt that accumulated in the suction header of this compressor, the total unseparated liquid was

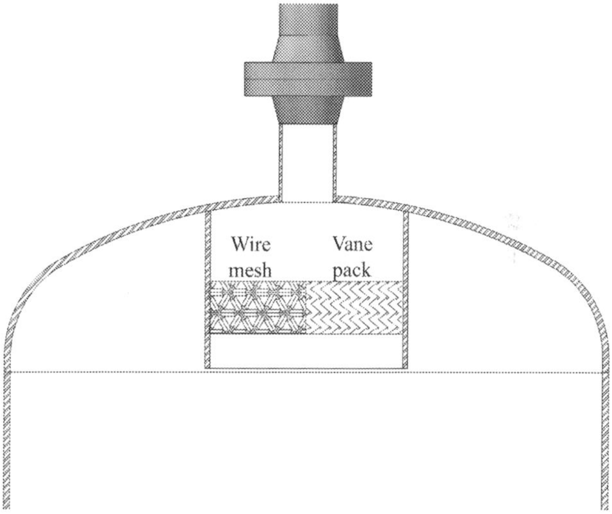

Figure 5.3 Separator mist pad.

Figure 5.4 Salted-up recip. *Personal collection of Dejan Ivanovic.*

less than 1 L/day (1 qt/day), but was enough to render this compressor unusable in less than 30 days. Velocity is straightforward:

$$v_{mist} = \frac{q_{mscf}}{\left(\frac{\pi}{4}\right) \cdot \mathrm{ID}_{mist}^{2}} \cdot \frac{\rho_{gasStd}}{\rho_{gasActual}} \tag{5.8}$$

Note that Eq. (5.8) uses gas parameters instead of full-flow parameters, this is because of the assumption that the separator works and the remaining mist is an insignificant portion of the total flow. Adjust the ID of the mist pad until you reach your target velocity. With a wire mesh pad, you can roll the plenum any size you want to since it doesn't attach to any manufactured fittings. With a vane pack you have to restrict your design to standard pipe sizes.

Eq. (5.9) is used to determine the minimum distance from the top of the mist pad to the entrance to the outlet nozzle.

$$d_{toNozzle} = \mathrm{ID}_{mist} - \frac{\mathrm{ID}_{nozzle}}{2} \tag{5.9}$$

5.4.4 Typical designs

Virtually all production units are either vertical two-phase, horizontal two-phase, horizontal three-phase, or combination units. Combination units include the HLP (discussed in the next section), and "L-shaped" units that include a vertical section for gas/liquid separation and a horizontal section for oil/water separation.

5.4.4.1 High–low producer

The most common production unit in conventional well-site use is a compound unit that goes by several names, one of them is "HLP." This configuration (Fig. 5.5) has a "T-shaped" high-pressure two-phase vessel (the "high side") inset into a horizontal low-pressure unit (the "low side"), but there are others. The high operating pressure of the high side keeps evaporation to a minimum and the low operating pressure of the low side enhances liquid–liquid separation.

The high side has a simple level float that opens or shuts to keep the end of the outlet stinger submerged in liquid (to keep high-pressure gas from entering the low side). The low side is a bit more interesting. Fluids enter the liquid end of the vessel and lighter fluids tend to float to the top of the body of liquid. As liquids accumulate in the liquid section, the oil will eventually skim over the "oil weir" into the oil section. The height of the "weir nipple" is calculated to provide a neutral monometer effect

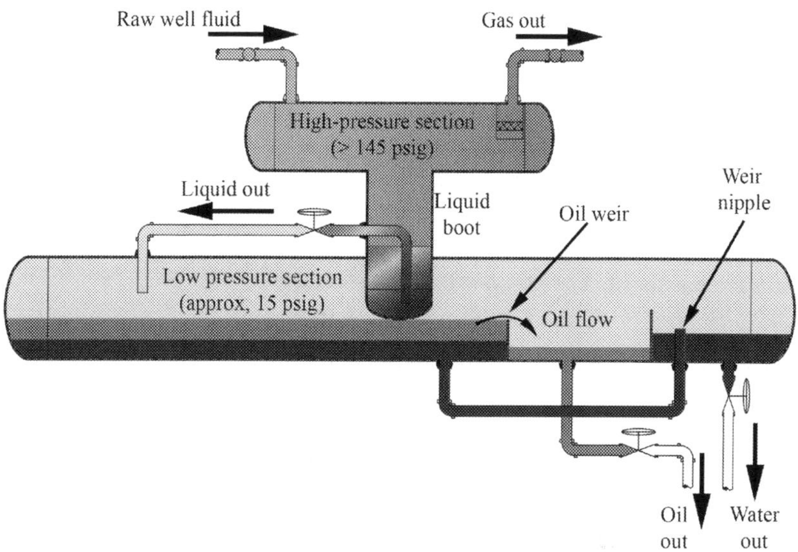

Figure 5.5 HLP separator.

with just water in the right-hand leg and a mixture of water and condensate in the liquid section up to the oil weir. The "weir nipple" will always be shorter than the "oil weir." Both the oil section and the water section have a simple level controller and a dump valve.

Effectiveness of the (small) top section falls off rapidly as pressures decline below 145 psig (1000 kPag). People operate them down to 60 psig (414 kPag), but generally notice significant decrease in gas production, and a significant increase in tank outgassing,

There are two other common configurations that are hydraulically similar to an HLP and are noteworthy: vertical high side and integral prescrubber. The vertical high-side unit goes by many names, but the basic design has the high-pressure gas section rising vertically from the liquid boot and everything else the same as shown in Fig. 5.5.

The integral pre-scrubber version (also known as a "compressor dome") has an additional low pressure, short, vertical vessel let into the low side. In this configuration, a low-pressure wellhead is piped into the prescrubber and then to a compressor suction, liquids are dumped into the low side. The compressor discharge returns to the high side.

5.4.4.2 Horizontal

If you add a raw gas inlet and a gas outlet to the low side of Fig. 5.5 you have a three-phase horizontal separator. If you either remove the liquid boot or set the level control for liquids in the horizontal shell of the high-side vessel, you have a two-phase horizontal separator (after you've adjusted the size to accommodate expected fluids at the expected pressures).

Two-phase horizontal separators are located at well-sites around the world in very large numbers. At high pressures they are able to do a better job with a smaller ID. At lower pressures, evaporation is such a big issue, and the coherent gas/liquid interface is so large that these units often put more water into the gathering system (as water vapor) than a similar size vertical scrubber without a mist extractor (and a similar size vertical separator will do a much better job).

5.4.4.3 Vertical

Two-phase vertical separators (top vessel in Fig. 5.6) rely on a change in direction of the incoming fluids (gas can change direction easier than liquids can and the initial direction change results in a significant number

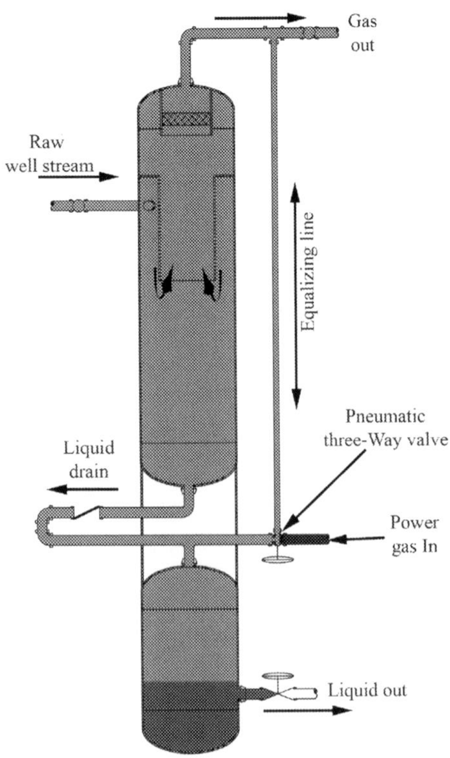

Figure 5.6 Vertical two-phase separator with blow case.

of droplet collisions) and a velocity change in the bulk vessel to separate gases and liquids. Normally, the bottom section of the vessel is used for liquid accumulation with appropriate level control and liquid-removal components located in that section. The vessel in Fig. 5.6 drains the accumulated liquid directly into an integral "blow case" (see Section 5.4.4.5) with the level controls and liquid removal relocated to the blow case.

Fig. 5.6 separator also has a unique innovation called a "top-hat diverter and tangential inlet." This combination acts as a "vortex tube" to cool the gas and warm the liquid. Vortex tubes are a unique bit of fluid mechanics that accept a warm fluid and split it into a hot fluid and a cold fluid without violating the Second Law of Thermodynamics (using Clausius statement the Second Law is "heat can never pass from a colder to a warmer body without some other change, connected therewith, occurring at the same time" (Wikipedia, 2)) by having an outlet stream that is warmer than the inlet stream (vortex tubes work because of the Law of Conservation of Momentum which requires the transfer of heat from two rotating streams in contact from the cold stream to the hot stream). The top-hat diverter units have been observed to lower the gas outlet temperature by up to 20°F (11°C) which lowers the outlet water content of gas that started at 80°F (26.7°C) and 20 psig (138 kPag) and exited the separator at 15 psig (103 kPag) at 60°F (15.6°C) from 1616 lbm/MMSCF (25890 mg/SCm) to 819 lbm/MMSCF (13120 mg/SCm) keeping 95 gallons/MMSCF (12.8 L/kSCm) of water out of the gathering system. Not all top-hat diverter separators provide this kind of temperature drop, but even if it is zero, the angular velocity in the annulus provides significant separation benefits.

Three-phase vertical separators use a displacer (often incorrectly called a "float") that is buoyant in water but not buoyant in oil so it floats on the oil/water interface. They have a second displacer of different material higher in the vessel that is buoyant in oil. The water-level controller is generally "continuous" (see Section 5.6.1) while the oil-level controller is usually "intermittent." These units work best when the process temperature is tightly controlled, and the inflow is maintained in a narrow range. In well-site use, nothing is ever tightly controlled and it is common for three-phase vertical separator to be converted to two-phase by changing the bottom displacer to one that is buoyant in oil (and the controller is usually converted to "intermittent"). These modifications are rarely done with the knowledge and approval of engineering staff.

5.4.4.4 Heated vs nonheated

It is common for the process of separation to be enhanced through some level of temperature control. For example, separators tend to work better if the water hasn't frozen and paraffin has not solidified. Some hydrocarbon liquids have a significant change in density with changes in temperature, often a much greater range than the change water density for the same change in temperature, so as you cool the liquid stream the density of the phases gets closer together and the tendency for the phases to stratify via gravity becomes weaker (i.e., it starts taking longer to happen, it still stratifies, but possibly not within the "liquid section" rather it happens in the "water section" of a three-phase separator). Adding heat can prevent freezing and paraffin formation, and can keep the hydrocarbon liquids as light as possible.

Burners on separators are called "fire tubes" which are actual pipes (called "burner tubes," often thin walled, but not always) rated for the external pressure applied by the vessel MAWP and include gas-fired burners in an open chamber protected by "flame arrestors" (Fig. 5.7) which are devices that act to remove the heat from a flame as it attempts to travel down narrow passages. The design of flame arrestors varies considerably from manufacturer to manufacturer and can either be classed as "flame arrestors" (typically installed with the unprotected side open to atmosphere) or "detonation arrestors" (typically installed in a line some distance from the end, Fig. 5.7). In either case these devices are intended to pass the thermal mass of a flame front to the physical mass of the arrestor which implies many small flow paths with a large surface-area per unit volume of flow. The multitude of small channels have an impact of differential pressure across the arrestor and care must be taken when specifying arrestor technology to ensure that there is adequate dP available satisfy the required flow rate of the arrestor.

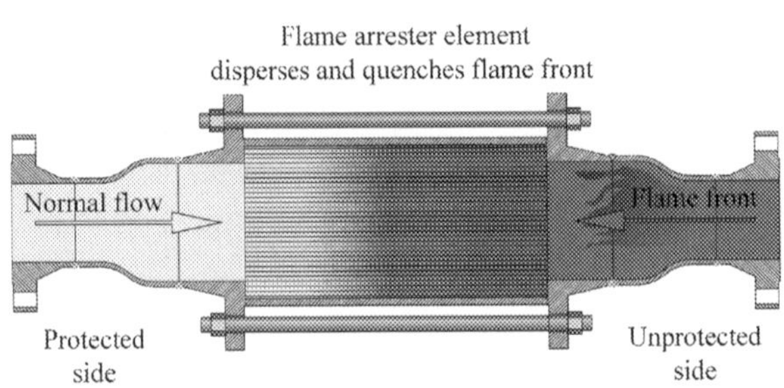

Figure 5.7 Flame arrestor.

"Indirect-fired" heaters connect a body of liquid (called a "water bath" in spite of the liquid rarely being water) to the pressure vessel wall and place the burner tube in the water bath. Since the water bath is at atmospheric pressure, the MAWP of the pressure vessel is not a factor in the wall thickness of the burner tube, and the tubes are much thinner walled. The burner heats the water bath and this energy is transmitted to the pressure vessel through the walls of the pressure vessel. This low-intensity heat source is adequate for freeze prevention, but not much else.

"Direct-fired" heaters have the burner tube within the pressure vessel and can apply very high heat directly to the process fluid. Much like placing an empty pot on a household range will quickly burn through the pot, if the burner in a direct-fired heater is not submerged in process liquid it will fail fairly rapidly and quite explosively. Direct-fired heaters are almost never used in vertical vessels and when they are used in horizontal units there are measures taken to maintain a higher liquid level than in unfired vessels such as setting the level control higher in the vessel and/or making the oil weir and weir nipple higher than in unfired vessels.

The low side of an HLP often has direct-fired heater elements to aid in separation and resist freezing. Typically direct-fired separators are called "heater-treater units" in oil fields, but gas fields have resisted that terminology. These units have a fire tube in the low side in direct contact with the produced liquids. With the liquids-rich shale plays like the Baaken and Eagle Ford, these heater-treater configured HLP units are very common (trying to keep paraffin mobile). Some of the liquids-rich wells put an unfired two-phase vertical scrubber in front of the HLP with the gas outlet of the vertical unit going to the high side of the HLP and the liquids dump going to the low side of the HLP.

5.4.4.5 Blow case

As wellhead pressures inevitably come down, the ability of a separator to move liquid is reduced. For example, if a separator is running at 5 psig (34.5 kPag), and is dumping into the top of a 400 bbl tank (i.e., 12 ft D × 20 H (64 m^3, 3.7 m D × 6.1 m H) it has to overcome nearly 9 psi (62 kPa) of hydrostatic head, but only has 5 psig to accomplish that task. When the dump valve opens, absolutely nothing will transfer from the separator to the tank. The most common solution to this ubiquitous problem is a "blow case." The bottom vessel in Fig. 5.6 is a blow case designed as an integral component of a vertical production unit. In this case, liquid drains from the vertical separator through a nonreturn (check) valve into the blow case. When the blow case is full, the "pneumatic three-way valve" repositions and the "power gas" (typically gas from the discharge of a well-site compressor) is directed into the blow case, shutting the check valve between the vessel and the blow case and increasing the pressure in the blow case. The "liquid outlet" valve (i.e., "dump valve") opens and the pressure in the blow case moves the liquid to its disposal location. It is common for the three-way valve to be replaced by a pair of independent control valves. Arguments for either design have merit, but it seems to be easier for people to visualize how the blow case performs using a three-way valve.

Commercial blow cases are readily available for accumulating environmental drains on process skids. These units are all in a horizontal configuration which requires a larger volume of gas for a given volume of liquid, but allows a very low profile (i.e., you don't have to bury them as deep for a skid drain). The design for a vertical blow case is easy to accomplish and they tend to be able to move a larger volume (due to the smaller liquid surface). The capacity of a blow case is made up of two components: (1) the volume shifted every cycle and (2) the length of time required for a cycle. The volume per cycle is easy to calculate as the difference between the start-of-cycle liquid height and the end-of-cycle liquid height.

The length of time required for each cycle is made up of four components: (1) time to reposition the three-way valve and pressurize the vessel; (2) time to open the dump valve and empty the liquid; (3) time to shut the dump valve and reposition the three-way valve; and (4) time to equalize pressure so the check valve can open. The first three components are reasonably quick and rarely are the controlling factor in overall capacity. The last one is crucial. Let's say that power gas pressure is 100 psig (690 kPag) and the vessel is at 5 psig (34.5 kPag). The "choked flow" arithmetic we talked about in the last chapter says that this process must be choked at some point, but where? At the three-way valve? That would require sonic velocity all the way down the equalizing line, which is not possible. At the intersection with the gas outlet line? Very possibly, but the flow from the blow case to the end of the equalizing line is anything but simple. If we assume that the location of the transitional standing wave (and therefore the choke point) is at the outlet line, then we can determine an initial mass flow rate (using the minimum upstream pressure for choked flow into 5 psig to calculate density). Iterate that mass flow rate every 1/10 second and you can get acceptably close to the required time to equalize the blow case. For pipe that is 1 in (25 DN) and larger this period is very short and not a major factor in the capacity of the blow case. Many compressor packagers use 1/8 in (3.1 mm) lines for equalizing and forego the three-way valve (accepting that the loss of power gas during the dump cycle is largely insignificant). This has a significant impact on the capacity of a blow case to move liquids.

The blow case in Fig. 5.6 is 24 in (600 DN) ID, and the level controller is 6 in (152 mm) from the weld seam on the top elliptical head (2.6 ft^3 (0.074 m^3) combined volume). When the level controller has been satisfied, the volume filled with gas is 4.2 ft^3 (0.119 m^3). Using the 100 psig power gas and 5 psig separator pressure above, an equalizing event must transfer 1.34 lbm (0.609 kg) of gas from the blow case to the separator outlet before the check valve will open. The fixed time to reposition valves, pressurize the blow case, dump the liquid, and reposition the valves is 14 seconds for all three cases in Table 5.6. The blow case in Fig. 5.6 dumps 11.75 gal/cycle (44.5 L/cycle).

Most compressor suction-scrubber installations will see much less liquid than 85 bbl/day (13.58 m^3), but the flow rate is never very steady. It is not at all uncommon for a well making 10 bbl/day (1.59 m^3) of liquid to make that volume in three to five distinct flow events per day, each less than 60 seconds in duration. Flowing 2 bbl (0.318 m^3) of liquid in 60 seconds is an instantaneous flow rate of 2880 bbl/day (458 m^3) which frequently results in a suction-scrubber high level kill on the compressor. Eliminating the cost of the three-way valve on compressor skids is a false economy.

The volume of liquid transferred every cycle of the blow case is constant. Inflow during the dump cycle is stopped by the power gas pressure holding the check valve between the separator and the blow case shut. This means that very accurate and repeatable volume flow rates can be

Table 5.6 Blow case capacity example

	2 in Equalizer (50 DN)	**1 in Equalizer (25 DN)**	**0.125 in Equalizer (3.2 mm)**
Initial mass flow rate	3.2 lbm/s (1.37 kg/s)	0.755 lbm/s (343 gm/s)	0.012 lbm/s (5.3 gm/s)
Equalize time	$\ll$ 1s	5 s	275 s
Cycles/day	9600	6646	305
Volume/day	2686 bbl (427 m^3)	1859 bbl (296 m^3)	85 bbl (13.58 m^3)

calculated by simply counting the change in state of the liquid outlet valve or the power gas three-way valve.

5.4.5 Wells with downhole pumps

An alternative to blow cases that makes the separator irrelevant to liquid transfer is to take advantage of the energy of downhole pumps to transfer liquid directly. In Section 3.4.3 we saw that pumping up the tubing while flowing up the annulus resulted in nearly all of the liquid falling out of the gas as it moved up the annulus, so the gas arrives at the separator about as dry as you are going to get it without chemical or thermodynamic dehydration. On the other hand there is only a small amount of gas in the pumped liquid. What we typically do with these two single-phase streams is recombine them at the wellhead and then separate them again in the production unit. Not terribly effective, especially since there is no connection between the pressure that the reservoir needs to maximize production and the pressure the pump needs to maximize pump efficiency.

If we abandon the idea that all well streams must go through a production unit, then we can pump directly into a tank or water gathering system and flow directly into gas sales, eliminating most surface equipment.

With zero gas in a pump, moving 20 bbl/day (1.59 m^3/day) from 3000 ft (914 m) into 5 psig (34.5 kPag) requires 0.548 hp (0.409 kW). Raising the tubing-head pressure to 300 psig (2068 kPag) increases the energy required to 0.674 hp (0.502 kW). Since it is rare to have a motor on or in a well that is less than 5 hp (3.72 kW) and 25 hp (18.6 kW) motors are common, this change is usually trivial. Adding backpressure also significantly improves the performance of downhole pumps in real service (i.e., with some amount of gas in the pumped stream). In order to maintain 300 psig (2068 kPag) upstream of a backpressure valve, it is immaterial if the downstream pressure is atmospheric, 5 psig (34.5 kPag), or 299 psig (2062 kPag). Pumping directly into a water gathering system is simply using energy that would otherwise be wasted across a backpressure valve, while saving the cost of transfer pumps and tanks.

Some amount of free gas will always accompany liquids into downhole pumps. Experiments show that pumped gas ranges from a minimum of 6 MSCF/day (170 SCm/day) to 245 MSCF/day (6.9 kSCm/day) with an average around 30 MSCF/day (850 SCm/day). This gas is certainly a nuisance in a water collection system and can create localized problems with explosive atmospheres in and around tanks. The conventional way to remove this gas comes from the irrigation industry and the equipment is designed to discard the gas (generally air in the irrigation industry) to atmosphere at low pressure. These float-operated devices nearly always carry some amount of liquid with the exhausted gas and have been a significant source of corrosion in aboveground piping and piping in underground vaults.

With gas prices around \$3 USD/MSCF (\$0.106 USD/SCm) the amount of gas discarded in the water system or tank could be the difference between profit and loss for many low-rate wells. Recovering this gas could be the difference between gas sales and plugging the well. Fig. 5.8 shows a device that has been very effective at removing the gas from a pumped liquid stream while retaining enough pressure to allow the gas to go to gas sales without added compression.

5.4.6 Liquid-storage vessels

Liquid-storage vessels are either "tanks" (above grade) or "pits" (set below grade). They can be made of steel, plastic, or fiberglass. Steel tanks have dimensions specified in "API 12F: Specification of Shop Welded Tanks for Storage of Production Liquids," glass-reinforced plastic (GRP) tanks

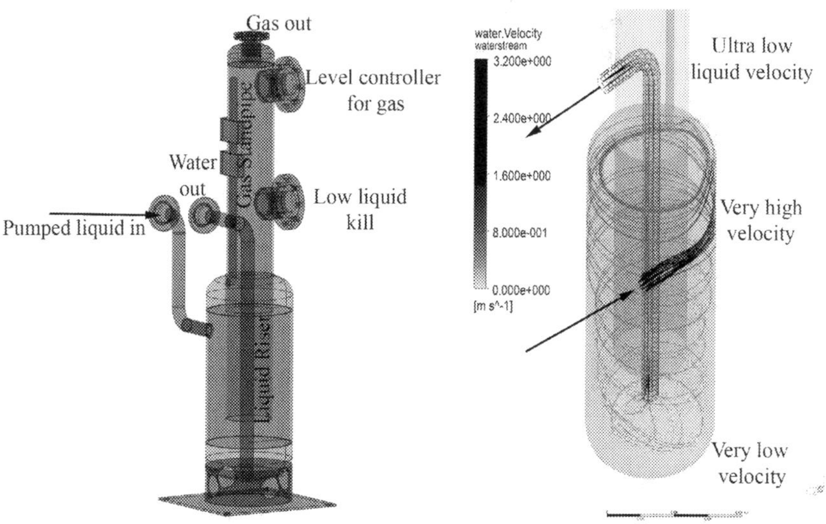

Figure 5.8 GasBuster (US Patent 8,439,999, AU Patent 2013302601).

Table 5.7 API 12F tank dimensions

Nominal size	Working capacity	Diameter	Height	Design press	Design vacuum
90	72 bbl 11.45 m^3	7 ft, 11 in 2410 mm	10 ft 3050 mm	1 psig 6.9 kPag	0.064 in Hg − 2.2 mbarg
100	79 bbl 12.56 m^3	9.5 2900 mm	8 ft 2440 mm		
150	129 bbl 20.5 m^3	9 ft, 6 in 2900 mm	12 ft 3660 mm		
200	166 bbl 26.4 m^3	12 ft 3660 mm	10 ft 3050 mm		
210	200 bbl 31.8 m^3	10 ft 3050 mm	15 ft 4570 mm		
250	224 bbl 35.6 m^3	11 ft 3350 mm	15 ft 4570 mm		
300	266 bbl 42.3 m^3	12 ft 3660 mm	15 ft 4570 mm		
400	366 bbl 58.2 m^3	12 ft 3660 mm	20 ft 6100 mm		
Skinny 500	466 bbl 74.1 m^3.	12 ft 3660 mm	25 ft 7620 mm		
Short 500	479 bbl 76.2 m^3	15.5 ft 4720 mm	16 ft 4880 mm	0.5 psig 3.45 kPag	0.032 in Hg − 1.1 mbarg
750	746 bbl 118.6 m^3	15.5 ft 4720 mm	24 ft 7320 mm		

are specified in "API 12P: Specification for Fiberglass Reinforced Plastic Tanks." These standards have a very specific and limited list of tank sizes (the tank sizes for API 12F are in Table 5.7, sizes for API 12P are slightly different). For tanks of other sizes we use "API 650: Welded Tanks for Oil Storage" for design standards.

The code has specific equations for calculating required wall thickness; depending on required corrosion allowance the wall thickness, typically API 12F tanks, have a wall thickness of either 3/16 or ¼ in (4.76 or 6.35 mm).

We will discuss valves and valve technology in the next chapter, but "tank ball valves" are an important topic to discuss along with tanks. A ball valve has a ball with a hole drilled through the center. When the bore is aligned with the axis of the pipe centerline, the valve allows fluid to flow. When the bore is normal to the axis of the pipe centerline it shuts flow off. Shutting the valve while flowing liquids traps a volume of liquid in the bore. Temperatures below freezing can turn this trapped water into a block of ice, increasing the volume it needs to occupy by about 8.3%. That small volume increase can create extreme pressure increases that can cause the valve body to fail—often emptying the tank onto the ground. To prevent trapping this volume of liquid, there are valves available with the tank side of the ball drilled through to the bore. In this configuration, freezing temperatures tend to push an ice block back into the tank instead of simply breaking the valve.

Tanks specified by the API 12 series can either be heated or unheated, and heated tanks are direct fired. Below-grade API 12 series tanks have evolved over time. When we first started setting them to accept environmental drains and other low-pressure sources, we backfilled the hole to help with freeze protection. Environmental considerations caused this to evolve away from API 12 series tanks by requiring double wall, double bottom, shorter tanks that could be checked for primary fluid containment. This further evolved to prohibit backfilling against buried tanks at all (which reduces the freeze protection significantly and raised the risk that the hole would fill up with water and float the tank like a battleship). Evolution continued further to cause many companies to ban buried API 12 tanks altogether. This ban has led those companies to either use blow cases on environmental drains (rare) or elevate well-site equipment (expensive).

GRP tanks have had a reputation on well-sites of accumulating static electricity which can create a hot enough spark to ignite any flammable vapors in the tank. This has been a major problem that can be solved by following manufacturer's and API 12P grounding instructions, but there have been many field installations over the years that have failed to provide adequate grounding. GRP tanks can be a good choice if properly installed and grounded.

The API specifications call out particular locations for truck loading, drain/sampling, and fill connections, but the end user has the ability to connect the piping in any way that he wants. You occasionally see a tank with the fill line tied into the drain/sampling connection. This can be a very bad idea since an upset like a separator dump-valve hanging open can blow the contents of a tank out of tank vents. The fill connection should always terminate in the vapor space of the tank to prevent an upset from disturbing the tank contents.

5.4.7 Vapor recovery units

There is always some amount of gas accompanying any reservoir liquid to surface. At the low end, gas dissolved in water is a very small quantity.

Entrained gas can be a significant quantity, especially in hydrocarbon liquids. If we don't extract that gas prior to putting it into a tank, then the tank will outgas to atmosphere. The industry would much rather capture that gas for sale than to vent it to atmosphere if there was a commercial or environmental incentive to recover it. The economics of recovering methane from a tank tend to be poor. Arguments by environmentalists that there is a strong environmental reason to recover methane tend to be self-serving nonsense that has been accepted by regulators without question. Sometimes regulations that carry zero scientific basis force actions like vapor recovery of methane from tanks, which is unfortunate but apparently unavoidable. On the other hand, some of the constituents of the outgassing of condensate streams are real pollutants that should not be released to the atmosphere. For example, BTEX (i.e., benzene, toluene, ethylbenzene, and xylene) are all carcinogens that cause both acute and chronic health problems when ingested (the exposure limit for benzene for employees in the United States is 1 mg/L of air for 8 hours or 5 mg/L of air for 15 minutes). Capturing these naturally occurring substances is a good thing.

"Live oil" vs "dead oil" is an indication of the amount of gas that will boil off of the oil at stock-tank conditions. The boiling point of iso-Pentane is 82.1°F (27.8°C). You can put iso-Pentane into a tank as a liquid at 70°F (21.1°C) and it will be a liquid. If a tank heater or the sun shining on the tank raises the temperature by 13°F (7.2°C), then the iso-Pentane will boil off. Normal butane has a boiling point at atmospheric pressure of 31.1°F (−0.5°C). If you raise the pressure to 32 psia (221 kPaa), the boiling point increases to 70°F (21.1°C). Putting 35 psia (241 kPaa) butane into an oil tank will lead to the butane boiling off. A "dead oil" will not have any components that pass their boiling point at storage conditions. The goal of vapor recovery is to provide dead oil to transport mechanism without releasing the vapors to atmosphere.

Equipment to recover the gases that are not stable in a liquid stream is lumped into a class of equipment called vapor recovery units (VRU). This equipment is often a system of components that include some sort of settling or vapor extraction and vapor collection. At its simplest, a VRU system includes sealing a tank to prevent air inflow and pulling the vapors off the tank with some sort of compression equipment. At the other end of the spectrum is a heated pressure vessel rated for vacuum service that is maintained at a low pressure and high temperature to force volatile gases out of the liquid before putting the resulting dead oil into the stock tanks. These units are the most expensive and the most effective.

Tank-based VRU are a very sensitive balancing act. A tank built to API specifications is a reasonably delicate structure. For example, take a 300 bbl (48 m^3) tank which has a diameter of 12 ft (3.7 m). Putting 5 psig (34.5 kPag) in the tank results in an up force on the lid of 81,400 lbf (362 kN)—eight times the up force allowed by API 12F. A small upset in the gas-extraction equipment can cause the lid of the tank to be explosively relocated into an adjacent field (usually to decapitate a rare pregnant racing steer and his yearling calf). At the other end of the pressure spectrum, a small overcompression can put similar forces working toward collapsing the tank. Well-site storage tanks are so sensitive to vacuum that it is reasonably common for painter's tape over the vents to create an environment where simply pumping liquids out of the tank will cause the tank to collapse. Further, a slight vacuum on the tank will cause the vents to ingest air which reduces the commercial value of the extracted vapors. These vacuum problems generally drive operators to maintain a slight positive pressure (on the order of 5 in of water (12.4 mbarg)) and an elevated temperature (usually at least 20°F (11.1°C) higher than ambient).

Pressure-vessel-based VRU have much more ability to operate away from local atmospheric pressure, but they still must have a means to get the liquid into the storage tanks. Common methods of accomplishing this liquid transfer are as follows:

- Maintain the pressure vessel at a high enough pressure to allow liquid to dump into the storage tanks. Unless you also maintain the temperature of the pressure vessel at a high value (i.e., on the order of 100°F (56°C) above storage tank temperature) this method results in the most gas in the tanks.
- Maintain a vacuum on a pressure vessel and install a pump. This option is the most obvious, but it has the problem that most pumps tend to introduce high shear forces on the liquid which can result in emulsifying the oil and lowering its value.
- Use a blow case on the pressure vessel (under vacuum). This option works fine, but many operators are concerned that the gas used for motive force will show up in their tanks. It won't, but that is the concern.
- Use a tall vertical vessel (under vacuum) that maintains a liquid level above the top of the storage tanks. This is the most effective operationally, but these very tall vessels are very expensive and adding heat

requires more equipment (e.g., an external heat exchanger and coils within the pressure vessel to transport heat-exchange liquid).

The other part of a VRU is the vapor collection. This is generally done with some sort of compression (see Chapter 8: Gas Compression). The most common units that we see on well-sites are:

- Oil-flooded screw compressors. These units are generally limited by designers to about 12 compression ratios, so with 10 psia (69 kPaa) suction, you can't reach much higher discharge pressures than 120 psia (827 kPaa). Since these units have compressor oil in contact with a high BTU gas, care must be taken in the selection of the compressor oil to make sure that it expected gases are compatible with the oil.
- Multistage reciprocating compressors. These units are limited by physical conditions to about four compression ratios per stage. A three-stage machine can do $4^3 = 64$ ratios and can take 10 psia (69 kPaa) suction to 640 psia (4.4 MPaa) discharge. These units are common, but keeping the stages balanced with changing suction temperature from day to night is often more effort than the field is able to commit and these machines are very prone to failure in second and third stage rods and valves.
- Thermocompressors. There are a number of commercial VRU that use an ejector (see Chapter 3: Well Dynamics and Chapter 8: Gas Compression) to perform the first stage of compression to allow a two-stage recip to start at (say) 20 psia (138 kPaa) and discharge at 320 psia (2.2 MPaa) with considerably fewer mechanical issues than a three-stage recip.

VRU can extract a very high energy-content gas from the liquids. Methane has an energy content of 909 BTU/SCF (33.88 MJ/SCm). It is common for tank vapors to have an energy content over 1500 BTU/SCF (55.89 MJ/SCm). Since natural gas is sold based on energy content, whenever the sales gas is analyzed it is in the operator's interest to ensure that the VRU is operating. If this high-energy gas is offline when the sample for the next quarter (or even year) is taken then the operator is giving a significant value to the gatherer without compensation.

It is normal for a significant amount of the vapors extracted in the VRU to condense to liquids in the gathering system due to cooling and normal pressure drop across the system. These liquids create operational difficulties (e.g., the only way to remove them is to pig the line), but if they can be successfully transported into a processing plant they have considerable value.

5.5 PRESSURE SAFETY DEVICES

The BPVC requires pressure safety devices (PSD) on pressure vessels under most scenarios. That word, "scenario" is very important to the understanding of PSD requirements. If you have a pressure vessel rated for 1 million psig (6900 MPag), and your most energetic pressure source is 100 psig (690 kPag) then you can apply engineering judgment and say that there is no need "in this scenario" to provide PSD, but there may be things that can happen (i.e., other scenarios) that do put the vessel at risk and require PSD.

The BPVC has generic PSD requirements that must of necessity apply to every industry. For Oil & Gas the API has published API 521, Sixth Edition, 2014 (earlier editions were dual stamped with an ISO code, but this release aggressively removes references to ISO). This standard draws heavily on the BPVC, but is specific to Oil & Gas. The scope of API 521 requires that you evaluate any vessel is built to a pressure vessel code for overpressure protection.

For the most part API 521 only applies to things designated by code as "pressure vessels." API 521 looks to piping design codes (e.g., ASME B31.8 for gas piping) for pipelines and pipeline accessories. There are some notable devices that API 521 would exempt from an API 521 analysis that regulations have specified must be considered for PSD protection. The most common regulatory add-on is that in many jurisdictions, pig launchers and receivers must be evaluated for overpressure protection. Because of the potential for a launcher or receiver to be isolated while liquid full, this is a quite reasonable requirement.

5.5.1 Credible scenarios

The API and ASME started moving away from "equipment-based sizing" to "system-based sizing" in about 1997. Equipment-based sizing looked at the size of a vessel and designated a relieving capacity based on that vessel size without considering what inflow rates were possible for a given location or service. Equipment-based sizing often resulted in overpressure protection that was significantly oversized.

System-based sizing is based on the capacities of the system where the vessel is installed. It looks at what can credibly happen and the volume requirement if that thing does in fact happen. API 521 lists 17 major categories that need consideration:

1. Closed outlets
2. Cooling-water failure to condenser

3. Top-tower reflux failure
4. Sidestream reflux failure
5. Lean-oil failure to absorber
6. Accumulation of noncondensables
7. Entrance of highly volatile material
   a. Water into hot oil
   b. Light hydrocarbons into hot oil
8. Overfilling
9. Failure of automatic controls
   a. Inlet control devices and bypasses
   b. Outlet control devices
   c. Fail-stationary valves
   d. Choke valves
10. Abnormal process heat or vapor
    a. Abnormal process heat input
    b. Inadvertent valve opening
    c. Check valve failure
11. Internal explosions or transient pressure surges (e.g., water, steam, or condensate hammer)
12. Chemical reaction
13. Hydraulic expansion
    a. Cold-fluid shut-in
    b. Lines outside process area shut-in
14. Exterior fire
15. Heat transfer equipment failure
    a. Heat-exchange tube rupture
    b. Double pipe
    c. Plate and frame
16. Power failure (steam, electric, or other)
    a. Fractionators
    b. Reactors
    c. Air-cooled exchangers
    d. Surge vessels
17. Maintenance

Many of these categories are never credible on a well-site (e.g., #3 Top tower reflux failure), some of them are always credible (e.g., #1 Closed outlets). One useful approach is to evaluate a site and select the categories that have a reasonable chance of being credible. Virtually every well-site needs to be evaluated for #1 Closed outlets, #9 Failure of

Table 5.8 PSD credible-scenario example

PSV number: 4	Equipment type: Suction scrubber	MAWP: 500 psig
PSV manufacturer: Mercer	PSV part number: 91-43F51V07P21	PSV size: 2 × 2 "F" orifice
PSV press setting: 100 psig	PSV calculated flow: 972 MSCF/day	Maximum credible flow: 116 MSCF/day
Recommended changes	None	
Possible scenarios	Closed outlet	A blocked outlet could put reservoir pressure on the vessel in excess of PSV set point (116 MSCF/day required capacity)
	Failure of automatic controls	No credible scenario
	Hydraulic expansion	Not credible. While a plugged drain line could result in the vessel becoming liquid full, someone would have to manually shut the inlet and outlet valves after the vessel filled, requires too many unrelated activities
	External fire	Not credible because there is no method of getting the vessel liquid full and isolated without failures that would be unrelated to the fire

automatic controls (the subcategories of that category are not often useful for well-sites), #13 Hydraulic expansion (mostly "a. Cold-fluid shut-in"), and #14 Exterior fire. An increasing number of well-sites also require you to analyze #16 Power failure. Table 5.8 is an extract from a credible-scenario analysis for a specific field.

The fire case seems to be the most controversial. The coefficient of thermal expansion of water is about 100 psi/R (1.2 MPa/K), so if a vessel is liquid full, even a low-grade external fire has the ability to rapidly raise the pressure within the vessel. Gas is a lot more forgiving. Raising the temperature of a vessel filled with gas (no liquid) from 60°F (15.6°C) at 20 psia (137.9 kPaa) to 61°F (16.1°C) would raise the pressure to 20.038 psia (138.2 kPaa), so it is clear that an external fire does not have the potential to rapidly increase vessel pressure. If there is some liquid, it will boil off which will raise vessel pressure, but when you do the volume/pressure calculations they rarely work out to having enough mass of liquid in the vessel to raise the pressure significantly.

The fire case is limited to pool fires (i.e., a standing pool of liquid is burning) since a jet fire directed at a vessel would cause the vessel to fail long before pressure could rise significantly from expanding fluids.

Not considering the fire case will often draw loud and extensive criticism of your work and your overall intelligence. This prejudice is as silly as any other prejudgment, but it is extremely common in our industry. Most of the "you always have a fire case" proponents will accept a reasoned analysis of the actual conditions, but not always. I've seen analysis where it was perfectly clear that the fire case was not a credible overpressure scenario because there were not even any flammable liquids stored on the location (to say nothing of it being impossible to isolate the vessel liquid full) be rejected because "the fire case is always credible." When confronted with that sort of unreasoning nonlogic it is best to pretend that there is a credible fire case and set the required capacity less than the highest flow of actual credible scenarios.

When your inflow source is a reservoir, people frequently do an absolute open flow (see Chapter 3: Well Dynamics) calculation and develop a flow rate based on the rate the reservoir would flow into atmospheric pressure. This is not what the reservoir is actually seeing. In the example in Table 5.8, the pressure safety valve (PSV) is set at 100 psig (690 kPag) and the well was flowing into a 30 psig (207 kPag) compressor suction. If you slam the outlet valve shut, the well is going to want to flow at 30 psig. As pressure builds-up, the flow rate will drop off at the difference between the square of reservoir pressure and the square of bottom-hole pressure. The proper flow rate to use is based on the reservoir flowing into the PSV set pressure.

5.5.2 Double jeopardy

The design of PSD is centered around "credible scenarios." "Credible" is a very important concept. While it is often credible that an operator might shut the outlet of a pressure vessel with a pressure source still able to come in through the inlet, it probably isn't credible that at the same instant in time the displacer fell off the level controller—the two events are unrelated and it stretches credibility that they would happen simultaneously. You do not have to consider multiple, unrelated events occurring simultaneously as long as they really are unrelated.

On the other hand, cascading scenarios may be very important. For example, a pool fire may burn through the gas-supply tubing to the automation equipment which would cause fail-closed valves to go shut. If a site has an emergency shutdown process then the "XV" would go shut on loss of control gas and now we have both an external fire and a closed outlet. That scenario is quite credible and should be considered in setting a required flow rate.

5.5.3 Set points

Vessel MAWP is set by the design and the design codes. Overpressure protection needs to allow adequate outflow to prevent pressure sources (e.g., inflow from the reservoir, or thermal expansion) from increasing the pressure to more than 110% of MAWP while the relief device is open (121% of MAWP for fire case). These limitations on pressure accumulation are described in the BPVC and have been removed from the current version of API 521. For vessels with multiple credible scenarios (each with its own flow rate) it would be reasonable to install a device with a small relieving capacity (sized for the smaller credible scenario) set at MAWP and a larger device set at 105% of MAWP with a combined flow rate large enough to prevent the pressure from building up. In the example in Table 5.8, the vessel MAWP is 500 psig (3.45 MPag) and the device set point is 100 psig (690 kPag) with a flow rate eight times larger than the maximum credible flow rate—certainly adequate to keep the vessel below 550 psig (3.79 MPag). The relief valve set point was set this low because the operators had elected to install a lower pressure accessory on the device which required re-rating the system. BPVC specifies a maximum pressure accumulation during an overpressure event; any combination of set point and flow rate that prevents exceeding this maximum pressure accumulation is acceptable.

Vessel designers frequently install multiple identical relieving devices in lieu of the customer supplying the result of a credible-scenario analysis. Multiple devices can be quite reasonable, but there needs to be separation between the device set points. Every type of pressure-relieving device has an uncertainty or dead-band value. These can be small, but they cannot be zero. Consequently two parallel devices with the "same set point" will have one that falls lower in its dead-band than the other device. In an overpressure scenario, the one with the lowest set point will lift, lowering system pressure. In most cases the second device will not open at all, but when it does it will tend to cause the other device to chatter between open and closed which can damage a device. Because of this dead-band issue, you can't take credit for the second device and each of the two devices must be capable of passing the entire inflow. In order to take credit for both devices their set point needs to be at least twice the magnitude of the dead-band difference (i.e., if the dead-band is 5 psi (34.5 kPa) then the set points need to be separated by at least 10 psi (69 kPa)).

5.5.4 Devices

There is a range of equipment designated as "PSD." These range from one-time use devices that fail catastrophically and cannot be reset to devices designed for such a small operating range that it is very common for them to lift and reseat (e.g., tank pressure/valve vents). In between these two ranges are devices which are intended to operate very rarely and their operation constitutes a significant safety event (as opposed to a normal "control event").

It is important to note that any device controlled by field automation is a "control device," not a "PSD." All PSD must be able to operate independently from program logic. Having a blowdown valve that is part of a process control may be a perfectly appropriate control feature, but it is not part of the overpressure protection.

5.5.4.1 Rupture disk

At the lower range of operational complexity is the "rupture disk." These devices are designed to have a ductile failure at a particular differential pressure. The "ductile failure" indicates that the metal will rip instead of shattering (and potentially sending projectiles downrange). When a rupture disk fails, it is unable to later close and stop flow, it simply becomes an open pipe that must be physically replaced prior to returning to service.

Rupture disks are very subtle engineering creations. Ductile failure of a metal doesn't usually happen at a precisely defined stress. If you take 100 "identical" pieces of steel and test them to failure, the failure point will exhibit a range of values. For rupture disks, that range must be exceedingly small. Consequently, rupture disks must be handled with great care. Any scratch or dent should obviously cause them to be discarded, but even the oil from a worker's hands can significantly change the failure point. The frequency of failures caused by mishandling of rupture disks has soured many organizations on their use and some companies have banned their use, therefore in cases where it seems that rupture disks would be a good choice it is useful to check company polices before committing much effort toward using them.

Rupture disks can have a secondary function of protecting conventional or pilot-operated PSVs from corrosive or toxic fluids. In that service, the failure point of the rupture disk is largely irrelevant (as long as it is below the setting of the PSV), and its reason for being installed is to protect the internal mechanism for the PSV from becoming fouled by process fluids. Installing a rupture disk below a PSV changes the predicted flow rate of the PSV which is accounted for in the flow calculation as given later.

5.5.4.2 Conventional PSV

Conventional PSV use spring tension to hold a disk on a sealing surface. When the net differential forces on the disk are greater from the process fluid than from the spring and exhaust pressure, the disk comes off the seat. These net differential forces are made up of: (1) process pressure applied over the (fairly large) surface area below the disk; (2) exhaust pressure applied over the (smaller) surface area above the disk; and (3) spring force above the disk. The set point of the PSV is determined with local atmospheric pressure above the disk (which is why you do PSV calculations in gauge pressure instead of absolute pressure). When a PSV is installed to exhaust into a flare header, imposed backpressure (say from other PSV lifting at the same time) shifts the set point upward.

In vacuum operations, all of the net forces are pushing the disk toward the seat, but the valves are designed to prevent vacuum and when exhaust pressure is higher than process pressure the PSV will pass exhaust pressure into the process. Conventional PSV can be used in vacuum service with rupture disks located under the valve.

5.5.4.3 Pilot-operated PSV

In a pilot-operated PSV, a secondary device senses process pressure against local atmospheric pressure. When the differential pressure is exceeded, the pilot opens and sends control pressure to the internal piston on the actual process valve. When this operation is trying to maintain an upstream pressure at a predetermined value, it is a "pressure regulator," when it is preventing an overpressure it is a PSV, the two pilots are very similar.

Pilot-operated PSV do not add any appreciable operational value when installed with the PSV exhausting to atmosphere with a very short tail pipe and were historically very rare on well-site equipment. With regulatory proscriptions on venting raw gas to atmosphere, flare headers have become much more common on well-sites and consequently the need for pilot-operated PSV has increased.

5.5.4.4 Tank pressure/vacuum vent

As we've seen, liquid-storage tanks are quite unable to operate with any significant pressure or vacuum. Consequently, equipment designed to protect tanks (Fig. 5.9) tends to operate far more often than other PSD equipment. Outgassing puts increasing mass into the vapor space, increasing pressure. Throughout the day, even an unheated tank will release pressure through the pressure/vacuum vent, and every time atmospheric pressure drops, more gas will leave the liquid. Throughout the night, most tanks will have some amount of condensation, and without the vacuum relief would risk collapsing the tank. Both valve paths on this device regularly have flow.

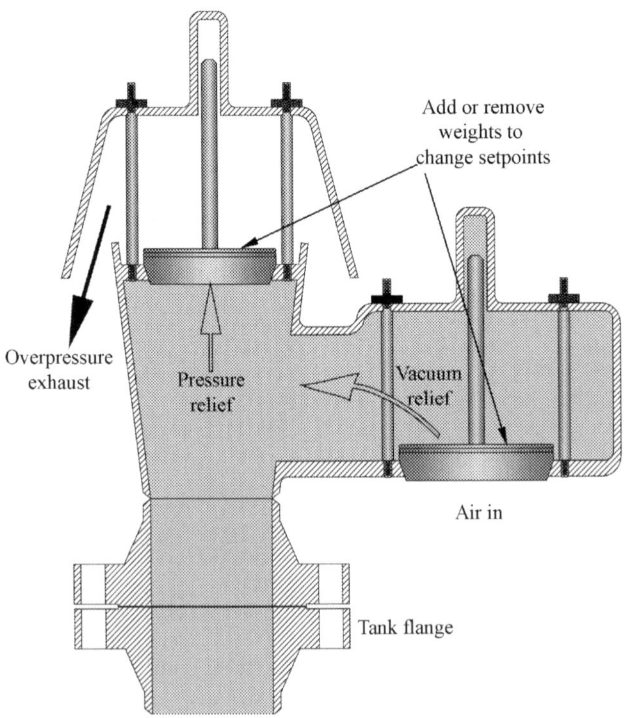

Figure 5.9 Tank pressure/vacuum vent.

5.5.5 Flow rate determination

All gas PSV are "critical flow devices" (i.e., they satisfy the requirements of Eq. (4.8)). This requirement makes the exhaust piping length and size to be of critical importance. If the friction losses in the piping to a flare header causes the pressure on the downstream side of the PSV to exceed the pressure allowed by Eq. (4.8) then the flow rate will be significantly less than the flow rate calculated as given in the next section. Confirming the friction loss in the piping when each PSV is flowing alone is only the first step. You also have to confirm that if all connected PSV lift concurrently, the flare header and the flare nozzle will allow enough flow to keep all of the PSV in critical flow. This calculation is a major reason that most well-site flare headers are inadequate. It is nearly always done properly in plants and rarely done adequately on well-sites.

5.5.5.1 Rate calculation

Flow rate is a function of the upstream (process) pressure, pressure (assumed to be less than Eq. (4.8)), sonic velocity, and the size of the opening within the valve:

$$C = 520 \cdot \sqrt{k \cdot \left(\frac{2}{k+1}\right)^{\frac{k+1}{k-1}}}$$
$$A_{orifice} = \frac{\dot{m}_{required}}{C \cdot P_{up_{psig}} \cdot K_d \cdot K_b \cdot K_c} \cdot \sqrt{\frac{T_1 \cdot Z_1}{MW}} \quad (5.10)$$

The constants are:

- K_d → Discharge constant, from manufacturer (for valve selection use 0.975)
- K_b → Backpressure constant, from manufacturer (for valve selection use 1.0)
- K_C → Rupture disk constant (use 0.9 if a rupture disk is present under the PSV, 1.0 if absent).

Once you've calculated the area required, round it up to a standard value and use Table 5.9 values (the first number is PSV inlet pipe size and the second number is the PSV outlet pipe size).

With a selected PSV, you can calculate a valve capacity with the actual orifice area and actual valve-specific constants from the manufacturer using Eq. (5.11).

Table 5.9 PSV orifice flow area for standard sizes

	Area (in²)	Area (cm²)	1×2	2×2[a]	2×3	3×3[a]	3×4	4×6	6×8	6×10	8×10
D	0.110	0.710	X								
E	0.196	1.265	X	X							
F	0.307	1.981	X	X							
G	0.503	3.245		X	X						
H	0.785	5.065		X	X						
J	1.287	8.303		X	X	X	X				
K	1.838	11.858				X	X				
L	2.853	18.406					X	X			
M	3.600	23.226						X			
N	4.340	28.000						X			
P	6.380	41.161						X			
Q	11.050	71.290							X		
R	16.000	103.226							X	X	
T	26.000	167.742									X

[a]Equal inlet/outlet combinations only available in threaded connections, flanged valves always have a larger outlet than inlet.

$$\dot{m}_{design} = A_{orifice} \cdot C \cdot P_{up\ psig} \cdot K_d \cdot K_b \cdot K_c \cdot \sqrt{\frac{MW}{T_1 \cdot Z_1}} \tag{5.11}$$

5.5.5.2 Exhaust forces

If a rocket can lift a mass into space, it stands to reason that a jet exiting a pipe at sonic velocity will exert some measurable force on the piping. That force is:

$$F_g = \left(\frac{K_d \cdot A_{orifice} \cdot (P_{up\ psig} + P_{atm}) \cdot K_f}{1.383 \cdot A_{outlet}} - P_{atm} \right) \cdot A_{outlet}$$

$$\text{If } F_g < 0 \Rightarrow F_g = 0 \tag{5.12}$$

$$F_r = \frac{C \cdot K_d \cdot A_{orifice} \cdot (P_{up\ psig} + P_{atm}) \cdot \sqrt{\frac{k}{k+1}}}{332.7} + F_g$$

The new constant (K_f) is a function of the adiabatic constant of the process gas (Table 5.10).

Table 5.10 API 520-2 reactive force Table 5.A1

k	K_f	*k*	K_f
1.01	1.15	1.55	0.95
1.05	1.13	1.60	0.94
1.10	1.11	1.65	0.93
1.15	1.09	1.70	0.91
1.20	1.07	1.75	0.90
1.25	1.05	1.80	0.89
1.30	1.03	1.85	0.87
1.35	1.02	1.90	0.86
1.40 (air)	1.00	1.95	0.85
1.45	0.98	2.00	0.84
1.50	0.97		

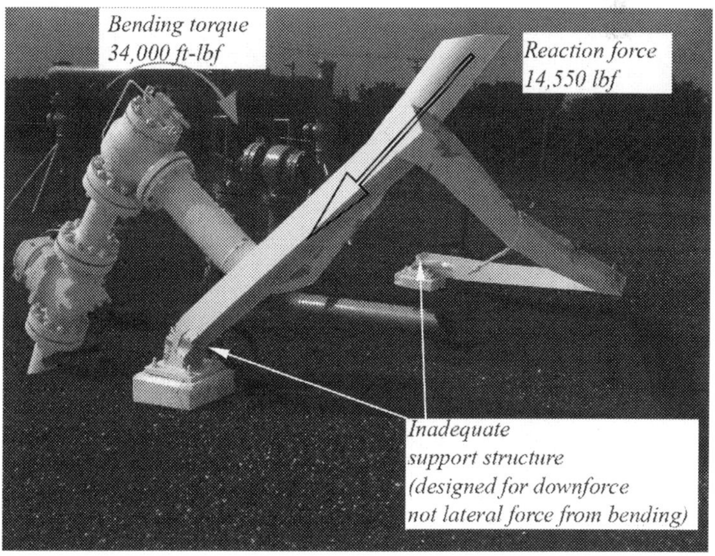

Figure 5.10 PSV forces. *US Department of Transportation, Failure Investigation Report—El Paso-Mojave GT 2012-5-2.*

The results of this calculation can predict a considerable force that must be countered with adequate support. It is common for the (unsupported) tail pipe on threaded PSV to unscrew from the valve and create a flying-debris hazard. Flanged PSV don't unscrew, but they can experience significant bending forces (Fig. 5.10).

5.6 WELL-SITE PROCESS CONTROL

Typical well-site process control is pretty simple. Separators need liquid-level controls. Separator heaters and many tanks require temperature control. Compressors often require suction-pressure control, discharge backpressure control, and/or recirculation control. All of these processes can be independently controlled without requiring a Program Logic Controller (PLC). In fact, the outcome of the control process is often far superior if there is not a PLC to act as a gatekeeper.

Historically the well-site control philosophy was one-controller-one-end-device. This says that a level controller output goes to an end device called a "dump valve," a suction controller uses upstream gas to operate a pneumatic control valve, a temperature controller sends an on/off signal to a burner supply valve, etc. This is the "old way" and the vast computing power in today's PLC is just aching to be used. Do your best to avoid the trap of using it.

I was recently asked to help a company troubleshoot a very expensive compressor skid that was not working properly. I found that they were using the on-skid PLC to control the suction control valve, the discharge backpressure valve, the recirculation valve, and even the control valves for the blow case. In addition to these control functions, the PLC had the assignment of gathering hourly performance parameters, connecting to the Internet, and sending the data to a database in the cloud. Unfortunately it was a simple-minded PLC that could not start a step until the last step finished. Since everything in the entire system reported on the top of the hour, it could sometimes take 2–3 minutes to accomplish the reporting task. When the well would experience a drop in flowing bottom-hole pressure at the top of the hour (maybe due to a dip in the liquid-level downhole), suction pressure would increase, which should cause the suction controller to go toward shut and keep the increased mass flow rate from tripping the engine on low rpm, but the PLC was sending data that no one ever looks at to a database that no one can find. When a slug of liquid hit the suction scrubber on the top of the hour, the blow case controls couldn't be told to work and the skid would go down on the high scrubber-level kill (which luckily didn't go through the PLC). When a slug of liquid hit the scrubber a few seconds before the top of the hour, the dump would open and then the PLC was too busy to close it and for several minutes the skid would blow a steady stream of gas into the tank, unfortunately the end of the pipe was submerged so it would blow the tank dry. There were 30 other scenarios that were interrupted by the data gathering subroutine. As a short-term "fix" they set a second PLC that only had reporting assignments and left the original PLC to do "Programmed Logic Control." Eventual solution was to replace all of the PLC control with local control (see later) and to not allow the PLC to control any of the compressor processes except engine rpm and engine load control. This was an extreme example and a particularly lame program, but it is by no means unique.

A control system is made up of: (1) a sensing element; (2) a control element; and (3) an end device. The sensing element can be something as simple as the float in a toilet tank. That system includes a flush handle (that is not part of the control logic) which drains the tank when depressed. The "sensing element" is the tank float that sees a decreasing tank level. The control element is the linkage from the float to the water-supply valve that is direct-connected to the float arm. The end device is the water-supply valve. It is easy to see that this system could have a radar-level sensor connected to a PLC that can open a solenoid on the water-supply line, but why in the world would you do that? It would only add complexity that does not add value.

The sensing element is a device that can discriminate the condition of a process variable and can communicate that condition to a controller. Sensing elements can be mechanical (e.g., a bimetallic temperature sensor has two metals with different coefficients of thermal expansion fused together, as temperature changes the two metals expand/contract differently which changes the curvature of the element), electrical (e.g., a "Mag flow meter" senses fluid velocity by evaluating the changes in the magnetic field generated by the flowing fluid), or hydraulic (e.g., a level float in a separator or tank).

The control element is anything that can cause an end device to change state in response to the input of a sensing element. A PLC is the most obvious example of an electric control element, it can take the input of (multiple) sensing element(s) and evaluate that input and change the state of (multiple) end device(s). At the other end of the spectrum is a temperature controller that can react to the curvature of a single bimetallic element to turn gas on or off to an end device controlling fuel flow to a burner.

The end device is simply something that can institute a physical change to the real world. End devices can be anything from a pneumatic actuator to open or close a gate, to your electric garage-door opener, to a valve that throttles process flow. In Oil & Gas the term "end device" is always a valve.

Process control systems can either be: (1) pneumatic (historically most common on well-sites), (2) electric, or (3) local. Hydraulic controls are common in other industries, but are very rare in well-site applications.

Local devices like the pressure regulator in Fig. 5.11 are autonomous units that have everything they need to function in a single package. There is no need to vent gas, there is no need for an external energy source or external program logic. There is also no need for the exhaust stream that has become such a large issue in regulatory oversight. Local devices are generally limited to pressure control (including pressure regulators that manage based on downstream pressure and backpressure regulators that manage based on upstream pressure) which can also be done either with pneumatic or electric control. Any job that can be done with local control probably should be done with local control.

Electric controls have historically been limited to small solenoid valves that started or stopped control gas. Government regulators have issues with using natural gas for process control because the gas is eventually released to atmosphere and that somehow is causing the climate to change (these overblown government actions in the name of controlling

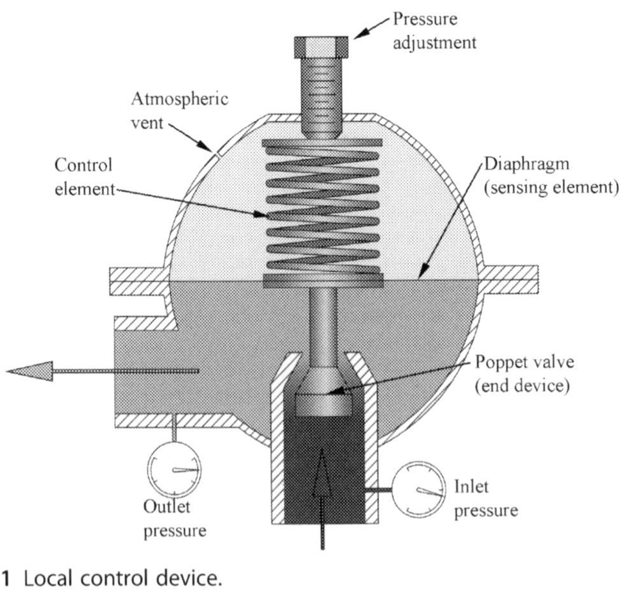

Figure 5.11 Local control device.

anthropogenic global warming have continued to get progressively more intrusive with progressively less impact on the actual environment). Consequently, operators are required to inventory the so-called greenhouse gases and to reduce the "emissions" of these gases year on year. Pneumatic controls using natural gas are being regulated out of existence. Over the last few years, some very high-performing linear operators using high-tech motors have come on the market. These linear operators are typically retrofit onto the bodies of valves that have previously had pneumatic operators. The new operators are characterized by very low power consumption and excellent, verifiable reliability. They really change the well-site control universe. For example, for a separator dump valve you can use an electric-level controller whose sensing element is indistinguishable from a pneumatic-level controller and has a simple switch to turn a linear electric motor on or off to reposition the end device, but that is not the only way to do this task. Linear electric-actuator technology allows you to explore technology that is not available for pneumatic controllers like radar devices, capacitance devices, and vibrating fork devices which all have potential for increasing reliability over a traditional float. Even with the apparent flexibility of electric control devices, the one-controller-one-end-device control philosophy should be continued to electric devices—adding control logic has a high potential for failure with very low potential for adding value.

5.6.1 Pneumatic control

The workhorse technology for well-site use is pneumatic controls, using well gas for control gas. Terminology in pneumatic control is confusing and simple understanding of terms goes a long way toward reducing confusion.

- Type of controller
  - *Continuous bleed*: Source gas always has an open path from the source to the end device, and end-device pressure is changed by adjusting the rate that gas released to atmosphere.
  - *Intermittent vent*: The controller determines whether control gas pressure should be directed to the end device or to isolate the control gas from the end device and the vent end device to atmosphere. No gas is purposely exhausted to atmosphere when the end device is "at rest."

- The defining difference is that a "bleed" controller does not have a mechanical barrier between the supply gas and the end device while the "vent" controller does have such a mechanical barrier.
- Type of service
  - *On/off*: When the input exceeds the set point, the controller sends pressure to the end device. When the controller senses that the condition has cleared it lowers pressure to the end device allowing it to go into its at-rest position.
  - *Throttling*: The controller is able to send a variable signal to the end device to allow it to maintain an intermediate position.
  - The defining difference is that "throttling" devices are able to maintain pressure in an intermediate position and an "on/off" device will send gas to the end device until the pressure on the end device is equal to the control gas pressure.
- "Snap vs proportional" is often used by manufacturers to further differentiate their equipment from the competitor's equipment. This language is used often enough for it to be helpful to explain the terms.
  - *Snap acting*: This means that the supply fully opens at a maximum signal and fully closes at a minimum signal.
  - *Proportional*: This implies that the supply valve will open proportionally with a change in input (e.g., as the process variable being controlled starts to go above the minimum value the supply valve cracks open and if the process variable keeps increasing, the controller opens further until it is fully open). Proportional control should not be confused with "throttling" since the supply valve simply opens or closes more in response to a value above minimum and even at a minimal opening the line to the end device will quickly reach control gas pressure in most systems.
  - It is common to think about "snap acting" as working like a traditional home light switch and "proportional" acting like a dimmer that ramped the light up to full power as it is left in an intermediate position. Snap vs proportional is not a defining characteristic of pneumatic controllers.
- Actuator condition
  - *At rest*: Valve position with the actuator vented.
  - *Actuated*: Valve position with some amount of gas pressure on the actuator.

Table 5.11 provides the defining parameters for a pneumatic controller.

Table 5.11 Categories of pneumatic controller

		Type of service	
		On/off	**Throttling**
Type of controller	Intermittent vent	Mechanical barrier between supply and end device, unable to sustain an intermediate valve position. Vents on de-actuation with emissions near zero between cycles	Mechanical barrier between supply and end device, able to sustain an intermediate valve position. Vents some gas pressure when valve needs to move toward closed
	Continuous bleed	No mechanical barrier between supply and end device, unable to sustain an intermediate valve position. Bleeds continuously, exhaust rate slows while process is "on," but average rate is about constant	No mechanical barrier between supply and end device, able to sustain an intermediate valve position. Bleeds continuously, rate varies slightly with actuation but average rate is about constant

5.6.1.1 Source gas

High-quality gas is required for process control. It should not have any free water, in fact it should have as little H_2O in any phase as possible. It should also be at a constant pressure.

There is a device in widespread use called a "fuel-gas dryer" which is neither a "dryer" (which would imply it has dehydration capability, which it doesn't) nor is it reliable. These units are typically made from 6 in Schedule 80 (150 DN) pipe with an ID of 5.761 in (146.3 mm) in a clumsy attempt to avoid requiring a code stamp on the vessel. It does not have a mist extractor or (generally) automatic level controls. Any liquid that it happens to remove from the raw field gas accumulates in the bottom of the vessel and must be manually drained (which should happen every couple of hours but typically happens every few months). Pneumatic control problems can often be traced to these inadequate vessels.

A cost-effective source of fuel gas that avoids most of the problems with "fuel-gas dryers" is called a "cold finger" (Fig. 5.12). This appendix

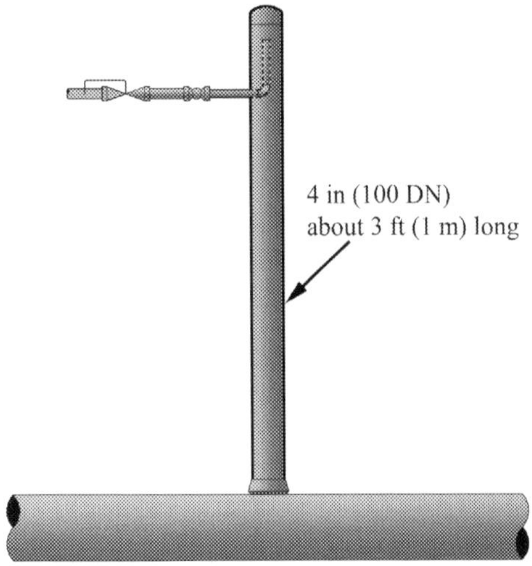

Figure 5.12 Cold finger.

on the separator outlet line is long enough, with enough exposed surface area to cause considerable condensation, and any liquid that does drop out of the raw gas stream will drop back into the process stream. I have used these devices on lines that regularly see ambient temperatures below −40°F (−40°C) without any tendency to freeze the control gas regulator.

When selecting a source-gas location for using raw well gas, it is important to spend a few minutes *thinking* about current and future rates, pressures, temperatures, and equipment. Will the well ever have a well-site compressor? If so you might want to include a compressor manifold on the separator outlet piping and put the fuel-gas source on the compressor discharge side of the manifold. If you are using a fuel-gas dryer and "can't" include automatic liquid-level control, is there anything you can do to keep it from standing full of liquid? Maybe drain it into a small commercial blow case with automatic level control. If you put a cold finger on the separator outlet piping on a vertical vessel it will be very high in the air, how do you plan to access the pressure regulator when it (inevitably) fails? There is not one answer to this question, but you need to have thought about the question.

5.6.1.2 Controller/sensing element

In pneumatic control all controllers fit in to one of the four boxes in Table 5.11. The API is currently developing a standard to help people quantify the emissions from pneumatic controllers that want to create many more categories, but you can determine what line in Table 5.11 a controller fits in by seeing if there is a way for the controller to block gas going to the end device. You can determine the column by seeing if the controller can hold the end device in an intermediate position. If a controller has a block between the source and the end device then it is intermittent vent regardless of what happens when the block is open (there are controllers on the market that have an isolation but once it is open they bleed continuously—this makes the controller an intermittent vent device).

It is normal in pneumatic service for the sensing element to be integral to the control element. The sensing element can be a float, diaphragm, bimetallic element, or a bellows. The control element is typically a rod or a series of rods that can transmit sensing element movement to alter the control signal to the end device.

5.6.1.3 End devices

Pneumatic end devices are nearly always valves on well-sites. In Fig. 5.13, the valve is set up in a "pressure to open" configuration. There is no way to tell from this figure if it is set up to throttle or be on/off since that is a

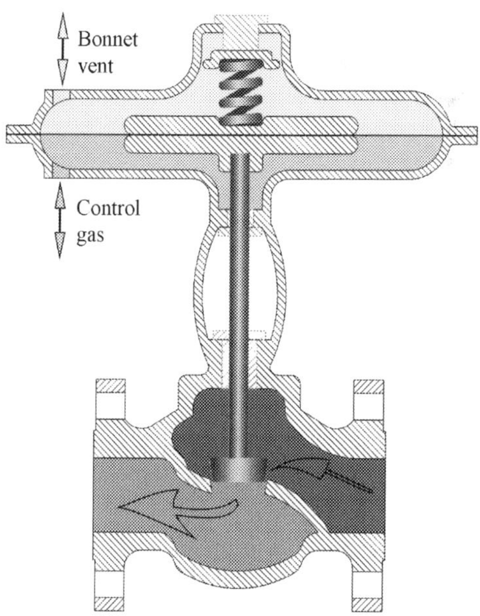

Figure 5.13 End device.

function of the controller, not the end device. On some manufacturer's valves, to change it to "pressure to close" you swap the bonnet vent and the control gas in connections and reverse the diaphragm carrier so that the spring pushes up instead of down.

For any pneumatic end device there will be a way to make supplied pressure act against a spring to move a stem connected to a valve seat. Just looking at a control valve it is impossible to determine the service it is put to. Pneumatic valves have different MAWP for the process body than from the actuator bonnet. The bonnet MAWP is usually less than 45 psig (310 kPag). The body can be any value needed for a process application. Many purchasers of these valves have confused the bonnet MAWP with the process MAWP and improperly rejected valves that were adequate for the service.

5.7 FLUID MEASUREMENT

All gas wells require volume information to be related to specific wells. Many regulatory agencies require monthly reporting by well. Many contracts require settlement by well. Reservoir analysis requires knowledge of the material removed from the reservoir. Reservoir fluids are measured multiple times: at the well-site for gas; at field-aggregation points for gas and oil (e.g., Lease Automatic Custody Transfer (LACT) meters for oil at the lease level; and field "Central Delivery Points (CDP)" for aggregated gas); at plant inlets; at each product outlet of a plant; at entry to transportation pipelines; at entry to local distribution systems and/or end-user facilities; and at final use. It is common to do a material balance for each of these subsystems and develop a "fuel, lost, and unaccountable" or "shrinkage" number. Since commercial transactions in gas are based on energy content rather than mass or volume, it is also common to do both a mass and an energy balance for field gas (commodity gas at the outlet of processing facilities has such a narrow energy per unit-mass range that mass balances and energy balances result in the same shrink so it is common to only do a mass balance on commodity gas). Any transportation agreement will have a shrinkage number in it, so it is very important for facilities engineers to understand both the magnitude of contractual shrinkage and whether that number is volume, mass, or energy. It is rare, but not impossible, for a contract to state a gas shrinkage in volume flow rate at actual conditions. Field shrinkage of oil is nearly always stated at actual volume. Both mass (stated through the surrogate of SCF or SCm) and energy are common for gas shrinkage.

If your gas has 20 volume percent CO_2 and 80 volume percent methane, then in the gathering system, midstream transportation, and compression stations you will get about the same shrinkage using either mass or energy. When that same gas goes through a sweetening plant, something like 55% of the incoming mass will not be present in the gas-sales stream, but something like 98% of the energy will still be present—in this case a mass-based shrinkage number would not be in anyone's interest, but an energy-based shrinkage is quite reasonable.

5.7.1 Key concepts

All of the hand-offs that occur between leaving a well-site and entering a burner require fluid measurement. The problem is that we have absolutely no way to "measure" velocity, volume flow rate, volume flow rate at standard conditions, mass flow rate, or energy flow rate. None. We can't do it. There are things that we can measure. We can measure the amount that a curved tube straightens in response to an increase in internal pressure. We can measure the electrical resistance change with respect to a known current at a known voltage through a specific metal in a changing temperature environment. With these inferred temperature and pressure values and measured pipe ID and orifice diameter we can infer a velocity through the orifice diameter. If the makeup of the fluids has not changed since the last time we analyzed a fluid sample then we can infer a mass flow rate and therefore a volume flow rate at standard conditions. But we didn't measure any dynamic flow property, and certainly didn't measure velocity. The key concepts of fluid flow measurement are "inference," "latency," "accuracy," "uncertainty," and "repeatability."

Inference: We use the few parameters that we can measure, many assumptions, and theoretical concepts to turn the scarce data into flow by using "inference." For example, a turbine meter has a vane pack on a wheel that rotates in response to a flowing fluid striking the vanes and transferring flow energy to rotational energy and then we count rpm. If we know the density of the fluid (and the fluid is homogenous without any included drops or bubbles) and the coefficient of rotating friction (including windage losses) and the mass of the vane pack, then we can infer a rate of momentum transfer which (using all the same values) lets us infer a velocity. If our density input is low, then the momentum of the fluid is low and since we have assumed a constant density the fluid velocity is reported as lower than it actually is.

Latency: Every sort of meter requires some time to "steady out." While the device is in transition from one state to another it is impossible to determine a flow rate because the underlying assumptions are not satisfied. Mechanical devices must "come up to speed," fluid devices must stabilize flow perturbations, and electrical devices must stabilize current flow before their underlying assumptions are valid.

As we will see later, a "Mag flow meter" infers velocity from a magnetic field which you would think can be communicated at the speed of light. Fig. 5.14 is an extract of data gathered at 1-second intervals used to evaluate a new type of vessel. As you can see from this data, the flow rate continued to increase for 1.75 seconds after the flow was stopped by shutting a valve. This shut valve took the actual flow rate to zero, but the Mag flow meter (bottom line in graph) still indicated a near-average flow for the entire 24 seconds that the valve was shut. At the end of the latency period, the flow rate increased as the pressure increased (and in this application, increasing pressure after the valve opened is an indication

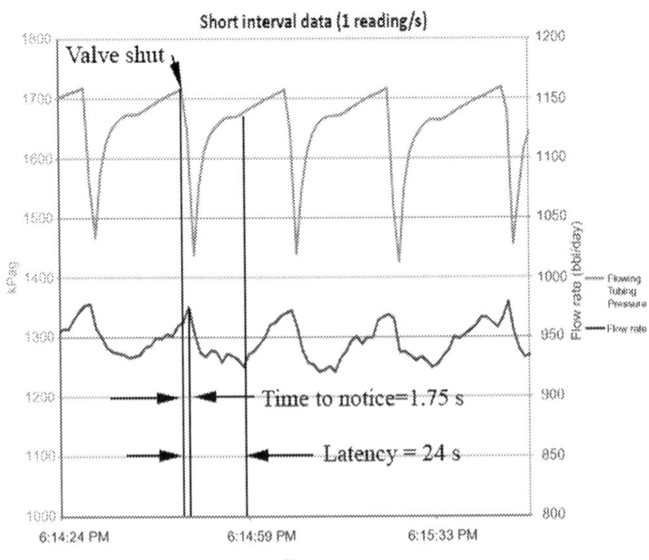

Figure 5.14 Latency example.

Table 5.12 Latency and inference

Technology	Inference parameter	Latency
Square-edged orifice	dP and temperature	4–20 seconds (depending on flow profile)
V-cone	dP and temperature	2–8 seconds (depending on density)
Vortex	Pressure changes from alternating vortices	10 seconds
Turbine	rpm	30 seconds
Coriolis	Vibration frequency	20 seconds
Ultrasonic	Doppler shift in sound frequency	45 seconds
Mag flow meter	Measured voltage	25 seconds
Pitot tube	dP	2 minutes

of increasing flow), which was verified by watching the measured liquid flow into a tank.

Table 5.12 provides what we are using to infer a velocity and how long it takes to steady out for some of the commonly available fluid measurement options. All of these technologies have a place where they would be a first choice, but none of them is universally the clear winner.

Accuracy: This is defined as "degree of conformity of a measure to a standard or a true value" (Webster), which should be an important part of fluid measurement, but isn't. There is no "standard" or "true value" to conform with. Any fluid measurement professional that uses the word "accuracy" most likely does not understand their topic and should be questioned closely prior to accepting their word on anything.

Uncertainty: Rather than "accuracy," fluid measurement professionals talk about "uncertainty." Uncertainty is the degree of not knowing. If you say that a pressure gauge "has an uncertainty of ±0.5% of full range," you are making a statement that can be verified against an objective standard. If you have a 0–100 psig (0–690 kPag) pressure gauge that has an uncertainty of ±0.5% then the uncertainty means that every reading you take from that gauge must be within ±0.5 psig (±3.45 kPag) from a value that would be determined by an objective standard. If a pressure reading with this instrument is "71 psig" then there is confidence that the real value is in the range of 70.5–71.5 psig. In flow measurement, you can take the uncertainty of every instrument, assemble those values in a way that honors their importance in the flow calculation (e.g., a parameter that only appears within a square root would have less impact on the total uncertainty than one that was squared) then you can determine a cumulative flow measurement uncertainty. An instrument that has a 0.1% uncertainty will give you data that is closer to reality than an instrument that has a 2% uncertainty, but will almost certainly cost much more. Balancing purchase cost vs operating cost vs uncertainty is a major part of the job function of a measurement professional.

Repeatability: "Uncertainty" speaks to the range of likely outcomes around a nominal value. Repeatability speaks to whether the nominal value reflects physical reality. For example, an expert marksman can place a number of shots into a very small pattern, but a small maladjustment of her sights can move that pattern away from her actual target. In that case she has good uncertainty and poor repeatability. This concept is called repeatability in fluid measurement because it is thought that if the marksman picked up a second weapon she would produce a second pattern and if there was no maladjustment in either weapon, the second pattern would overlie the first. When we think about fluid flow, any high-repeatability instrument would yield the same nominal value in the same flow stream every time. A low-repeatability instrument would yield different nominal values from one instrument to the next.

5.7.2 Makeup of a flow measurement system

A measurement system is made up of three classes of components: (1) primary element, (2) sensing element, and (3) recording element.

Primary element: experiences pressure, temperature, and flow velocity, but has no electronics or recording capability, it is mostly just a carefully designed lump of metal and maybe plastic.

Sensing element: experiences the parameters being used to infer flow (e.g., pressure, temperature, rpm, and number of pressure surges), but is not able to sense flow. It is made up of electronics and exotic metals.

Modern electronic analog sensing elements communicate to the recording element using either a 4–20 mA, 1–5 VDC, or Modbus. With 4–20 mA far more common on well-sites. When connecting a sensing element using a 4–20 mA control loop you tell the recording element that 4 mA means zero. Then you tell it what 20 mA means (say 100 psig (690 kPag)), then the recording element is able to convert any intermediate reading into a pressure (e.g., a 12 mA signal with a 0–100 psig calibrated range would be 50 psig (345 kPag)). Most pressure transducers are able to be adjusted to indicate 20 mA at any value between zero and their inherent maximum value. It is often useful to change the calibrated range of a dP transmitter on a square-edged orifice meter installation to allow precise measurement under changing conditions without changing the orifice plate (you simply do a calibration, apply your new maximum dP to the transducer, and adjust the output to read 20 mA). It is important (but frequently overlooked) to also change the maximum value in the recording element.

Recording element: is either electronics or pen and ink and sees electronic signals or pen movements. Historically the recording element was a circular chart and pens connected to pressure/temperature sensors that marked a trace on the chart that could be "integrated" by having a person follow the line with a stylus that digitized the traces and allowed the pressure/temperature data to be converted to a volume.

Today the recording element is usually a remote terminal unit (RTU) which is a small computer (typically less computing power than a smart phone) that has inputs for all of the sensing elements and outputs for some (small) number of digital control outputs. RTU equipment often has some specialized data storage for the parameters necessary for flow calculations (e.g., gas analysis, liquid density/SG, meter tube ID, orifice size, and volume/pulse). It will always include specialized subroutines that convert the output from the sensing element to a flow rate (either volume flow rate at standard conditions or mass flow rate). Finally the RTU often has the ability to transfer the accumulated flow data to a central repository (that is increasingly on the Internet instead of a local database). A single RTU can accept between 2 and 20 flow measurement devices.

5.7.3 Water measurement

A significant quantity of liquid measurement (much water and virtually all condensate) is measured on well-sites using tank gauges and truck "run tickets." This measurement technique is eventually very accurate, but the flow of paper and the ability to relate a run ticket to a particular tank can be challenging. When we try to meter liquids into tanks, on every timescale, the flow measurement is higher than run tickets, sometimes significantly higher. That discrepancy is largely due to the dump event being shorter than the latency of the meter. For example, when a turbine meter is shut off, the liquid will tend to be replaced with gas to some greater or lesser extent. When the dump valve opens, this gas flows very fast and over-speeds the turbine meter. By the time the gas is gone and the dump valve is closing, the meter has not yet slowed to a speed consistent with actual flow rate, so for the whole dump cycle the meter is recording a higher flow rate than is actually moving. This phenomena has caused no end tail-chasing work and finger pointing.

When liquids (almost always water) are sent directly into a water gathering system it is usual to try to measure it. Many technologies have been used for water measurement, this section reviews a few of them.

5.7.3.1 Turbine meter

The momentum of a flowing fluid has the ability to do work on a structure located in a flow stream. In the case of a turbine meter, the fluid is spinning a wheel and the assumption is that the work that the fluid does on the wheel is proportional to fluid velocity. In a steady (i.e., no start/stop latency issues) flow, water measurement with a turbine meter can be pretty good. It is not so good with gas flow. If you define fluid momentum $= \rho \times v^2$ then pure water with a velocity of 10 ft/s (3.05 m/s) and slightly above atmospheric pressure would have 13 times more momentum than a methane stream at 100 psia (690 kPaa) and 60°F (15.6°C) flowing at 40 ft/s (12.2 m/s), which gives the water flow much greater resolution to see small changes in velocity reflected in the angular velocity of the wheel.

5.7.3.2 Vortex meter

We've all seen a flag waving in the breeze. It moves like it does because when a flowing fluid impacts on a bluff body in the flow stream it sheds small vortices to either side of the bluff body. These small swirls are called "von Karmen Vortex Streets" and as you can see by the movement of a flag they have nonzero energy. The magnitude and frequency of the swirls is proportional to the fluid velocity. Vortex meters were proposed many years ago, but the electronics required to differentiate individual swirl-pulses (as pressure changes) and to accumulate them wasn't available until the 1990s. The Oil & Gas industry is not quick to adopt new technology, so these meters are not terribly widespread as of this writing, and efforts to use them in gas by isolated individuals have not met with a lot of success. There have been some very successful applications of this technology measuring the flow rate of liquid from separator dumps in unconventional gas fields due to the relatively short latency period.

5.7.3.3 Mag flow meter

Faraday's law can be stated as "generated voltage is proportional to the product of fluid velocity, magnetic field strength, and the length of the conductor." You can measure voltage, field strength, and the length of the conductor, so you only need to establish the constant of proportionality to determine velocity from parameters you can measure. This technique requires that the pipe be full of a conducting liquid—they don't work in distilled water or any gas.

5.7.3.4 Coriolis meter

Proponents and manufactures of Coriolis meters claim that these meters "directly measure density" and "directly measure mass flow." Both claims are utter nonsense. The device has a bent tube located within some very sensitive sensors. The frequency of vibration of the tube in response to excitation is said to be a direct measure of density. It is actually a direct measure of vibration frequency. A slug of gas will change the vibration frequency, but latency in the measurement of the frequency will cause the indication to lag the change and for a time that can be as much as 2 minutes, the density used in the calculation will have no relationship with actual density. There are other normal conditions (e.g., intermittent road traffic and vibration frequencies imposed on the pipe by compression, etc.) where a change unrelated to changing density can change the vibration frequency.

To measure flow rate, the sensors look at the amount of displacement that the tube experiences from the momentum of the flowing fluid. Compare this displacement at the beginning of the tube to the displacement at the end of the tube and the transit time of a given wave to determine a flow rate. It is a direct measurement of pipe displacement, not of flow.

Coriolis meters do a good job of measuring liquid (I've had less success using them in gases) and their uncertainty and latency are acceptable. My biggest objection to them is the hype of the marketing material. They are not magical, they are simply decent technology that does an acceptable job and are kind of expensive.

5.7.3.5 Ultrasonic meter

The speed of sound in a given fluid at a given temperature can be determined with high accuracy. Speed of sound is fast, but it is only one to three orders of magnitude faster than typical fluid velocity in commercial applications. This difference is small enough that the flowing fluid can impact the transmission of a sound pulse from a sender to a receiver in a measurable way. There are two distinct technologies used in ultrasonic meters: (1) time of flight and (2) Doppler shift. Time of flight meters have two senders and two receivers—one pair has the sender downstream and the receiver upstream, the other pair has the sender upstream and the receiver downstream. The time of flight of the sound pulse from the downstream sender to the upstream receiver is delayed by the flowing fluid retarding the pulse. The time of flight from the upstream sender to the downstream receiver is shortened by the fluid carrying the sound with it. Using this information, software can be developed to convert the time of flight of the two pairs into a velocity that can be converted to a flow rate using known fluid properties and measured temperature.

Doppler shift is the change in the apparent magnitude of a frequency or wavelength for an observer relative to a moving source. In astronomy, scientists look at distant objects moving at a significant fraction of the speed of light and measure the "red shift" to determine how fast the object is moving away from us. Commercial fluids are moving far too slowly for this technique to be useful with light, but those fluids are moving at a significant fraction of the speed of sound and that frequency shift is measurable. This is like standing next to a train track and noticing that the sound of a train going away from you sounds different from the sound of the same train coming toward you.

Eq. (5.13) makes it clear that the Doppler-shift functionality of an ultrasonic meter provides a reasonable way to determine a Mach number. If you have an accurate (and constant) fluid composition and temperature then this equation yields a fluid velocity.

$$\begin{aligned} f_{apparent} &= \left(\frac{v_{sonic} + \cancel{v_{receiver}}}{v_{sonic} + v_{source}}\right) \cdot f_{emitted} \\ v_{souce} &= \left(\frac{f_{emitted}}{f_{apparent}} - 1\right) \cdot (v_{sonic}) \end{aligned} \tag{5.13}$$

There are uncertainties in both the time-of-flight calculation and the Doppler shift calculation which leads manufacturers of many ultrasonic meters look at both techniques and compare the results in a proprietary weighting that varies by manufacturer based on their assessment of the best way to minimize uncertainty.

5.7.3.6 *Blow case dump counter*

The simplest liquid measurement on a well-site, with oddly enough the lowest uncertainty, is to put a counter on a separator blow case. The blow case fills to the same point every cycle and is isolated from the main vessel during the outflow portion of the cycle which prevents transient volume from being an issue while the vessel is drained to the same level every cycle (same volume transferred every time). Most well-site RTU have a "pulse input" that is designed to be used with a turbine meter (every pulse is a known volume) that is ideal for a pulse sensor connected to the control gas on the blow case dump. Where these have been used on well-sites with on-site liquid tanks, the comparison between recorded volume and run tickets has been excellent.

5.7.4 Gas measurement

A molecule of gas will pass through gas-measurement equipment between four and eight times between the reservoir and the burner tip, with the average probably around five times. World natural gas production in 2015 was on the order of 335 BSCF/day (9.5 GSCm/day). Commercial transactions vary widely both in units, currency, and care in recording around the world, so any value you place on this activity is the result of many assumptions of widely varying applicability. My guess is that wellhead sales are on the order of $1 billion USD/day; sales to final users would be on the order of $2.5 billion USD/day. Commercial transactions based on gas-measurement equipment are a significant factor in the world's economy.

5.7.4.1 *Square-edged orifice meter*

A huge proportion of the equipment used in gas measurement is done using square-edged orifice meters as defined by API 14.3/AGA3 (Fig. 5.15). This type of measurement starts with the Bernoulli equation (see Section 0.7.2.2) and "adjusts" it for real flows. If you'll remember, the assumptions that made the development of the Bernoulli equation possible were as follows:

- Inviscid (i.e., viscosity and friction are zero)
- Incompressible
- Irrotational (neither vorticity or rigid body rotation)
- Reversible
- Isothermal
- Isentropic (flow not a function of position)

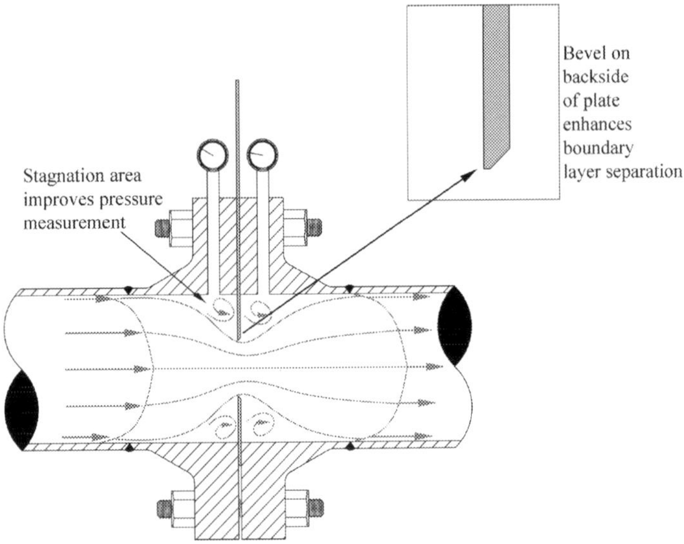

Figure 5.15 Square-edged orifice meter.

- Adiabatic (no heat is gained from or lost to the environment)
- There is no work done on or by the fluid
- No fluid is added or removed from the control volume being investigated

To that list, gas measurement adds:

- Tube is long, straight, and level (removing the elevation term from Bernoulli).
- Flow profile is fully developed (i.e., it matches the power law shown in Fig. 0.1).
- Single-phase flow.
- Fluid properties known and constant.
- Enough friction to dampen swirl.
- Tube roughness in a very narrow range.
- Condition of the flow restriction meets published specifications.
- Meter tube ID is in the range 2–36 in (51–914 mm)
- Ratio of orifice diameter to tube diameter (i.e., "β-ratio") must be in a range defined by Reynolds number (Miller, 1989, Table 9.54). The values below were developed by a committee that had quite diverse priorities. My experiments have found repeatability to suffer when β-ratio is less than 0.3 (more at low Reynolds numbers than at high Reynolds numbers) and I will not use plates in the 0.2–0.3 range by

personal preference. I also found that β-ratios greater than 0.72 degrade the repeatability more than any other parameter (you can usually achieve the same flow result by increasing the calibrated range of the dP instrument with a smaller impact on both uncertainty and repeatability). It is reasonable to design meter tubes with β-ratio in the 0.3–0.72 range for all valid Reynolds numbers.

- $Re < 6 \times 10^3 \rightarrow$ Orifice measurement invalid regardless of β-ratio (flow must be in turbulent region of Moody diagram)
- $6 \times 10^3 \leq Re < 10^4 \rightarrow \beta$-ratio in the range of 0.20–0.75 (uncertainty $\pm(0.6\% + \beta/100)$)
- $10^4 \leq Re < 10^7 \rightarrow \beta$-ratio in the range of 0.20–0.60 (uncertainty $\pm 0.60\%$)
- $10^4 \leq Re < 10^7 \rightarrow \beta$-ratio in the range of 0.60–0.75 (uncertainty $\pm(0.6\% + \beta/100)$)
- $Re \geq 10^7 \rightarrow$ Orifice measurement invalid for all β-ratios

It is obvious that a flow cannot be both inviscid and "have enough friction to dampen swirl" at the same time, so how in the world can we tie up billions of dollars/day based on arithmetic that is patently wrong? The industry has developed a database of millions of records representing the preponderance of the possible outcomes from within the precisely defined flow envelope and has created adjustment factors to mask the shortcomings of the underlying arithmetic and provide very good results. The key learning from this discussion is that the only reason that the measurement from a square-edged orifice meter has any relationship to physical volume is that the equipment is strictly limited to very proscriptive design parameters. Outside of these design parameters the discrepancy between measured flow and physical flow quickly becomes too large to use.

It is rare for the flow profile to match Fig. 0.1 for any number of reasons (primary among them is swirl created by upstream piping configuration). The original method of damping the swirl was to make the upstream piping longer and longer but again, if there is no friction (or even minimal friction) this doesn't work. The next approach was to install a "straightening vane" as shown in picture #3 in Fig. 5.16. These tube packs seem to address rigid body rotation, but don't do much with swirl. In the 1980s considerable research was conducted to create a low dP element that could force flow into the desired pattern. A number of commercial products were developed to provide "flow conditioning" using a perforated plate, and all of them worked better than the tubing pack that we called "straightening vanes." They all had varying sized holes arranged in a proprietary pattern and each had a plate thickness that was also proprietary. By the mid-1990s products with a combination of a vane pack and a perforated plate had taken over the market and were trying to differentiate themselves on price.

There are technical and regulatory requirements to "calibrate" the sensing elements on a rigid schedule to a rigidly defined standard. When the sensing elements were integral to a pen-and-ink recording device, these periodic calibrations were necessary because these instruments had a tendency to drift with time and needed to be brought back into proper compliance with an objective standard. To perform the calibration, a tech uses a comparison device to apply a specific value (of pressure for example) to the instrument and records the reading from the instrument being calibrated in the "as found" column for each of a predefined list of test values. Then the tech adjusts the instrument and runs the same test values again, recording the "as left" values for each. An instrument that cannot be adjusted to very close agreement with the standard over the entire range must be discarded and replaced. When the typical sensing element was an integral part of the pen-and-ink recording element, it was normal for the "as found" to be considerably different from the "as left" column and this often resulted in adjustments to the historical settlement values. With the advent and evolution of electronic sensing elements, this tendency for instruments to drift has largely been eliminated. The last time I saw a calibration report with the as-found column different from the as-left column (without a notation "replaced instrument") was at least 20 years ago. However, the calibration requirement remains in the standards and regulations and must be performed.

Plate holders: The simplest kind of plate holder is the "orifice flange union (OFU)" as seen in Fig. 5.15 which is simply a pair of flanges with gauge taps located precisely 1 in (2.54 cm) from the upstream and downstream faces of the orifice plate. OFUs use a "paddle plate" as shown in photo #1 in Fig. 5.16. It can take considerable time and effort to inspect a plate in an OFU, so the industry developed a single chamber plate carrier where you have to isolate and depressurize the meter tube to open the quick-release plate carrier. A dual-chamber fitting is available for an increased price that allows the operator to crank the plate carrier up into a second chamber and then shut an isolation port to allow plate inspection without depressurizing the meter tube.

Primary element issues: On the other hand there is no requirement to ever look at the primary element, where the actual fluids reside, flow, and

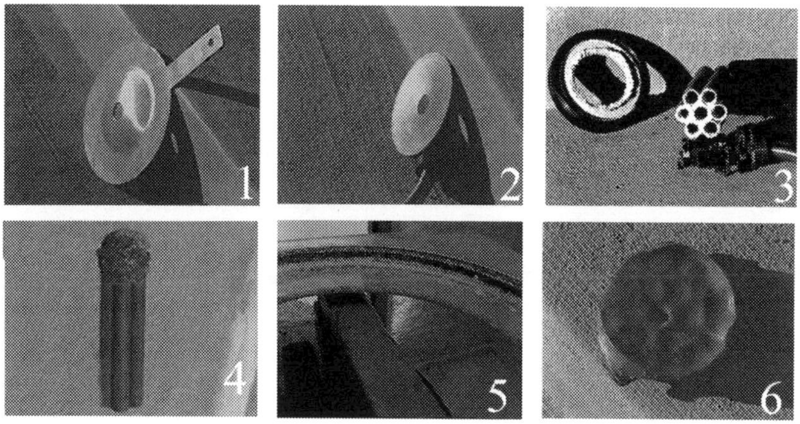

Figure 5.16 Primary element issues (1) bent plate; (2) bent plate; (3) phase-change scale; (4) mist extractor unwound; (5) β-ratio; and (6) pipe protector left in pipe.

interact. Fig. 5.15 shows some examples of why this oversight in the rules may not be appropriate:

1. Bent plate. This orifice plate was struck by a large chunk of hydrate "ice" moving at high velocity. The result was a significant deviation from the flatness requirement of the spec. This tube also has a β-ratio significantly less than 0.3, so it probably wasn't doing a very good job prior to being struck.
2. Bent plate. This plate was also struck by a hydrate chunk. This one shows that a secondary ice chunk made it through the small orifice and struck the thermowell making the temperature probe very difficult to remove.
3. Phase-change scale. The straightening vanes (which are described in API 14.3/AGA3 and were properly installed and properly located by the standard) and the orifice plate both show significant accumulation of scale. The straightening vanes are located in the straight pipe directly upstream of the orifice plate, and this plugging (which actually closed off two of the holes completely and reduced the diameter of the others) resulted in the flow reaching the plate in four independent jets. The scale on the plate changed the flatness, changed the quality of the "square edge," changed the shape of the downstream bevel, and even changed the OD of the orifice. The cumulative error on this plate was on the order of 45% low (i.e., the well was being paid for 55% of the gas actually produced).

4. Mist extractor unwound. Separator mist pads are very reliable items and rarely have a problem that causes them to come apart. When this one did come apart (probably due to a manufacturing defect), it filled all of the piping between the separator and the straightening vanes. Having a mist extractor that was several meters of small diameter pipe reduced the flow rate until this meter had a Reynolds number that was outside the range allowed in the spec.
5. β-Ratio: This meter had a 7 in (178 mm) plate in an 8 in (203 mm) line or a 0.875 β-ratio. When we replaced this plate with a plate in the acceptable range, the flow rate through the meter increased by 34%.
6. Pipe protector left in pipe: When straightening vanes are shipped from the manufacturer they have pipe protectors on both ends. In this case the meter fabricator neglected to remove the protectors prior to installing the vanes, and the four inspectors that signed off on this tube in the fab shop and field didn't catch it. This piece of red plastic was in the meter tube for nearly 4 years before someone questioned the very high differential pressure between the wellhead and the meter run and ordered a meter tube inspection. After the red plastic was removed, the well changed from a "dog" to an "above-average" well for this field.

The moral of this story is that while square-edged orifice measurement is the most common form of gas measurement used in the world by a very long margin, it cannot be treated with neglect and when calibrations are done, it is a good idea to spend a little while taking a look at the plate and tube. As engineers designing well-sites, we need to facilitate this examination by configuring the tubes such that someone can look inside them without removing them from the pipeline by entering the meter run with an "inspection tee" instead of a 90-degree elbow. An inspection tee has the flow in and out of the tube through the branch and the flow out one end. The other end has a threaded plug (for 2 in (DN 50) tubes) or a blind flange for larger tubes. In tight spaces and in tubes that are likely to need to be inspected more often you can consider specifying pig-closure caps that can be opened for inspection quickly.

Table 5.13 provides some of the basic equations included in API 14.3/AGA3 in the middle column. If you have trouble seeing the Bernoulli equation in the mass flow rate equation you are not alone. The derivation of that equation is dozens of steps, many of which are quite counterintuitive.

Table 5.13 Differential producer equations

	AGA3	V-cone
Mass flow rate ("d" is "orifice diameter" and "D" is "pipe diameter")	$\dot{m} = N_1 \cdot C_d \cdot E_v \cdot Y \cdot d^2 \cdot \sqrt{\rho_{(P,T)} \cdot \Delta P}$	$\dot{m} = F_a \cdot C_d \cdot E_v \cdot Y \cdot D^2 \cdot \sqrt{\rho_{(P,T)} \cdot \Delta P}$
Unit conversion/thermal expansion factor	N_1 is a unit conversion	Fa = 1.0 if flowing gas temperature < 100°F (37.8°C) (unit conversions are in E_v and Y)
Velocity approach factor (the "U" terms in the V-cone column are unit conversions)	$E_v = \frac{1}{\sqrt{1-\beta^4}}$	$E_v = \left(\frac{\pi \cdot \sqrt{2 \cdot U_3}}{4 \cdot U_2}\right)\left(\frac{D^2 \cdot \beta^2}{\sqrt{1-\beta^4}}\right)$
Expansion factor (Y)	Complex, multistep calculations (see AGA3 Part 1)	$Y = 1 - (0.755 + 6.78 \cdot \beta^8) \cdot \left(\frac{U_1 \cdot \Delta P}{k \cdot P}\right)$
Diameter ratio	$\beta = \frac{d}{D}$	$\beta = \sqrt{1 - \frac{d^2_{coneOD}}{D^2_{tubeID}}}$
Example for $\beta = 0.5$ in 4 in pipe	$\beta = \frac{2}{4} = 0.5$	$\beta = \sqrt{1 - \frac{3.464^2}{4.0^2}} = 0.5$

5.7.4.2 V-cone

McCrometer's V-cone meter addresses a major issue with API 14.3/AGA3 meter tubes, specifically: the equation assumes fully developed flow (Fig. 0.1) with no rotation and then doesn't have a good way to force the flow into that configuration. Most of the "flow conditioners" on the market since the mid-1990s do a reasonable job of simulating fully developed flow at a limited range of Reynolds numbers, but the big issue is that we install them and then never think about what flow conditions are appropriate to properly develop the flow ever again.

Fig. 5.16 is a cross section of the V-cone solution to this issue. The flow is gradually squeezed down into the annular area between the cone and the pipe walls, and any flow discrepancies are evened out by the fluid moving from side to side to take advantage of portions of the space that have less gas than other portions. At the exit to the cone, the flow is a very close approximation to fully developed. This results in significantly reduced uncertainty and significantly improved repeatability.

Measurement techs seem to hate V-cones and often say that they do not work. Some of the things I've found when evaluating why measurement with V-cones is "bad":

- One site claimed that their system balance was off by 30%, "V-cones are junk." It turns out that the calculation routine and they were using the equations from the center column of Table 5.13 instead of the right-hand column. After the proper equations were loaded to the RTU, the system balance came into line with expectations and instead of the contractual 2% system shrinkage they were experiencing 1.1%—and able to sell the 0.9% for their own account.
- Another site claimed that their system energy balance was "way off" after installing V-cones on about 12% of the wells—they had left the default gas analysis in the RTU and it was for AIR. Putting in the proper gas analysis brought the energy balance into values well within contractual limits.
- At another site the meters were registering zero flow when you could hear flow in the pipe. When they moved the high-pressure instrument to upstream of the V-cone and the low-pressure instrument to downstream it magically started registering flow properly.
- A V-cone on a gas-lift system was registering 1.6 MMSCF/day (45.3 kSCm/day), but the well was only returning 1.2 MMSCF/day (34 kSCm/day). The meter was disassembled and it was found to be packed full of emulsified compressor oil and gravel (actual gravel from

the road) which had resulted in a no-flow dP that was consistent with the reported flow rate.

- Another gas-lift system used V-cones for gas-lift gas, but these were multipad wells and they had very poorly designed orifice meters at the pad level—the sum of the individual wells was never even close to the pad-level meter. They had a high-quality orifice meter at the compressor station and upon comparing the sum of all the well's V-cone readings to the station meter, the difference was within the uncertainty range of the orifice meter (i.e., the sum of the V-cones was the same number as the orifice meter). The recommendation from this analysis was to discard the junk orifice meters and do their engineering based on the V-cones, they didn't follow it.
- Yet another site decided that there was serious money to be saved by specifying that all gas flow meters would be exactly the same (Fig. 5.17) to save on spare parts. With a little digging, you can find from the model number that they settled on was ASME 16.5 Class 150 flanged 4 in (200 DN) schedule 40 carbon steel with ½ in (12.7 mm) NPT instrument connections with a 0.5516 β-ratio. Normal operating pressure on most of their wells is 20 psig (138 kPag) and the dP instruments are calibrated 0–100 in H_2O (0–187 torr). This meter with this calibrated range can measure 0–1745 MSCF/day (0–49.4 kSCm/day) (the "7MMSCFD" designation of the meter is at MAWP of 280 psig (1930 kPag) and maximum calibrated range on the dP instrument of 150 in H_2O (280 torr)). This meter in this configuration has a total meter uncertainty of ±36 MSCF/day (1 kSCm/day). The meter manufacturer recommends a minimum dP of 10 in H_2O (18.7 torr), so the valid range of the meter is 600–1745 MSCF/

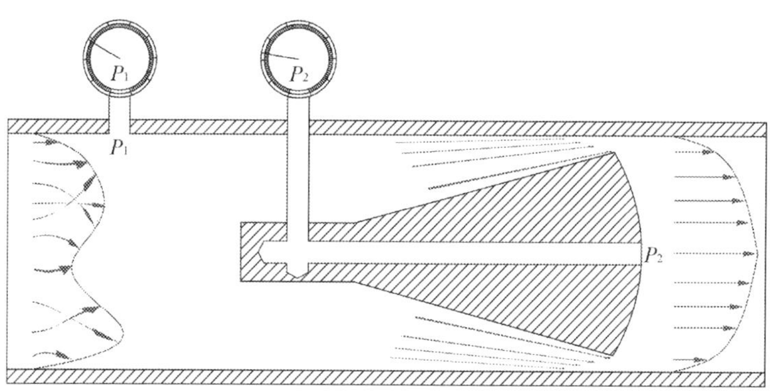

Figure 5.17 V-cone flow.

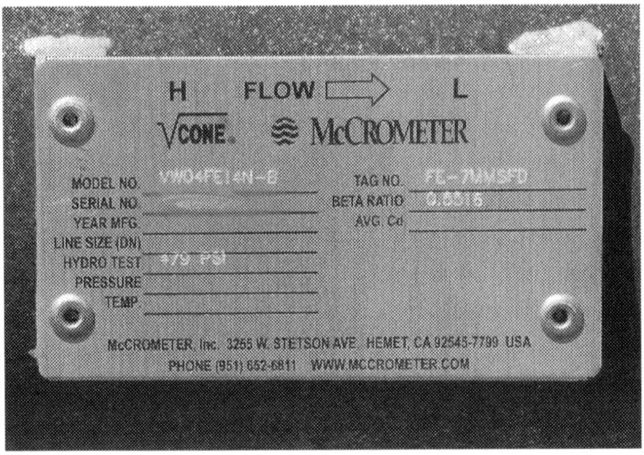

Figure 5.18 V-cone meter nameplate.

day (17–49.4 kSCm/day). Approximately 75% of the wells in this field produce less than 600 MSCF/day (17 kSCm/day), and 10% of the wells produce more than 1745 MSCF/day (49.4 kSCm/day), so they have valid gas measurement on about 15% of the wells. The meter manufacturer would be delighted to sell them a 2 in (50 DN) meter with a smaller β-ratio that has ASME B16.5 Class 150 4 in flanges and spacers to fit into the space that this meter occupies, but that would violate the "standardization" strategy (Fig. 5.18).

V-cone meters yield consistently better measurement than small orifice meters, they are less prone to primary element problems, have primary element problems that are easier to repair, and they cost less. However, in oil-field terms they are "new" (entering only their third decade of excellent service) so measurement techs resist their use. This resistance is worth the effort to combat.

5.7.4.3 Other measurement technologies

Most of the technologies described earlier for liquid measurement have been used for various gas-measurement applications. None of them have the adaptability, repeatability, or uncertainty of a V-cone, and several of the technologies provide terrible results in gas (e.g., vortex meters are often not even compensated for pressure and temperature and cannot provide either a mass flow rate or a volume flow rate at standard conditions).

People occasionally apply pitot tubes to well-sites. These devices look at the pressure differential between the upstream side of a bluff body in the flow and the downstream side. These meters have worked well in laboratory conditions with well-known and constant fluid properties, but for well-site application the variability has proven to be far too great for these meters to be reliable in the real world.

5.8 WELL-SITE EQUIPMENT SPACING

Companies (especially large companies) have had a near-obsession with "Standard Well Site Equipment Spacing" for decades. Over the years I've seen hundreds of "standard" designs on paper (that took millions of man-hours to develop) and nearly zero on the ground. Things like the location of roads, the location of pipelines, topographical features, and varying equipment requirements have conspired to make this goal largely unreachable. In my estimation that is a very good thing. Remember, "Every well has a personality, most are unpleasant, and many are just awful." Trying to put "standard" clothing on a petulant child is rarely a delightful experience and trying to force a well into a "standard" is just as unpleasant.

Rather than an immutable "standard layout," it is more effective to approach a well-site as a system:

- Draw a scale diagram of the site you have purchased the rights to use. Include the wellhead, roads, pipelines, and any topography that is important (e.g., is there an ephemeral drainage that runs water 3 days/year? For a well-site that is going to be in place for 75 years it might make sense to account for it in your design).
- Draw an exclusion area around the wellhead to include an area to lay down any future workover rigs and room to stage necessary rig equipment (e.g., you may need a pipe rack when you pull tubing and you may need a place to put an air pack for clean-outs)
- What kind of deliquification equipment is likely? If it is nodding donkey SRP then you need considerable room for the surface equipment. Other kinds of deliquification require space as well (e.g., gas-lift requires a compressor or a gas-lift line from a central location).
- Will you need grid power? What route will it use to come onto location? That route should not require moving equipment. If no grid

power is anticipated, but if you likely need power, where are you going to put a genset?

- Think about what equipment you will need and how will it need to be accessed.
  - Will you need tanks? Tanks need secondary containment and access for trucks, truck access should not require backing, and turnarounds shouldn't encroach on the rig exclusion zone. Will you require low-pressure tankage (i.e., buried tanks or elevated equipment)? Buried tanks take more space than you expect (and need truck access like aboveground tanks).
  - Will you need a transfer pump? Can it go in the secondary containment? If not, where will it be, if so is the secondary containment big enough?
  - What sort of production equipment will you need? Will it need to be elevated for drainage? How will you elevate it if it needs to be elevated?
  - What measurement equipment do you need? How much space does it need? Can a tech get his truck up to it for calibrations?
  - Will the well-site ever need a compressor? Most do at some point. Where are you going to put it and how will you tie it in? Installing a tee and blind flange above and below the separator outlet valve on the production-unit outlet piping can save significant future costs. These tees allow for tying-in future compressors without having to cut out the installed piping, for minimal cost.
  - Where will the components of an automation system reside and how will you get wires between components? It is sometimes a very good idea to share ditches between pipe and wire, but not if the route of the pipe adds significant distance to the wiring. Today's trend is toward more wireless end devices so this may not be a consideration for much longer. Even with the current generation of wireless being significantly superior to early generations, there are still battery-life issues that too often addressed by reducing polling frequency without a place in the database to indicate whether a particular reading is live or a holdover from the previous period—an oversight that has led to erroneous conclusions.
  - Layout the equipment (and access routes) on the scale drawing.
- Select a size for the pipe from the wellhead to the production equipment/tanks and layout a route. I've found that transitioning from the

wellhead equipment to the piping should happen at ground-level flanges located at least 6 ft (2 m) from the well bore. This allows you to pull the wellhead piping and put a blind flange on the flow lines to protect them from rig activity. Putting the start of the piping closer to the wellhead too often results in the line being damaged, delaying the time required to start back up after a workover. Add the pipe route to the scale drawing.

It is common for companies to agonize over separation distances between equipment and the wellhead and between pieces of equipment. This agony is truly pointless. Whatever distances you select will ultimately be arbitrary and will frequently require exceptions caused by location topology and well-pad size (which tends to get smaller with every project). Repeated dispersion analysis for non-sour wells show that the risk of one fire igniting other well-site equipment is very low, and the consequences of such a sympathetic fire are really low. We tend to apply electrical area classification distances to nonelectric applications and overstate the electrical area classifications. These are pretty expensive extravagances.

5.9 CONTROL ROOMS

The growing tendency in many companies is to combine a "control room" with "dispatched pumping." The control room part is fairly self-explanatory, it is a place (sometimes a room, sometimes an entire floor of a large building) where people sit in comfortable chairs, in conditioned air, out of the wind, rain, and snow and look at computer screens to determine where there are issues that should be addressed.

The "dispatched pumping" concept is one of the most wrong-headed ideas of the 20th century. Under this foolish concept, the control room is responsible for "looking at" all the wells in the field and developing a prioritized schedule of sites for the pumper to visit that day. This schedule is often ignorant of driving routes so a pumper will visit a well, drive past another well on her list to the next priority on the schedule, and then (eventually) spend considerable windshield time returning to the well that she drove past earlier. At the most extreme implementation of this silliness a pumper can be fired for visiting a well not on her list. The big problem with this concept is that most things that go wrong on well-sites happen gradually, and routine visits frequently find situations that can be repaired easily without the control room ever seeing that there was something to repair.

Things that can go wrong on a well-site generally fit into one of four categories:

- "Anomalies" are some "small" thing that is not right.
- "Issues" are anomalies that have been noticed and reported (as opposed to having been fixed).
- "Problems" are issues and anomalies that have escalated rather than being resolved.
- "Failures" are problems that were not fixed in time.

The way we deal with the event continuum changes by who we are:

- A lease tech on site can detect, diagnose, and repair an anomaly before it becomes an issue or a problem.
- An engineer on site can detect an anomaly and prioritize response to observed anomalies.
- Someone in a control room can detect and dispatch people to failures.
- An engineer in a remote office can go to meetings.

The single most powerful tool available for well-site operations is "boots on the ground." Automation and control rooms have evolved to significantly reduce the effectiveness of field staff without providing much effectiveness themselves. High-quality field data is a diagnostic tool without peer, but it has to be placed in the hands of people with the latitude *and proximity* to act on it.

There is a timing progression for well-site activities:

- If a well has an anomaly or issue, then the control room will not see it and disallowing routine visits by field techs will ensure that it will escalate into a problem or failure.
- If a well has a problem, then it will also rarely be seen by the control room and without routine visits it will certainly escalate to a failure.
- Failures are seen by control rooms and dispatched field techs will generally find expensive problems that could have easily been fixed by simply adjusting lubrication or proactively replacing a diaphragm that was beginning to leak or scheduling the replacement of a leaking PSV.

While control rooms and dispatching systems make office-types feel that they are contributing, they are nearly always counter to the needs of an operating reservoir—any event that can wait for a work order cycle to dispatch a field tech probably doesn't really need to be addressed at all.

5.10 PROCESSES VS DECISIONS

The intent of processes like dispatch systems, management of change processes (MOC), and work orders is to drive decisions toward a "standard" answer. This goal ignores the fact that no well is "standard" no matter how desperately we want them to be. A process that maximizes production on one well can (and has) kill(ed) the next well.

Effective field operations:

- Build real disincentives for failing to make a decision.
- Don't demonize someone for making a decision that turns out to be suboptimum or even horrible (focus on damage control and learning).
- Clearly define delegations of authority (i.e., who is accountable for each class of decision) and then let the appropriate person make the decision. Delegations of authority should be pushed as far down the organization as possible.
- Minimize processes and procedures to those (few) required by law (such as lockout/tag-out).

There is no "one way" to operate a gas field, and nearly every possible organization and delegation of authority have been tried, probably most of them have been tried within any given organization at one time or another. In Chapter 10, Integration of Concepts, we'll discuss the concept of people "owning" the operation they work in. If the organization does not instill ownership in the people working there, then it simply cannot succeed to the degree that is possible.

REFERENCES

ASME B31.3. Process Piping: ASME Code for Pressure Piping, B31 2014.

ASME B31.8. Gas Transmission and Distribution Piping Systems, B31, 2014.

BPVC. ASME Boiler and Pressure Vessel Code, Section VIII, Division 1, 2013.

Miller, R. W. (1989). Flow Measurement Engineering Handbook. New York, NY: McGraw-Hill.

Webster. (n.d.). Retrieved April 21, 2016, from http://www.merriam-webster.com/dictionary/biogenic (other dictionary references replace "biogenic" with the referenced term).

Wikipedia, 1. ASME Boiler and Pressure Vessel Code (BPVC), 2015. Retrieved October 19, 2016, from https://en.wikipedia.org/wiki/ASME_Boiler_and_Pressure_Vessel_Code_(BPVC).

Wikipedia, 2. Second Law of Thermodynamics, 2016. Retrieved October 20, 2016, from https://en.wikipedia.org/wiki/Second_law_of_thermodynamics#Various_statements_of_the_law.

NOMENCLATURE

Symbol	Name	fps units	SI units
A	Flow area	ft^2	m^2
$A_{orifice}$	Flow area of PSV orifice	in^2	cm^2
A_{outlet}	Flow area of PSV discharge pipe	in^2	cm^2
$E_{tableA1B}$	Velocity approach factor	decimal	decimal
ID_{pipe}	Pipe inside diameter	in	mm
f	Frequency	1/s	1/s
F_g	Force at outlet of exhaust pipe	lbf	N
F_r	Reactive force on PSV	lbf	N
k	Adiabatic constant (ratio of specific heat)	decimal	decimal
K_b	Backpressure constant for PSV	decimal	decimal
K_c	Rupture disk constant	decimal	decimal
K_d	Discharge coefficient for PSV	decimal	decimal
K_f	Reactive force constant	decimal	decimal
K_s	Constant for vessel sizing	ft/s	
$\dot{m}_{design}$	Result of design calculation after PSV is selected	lbm/h	kg/h
$\dot{m}_{required}$	Result of PSV design calculation for credible scenario	lbm/h	kg/h
MW	Molecular weight	lbm/lb-mole	gm/gm-mole
OD_{pipe}	Pipe outside diameter	in	mm
P	Pressure	psia	kPaa
P_{atm}	Local atmospheric pressure	psia	kPaa
P_{mawp}	Maximum allowable working pressure	psia	kPaa
P_{gauge}	Pressure relative to local atmospheric pressure	psig	kPag
P_{up_psig}	Upstream pressure	psig	kPag
$S_{tableA1}$	Basic allowable stresses from Appendix A	ksi	MPa
t	Thickness	in	mm
t_{corr}	Corrosion allowance	in	mm
v	Velocity	ft/s	m/s
T	Temperature	R	K
W	Weld joint strength reduction factor	decimal	decimal
Y	Metallurgy derate	decimal	decimal
Z	Compressibility	decimal	decimal

(*Continued*)

(Continued)

Symbol	Name	fps units	SI units
Subscripts			
1	Upstream value		
2	Downstream value		
avg	Average		
disch	Condition at the discharge of a pump or compressor		
gas	Gas		
liq	Liquid		
std	Standard condition		
suct	Condition at the suction of a pump or compressor		
tbg	Tubing		

EXERCISES

1. A well-site needs a new flow line, which pipe option in Table 5.14 would satisfy API 31.3?
2. Select a pipe size and material and determine upstream pressure for:
   - Type gas → CBM
   - Gas rate → 350 MSCF/day (9.9 kSCm/day)
   - Water rate → 15 bbl/MMSCF (84.2 L/kSCm) (SG = 1.03)
   - Downstream pressure → 20 psig (138 kPag)
   - Gas temperature → 70°F (21.1 °C)
   - Pipe length → 185 ft (56.4 m)
3. For the conditions in #2, the vessel fabricator recommend a 24 in (600 DN) vertical two-phase separator. What diameter should the mist pad be?
4. What PSV would be recommended for the vessel in #3?
5. The pneumatic device in Fig. 5.11 is labeled "local control device." What about this valve makes it "local"? When the end device has to move toward "closed," where does the excess gas go?
6. What is the difference between a "separator" and a "scrubber"?

Table 5.14 Data for exercise 1

	fps	Both	SI
MAWP	600 psig		4137 kPag
Design temperature	100°F		37.8°C
Nominal pipe	10 in		250 DN
Type test		Pneumatic	
Test pressure		150% of MAWP	
Corrosion allowance	3/16 in		4.8 mm
Normal op pressure	150 psig		1034 kPag

Pipe options

	Sched 20	Sched 40 (std)	Sched 60 (X)	Sched 80
OD	10.750 in (273 mm)	10.750 in (273 mm)	10.750 in (273 mm)	10.750 in (273 mm)
ID	10.250 in (260 mm)	10.020 in (255 mm)	9.750 in (248 mm)	9.562 in (243 mm)
t	0.250 in (6.35 mm)	0.365 in (9.27 mm)	0.500 (12.70 mm)	0.594 in (15.09 mm)

Table A1

	SMYS (psig (MPag))	Basic allowable stress ($S_{tableA1}$ psig (MPag))	Maximum temperature (°F (°C)) before derate
API 5L Gr B	35000 (241)	20000 (138)	400 (204)
API 5L X42	42000 (290)	20000 (138)	400 (204)
API 5L X60	60000 (414)	25000 (172)	400 (204)
API 5L X70	70000 (483)	27300 (188)	400 (204)

Table A1B

	Description	$E_{tableA1B}$
API 5L	Seamless pipe	1.00
API 5L	Electric resistance welded pipe	0.85
API 5L	Electric fusion welded pipe	0.95
API 5L	Furnace butt welded	0.95

ASME B31.3 Table 304.1.1 (valid for $t < OD/6$)

	Temperature range	Y
Ferritic steels	<900°F	0.4
Austenitic steels	<900°F	0.4
Other ductile materials	<900°F	0.4
Cast iron	<900°F	0.0

CHAPTER SIX

Gas Gathering Systems

Simpson's Fourth Postulate: Gathering systems and compressor stations are "tools of reservoir management". Thinking of them as a "sales tool" will reduce ultimate recovery.

Contents

Practical Onshore Gas Field Engineering.
DOI: http://dx.doi.org/10.1016/B978-0-12-813022-3.00006-7

6.1 OVERVIEW

Before the 1990s, virtually all gas in the United States was purchased at the well-site by gas gathering companies for contractually fixed prices. All potential profit and loss from price fluctuations fell to the gathering companies. Under that model, the gathering companies had a significant profit motive for operating their piping systems, compressor stations, and plants in a way that maximized production without a lot of concern for costs. The Federal Energy Regulatory Commission (FERC) issued FERC Order 636 in early 1992 which ended this economic model. FERC 636 required pipelines to "unbundle" (i.e., separate) their sales services from their transportation services. This meant that the producers were now free to negotiate gathering/processing fees with their old gas purchasers, and to sell their commodity products to end users at the outlet of processing plants. This rule shifted the potential for profits and losses related to price fluctuations to the producer and removed it from the gathering company (who now move and process the gas based on a fixed fee) and shifted their profit motive away from maximizing production to minimizing costs.

The other major shift was due to the US Congress passing the Section 29 Tax Credits as part of the so-called Windfall Profits Tax on energy in 1980. The Windfall Profits Tax proved to be a law that was very expensive for producers to capture necessary data and generated approximately zero revenue for the Treasury, but the Section 29 Tax credits were largely overlooked by the Oil & Gas industry until the late 1980s. Eventually the industry realized that in a \$1/MMBTU (\$0.948 USD/GJ) sales market you might be able to make a profit selling 900 BTU/SCF (33.5 MJ/SCm) gas if it allowed you to deduct another \$1/MSCF (\$0.035/SCm) from your final tax bill—losing a bit of money on the actual coalbed methane (CBM) production looked good if it could offset taxes owed on other operations. Remember that prior to the development of CBM, the gas-sales market was so anemic that no one developed "dry gas," only gas with a reasonable expectation of producing hydrocarbon liquids was ever developed. CBM had no expectation of hydrocarbon liquids and had a high potential for producing unacceptable levels of both water and CO_2. The industry had no infrastructure for dealing with either the water or the CO_2, and the traditional gas gathering companies wanted no part of the effort to create this infrastructure—there just wasn't enough money in it for a common carrier.

Producers were in a dilemma. They wanted the tax credit and to sell the gas they had to get the gas to market, but all of the taxis were on strike. The decision by some operators was to build gas gathering, transportation, and sweetening infrastructure themselves. Other operators banded together and offered very favorable fee structures to third parties to build midstream and processing infrastructure. These producers had to accept that they would have to build primary gathering systems to aggregate their gas at central delivery points (CDP) because with FERC 636 the gathering fees that were demanded by third parties were simply too high to be attractive even with the Section 29 Tax Credits. This leaves us with three infrastructure models (these names for the models relate to the major players in the San Juan Basin in the early days of CBM):

- *El Paso model*: Third party picks up raw wellhead gas at the well-site and delivers commodity products to the purchaser at a plant tailgate.
- *Williams model*: Producer aggregates raw wellhead gas and brings it to CDP for compression, sweetening, and dehydration; and the third-party delivers commodity gas to the purchaser at the plant tailgate.
- *Burlington model*: Producer delivers commodity gas to the purchaser at the tailgate of producer-owned processing plant.

The El Paso model can have reasonable economics in liquid-rich gas with low contaminants where the third party keeps a portion of the hydrocarbon liquids as a processing fee. Absent hydrocarbon liquids, this model tends to have terrible economics for the third-party and poor economics for the producer.

In dry gas both the Williams model and the Burlington model have proven to be very attractive. These models return the potential benefits of gathering-system improvements to the people with an incentive to realize those benefits—if a line looping project is expected to increase production by 1 MMSCF/day (28 kSCm/day), it has a much shorter payout at

$1000/day plus a $365,000/year reduction in taxes for the producer than it would have had for a $100/day increase in taxable revenue for a third party. The same is true for gathering operations like running pigs, if running a weekly pig gets you a production increase of 50 MSCF/day (1.4 kSCm/day), then that would be an increase in the producer's after-tax revenue of $21,000/year as opposed to an after-tax revenue for a third party of $1200—if running the pig costs $200/week in labor, the producer makes money on pigging and the third party loses money on that activity. In other words, these models align gathering expenditures with revenue drivers. Third-party pipeline techs will operate a gathering system to minimize operating cost, producer pipeline techs will operate a gathering system to maximize production, and the result looks very different.

All gas wells perform better if they see delivery pressures that are consistent with the needs of the reservoir and at the same time are very consistent. To accomplish those complementary goals the primary gathering system must be properly maintained and operated, which means that it needs to be operated by people with an interest in maximizing ultimate recovery.

While anecdotes never prove anything, they can illuminate a point. Looking at Fig. 6.1, these three wells were located adjacent to each other,

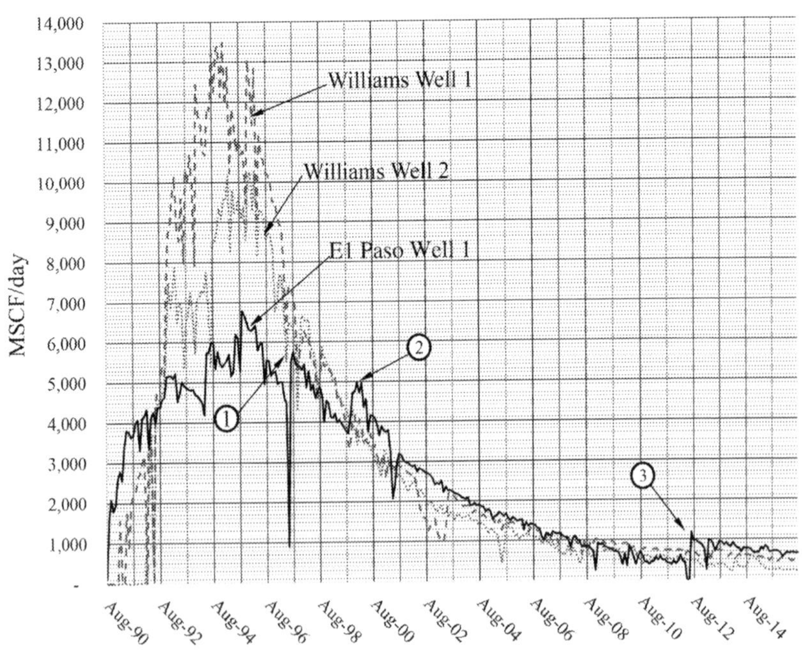

Figure 6.1 Gathering choice example.

and each had the opportunity to take gas from the ultimate recovery of the other two. The well labeled "El Paso Well 1" was connected to third-party gathering pipe for the first 7 years of production, and the other two wells were on a gathering system operated by the well's lease techs. The labeled points on the graph are:

1. The operator of El Paso 1 set a 1000 hp, two-stage reciprocating compressor to pull the wellhead flowing pressure down to 5 psig while still connected to the third-party system.
2. The well was moved from the third-party system to the producer-operated gathering system. Wellhead flowing pressure was increased to 16 psig; the compressor was replaced with a 500 hp oil-flooded screw compressor.
3. The well was equipped with a linear rod pump.

It is easy to see from the late-life data that this well likely started with similar potential as the other two wells. As of this writing, the well labeled "Williams Well 1" has recovered 32 BSCF (0.906 GSCm), and El Paso 1 has recovered 25 BSCF (0.708 GSCm), and including the tax credits the difference in after-tax revenue between these two wells has been $22 million USD in current dollars (the difference between Williams Well 2 and El Paso Well 1 is only $6 million, which is the expected total revenue of four conventional wells in this field). The two wells on the producer-operated gathering system have recovered over 95% of their original gas in place (OGIP), the well that started life on the third-party system wasn't able to take advantage of the early high rate period (which happened during the Section 29 Tax Credit window) and has "only" recovered 82% of OGIP. Current flow rates in the field indicate that it will not be economic to operate the CDP for much longer and ultimate recovery will likely be limited to 96% of OGIP for the wells that have been on producer-operated gathering from the beginning and 90% for the well that entered the game late. Ultimate recovery from the original 22 wells will be on the order of 300 BSCF which added over $1 billion USD, the producer's revenue stream over 27 years.

Nothing I've said about the benefits of producer-operated gathering system applies to affiliate-operated gathering systems. In the interest of "operating efficiency" many companies have accepted that they need to build their own gathering systems, but then put into place an organization that moves the management and operation of those systems away from the lease operators to a dedicated team. This "midstream organization" is actually the worst of all possible worlds.

- Gathering group has different goals than the production group, and their goals have nothing to do with reservoir performance.
- Production group has no recourse for nonperformance (legitimate complaints sound like "whining" and you can't invoke a gathering agreement as the final authority).

Having the two groups working for the same boss has not proven to be improvement because most organizations look at production groups as "profit centers" and set revenue goals and gathering groups as "cost centers" and set cost-control goals—the benefit of producer-operated gathering is that the operation of the gathering system is done within the context of "maximizing reservoir profitability," not minimizing the (tiny) costs of operating a gas gathering system. The two have very different realities and focusing on gathering without considering the reservoir performance often results in developing a rigid work schedule that cannot be modified by exigencies of field operations and subpar reservoir performance.

The poor average performance of affiliates has only been true when talking about raw reservoir gas. Once the gas is in a CDP, pressure raised to midstream levels, and dehydrated to 7 lbm/MMSCF (110 mg/SCm) it becomes quite appropriate to treat the system as a cost center since at a plant inlet pressure of 900 psig (6.2 MPag) and 60°F (15.56°C) it is at 36% relative humidity and the chance of accumulating condensate is acceptably low.

Sometimes acreage dedications require that a particular group of wells is committed to a particular third-party gathering system. You can have success in building "micro gathering systems" where the producer takes responsibility for the flow lines from the well-site to the trunk, the gas measurement, and removes any well-site dehydration equipment. Then the producer builds a new trunk parallel to the third-party trunk and ties the wells into it. This short trunk then terminates in a small compressor station with any necessary dehydration and a single meter for the third-party to assume responsibility for. The third party gets the benefit of going from multiple flow meters to maintain to one, and often from multiple dehydration units to zero. The producer gets all the benefits of producer-operated gathering and has compression horsepower between the wells and the vagaries of third-party gathering pipe. The first micro gathering system I built had seven wells, each with a well-site compressor (total of 400 hp), well-site dehydrator (on the suction side of compression, operated by the third-party gatherer), and well-site measurement.

When we finished the project in 2004 the gas sales from the seven wells increased by 30% and the decline dropped from over 12% to under 4%—12 years later the wells are still making more gas than they were making prior to the project.

A facilities engineer needs to think about what she needs to know before assuming the task of designing, building, commissioning, and operating a gas gathering system. It is different for each of the three models:

- El Paso model
  - You need to know what the gatherer is obligated to do by the contract.
  - You need to know how to influence the gatherer when they don't do what you see as best for your reservoir.
  - The phone number of your attorney for when you can't reach a meeting of the minds.
- Burlington model
  - You need to know everything there is to know about primary gathering.
  - You need to know everything there is to know about regulated compressor stations.
  - You need to know everything there is to know about midstream pipelines.
- Williams model
  - You need to know everything there is to know about primary gathering.

But what the heck does "everything there is to know about ..." even mean? It includes: (1) selection, specifying, and purchasing of gathering piping, accessories, and equipment; (2) acquiring rights of way (ROW) and government permits; (3) construction; and (4) operation. In reality no one knows "everything" about any of these topics, and few people have even been able to keep up with changes in regulations, technology, and safety/environmental concerns in any given portion of the continuum. I've always found it adequate to be aware of the broad scope of these topics so that I could tell when a narrow "expert" was blowing smoke up my pants legs. Before I was willing to contract a system design out to an engineering firm, I designed three major projects. Before I was willing to allow Supply Chain Management to "help" me on a project I actually did the purchasing on a project. I spent time walking with surveyors and archeologists to understand their process well enough to be able to tell when I was getting value for my money. I spent time with construction crews until I could tell the difference between competence and incompetence. I dug a hole with a track hoe so that I could see for myself how to set realistic expectations for progress (it turned out to be a lot harder to accomplish than I had expected). I spent time with a welding inspector to understand what his issues were. This chapter is not going to make anyone an expert on anything. It is not intended to do that. The intention is to help with the question "what haven't I thought about."

6.2 PROJECT LIFE CYCLE

Increasingly, companies are spending an ever larger portion of project budgets on life cycle management activities. These activities are intended to reduce project costs from rework and repeating activities, and they sometimes accomplish that. Other times these rigid structures add more oversight and communications costs than they remove. An organizational structure should serve the project instead of the project being force fit into an inappropriate structure.

There are many versions of these project life cycle models, but they all have similar functionality. We're going to discuss one that is made up of: (1) planning phase; (2) engineering, procurement, and construction (EPC) phase; and (3) operations phase.

6.2.1 Planning phase

In the planning phase the project owner describes the project in enough detail get preliminary funding. This phase starts out with broad statements like "we will drill xx wells for $yy per well, well-site and gathering facilities should cost $zz per well and midstream costs are expected to be $aaaa so the project ultimate scope should be in the region of $bbb billion." No one is going to fund a project with this little definition, but that is often enough definition to get the money to refine the estimate.

This phase compares expected hydrocarbon resources to the amount that can be recovered, the timing of the recovery, and the confidence that the project will meet economic expectations.

Often the result of the planning phase is a document that flows into the Front End Engineering Design (FEED) and is usually called something like "Pre-FEED." The Pre-FEED document is really intended to focus the subsequent steps to avoid as many dead-end analysis as possible (e.g., if the Pre-FEED describes the reservoir as "CBM with 6% CO_2 and 94% CH_4, 1 MMSCF/day gas production, 20 bbl/day of water, and no hydrocarbon liquids" and goes on to indicate that water rate should decrease with time and CO_2 was likely to increase toward 25% over 30 years of production, then the FEED process should not consider three-phase separators in their engineering design).

6.2.2 EPC phase

The EPC phase is often turned over to a contractor without much input from the owner beyond the Pre-FEED document. This can be a major mistake, and companies know that it is a mistake, but have often been unable to change the process to allow the owner's engineers to take a significant role in the EPC phase.

The EPC phase generally includes: (1) FEED (which is sometimes prepared by a different engineering firm than the rest of the EPC effort); (2) detailed engineering; (3) procurement; and (4) construction.

FEED. The FEED is intended to lock in many design elements (like pipe size, pipe material selection, valve locations and types, and pigging equipment). Ideally, ROW would be purchased and permits issued prior to the end of FEED, but there are too many elements of a permit that use the final information from the engineering design for the permitting effort to begin in earnest prior to the completion of FEED. Consequently, the FEED is never a complete engineering document.

At the end of FEED, the cost estimate is refined to something closer to $\pm 25\%$ than the wild guess from the planning phase.

Detailed engineering. At the end of the FEED, management has to decide if the project is still viable and economically sound. If it is sound, then funds are authorized for detailed engineering (on a large project this can be millions of dollars), but generally this does not release the required funds for procurement or construction. This stage is typically done by one of the growing number of engineering firms that provide complete EPC services.

Detailed engineering is intended to build on the FEED, not to redo the efforts. If the FEED was based on the project being built out of steel pipe with a maximum allowable working pressure (MAWP) of 600 psig (4140 kPag), it is inappropriate for the detailed engineering to design the system based on using high-density polyethylene (HDPE) with an MAWP of 100 psig (690 kPag), but it happens. Communication between the EPC team and the FEED team is not encouraged (generally if the FEED was done in-house the team has moved on to the next project, if it was contracted out, that contract has ended and cannot accept new charges).

At the end of detailed engineering, project economics have been developed to ±10%; environmental assessments have been completed; pipeline routes have been laid out; necessary ROW have been acquired; necessary government permits have been approved; drawings are completed; if required (it usually is), bid packages have been prepared and potential bidders have been identified; and materials/equipment lists have been prepared in adequate detail to prepare a purchase order. The end of detailed engineering is also the final opportunity for management to sanction or kill the project (on most approved projects this is the point where serious money gets allocated).

Procurement. Purchasing has become a very specialized undertaking and it is generally done by specialists who only do purchasing. Unfortunately, communication between detailed engineering and procurement tends to be quite limited. If the engineer specifies "35 blue widgets" and the manufacturer says "we can do that, but the yellow widget exceeds your design specifications and costs one-fourth the price," the purchasing agent will almost always say "send the blue widgets" while the design engineer might have looked at the yellow widget specifications and approved the substitution.

There were certainly inefficiencies in the procurement process when engineers did their own purchasing, but there was also the capability to adapt the system design to the realities of the marketplace. Early in my career I designed a line with 18 in (450 DN) steel pipe (which turns out to be fairly rare). I felt like this size was proper to reach my target fluid velocities and I tried to order the material—fittings were going to take 20 months for delivery, valves would be 2 years out, and pipe was at least a year for delivery. As both the design engineer and the purchasing agent, I was able to ask about the availability of 20 in (500 DN) and was told that they had all my valves, fittings, and pipe in one of several of their warehouses and they could have everything on the ground at my job in 3 weeks. I've never had a dedicated purchasing agent ask me about alternatives when something was long lead time or very expensive.

At the end of the procurement stage material is on order; a materials-delivery schedule has been prepared; a construction contractor has been selected; and a construction schedule has been prepared.

Construction. The construction stage results in dirt moving and pipe being attached to other pipes. Engineering involvement in this stage varies widely from company to company, from project to project, and even from engineer to engineer. My preference was to manage the construction project and have all of the inspectors report to me. That had an opportunity cost (if I was standing on the pipeline I wasn't designing the next project) that my company eventually decided was too high. After that I had a construction manager who reported to me and the inspectors reported to him. It worked well. Other projects have a construction manager who is an employee of the EPC and will either be a senior construction operator or a junior engineer. Other projects have other organizations.

6.2.3 Operations phase

When the project is built, it is turned over to operations to actually sell gas. Many major projects remain open for several months of operation to allow for repair of identified problems to be paid for by the project. Other projects are closed the day the last piece of equipment is loaded onto the truck.

6.2.4 Cost estimating

Building a gathering system is an expensive undertaking and the costs have to be authorized. There is typically no way to get funds committed without a cost estimate, in fact you will have to estimate costs several times during the development of the project:

- You will need to do a gross-level scoping evaluation to even start
  - Get a scale map with the approximate well locations on it and draw in where you expect to put the central points of delivery (CDP) (it doesn't really matter how close you are to the final location, so don't spend inordinate time on siting the CDP at this time).
  - Draw a pipeline network and make a guess at pipe sizes. Be careful to minimize highway, railroad, and river crossings, they are all very expensive.
  - Measure your pipe segments and aggregate by pipe size.

- Sum the length of each size (usually done in miles) and multiply the sum times the expected pipe size (usually done in inches). Add all the products together to get an "in-mile" number.
- Apply your company's "cost/in-mile" factor. This number varies widely, I've seen it as low as $800 USD and as high as $2500 USD. This gross-level number should be all-inclusive and can easily be ± 50% for your particular project. Experienced piping designers will tend to adjust the factor so that the answer is +0%, −50% because it is always easier to go to your boss and say "steel prices went down and that $25 million project should be closer to $15 million," than to tell her that the project is now $60 million.

- You will need to do an authorization-for-engineering cost estimate
  - You have picked an MAWP.
  - You have laid out the piping network and modeled the network to get reasonable pipe sizes, lengths, schedules, and grades.
  - You have verified that there is a reasonable chance that you can acquire all of the ROW you need.
  - You have thought about what pipeline accessories you need (e.g., block valves, pigging equipment, line drips, and any automation)
  - Gathering-system costs are quite variable with changing materials costs, changing labor rates, and changing transportation costs. The only thing you have a good handle on is the pipe. You can call a couple of vendors and get today's price for each pipe size/material you are planning to use. Table 6.1 has factors that have tended to work well over many years and many locations. Using Table 6.1 factors, you divide today's pipe price by the factor in the last column to get a good handle on the final cost of the project.
- Project authorization cost estimate
  - Developing this number is a part of the detailed engineering and is done with detailed materials lists and competent ROW and permitting information.

Table 6.1 Rule of thumb cost estimating

	Material cost	Material part of job (%)	Rest of project (%)	Total cost
Steel	Pipe cost ÷ 0.80	40	60	Pipe cost ÷ 0.32
HDPE	Pipe cost ÷ 0.20	5	95	Pipe cost ÷ 0.10
RTP	Pipe cost ÷ 0.85	90	10	Pipe cost ÷ 0.77

6.3 GATHERING EQUIPMENT SELECTION (FEED)

The first decision that you have to make in a gathering project is pipe. This decision sets the stage for the entire design process and can have ramifications on operations for decades. The pipe decision is trying to find an "optimum" point balancing cost, capacity, and longevity. A heavy-wall steel pipe will last longer than a thin-walled pipe, but can cost a lot more. Stainless steels can resist many types of corrosion better than mild carbon steel, but cost more. HDPE pipe has a slicker surface and has a higher flow capacity than steel (in the same size), but has limited MAWP (pressure capacity).

The most important thing for a pipe designer is that they understand that there is no way to "get it right." That cannot be done. Remember that every well has a personality. That means that the flow rate and ultimate recovery vary widely from one well to the next, and while you may be given a well count and a predicted maximum flow rate per well, any well that conforms to that flow rate is an absolute coincidence. I've seen the reservoir guys get the field-maximum flow rate pretty close (not often, but it has happened), but not a single well made the rate described by field-maximum divided by well count. I operated a gathering system that the design flow rate was 38 MMSCF/day (1122 kSCm/day) with 46 wells, about 820 MSCF/day/well (23 kSCm/day/well). The designer of the system built his model based on 2 MMSCF/day/well (57 kSCm/day/well) and laid out the piping to reach his target velocities throughout the system. When I took over operations there were two wells that were pretty close to the original average, but the range on the other wells was 0.050–14 MMSCF/day/well (1.42–396 kSCm/day/well) and the total system was making over 100 MMSCF/day (2830 kSCm/day) with many parts of the system seeing very high line pressures, while other parts of the system had pipe velocities approaching separator velocity and significant accumulated liquid. I think that the system designer did a very good job with what he had to work with. This development project was done in three stages and for the subsequent stages the designer asked the reservoir engineers to provide estimated flow rates by well—the resulting error was very similar in these systems to the system-wide average approach, the reservoir engineers simply cannot predict the vagaries of drilling effectiveness or the risk of drilling into a small-scale discontinuity. The point of this discussion is to illustrate that any gathering-system design needs to expect that it will have weak places that will require fixing in subsequent years and to not treat the need to loop lines or add water-removal equipment as a failure of initial design.

6.3.1 Design standards

A gas gathering system is a pipeline that carries pressurized fluid. The system can be regulated by local, state/provincial, and/or federal regulations. These systems often have public health and safety ramifications. Consequently, an objective standard is required to provide a basis for decision making. These objective standards can be ASME standards, ISO standards, government regulations (codes), API standards and recommended practices, and company policies. Most often a mix of these sources will be required to be layered atop one another. Government regulations generally set the minimum performance level, but it is very common for the code of federal regulation to include an industry standard by reference and add requirements to the industry standard (e.g., the standard says that you have to do "A" or "B" and the code says you have to do "A" *and* "B").

Finally company standards are layered onto the design. Company standards are intended to clarify items that the codes and standards are silent on, or remove options that are allowed in the code. One very common and quite irrational addition by company codes is a proscription against threaded pipe. The standards include extensive language about how to use threaded pipe safely and how to understand physical design limitations. The codes are mostly silent on threaded pipe. Many company policies simply say that tapered pipe threads will not be used. This company decision actually increases the risk of personnel injury and environmental releases (small butt welds are very hard to do properly and the rejection rate is many times the rejection rate of butt welds on bigger pipe, the result is that more bad welds are put in service on small pipe that would have historically been threaded). As silly as some of the company additions can be, your choices are to follow them or get them changed and the process to change company policy can be challenging and can require years of work.

We've talked about "ASME B31.3: Gas Processing Plants" as the design code for well-sites. There are companies that use B31.3 for gathering lines even though these codes do not have any way to indicate population density or to account for high-risk traverses like crossing a major highway or railroad tracks or rivers. Using B31.3 a designer would use the same safety factors moving across farm country in Pennsylvania as they would use crossing the East River into Manhattan and running under Trump Tower. This seems like either an excessive safety factor for the agrarian pipe or a grossly inadequate safety factor for a 58 story residential/retail building.

"ASME B31.8: Gas Transmission and Distribution Piping" does include methods to evaluate varying risk profiles as you move down the pipe. Not every jurisdiction uses ASME B31.8 (e.g., in Canada you would use CSA Z6662-2007, and in countries using ISO standards it would be EN 14161:2003), but the processes and equations are very similar from standard to standard. The most significant difference tends to be in the safety factors that are required and when they are required.

The scope of ASME B31.8 is similar to the other piping codes. It is stated as: "This Code covers the design, fabrication, installation, inspection, and testing of pipeline facilities used for the transportation of gas. This Code also covers safety aspects of the operation and maintenance of those facilities." The code does not apply to:

- Pressure vessels
- Piping above 450°F (232°C) or below −20°F (−29°C)
- Piping beyond the outlet of the customer's meter set assembly
- Piping in oil refineries or natural gasoline extraction plants, gas treating plant piping
- Vent piping
- Liquid petroleum piping systems
- Liquid slurry piping systems
- Proprietary items of equipment, apparatus, or instruments
- The design and manufacture of heat exchangers
- CO_2 transportation piping systems
- LNG piping systems
- Wellhead assemblies, including control valves, flow lines between wellhead and trap or separator, offshore platform production facility piping, or casing and tubing in gas or oil wells.

The ASME B31.8 process for determining minimum wall thickness for piping is very different from what we discussed in Chapter 5, Well-Site Equipment, and is specific to pipe material. These equations will be presented as follows.

6.3.2 Pipe selection

When you are selecting pipe, you have to be concerned with balancing cost, flow capacity, and longevity. Generally one of these factors will dominate the decision. For example, when CBM was being anticipated, we knew that the gas had significant CO_2 and that we were not going to perform well-site dehydration so we would have significant condensed water in the presence of acid gas which was known to be a recipe for high corrosion rates which made longevity the primary consideration and we preferentially ran high-cost glass-reinforced plastic (GRP) (fiberglass) piping often in sizes smaller than we would have chosen for steel (because of limits on available sizes in the pressure ratings we needed).

6.3.2.1 Size selection

If you are designing a brand-new system or a simple loop line to alleviate a high-pressure section of the system, you need to have some sort of pipeline model. Choices are:

- A manual model. Using the arithmetic from Chapter 4, Surface Engineering Concepts, you have all of the equations and you can solve the model one segment at a time. Since the segments are interrelated, this iterative solution can be quite cumbersome and time consuming.
- A plant model. Generally plant models like Hysys or AspenONE have very robust and rigorous piping models that do many things that you don't need and not many things that you do need (looped lines and compressor processes are much weaker in these plant models than they are in pipeline models).
- Commercial pipeline models. Pipeline models all have extensive capability to model new lines in existing systems (from a simple twinning to a line that takes off the existing system picks up new gas and returns to a different part of the system creating an effective cross connect that can flow either direction), but they all have significant weak features. For example, we had an engineering student working for us one summer and assigned him an evaluation of a very-high-pressure portion of a system. This system was a simple wishbone layout with a north lobe and south lobe that came together and ran into a CDP, the bottleneck was on the north lobe. The summer-intern analyzed the problem with a commercial system and recommended a simple twinning of the north loop with the same pipe sizes as the original design. I asked

why he didn't just cross connect the north loop to the south loop to take advantage of the underutilized pipe in the south loop. He said that he had analyzed that and it resulted in a bigger problem that what he had started with. His answer seemed counterintuitive to me so I built a simple model in another software product and that second program liked the cross connect (we later built it my way and got results very close to the pressures predicted by my model). We spent the rest of the summer trying to find out what went wrong with his model. We finally determined that the model had a built-in bias to flow from low-numbered nodes to high-numbered nodes, and simply renumbering the nodes changed the program recommendations from cross-connect-bad to cross-connect-great. From that day on, we all knew that if we ran that particular program we needed to renumber the nodes of the final model and see if we got the same answer. It was a widely used (and very expensive) program in the industry with a blind spot that was difficult to even see is there.

- Simple chart. Many problems can be addressed (certainly initial sizing) using a simple chart like Fig. 6.2. Note that in the title, a very specific set of conditions are listed. Your conditions may (will) vary. For example, for 12 in (300 DN) and smaller projects a reasonable limit is 15 psi/mi (64 kPa/km), but for 12 in (300 DN) and bigger projects it is common to limit the pressure drop per unit length to 5 psi/mi (21 kPa/km) because of the much larger volume in the bigger pipe you want to minimize the number of large compressor stations required to replace pressure lost to friction. Also note that there is a 100 ft/s (30.5 m/s) line on the bigger pipe in Fig. 6.2. This limitation is there to keep the calculation within the incompressible limitation (as defined as downstream density $\geq$90% of upstream density). This velocity is just under 0.1 Mach which experiments have shown is a minor cusp in the velocity vs density curve. Since expected conditions rarely match this chart, Fig. 6.3 shows how the equations change with various parameters. Best-to-worst comparison is in the upper left corner. It shows that pipe roughness always matters. Temperature sensitivity is never very high, but is higher at higher pressures. Specific gravity sensitivity goes the other way and decreases with increasing pressure. Elevation (i.e., local atmospheric pressure) is a big deal below about 29 psig (2 barg), but above that assuming sea level should be fine for pipe capacity calculations.

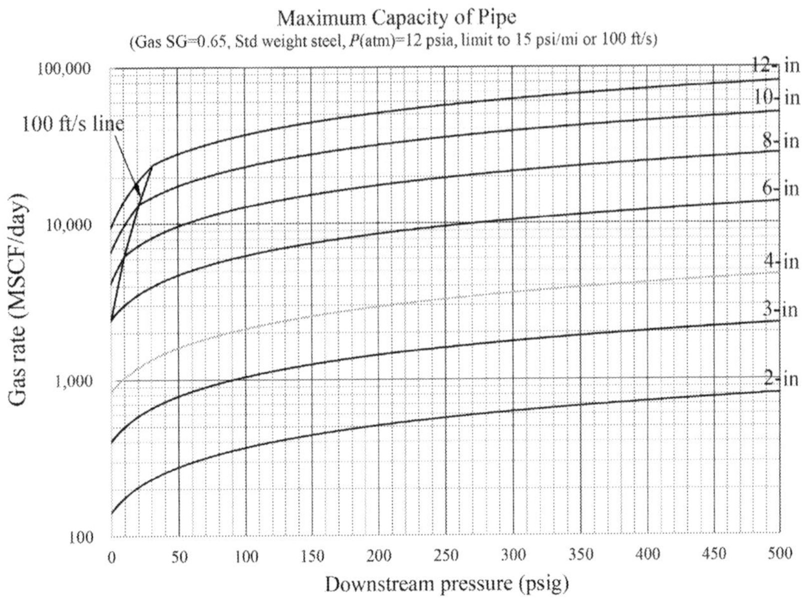

Figure 6.2 Capacity vs pipe size.

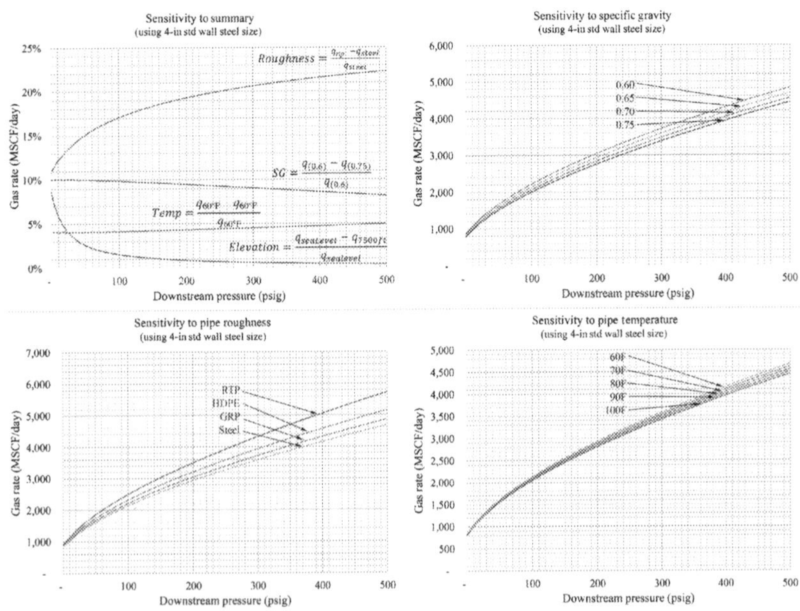

Figure 6.3 Pipe capacity sensitivities.

There are not many "never ..." in piping design, but there are some. In the interests of successful operations, the piping designer should never:

- Change pipe diameter underground (this transition will create confusion and operational difficulties for the life of the system, and the savings are inadequate for the problems).
- Design nonpiggable lines in a raw well-gas system.
- Allow the piping selection to limit system MAWP (i.e., piping MAWP should always be greater than the ASME B16.5 class of the flanges in the system, which should be the same or greater than the MAWP of the supply and delivery systems).

6.3.2.2 Material selection

Gathering-system materials choices have changed considerably in the last 30 years due to (largely unfounded) corrosion concerns with high-CO_2 gas, and the requirement to move very large volumes at very low pressure. Many operators mistakenly assume that since CBM and Shale gas will spend a large portion of the production life at low pressures that it is reasonable to build very low-pressure gas gathering systems. The problem with this choice is that it eliminates the ability to actively manage reservoir pressure during the high-productivity early production period. Another problem is that it requires major capital expenditures on day 1 to make up the energy discarded across wellhead chokes. As a gathering-system designer, it is a good idea to have a system with at least the pressure performance of the well-site equipment (assuming that the well-site equipment has been designed to optimize the ultimate recovery from the reservoir).

I was on a well a few years ago that the wellhead choke made so much noise that you couldn't talk on the location. The flowing data showed that the well was making 16 MMSCF/day (453 kSCm/day), reservoir pressure was 1400 psig (9.65 MPaa), wellhead pressure upstream of the choke was 450 psig (3100 kPag), and separator inlet pressure (downstream of the choke) was 250 psig (1724 kPag), because they had a very low-pressure HDPE gas gathering system, they further cut the pressure out of the separator to 10 psig (68.95 kPag), the gas then went 3 miles to a compressor station where the 2 psig (13.8 kPag) gas has to be boosted to 300 psig (2070 kPag)—with a rational system design the multimillion dollar compressor station likely could have been delayed 5–7 years just by eliminating the dP across the two well-site chokes.

Piping material must be compatible with design pressures, expected temperatures, and expected chemicals. It is useful to check with local contractors and find out what materials they have experience with. If they've installed the material and like it, then your installation may be able to benefit from a learning curve that someone else paid for. If they've used it and hate it, then you might be able to avoid some hidden shortfall in the material (or you may talk to a different contractor, your choice). If they've never used it then you need to budget for a steep learning curve.

Steel. Historically all gathering-system pipe was steel, mostly API 5L Grade B Seamless. Coatings have evolved from painted-on tar, to coal-tar-epoxy, to fusion bonded epoxy (FBE). The modern FBE coatings are orders of magnitude tougher and adhere to the pipe far better than older FBE designs.

In the early 1980s all gathering pipe was seamless because the rolled-plate pipe like spiral wound and electric resistance welded (ERW) pipe were truly junk. The manufacturing process on this pipe created so many stories of piping coming "unzipped" that the industry refused to purchase longitudinally welded pipe at all, and many were very reluctant to purchase pipe made in the United States. The U.S. Steel upgraded their facilities at Fairfield Works in Alabama in the early 1980s and reversed a decades-old decline in the quality of steel made in the United States. By the mid-1980s U.S. Steel and Lonestar Steel (later purchased by U.S. Steel) had straightened up their performance to the point that both ERW and steel produced in the United States were at least as good as any pipe manufactured in the world. With those changes, ERW pipe moved to the mainstream and it is many people's (including mine) preferred choice when the design choice is steel, but it is important for the designer to specify that the longitudinal weld not be installed on the bottom of the pipe (the heat-affected zone of the weld seems to be more prone to damage from MIC (corrosion) than the rest of the pipe or the girth welds).

A current trend in piping design is to use higher grades (i.e., higher specified minimum yield strength or SMYS known colloquially as "X Grades") to allow thinner walls for the same MAWP (or higher MAWP for a given wall thickness). The higher grades are stronger, at the price of being more brittle with higher internal stresses. I am concerned that the higher internal stresses will increase the incidence of "stress corrosion" modalities that tend to be a higher risk of these sorts of cracks. There are times that the X Grades are the best choice, but it is a bad idea to just use them because you have them. Wall thickness in ASME B31.8 is in Eq. (6.1).

$$t_{\min} = \frac{P_{mawp} \cdot OD_{pipe}}{2 \cdot P_{smys} \cdot F_f \cdot E_f \cdot T_f} \tag{6.1}$$

Most of the terms in this equation are obvious, but there are some new ones. E_f is a longitudinal joint factor that is 1.0 for seamless and ERW (remember in ASMB B31.3 Table A1B has a derate factor of 0.85 for ERW). T_f is the temperature derate factor which is 1.0 for any fluid operating at less than 250°F (121°C). The interesting new term is F_f which is called the "design factor" used for matching the safety factor of the pipe to the population density (Table 6.2).

Table 841.1.6-2 in ASME B31.8 contains a list of specific situations that require these factors to be lowered (e.g., fabricated assemblies in Loc 1 Div 1 must be designed to a design factor of 0.60 instead of 0.80).

The "1 mile section" is important. We built a gathering system on farmland (Loc 1 Div 2) and over time one of the farmers subdivided his property and built houses. Now a portion of the system is Loc 3 and the pipe wall thickness is no longer adequate for the system MAWP. We took several actions to render this part of the gathering system appropriate for the new population density. First we lowered the MAWP on the entire system (but we did not change the location class of the entire system) and reset all pressure safety valve (PSV) on the system to values consistent with the new MAWP. Second we installed new block valves on the trunk passing the housing, we were careful to place the valves far enough back to minimize the risk of further encroachment. Finally we actuated the

Table 6.2 ASME B31.8 design factor

Location class	Description	Design factor (F_f)
Loc 1 Div 1	1 mile section of pipe with ≤ 10 buildings intended for human occupancy that is operated at >72% of SMYS, static tests must be done with water	0.80
Loc 1 Div 2	A Loc 1 Div 1 pipe that is operated at <72% of SMYS, can be tested with water, air, or natural gas	0.72
Loc 2	1 mile section of pipe with $10 < x < 46$ buildings intended for human occupancy	0.60
Loc 3	1 mile section of pipe with ≥ 46 buildings intended for human occupancy, but that is not Loc 4	0.50
Loc 4	Location where multistory buildings are prevalent	0.40

new block valves with high-reliability actuators to shut off the piping through the housing development on either high pressure or low pressure. This resulted in a short section of an extensive gathering system legitimately having a different design factor than the rest of the system.

6.3.2.2.1 Steel pipe wall thickness example

For the conditions in the example in the last chapter, see Table 6.3.

If you assume Grade B (P_{smys} = 35,000 psi) ERW pipe, then Eq. (6.2) shows that standard weight (0.365 in (9.27 mm) wall thickness) would work where in the ASME B31.8 calculations Schedule 80 (0.594 in (15.09 mm) wall thickness) pipe was required.

$$t_{\min} = \frac{600\ \text{psi} \cdot 10.75\ \text{in}}{2 \cdot 35000\ \text{psi} \cdot 0.72 \cdot 1.0 \cdot 1.0} + \frac{3}{16}\ \text{in} = 0.315\ \text{in} \tag{6.2}$$

Now we need to calculate stresses using Eq. (6.3) (called "Barlow's formula") we see that at MAWP the hoop stress is 51.9% of SMYS, at test conditions it is 77.9% of SMYS, and normal operating pressure it is 13.0% of SMYS.

$$S_H = \frac{P_{psig} \cdot OD_{pipe}}{2 \cdot (t - t_{corr})} \tag{6.3}$$

HDPE. Polyethylene is a "thermoplastic" material that does not change physical characteristics after a phase change (i.e., you can melt it and let is solidify without changing its physical properties). "High density" refers to material whose density is between 0.941 and 0.965 gm/cc.

HDPE was invented at the Kaiser Wilhelm Institute in 1953, 2 years later it was first formed into pipe. HDPE pipe has been used in drainage and irrigation since the 1960s. Over time the "long-term hydraulic strength" (the plastic industry's replacement for the steel industry's use of

Table 6.3 Pipe thickness example data

	fps	Both	SI
MAWP	600 psig		4140 kPag
Design temperature	100°F		37.8°C
Nominal pipe	10 in		250 DN
Type test		Pneumatic	
Test pressure		150% of MAWP	
Corrosion allowance	3/16 in		4.8 mm
Normal operating pressure	150 psig		1034 kPag

SMYS, it is the applied pressure where material creep begins) improved dramatically, and people started using HDPE in gas service in the 1980s (the long-term hydraulic strength had increased by a factor of 4 compared to the 1960s value). HDPE pipe comes in short spools (limited to small diameter) and sticks. Lengths of pipe are melted together using specialized welding equipment.

All of the calculations in HDPE are a mixture of industry association documents, engineering standards, and regulations. Some jurisdictions actually refer to association documents in regulations, giving them the standing of a law. ASME B31.8 limits unreinforced thermoplastic pipe in natural gas service to a maximum pressure of 100 psig (689 kPag) under all conditions. One industry association document (that is referenced in regulations) does not differentiate gas service from liquid service and allows HDPE to be used in pressures substantially higher than ASME B31.8 allows. One equation that all sources seem to agree on is the definition of dimension ratio (DR):

$$\mathrm{DR} = \frac{\mathrm{OD}_{pipe}}{t}$$

which leads logically to: (6.4)

$$\mathrm{ID} = \mathrm{OD}_{pipe} \cdot \frac{\mathrm{DR} - 2}{\mathrm{DR}}$$

Some DRs have been standardized to be widely available, and these DRs are designated "SDR" for "standard dimension ratio."

From ASME B31.8 paragraph 842.2.1 (modified to use terms consistent with this document) Eq. (6.5) is used to determine the MAWP for HDPE. Material specifications for HDPE are contained in ASTM D 2513 and most commercial HDPE is classified "PE 3408." P_{lths} is specified by material, and for PE 3408 it is 1600 psi (11.0 MPa). The F_{design} term is a factor that distinguishes natural gas service from water service, for natural gas it is 0.32, for water it is 0.50.

$$P_{mawp} = 2 \cdot P_{lths} \cdot \frac{1}{\mathrm{DR} - 1} \cdot F_{design} \tag{6.5}$$

HDPE gas gathering systems are very common around the world, and they seldom fail to be a bad idea. In nearly every case they require significantly choking the reservoir pressure to protect the pipe from overpressure. When the goal is to protect an artifact instead of achieving an optimum set of conditions for gas flow, then the ultimate profitability of the reservoir will always be compromised. That is a lot of "never" and "always," and I strive not to make value judgments, but I am yet to see a case where an operation achieved better results with HDPE than they would have gotten with steel or RTP.

Reinforced thermoplastic pipe (RTP or Spoolable composite pipe). Due to both corrosion concerns and the cost of laying offshore flow lines, the Oil & Gas industry got serious about finding viable alternatives to steel pipe in the early 1980s. All of the so-called Majors and many of the Second Tier operators invested heavily with both university research and providing venture capital to start-up companies to develop pipe that met the difficult performance requirements for an installed cost that was competitive with steel. Like any emerging field, there were a number of products that proved to be unable to compete in the marketplace, but several companies have emerged with products that are proving to be superior to steel. All of the RTP commercial products have a few things in common:

- They are made up of a sandwich of dissimilar materials
  - Inner layer of material resistant to corrosion and scale formation. This layer is usually HDPE or cross-linked polyethylene (PEX, used for higher temperature applications).
  - Intermediate reinforcement layer. The reinforcement layer varies widely by manufacture from steel braid (Flexsteel) to "glass fiber-reinforced epoxy" (Fiberspar), to a rope-like aramid fiber (Soluforce). The Flexsteel product was developed specifically for offshore applications and was designed to withstand being hooked by a super tanker's anchor in a hurricane. The Soluforce product was designed to have a very low installation weight and has proven to have an advantage in installation costs.
  - Outer strength layer is almost always HDPE.
- Scope of original material-design projects was heavily focused on:
  - Downhole tubulars, those products have not met with much market success.
  - Multiple flow paths in a single diameter, again those products have had limited market success.
  - Offshore pipelines, one product (Flexsteel) has had significant market success.
  - Onshore gathering systems, several products have had significant market success.

$$P_{mawp} = 2 \cdot P_{lths} \cdot \frac{1}{OD_{pipe} - t_{\text{reinforced}}} \cdot F_{design} \tag{6.6}$$

Long-term hydraulic strength in Eq. (6.6) is 11,000 psi (75.8 MPa) (from ASTM D 2517). The thickness used in this equation is the thickness of the inner layer (referred to in manufacturer's literature as "reinforced thickness") as opposed the half the difference between the outside diameter and the inside diameter. The outside diameter is the ultimate OD of the sandwich. The design factor is the same as HDPE.

This material tends to cost more per unit length than steel (and considerably more than HDPE). Cost of installation is a fraction of the cost of installing steel or HDPE (it has been reported when plowing RTP into the ground without open ditch it has been installed at the rate of 1 mi/h (1.6 km/h) as opposed to an average of 0.5 mi/week (0.85 km/week) or 0.01 mi/h (0.016 km/h) for the same size steel in the same location). "Your results may vary," and I've never achieved 1 mile/h, but I've found it common to install RTP for less than 1/10th the cost of installing steel. It is common for the total installed cost of RTP to be 20%–40% less than steel (depending on how much open ditch the project needs).

Glass Reinforced Plastic (GRP or fiberglass). Fiberglass pipe is not included in ASME B31.8, which implies that the ASME does not encourage its use in gas service. It is included in ASME B31.3 as "Glass-Reinforced Thermosetting Resin (Glass RTR)," but ASME B31.3 does not discuss the procedure or equations for determining the piping MAWP. "ASTM D 2992, Practice for Obtaining Hydrostatic or Pressure Design Basis for 'Fiberglass' (Glass-Fiber-RTR) Pipe and Fittings" lays out the (lengthy) procedure for setting an hydraulic design basis and pressure design basis for GRP and at the end of the test the pipe pressure rated is published by the manufacturer, the user of this pipe selects their MAWP and selects pipe which fits that need. ASTM D 2992 is silent on the fluids that GRP is appropriate for.

As mentioned earlier, in the early days of CBM, GRP was the pipe of choice. It was available in pressure ratings consistent with desired design pressures and was very corrosion resistant. A considerable amount of it was installed in the San Juan Basin, and crews became very adept at its installation. We did not know the extent of damage that could be done by bruising the pipe (e.g., when a rock falls from the edge of the ditch onto the pipe it makes a mark on the pipe). These bruises did not impact the ability of the pipe to pass a static test, but over time the damaged fibers rubbed together and eventually created a path for gas to escape. None of these leaks created the typical "anthill" as shown in Fig. 4.14, but they created a nuisance.

6.3.2.3 Pipe summary

The typical values and the pipe material and pipe comparison are given in Tables 6.4 and 6.5, respectively.

6.3.3 Ditch

Buried pipelines must be, well, buried. With steel and HDPE lines and any line in hard-rock, burial means an open ditch. In most soils RTP can be "plowed-in," meaning that it is put into the ground and backfilled with a single-step process.

6.3.3.1 Open ditch

Historically the only way to get pipe underground was to dig a ditch, lower the pipe into it, and then backfill the ditch over top of the pipe.

Table 6.4 Pipe material typical values

Material	Largest size typically used	Typical MAWP in gas	Comments
Steel	20 in (500 DN)	600 psig 4140 kPag	Requires external coating, internal surface subject to corrosion
HDPE (SDR-7)	16 in (400 DN)	100 psig 690 kPag	All sizes of an HDPE SDR number have the same MAWP (limited by maximum pressure in ASME B31.8)
HDPE (SDR-13.5)	16 in (400 DN)	82 psig 565 kPag	Much thinner wall than SDR-7
GRP (stick)	6 in (150 DN)	1440 psig (10 MPag)	Delicate to install
SoluForce	6 in (150 DN)	1440 psig (10 MPag)	Spool, HDPE in contact with fluid, strength layer aramid fibers
FlexSteel	8 in (200 DN)	1440 psig (10 MPag)	Spool, HDPE in contact with fluid, strength layer steel
FiberSpar	6 in (150 DN)	1440 psig (10 MPag)	Spool, fluid in contact with HDPE or PEX (high temperature), strength layer glass-reinforced epoxy

Ditches are made using a combination of "wheeled ditching machines," "rock saws," and "track hoes" depending on the terrain, length/size of the line, and ground condition. Open ditch is a trade-off—the more ditch that is open the longer continuous pipe section can be lowered in at one time and the fewer times you have to redeploy specialized equipment, but open ditches can collect a lot of water in a rain event and are a danger to farm animals and wildlife (Fig. 6.4, don't worry, the track hoe in the picture was used to dig a walkway out of the ditch and the bull and calf were able to walk out of the ditch unharmed). Different engineers have different risk tolerances and personally I don't allow a ditch to be open more than 3–5 days.

We'll talk more about how to make a ditch later, but for now let's talk about considerations for a ditch.

- First, the ditch must be deep enough to allow for adequate "cover" (i.e., the amount of dirt between the top of the pipe and the finished grade). The amount of cover is a combination of your choice of a

Table 6.5 Pipe comparison

	Steel	HDPE	RTP
End connection	Welded	Welded	Press fit
Distance between interventions	30 ft (9.14 m)	30 ft (9.14 m)	2000–5500 ft (910–1700 m)
Crew size	11 workers (most must be highly skilled)	6 workers (half highly skilled)	Three workers (moderate skills)
Ditch bottom	Flat	Flat	No requirement for flatness (also, can be plowed-in without an open ditch)
Maximum pressure in gas	2500 psig (17.2 MPag)	100 psig (690 kPag)	2500 psig (17.2 MPag)
Minimum radius for field bend	40 × OD	25 × OD	10 × OD
110-degree bend	Fitting plus field bend	Fittings	Rope pipe (no fitting needed)
Weight	Very heavy	Heavy	Light
Material cost	High	Very low	Highest
Total Cost	High	Highest	Lowest

Figure 6.4 Moose in open ditch.

standard (ASME B31.8 has different minimum cover for different location classes, the more population density the deeper the pipe must be), any applicable regulations, and company policy. It is important to review all three and ensure that you have met the requirements of the most restrictive.

- You need to pick a "working side" and a "spoil side." Every contractor has a preference for working side, and it is worthwhile to follow that preference (after the decision is made it no longer matters and if you force your preference on the contractor he will make you pay dearly for that demand in a thousand small inefficiencies). The pipe is strung out and joints are welded together on the working side. The dirt removed from the ditch is piled on the spoil side.
- You need enough ROW for the activity. For larger diameter pipe, you have to dig a wider hole deeper, which creates a big spoil pile. If any part of the spoil pile goes off of the ROW then bad things can happen (from being required to pay for additional ROW to having your job shut down for trespass). Bigger pipe is heavy and the equipment required to handle it for welding and for lowering it into the ditch is bigger than you need for smaller pipe so you need more space on the working side too. If your ROW is limited (as it often is on federal land), then if worse comes to worst it is possible to "work on the spoil" where you put your ditch adjacent to one edge of the ROW, put the spoil on

the other side, level the spoil pile and string the pipe on top of it. This is never ideal, and it creates some additional risks from people working on the unstable spoil pile, but it can be done if done with care.

- Ditch bottom is important. It needs to be reasonably consistent and free of larger rocks. The ditch bottom should be approximately parallel with the finished grade which means that it is going to undulate to some extent. The pipe is surprisingly flexible and will naturally sag into most undulations. If the pipe won't fit a particular section of the ditch (e.g., overtopping a hill or crossing a valley) then you have to "overbend" the pipe to fit the ditch. It should never be acceptable to the project engineer to "build up" the ditch bottom with sandbags in lieu of bending the pipe—sandbags simply are inadequate structural support for the long term. Some projects require that pipe be laid in a "bedding material" which can either be imported sand or sifted spoil. The goal of bedding is to ensure that the pipe is resting on the bottom of the ditch without gaps and is not resting on rocks that can damage the pipe coating. Bedding was specified much more often before the current generation of pipe coatings—today's pipe coatings are tough enough that adding bedding material is largely pointless, but is still required in some jurisdictions and some companies.
- Once the pipe has been lowered in, you have to decide what the first material to put on the pipe is. Again this decision has more to do with "policy" than actual need. Some companies require bringing in sand for pad, others allow sifted spoil, others don't specify and blown spoil is used. When the pipe is covered with appropriate fill, then the rest of the dirt is pushed into the ditch and "driven down" (i.e., construction equipment is driven over the ditch multiple times in multiple directions) to compact it enough to keep it from settling.
- Finally, the ROW has to be returned to as near pristine conditions as possible. The contour must be returned to approximately original, barricades must be put in place to keep it from becoming a new roadway, appropriate seed must be sown to grow native plants, pipeline markers must be placed to show the pipeline route, and the ROW must be monitored for erosion until it has reached some percentage of coverage from new growth.

6.3.3.2 Plowed-in

For many years, the power-distribution industry has used equipment to "plow-in" buried cables. This process starts with the cable on a spool that is associated with a "sword" that is inserted into the ground at the required burial depth and clears a path through the earth while positioning the cable along the path. Cable is very flexible and this process has been largely perfected for underground cable installation. The advent of RTP has led several companies to produce equipment to adapt this installation technique to spooled pipe. The pipe is considerably stiffer than cable and required some significant adaptations, but those adaptations were accomplished and it is very common for RTP to be plowed-in without an open ditch.

Looking at the concerns for open ditch, this technique has some significant benefits:

- The depth of the pipe is set by positioning the sword and can be much deeper than the minimums.
- There is no need to specify a working side/spoil side since only minimal work is done on the ROW.
- There is no spoil pile, so the ROW only needs to be wide enough for the plow with occasional sites to allow equipment to reverse direction.
- Ditch bottom is irrelevant and bedding is not an issue.
- Padding is irrelevant.
- Finally, the ROW is only disturbed a few inches that can be largely returned to pristine conditions by running the tire of a piece of equipment (like a rubber-tire hoe) down the gash. It is important to place pipeline markers within a few weeks of running the pipe or you may not be able to find the "disturbance."

6.3.4 Pipeline obstructions

As a pipeline progresses down a ROW, it will encounter obstacles. Standards, codes, and company policy frequently have something to say about crossing these obstacles. For example, ASME B31.8 specifies that putting your pipe on a bridge over a river requires a maximum design factor of 0.6. It also increases requirements for nondestructive testing around crossings (generally increases x-ray percentage). All crossings should be as close to perpendicular to the obstacle as reasonable.

Some obstacles like farmer's fences should have been discussed in the ROW agreement and it is clear (from that contract) that you have to stop the process of making up the pipe outside of the ditch at the fence and start it back up after the fence, and then remove the fence while putting alternate guards against cattle movement while you connect the two sections in a "bell hole" (see Section 6.7.5). Other obstacles may not be as cut and dried.

Rivers. There are basically four ways to cross a river: (1) directional bore, (2) bridge, (3) river cut, and (4) plow-in.

Directional bore. As directional boring equipment has improved over the last few decades, this option has gotten progressively less expensive and more precise. Thirty years ago, this option cost 4–5 times more than other options and you never really knew where the line was going to emerge from the bore. Today the costs are only slightly higher than other options and the pipe emerges from the ground within a footstep or two from the target. The design of most modern directional bores put the pipe deep enough that the pipe does not tend to reside in standing water under the river and adding weight to the pipe to maintain a negative buoyancy is generally not necessary. Directional bores can either be cased (i.e., a thin wall pipe is driven through the bore hole when the drill is withdrawn and the pipeline is pulled through the casing) or allowed to collapse around the pipeline.

Bridge. If you will look, you will see that many railroad and highway bridges include pipes under them. Since the static load of a pipeline (or several) is tiny compared to the dynamic load or a train or truck, the use of transportation bridges is a common way for pipes to cross rivers. Permission and structural analysis is required, and the permission is not always granted. Requesting necessary permits prior to finalizing route is a prudent approach.

There are also dedicated pipe bridges (Fig. 6.5) that it may be possible to add your pipe to. The bridge in Fig. 6.5 was owned by a third-party

Figure 6.5 Pipe bridge.

gatherer and its use required: (1) detailed review by a structural engineer; (2) detailed review by a geotechnical engineer; (3) reinforcement of the pier foundations using helical piers; (4) I-beam supports along the length of the piers; (5) a new carrier cable (located above the cable for the existing pipe); and (6) a support arrangement that allowed the new upper cable to carry the new lower pipe. After 3 months of negotiations and engineering analysis, the owner of the bridge allowed us to use the bridge without a fee.

River cut. On very small rivers it is sometimes allowed to stop the flow of the river long enough to dig a ditch, lower the pipe in, and backfill it before releasing the temporary dyke. This approach is rarely allowed and requires: (1) permits from federal and state agencies; (2) explicit permission from all impacted upstream land owners (and usually a temporary use fee); (3) downstream water monitoring both before and after the job; (4) a way to prevent the pipe buoyancy from causing it to float to the surface (usually concrete river weights on open ditches); and (5) a mitigation plan if the turbidity of the downstream flow is impacted by the job (can be very expensive). On larger rivers (where damming the entire river is impractical), it is normal to do a river cut in stages by setting a cofferdam in part of the river and allowing the full flow of the river in a narrowed channel followed by moving the cofferdam to the next section. The logistics of doing a river cut in stages are extremely difficult and subject to intense scrutiny by regulators (primarily the Corps of Engineers in the United States).

Plow-in. With the advances in RTP, it is tempting to just ignore that there is a river in the way and continue plowing the pipe right through the river. This may be a great idea (virtually no disturbance in water quality and no obligation to stop or divert flow), but you have to be concerned with the pipe staying where you put it. For example, 4 in (100 DN) Soluforce weighs 3.3 lbm/ft (4.91 kg/m). This means that 3.3 lbm (1.5 kg) of pipe will be displacing 6.9 lbm (3.1 kg) of water, a significant up force on the pipe. A heavier product like Flexsteel is still two-thirds the weight of water and will still float. It could be that the tendency of the sliced path is small enough to restrict the amount of water that is installed with the pipe, but this would still make me nervous. Check with your pipe manufacturer for case studies prior to using this option.

Railroads. Other obstacles are easier to describe, but still very difficult to actually cross. The other obstacles include railroads, roadways, and foreign pipes. Railroad crossings are a major issue. The roadbed is subjected to extreme forces from loaded trains and the rail companies guard the integrity of their roadbeds very closely. It is difficult to get a railroad company to even talk to you, and when they do respond to inquiries they have extensive restrictions. As directional boring technology has improved that is becoming the only option being allowed. In previous decades you could often dig a bell hole 6–10 ft below the bottom of the constructed roadbed and do a straight bore that had to be lined with a heavy-wall casing that you pulled your pipe through. Today you can do an uncased bore, but the minimum depth is much deeper (one railroad demanded 60 ft (18 m) below the surrounding ground level, which is achievable with a directional bore, but requires offsetting the start/end of the bore a considerable distance).

Roadways. Pipelines frequently have to cross roads. "Roads" run the gamut from a "two-track" path that a farmer uses several times a year to access remote fields to elevated, limited-access highways. Each has its own issues. The main issue is the risk of road traffic damaging the buried pipe. This risk will be somewhere between zero and significant. You design road crossings to keep the risk within your comfort zone.

Cased roadway crossings. A roadway casing is a culvert-like structure that is placed under the roadway and the pressure-containing pipe(s) is/are pulled through the casing. Standards, codes, and most companies leave the case/uncase decision up to each project. While regulations are generally not helpful on the case/uncase decision, if you decide to case a roadway crossing the regulations tend to be very specific on the casing details. A roadway casing generally needs to be sealed on both ends and the sealed casing needs to be vented so that a pipeline leak won't overpressure the casing (or you must design the casing and end caps to withstand pipeline test pressure).

The regulations suggest that the casing must be vented, but they don't usually specify that the vent should be on one end or both. Companies that vent both ends have reported creating a "thermal syphon" that allows cold air to drop into the casing and be warmed by the pressurized pipe to rise up the other vent, in extreme cases this airflow has created pipeline freezes (that can be prevented by removing one of the vents). Some operators have had good success filling the pipe/casing annulus with paraffin or beeswax to reduce corrosion risk and prevent airflow.

Uncased roadway crossings. Some engineers will always install a roadway casing on every roadway they cross. Others (like me) will resist roadway casing at all costs. The upside of casing is that even a thin-walled casing will prevent road traffic from damaging pipe. The downside is that the casing (even though the ends are sealed with elastomer boots) will almost always be full of water and it is difficult to pull a pipe through a casing without damaging the pipe coating—the combination of standing water and damaged coating is a recipe for accelerated corrosion that can be very difficult to remediate.

At the end of the day it is difficult to justify not casing roadway crossings of paved roads. For public unpaved roads, if you can bury the pipe with at least 8 ft (2.4 m) of cover there is a good argument for not casing. For private unpaved roads (including the farmer's two-track) you can make a good argument for continuing with your normal pipe depth.

Foreign pipe. "Foreign pipe" is any pipe that is not part of your project. It can be owned and operated by your company, or by some other entity (in Fig. 6.6 three of the four foreign pipes are operated by the company laying the new pipe, the fourth is owned by a third party).

Crossing foreign lines can be contentious. You must always notify the foreign pipe owner any time you are disturbing the soil near their pipe, they have the right to be present during the excavation, but their

Figure 6.6 Foreign pipe crossing.

presence needs to accommodate your construction schedule, if they don't show up at the appointed time, proceed without them.

ASME B31.8 specifies minimum clearance to an "underground structure" of 6 in (152.6 mm) (only 2 in (50.8 mm) from gas mains oddly enough), but they don't specify if you need to go over or under the structure. The company that owns the foreign line will often have strong opinions concerning what they will "let" you do. In most jurisdictions, they don't have the right to specify anything about the crossing. In the interests of community relations, if you can meet their expectations then you should.

I had one crossing where the foreign pipe was laying on solid rock and the top of their 16 in (400 DN) 1000 psig (6.8 MPag) gas pipeline was buried with just over 3 ft (1 m) of cover. I was laying a 6 in (150 DN) GRP line. The foreign pipe owner demanded that I go under his pipe and provide no less than 4 ft (1.2 m) of separation. Going under would have required the use of explosives adjacent to this high-pressure gas line. I told them that I was going over their line and providing 6 in (152 mm) of separation. When the lawyers got finished with their (very expensive) review of law and precedent, the owner of the foreign line was happy with my original plan and I went over their line. Accommodating reasonable requests can come back to you in goodwill, refusing unreasonable requests is your right, just make sure that you learn the difference.

It is common for steel lines in close proximity to have a test lead "cad welded" (also known as "exothermic welding") on both pipes. These test leads are run to the surface so that a technician can confirm that the cathodic potential of the two pipes has not gotten cross connected so that one pipe acts as a sacrificial anode for the other pipe. The cad weld process generates very high heat that has the potential to damage marginal pipe (i.e., pipe that has been in the ground too long and the corrosion allowance was used up decades ago). For this reason never allow your construction crew to attach a cad weld to a foreign pipe. You can (and should) provide the equipment, but the owner of the foreign pipe needs to have their contractor or employee apply it. I had one situation where the cad weld burned a hole in the foreign pipe (the pipe was 1950s vintage, had been in the ground for 45 years and was "coated" with a field-applied asphalt paper and tar wrapping, and was well past its 20-year design life). Had my crew done the weld I would have been liable for the cost of the repair, the lost gas, and probably the lost production. In the actual case, we left the ditch open for them to do their own repair and we finished our job.

6.3.5 Liquid in gas gathering systems

Gas leaves the well-site equipment at the highest pressure it will see until it leaves the gathering system, often much warmer than the ground temperature, and fully saturated with condensable vapors. As the temperature drops, the ability of the gas to carry these substances as a vapor diminishes. As the pressure drops, the ability of the gas to carry these substances as a vapor increases. Consequently, as the gas moves down the gathering pipe you will have many condensation/evaporation events happening at different locations. The result is standing liquid in virtually all gas gathering systems. This standing liquid can create corrosion cells, and it will increase the pressure drop in the system.

Every sag and low point in the system (and there are dozens of sags and low points in every mile of pipe, Fig. 6.7) has the potential to collect liquids. Liquid-management strategies that use "line drips" (see later) to collect the liquid at significant low points (and designers have a wide range of definitions of "significant") will collect any liquid that happens to drop out uphill from the drip, but that is only going to be a fraction of the total liquid accumulation.

When designing a gathering system it is important to think about how you are going to manage accumulated water. One common strategy is to simply use nonmetallic pipe (HDPE usually) and pretend that it doesn't matter. Even if corrosion does in fact become a nonissue, gas flowing over a coherent liquid has to do work which must be paid for in terms of increased pressure drop and increased incidence of slugs.

Liquid slugs are a much larger issue than many designers accept. I operated a gathering system that had a piggable (but not pigged) subsystem with 16 wells making a cumulative 85 MMSCF/day (2400 kSCm/day). The system had three highly efficient piggable drips with automated dumps (and dump counters so we knew how much liquid was being removed) that were on average removing 21 bbl/day (3.4 m^3/day) so we expected that everything was just fine. One cold Thursday in December,

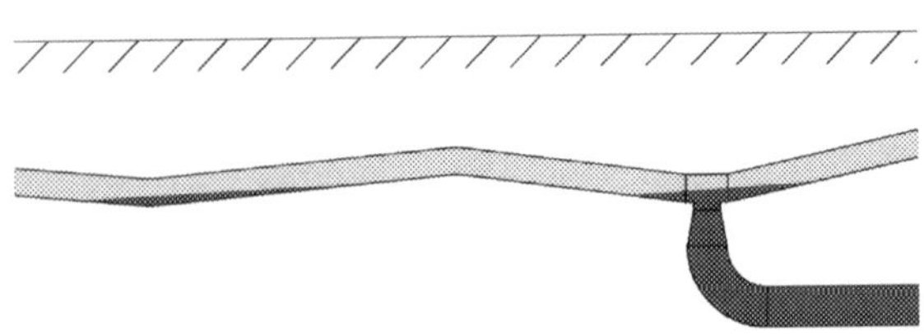

Figure 6.7 Pipe sag.

the water that was being accumulated in the sags and low places "launched" into a slug. All three drips were overwhelmed and unable to keep up with the inflow. The slug overwhelmed the compressor-station suction scrubber and did major damage to filters, compressors, and dehydrators. We recovered over 1000 bbl (159 m^3) of liquid from inside the compressor station, which was certainly a small portion of the total slug size since the automatic dump valves on the drips and suction scrubber were all wide open for several minutes. After we cleaned up the mess and fixed the broken components we started pigging the system weekly. After we started pigging the recovery went up to 36 bbl/day (5.7 m^3/day) and the line pressure at the farthest well dropped 30 psig (207 kPag). We never did learn what the "trip event" that caused the liquid slugs to start moving, and once we started regularly pigging the system we never had another occurrence.

6.3.5.1 Line drip

Condensed hydrocarbons in a gathering system have always been called "drip." A more accurate name for this category of equipment would be "line drip trap," but it is what it is. Drips are never the only answer for removal of condensed liquid, but they can be a component of the solution set. Since humans have a desire to think about fluid flow in human terms, it is easy to see why you would put a trap to catch condensed liquids in the low point of a gathering system, in fact these devices are often called "low point drains," but fluid flow and pipeline topography are far more complex than that. Often the low point will occur after the system pressure has dropped the vapor content far below 100% RH and which tends to inhibit condensation. The drip is installed at a low point, but not at a point that is conducive to condensation so it is useless.

I'm going to talk about four subclasses of line drips: (1) side-stream drip, (2) insert drip, (3) single-line piggable drip, and (4) dual-line piggable drip. In all of these I will describe a device with a nearly horizontal barrel (they need to have a slight slope toward the blowdown valve stinger). This is not the only possible configuration—many operators build simple vertical barrels that either have to be buried very deep or have limited volume, but the hydraulics are the same as described later. The volume capacity of the barrel should be based on the largest slug that can be expected to develop. For piggable lines it is the volume that the expected available dP can push up the largest hill it must traverse (i.e., if you have designed a compressor-station inlet to be 50 psig (345 kPag)) and your gathering system is HDPE rated at 100 psig, then you only have 50 psig available. If the line ID is 10 in (254 mm) and your largest hill is more than 116 ft (35.4 m) of elevation change, then your drip barrel should be able to hold 11.3 bbl (1.80 m^3). Stronger pipe would result in a larger available dP and you could tolerate a higher hill and would need a larger barrel.

Side-stream drip. The drip in Fig. 6.8 is the result of generations of wishful thinking. The hope is that liquid will condense from the gas in the piping uphill from the tee that starts the drip. It does sometimes. Other times it drops out before the last hill or after the next hill.

The side-stream drip in Fig. 6.8 has a barrel length of 8 ft (2.4 m) and a volume of 4.2 bbl (0.7 m^3). If it didn't have the elbow, the ditch would have to be nearly 25 ft (7.6 m) deep to achieve the same volume.

Note that the blowdown line is offset from the pipeline, this is achieved by rotating the small side of the concentric reducer 1 degree in either direction. The slope of the barrel is achieved by installing the tee in an orientation where it is at least 3 degrees (but not more than 5 degrees) out of level.

Side-stream drips need to have "pigging bars" welded into the branch on the tee so that pigs won't fall into barrel. Without the bars, this style of drip renders the line nonpiggable. Note that Fig. 6.8 shows the drip totally full of liquid, in my experience that is the normal state of these devices, people blow them down several times a year, but it only takes a few hours for them to fill back up. There are some side-stream drips that

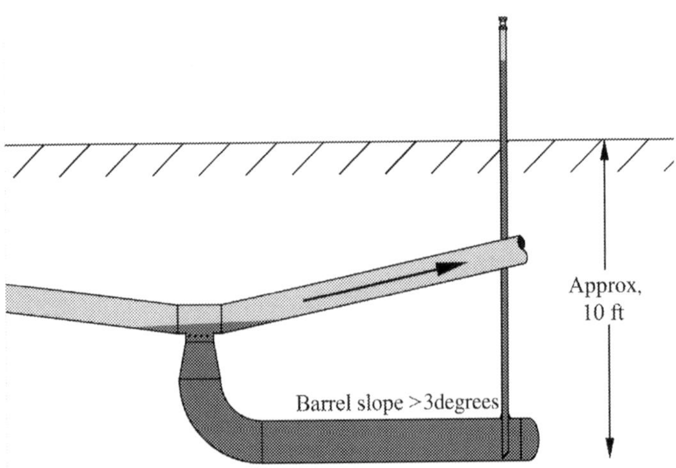

Figure 6.8 Side-stream drip.

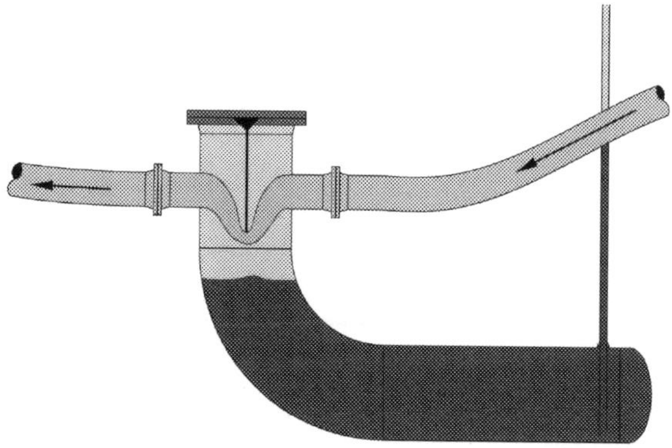

Figure 6.9 Insert drip.

have automated level control, and they will typically dump 4–5 times/day.

Insert drip. A significant step up from the side-stream drip is the insert drip (Fig. 6.9). This innovation has a full-width plate attached to a blind flange and lowered into a length of pipe. The insert plate extends at least one inlet-pipe diameter below the bottom of the inlet pipe. The plate forces the stream to change directions and accelerate under the plate. Just like in a separator, any droplets in the flow will tend to be thrown out of the flow stream and tend to coalesce into a puddle in the drip barrel. If insert drips are not manually drained often, the liquid level can rise until the flow tends to blow mini-slugs into the downstream piping and actually do more harm than good. There is no real impediment to designing insert drips with automatic level controls, but it is rarely done even though automatic level control would improve the function of the device by several orders of magnitude.

Insert drips are not piggable in any configuration. Just like the side-stream drip, the inlet/outlet piping centerline should be a degree or two off the centerline of the barrel to allow the blowdown piping to bypass the inlet (or outlet, there is no preferential flow direction) piping. The impingement force on the insert plate can be significant, so it is important that the plate be thick enough to prevent buckling and the attachment be robust enough to prevent detachment.

Single-line piggable drip. The single-line piggable drip in Fig. 6.10 is similar to a side-stream drip with pigging bars with two major differences: (1) the slots go all the way around the circumference of the pipe (allow capturing any liquid in annular flow and improves the slug-handling

Figure 6.10 Single-line piggable drip.

ability several fold); and (2) this design always has automatic level control (I can say that because so far every drip made to this configuration has been on my projects to my design, someone may eventually adopt this design and eliminate the level-control line, but it would be a mistake). The slots go nearly the entire length of the barrel diameter so there is considerably more flow area from the flow line into the drip than you get with a tee, which allows more of a slug to be diverted into the barrel instead of continuing down the line.

Through trial and error, the most effective slot width has been found to be about 2.5 times the gathering pipe wall thickness. The space between the slots should be approximately equal to the slot width. There is little room for error in the slot width, length, or spacing, and cutting the slots in the field or in a welding shop has resulted in a poor job; it is far better to have the slots cut in a machine shop. Since the radius on the ends of the slots will rarely be a standard drill diameter, programming the radius on a numerical control milling machine is far easier and better job than trying to drill them.

Total open area in the slots should be at least twice the pipe flow area, so the minimum length of the slots can be calculated using Eq. (6.7). The minimum length determines the smallest barrel that can accommodate the pipeline (Fig. 6.10). Solve this equation and round the result up to the next readily available pipe size (e.g., if Eq. (6.8) were 16.9 in (455 mm), a prudent person would expect that a barrel size of 18 in (450 DN) would be difficult to fabricate and possibly difficult to acquire end caps and elbows and would call it 20 in (500 DN)). Generally the nozzle reinforcement calculations in ASME B31.8 do not require reinforcement, but I generally include full saddle reinforcement due to the (admittedly small) risk of the weight of the barrel full of liquid hanging from a pipeline whose external coating in the vicinity of the drip was field applied.

For example, a standard wall 12 in (300 DN) pipe would be configured like:

- Pipe outside diameter: 12.750 in (324 mm)
- Pipe wall thickness: 0.375 in (9.53 mm)
- Pipe circumference at the OD: 40.05 in (1017 mm)
- Pipeline MAWP: 600 psig
- Estimated slot width: 0.9375 in (23.8 mm)
- Number of slots (circumference ÷ slot width ÷ 2): 21.36 (round down to 21)
- Adjusted slot width: 0.954 in (242.3 mm)
- Slot centerline length: 12.75 in (324 mm).
- Minimum barrel size: 18.031 in (458 mm)

$$L_{slots} = \frac{\pi \cdot OD_{pipe}^2}{2 \cdot W_{slot} \cdot n_{slot}} \tag{6.7}$$

$$ID_{barrelMin} = \frac{OD_{pipe} \cdot \sqrt{\pi^2 \cdot OD_{pipe}^2 + 4 \cdot W_{slot}^2 \cdot n_{slot}^2}}{2 \cdot W_{slot} \cdot n_{slot}} \tag{6.8}$$

$$L_{avail} = ID_{barrel} \cdot \sqrt{1 - \frac{OD_{pipe}^2}{ID_{barrel}^2}} \tag{6.9}$$

- Specified barrel: 20 in (400 DN) Sched 30 (0.5 in (12.7 mm) wall thickness) is the smallest size that will accommodate the minimum slot length and the MAWP. Actual barrel length should be calculated based on expected water influx during pigging.

Dual-line piggable drip. These devices (Fig. 6.11) serve two purposes: (1) they connect two flow lines in a manner that allows two pig launchers to share a single pig receiver; and (2) removing liquid from gas flow lines. The inset in Fig. 6.11 shows why these units are particularly effective; the two streams trying to join into a single space introduces significant angular velocity that tends to throw any liquid droplets to the slots. While side-stream drips are only effective with liquids in the bottom of the pipe,

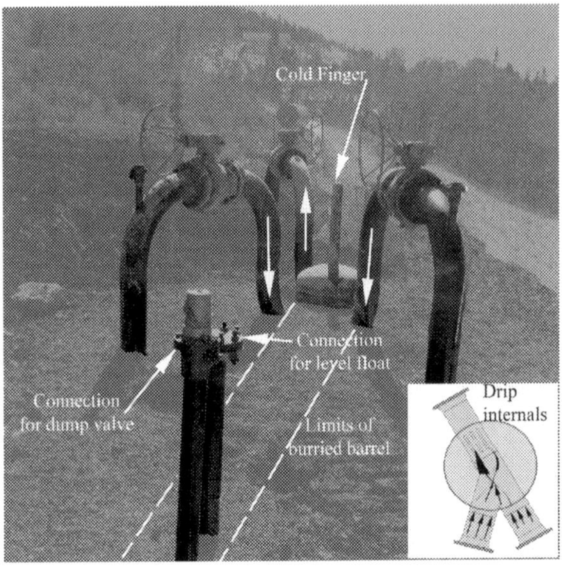

Figure 6.11 Dual-line piggable drip.

and single-line piggable drips are effective with liquids in the bottom of the pipe and annular flow, dual-line piggable drips are effective with stratified flow, annular flow, and mist flow. These units have proven so effective, that the usual way to install them is to bring the pipe above grade to install block valves and then back down to the drips installed below grade like in Fig. 6.11.

The interaction of the two streams allows the total area of the slots to be reduced to 1.25 times the pipe flow area (from 2.0 for single-line piggable drips). The inside diameter of the barrel must now accommodate both the slots and the geometry of the junction of the two lines. To ensure the integrity of the barrel connection, the holes cut into the barrel on the inlet side must be separated by at least 3 times the wall thickness of the barrel. In most cases, making the barrel equal the MAWP of the pipeline will require 0.500 in (152 mm), so that is a good first guess for hole spacing and Eq. (6.10) is the straight-line distance from the centerline of the inlet hole (which lines up with the outlet hole) on the ID of the barrel and the centerline on the ID of the barrel of the branch hole. The two inlet holes are so close together that full-encirclement saddles interfere with each other. I have addressed this by trimming both saddles and welding them together while welding the rest of the saddle to the barrel.

$$d_{holeCenterline} = 3 \cdot t_{barrel} + \mathrm{OD}_{pipe} \tag{6.10}$$

$$r_{lateral} = \frac{d_{holeCenterline}}{2 \cdot \sin\left(\frac{\pi}{8}\right)} \approx \frac{d_{holeCenterline}}{0.7654} \tag{6.11}$$

$$L_{slot} = \frac{5 \cdot \pi \cdot \mathrm{OD}_{pipe}^2}{16 \cdot W_{slot} \cdot n_{slot}} \tag{6.12}$$

$$\mathrm{ID}_{barrelMin} = r_{lateral} + L_{slot} + \frac{\mathrm{OD}_{pipe} \cdot (\sqrt{2} - 1)}{2} \approx R_{dual} + l_{slot} + 0.2071 \cdot \mathrm{OD}_{pipe} \tag{6.13}$$

To revisit the example in the previous section, two standard wall 12 in (300 DN) pipes would be configured like:

- Pipe outside diameter: 12.750 in (324 mm)
- Pipe wall thickness: 0.375 in (9.53 mm)
- Pipe circumference at the OD: 40.05 in (1017 mm)
- Pipeline MAWP: 600 psig
- Estimated slot width: 0.9375 in (23.8 mm)
- Number of slots (circumference/slot width/2): 21.36 (round down to 21)
- Adjusted slot width: 0.954 in (242.3 mm)
- Slot centerline length: 7.97 in (202 mm)
- Minimum barrel ID: 30.208 in (767 mm)
- Specified barrel: 36 in (900 DN) Sched 40 (0.75 in (19.1 mm) wall thickness) is the smallest size that will accommodate the minimum slot length and the MAWP. Actual barrel length should be calculated based on expected water influx during pigging.

The actual piping within the barrel is more complex than it looks. There is a fitting called a "lateral" that connects two pipes in a 45-degree configuration. These fittings have a bad reputation in Oil & Gas because the crotch is high stress, and have been subject to stress-related corrosion issues. Inside the barrel of the drip, the stresses are shifted to the straight piping and the drip barrel, not the crotch and this issue does not exist in this application. It turns out that cutting slots in a lateral is quite expensive and prone to error. At the end of the day it is more economical to have a machine shop cut the slots and the "birds mouth" on the straight pipe and the profile on the branch on an NC milling machine and then weld the two pieces together prior to assembly (Fig. 6.12).

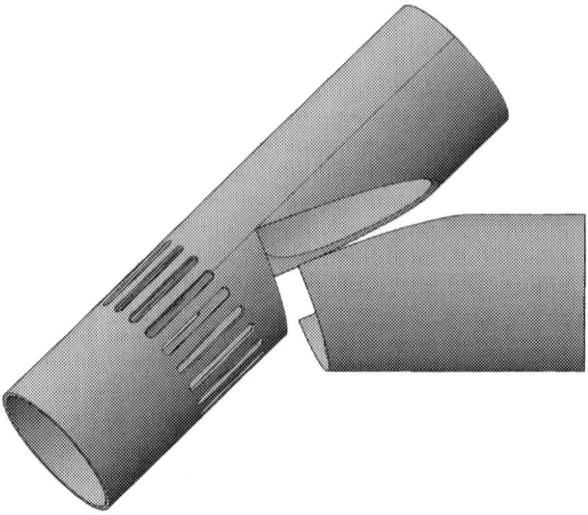

Figure 6.12 Piggable drip internals.

6.3.5.2 Pigging equipment

As we discussed in the last chapter, the only way to manage the risk of internal corrosion in steel pipes is by removing any standing water. It should have been clear in the last section that drip traps are only effective at removing the liquid that happens to arrive at the trap. Liquid that accumulates away from drip traps is a serious corrosion risk and will often have detrimental impacts on your ability to control operating pressure at well-sites. Finally, accumulated liquid can become mobile for reasons and at times not of our choosing, and the resulting slugs can do real harm.

It is critical that we manage liquid accumulations as part of an ongoing, carefully considered plan. The only way to manage these liquid accumulations is to run a device through the line to displace the liquid toward some piece of equipment that can capture it. This device is called a "pig." There are many apocryphal stories about where the name came from. It has been proposed that it is an acronym, and it has been proposed that the name comes from the squealing sound it makes as it travels down the pipe. The acronyms are sometimes humorous (such as "pipeline inspection gadget"), but incorrect. The persistent story is that early pigs were made of straw wrapped in wire and the sound was distinctive. The sound of a pig traveling down a pipe is rarely distinguishable from the background flow sounds and when it is noticeable it doesn't sound much like a "squeal." The actual genesis of the name is that early pigs were round, fat, and the end tapered toward something that looks slightly like a pig's snout. That was the limit of thought that went into naming this ubiquitous device.

Pigs are used for many specific tasks. Pigs to remove liquid are fairly simple and increasingly the industry is using "Turbo Pigs" ("Turbo Pig" is a registered trademark of Girard Industries) which are reusable resilient plastic with a central core and a number of cups or wipers. See the pigs toward the right-hand side and the small pig in front on the left-hand side of Fig. 6.13, but many people still prefer the traditional "bullet" or porcine shape. If you need to clean the pipe walls of solids accumulations, the pigs with brushes are available in both the Turbo style and the traditional porcine shape.

Today's pigs are available in the traditional porcine shape, with or without tough polyethylene coatings, with or without wire brushes. Turbo pigs are available with a series of disks, cups, or wipers that each do a specific job. You can buy spheres, or you can purchase a mandrel that allows you to stack the equipment that makes sense to a particular job. There are pigs with instrumentation to evaluate pipe condition (called smart pigs). Different goals point you to different equipment.

One lesson that must be relearned every couple of years is that spheres require that the fluid behind them be similar in density to the fluid in

Figure 6.13 Range of pigs. *Courtesy of Girard Industries (www.girardindustries.com).*

front of them—they are a good choice to put between two different hydrocarbon liquids in a liquid pipeline, but are not effective with liquid on one side and gas on the other side.

Pig runs are initiated from "pig launchers" and terminate in "pig receivers." These devices need to:

- accept the pigs that are required to be run,
- facilitate batch chemical treatment,
- allow for disposal of the liquids that come in with the pigs,
- operate quickly.

Launchers fit into two categories: (1) gravity launch and (2) pressure launch. Gravity launchers operate by placing the pig in the device, sealing the chamber (called a "barrel"), and opening the chamber to the process pipeline and allowing the pig to fall down the inclined line into the flow. These devices are not terribly effective and it can take considerable time for the pig to actually fall into the flow.

Pressure launchers on the other hand require altering the flow path in the pipeline to place the flow behind the pig. There is a class of pressure launcher called a "pigging valve" that is very effective in many situations, and they will be discussed under "trunnion ball valves" later.

The launcher and receiver in Fig. 6.14 represent nearly 30 years of my evolution in designing pigging equipment. When I look at the equipment that I did on my first project I dearly want to "fix" it, but it works and I'm not going to apologize for designing this equipment based on designs in common use in the industry. As usual the "designs in common use" are the result of compromises between fabrication complexity, cost, and operability (with limited input from the operators), and are not as good as they could be. I've built upward of 20 launcher/receiver pairs to the exact

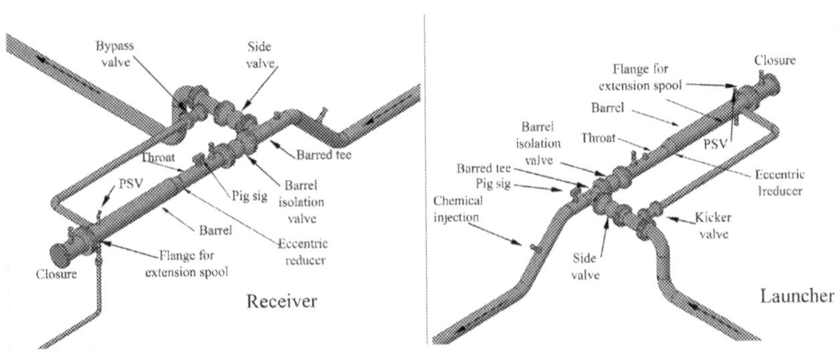

Figure 6.14 Pressure launcher and receiver.

design represented by Fig. 6.14 and it has been several years since I was last tempted to tweak it.

One of the most important learnings has been that there is no valid reason for a launcher to be significantly different from a receiver. I learned this lesson when our gas marketing department found a new market for our gas that required taking gas off the opposite end of the gathering system than we designed the system to provide. This required all launchers to become receivers and vice versa. Some of the auxiliary equipment was in awkward positions for the new functionality and required adaptations in processes, but it worked after a fashion. The current design would work well period.

We frequently have difficulty with terminology in communicating procedures and instructions between peers. It is worthwhile to describe the labeled items in Fig. 6.14.

Closure. The closure on a launcher or receiver should:

- Operate reasonably quickly.
- Have a pressure telltale as an integral part of the seal mechanism. This means that you cannot open the closure while the equipment is under pressure without there being a warning sound.
- Be supported by hinges or davits that would prevent the closure from becoming a projectile in an opened-underpressure situation.
- Be able to release pressure while still captured.

For many years all closures were "Huber-type" (lower image in Fig. 6.15) which have ears on the outer circumference that are intended to be hammered off and back on. These closures fail all of the criteria given in the previous list except for the first one. I have removed a Huber-type closure under pressure (the vent line plugged with paraffin after blowing down for 30 seconds) and it was very exciting—luckily I was standing out of the line of fire and when it finally opened it swung away from me instead of breaking my body.

The industry has largely transitioned away from hammer closures in favor of Yoke-type closures which have a tapered ring (the "yoke" in Fig. 6.15) that hold the door tightly against the flange. The two halves of the yoke are locked together with the yoke lock which is held in place with a "telltale" nut on a drilled stud which noisily releases any trapped gas when loosened. As the jacking bolts move the yoke halves apart, the door can come off the flange if there is still trapped pressure, but it is still captured by the yoke so it can't swing. When the yoke is fully open the door can open. These closures satisfy all of the criteria given in the previous list.

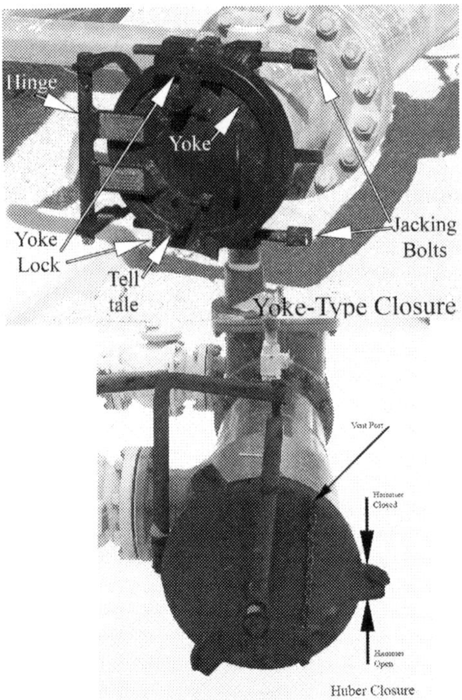

Figure 6.15 Pig trap closures.

Barrel. The barrel is the pipe between the closure and the eccentric reducer and it should:

- Be one standard pipe size larger than the pipeline. Typically "the next pipe size" for 16 in (400 DN) pipe is 20 in (500 DN) since it can be difficult to find an 18 × 16 (450 × 400) eccentric reducer or an 18 in (450 DN) closure.
- Be long enough. Barrel lengths have traditionally been fairly short (on the order of 5 ft 7 in (1700 mm) on a 20 in (500 DN) receiver) which limits the equipment that can be run. The minimum length for a 6 in (150 DN) launcher barrel should be 6 ft (1.83 m). For each pipe size above 6 in (150 DN) add 1 ft (0.3 m) (i.e., 8 in (200 DN) should be 7 ft (2.13 m), 20 in (500 DN) should be 12 ft (3.66 m), etc.).

Flange for extension spool. This flange on the barrel is an insurance policy against ever needing to run a very long pig. Regulations have begun to require periodic pipeline inspections using smart pigs. Smart pigs can be very long depending on how many evaluations are ongoing concurrently. Many operators are having to cut up launcher/receiver barrels to add piping to accommodate the long pigs, and then in many cases the longer barrel is in the way of normal traffic. Note in Fig. 6.14 that there is nothing connected between the closure and this flange. Having this flange on the barrel allows you to drop in a spool piece of any length without welding or disconnecting any piping.

Eccentric reducer. Use an eccentric reducer (with the flat side down) between the barrel and the throat to facilitate shoving the pig into the throat (with a concentric reducer or an eccentric reducer with the flat side up it can be very difficult to get a heavy pig to engage in the throat).

Throat. The throat is the same size as the pipeline. When loading a pig you try to engage the pig into the throat to seal the throat so that kicker gas will not bypass the pig. The throat should be at least twice as long as the maintenance pigs that you plan to run.

Pig signal. These devices (also called "pig indicator") are used to inform the operator that a pig has passed. They can be intrusive (i.e., an arm reaches into pipe and a pig passing trips the arm) or nonintrusive (i.e., they have the ability to sense the pig passage from the outside of the pipe). They can be unidirectional (less expensive, but will break if you run a pig through backward, rarely a good investment) or bidirectional. Mechanical or electronic. This technology is changing rapidly and prior to deciding on a device you need to see what is currently available. A pig signal should be located to indicate that the pig has passed the barrel-isolation valve. For launchers this is on the barred tee. For receivers it is in the throat. Since we need to plan for someone requiring a change in flow direction, it is prudent to either use a nonintrusive pig signal or to use 4 pig signals for a launcher/receiver pair.

Process valves. All of the process valves should be full-port, trunnion ball valves (see later for valve descriptions). The valves we are concerned with are as follows:

- Barrel-isolation valve. A normally shut valve that is the same size as the pipeline.
- Side valve. A normally open valve that is the same size as the pipeline (this valve can be reduced port if there is a valid reason, but I don't do it).
- Kicker/bypass valve. Both valves should be sized to provide less than 0.13 psi/ft (2.9 kPa/m) at 100 psig (690 kPag) using Table 6.6. The location for source/return gas has evolved over time. I have found that with the current location I can build the entire launcher/receiver in a shop and ship it bolted together which has resulted in cost savings.

Table 6.6 Kicker line size

Pipeline size	Kicker size	Pipeline size	Kicker size
4 in (100 DN)	2 (50 DN)	16 (400 DN)	6 (150 DN)
6 in (150 DN)	3 (150 DN)	18 (450 DN)	6 (150 DN)
8 in (200 DN)	4 (100 DN)	20 (500 DN)	8 (200 DN)
10 in (250 DN)	4 (100 DN)	24 (600 DN)	8 (200 DN)
12 in (300 DN)	6 (150 DN)	30 (750 DN)	10 (250 DN)
14 in (350 DN)	6 (150 DN)	36 (900 DN)	10 (250 DN)

The kicker/bypass line should tie into the barrel about one barrel diameter away from the flange for extension spool.

Barred tee. This tee needs to have pigging bars to make sure that you don't lose control of the pig.

PSV. This safety valve is required by many jurisdictions, and it is a really good idea. It is possible (though not a good practice) to isolate the launcher/receiver barrel full of liquid. In that eventuality any increase in temperature will overpressure the barrel. Even if the barrel is just isolated and drained, it doesn't take much of a leaking valve to put gas into the isolated barrel and in gas gathering systems that gas will be saturated with water vapor. A launcher that is opened once a quarter will always have some amount of liquid in it they are occasionally full. A very small thermal relief will prevent this liquid accumulation from damaging piping and/or valves.

Chemical injection port. I put a 1 in (25 DN) valve on the sweep on the pipeline side of the barrel-isolation valve to put chemicals directly into the line in front of a pig. My issues with the ineffectiveness of most chemicals are based on the inability of a flowing gas stream to keep them mobile and transport them through the system. A pig can accomplish this task quite effectively and facilitating their use is prudent. One chemical that I have had good success with is called a "gel pig" which is a fluid polymer that can be pumped into a line and then chased with a pig. The polymer tends to do a great job of aggregating scale, sludge, and slime into its matrix and can significantly improve the function of a line clogged with solids. The chemical injection port makes this evolution much easier. I put a chemical injection port on the receiver as well, but this valve is only there in case the line needs to change directions (or you need to run a pig backward to enhance cleaning).

Vents/drains. I put the barrel vent very close to the barrel-isolation valve. This location was selected in response to several leaking barrel-isolation valves one after the other. When the barrel-isolation valve has a small leak, the launcher is still usable, but when you seal the throat with the pig the leaking gas will spit it out of the throat before you have the closure shut, then opening the kicker valve will just bypass the pig. One inventive operator cut the handle off of a shovel and braced the handle between the pig and the closure, this allowed the pig to launch, but the stick had penetrated the pig and became the pig's "tail" on the cleaning run. When it arrived, the tail prevented the receiver barrel-isolation valve from closing and the line had to be blown down to remove the pig.

There is also a vent on the spool between the "flange for extension spool" and the closure. That never gets used on the launcher. On the receiver it is used to shift the pig out of the throat for retrieval (using the other vent will stall the pig in the throat and it may not be possible to extract it without opening the barrel-isolation valve with the closure open, a practice that should be discouraged).

I put a drain on both launchers and receivers. On receivers the drain is piped to some disposal container that can hold the liquid from the pig run. On launchers I generally build a small containment area to drain condensation. Some operators leave this drain open between pig runs which adds to the surface corrosion inside the barrel and throat, but only minimally and has not been a problem.

Sweeps. Long pigs can have difficulty traversing a "long radius 45-degree elbow" which is the proper designation of the most common fitting used in gathering piping. By "long radius" ASME B16.7: Factory-Made Wrought Buttwelding Fittings means that the radius of the bend at the centerline of the pipe is 3 times the outside diameter of the pipe. For launchers and receivers it is better to use a "hot bend" which is a length of pipe that has been heated and bent to a specific bend radius. The most common bend radius specified for launchers and receivers is 6D, but I have seen 9D specified, the bigger the radius, the longer the pig that will traverse it. When smart pigs first came into the industry it was common for smart pigs to require 42D bend radius, but that excluded virtually all pipelines from using these tools. The technology has evolved and now most can pass a long-radius fitting, and all can pass a 6D sweep. When specifying the fabrication of hot bends (also called "induction bends" or "sweeps"), it is important to specify a minimum tangent length. The bend starts with a length of pipe (e.g., a 12 in (300 DN) pipe would have 60 in (152.4 mm) included a 6D 45-degree bend, the fabricator will start with a pipe a bit longer if you don't specify a tangent), no matter how careful the fabricator is there will be some amount of ovality in the bent pipe, welding this out-of-round pipe to straight pipe can be very difficult and will frequently cause the weld to fail inspection. Adding pipe to the bend solves this. I always specify a minimum of 18 in (457 mm) tangents. It is common to get one tangent, i.e., 15 in (381 mm) and the other 21 in (533 mm), but both will be round pipe.

6.3.5.3 Compressor-station piggable bypass

Late in the life of a field we often find it necessary to set a compressor station on lines that sit between launchers and receivers. This usually results in the line being declared "nonpiggable" or setting a receiver at the compressor suction and a launcher at the compressor discharge. This works, but is labor intensive to shift the pig. An approach that has worked very well and that lends itself to automating is shown in Fig. 6.16. Valves V5, V6, V7, and V8 are sized using Table 6.2. The distance between V5 and V6 should be about twice the length required for a launcher barrel given in the figure.

In normal operations the valves are in the position shown (i.e., V1, V2, V3, V6, and V7 are open; V4, V5, and V8 are shut). When you decide to run a pig:

- Ensure that the pig sig is reset
- Shut V1
- When the pig arrives
    - Open VI
    - Open V5
    - Shut V6
    - Shut V3
    - Shut V7
    - Open V8

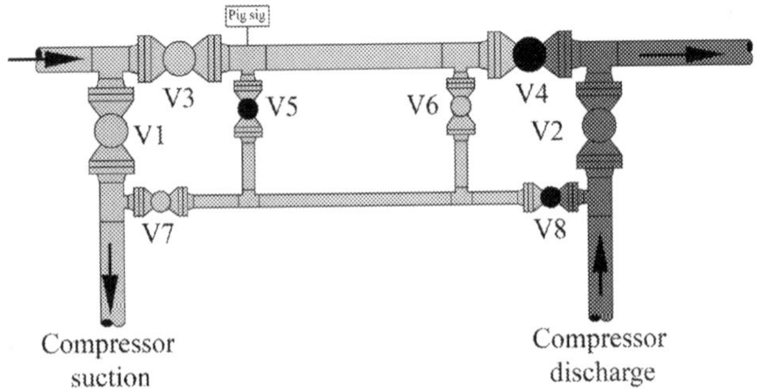

Figure 6.16 Compressor-station piggable bypass.

  - Open V4
  - Shut V2
- After enough time for the pig to pass V4:
  - Open V2
  - Shut V4
  - Open V6
  - Shut V5
  - Shut V8
  - Open V7
  - Open V3
  - Reset the pig sig.

This sequence never puts the compressor station in recirculation and minimizes the time that the header is isolated. We had one of these assemblies on a line with six launchers sharing a receiver—all of the pigs were run once or twice a week so a launcher/receiver pair would have been very manpower intensive. After some learning curve we finally reached the point where the compressor-station operator didn't even have to attend his compressors during pig runs because this process became very routine.

6.3.6 Gathering-system valves

When you look at the final costs of a gathering system you usually see that the cost distribution is something like Table 6.7. With any of these materials, valves represent a significant cost. Selecting valve technology can be confusing. The three most important decisions about valves are: (1) what kind of valve to use; (2) which valves will be actuated; and (3) will block valves be in cans or on the surface.

Almost every sort of valves has places where it should be chosen. Some valve choices are "Ford vs. Chevy (Holden for the Australians)," but most are not. On my first project I told the supply house that I needed "twenty 8 in valves," and I got what I deserved. Eighteen of them were ball valves, two were gate valves. I didn't specify that the valves should comply with the dimensions of API 6D, and found that if you don't specify a length then all of those valves that are on back shelves that don't

Table 6.7 Valve costs

	Labor	Pipe	Valves
Steel	Highest	Second	Third
HDPE	Highest	Third	Second
RTP	Third	Highest	Second

comply will come to your job, valve manufacturers tended to be very creative about valve lengths when no one specifies a standard. There were at least 10 different flange-to-flange distances. The significant cost reduction received for ridding their shelves of unsellable valves was more than offset by increased fabrication costs (e.g., we built one fabrication in the shop using valve "A" dimensions and took it to the field and tried to put valve "B" in a spot that was 2 in (50.8 mm) too short). Always make technology choice and an objective standard part of your valve specifications.

It was once common for valve manufacturers to provide a plot of valve position vs flow. The closer this plot was to a straight line, the better the throttling characteristics. This measure was called "linearity" and it was very useful. Something like a "needle valve" had excellent linearity in that from nearly any starting point in the valve travel, a 1% change in valve position would result in very close to a 1% change in flow. On the other hand, a gate valve will flow nearly 100% of maximum flow when the valve is 12% open, and even within that 12% of travel you get a different change in flow as you move 1% toward open than you do for the same change in valve position going toward shut. It is rare to see this plot any longer, but it can still be useful to talk about a valve's linearity.

Fig. 6.17 demonstrates an important concept in valve performance known as "linearity." The most common valve designs are:

- Linear: Anywhere in the valve travel, a change in valve position by some percent of full travel will yield about the same percentage change in flow (relative to maximum flow). While this design is very effective for manual operations it is reasonably difficult to program a PLC to utilize linear valves.
- Equal percentage: Once flow is established a change in valve position (as a percent of full travel) will provide the change in the flow (relative to the initial flow) that is the same magnitude of change across valve travel. For example, if increasing the valve opening from 10% to 20% increases the flow to 170% of the rate that the valve could pass at 10% open, then increasing from 50% to 60% would increase the flow by 170% of the 50% value. This concept is difficult for most of us to get our heads around, but it is very easy to program into a PLC.
- On/Off: Flow is largely unrelated to valve position after a very short start-up period. On/Off valves will generally pass over 75% of maximum flow at less than 10% open, and nearly 100% of maximum flow rate at something less than 30% open. On/Off valves in throttle service are nearly impossible to program into a PLC successfully.

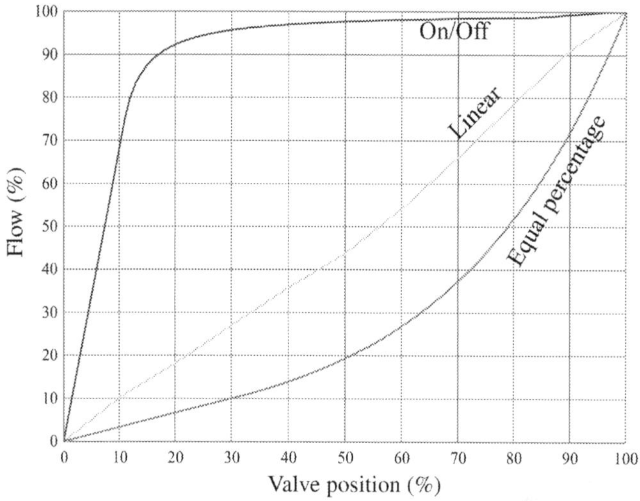

Figure 6.17 Valve linearity.

Historically, valve manufacturers regularly provided charts of valve position versus flow, but that concept has become rare in manufacturing literature. Today it is more common for valve manufacturers to provide a table of valve position vs "flow coefficient" (usually designated "c_v"). One particular manufacture provides a table of c_v with a value for each 10% of travel that can be used to generate a linearity curve ("linear" trace in Fig 6.17). This same data is used to calculate flow rate. One version of this calculation (there are many, confirm that the equation you use is compatible with the valve and the fluids you are calculating) is presented in Eq. (6.14) (Fisher). For valves with markedly nonlinear performance it can be very misleading to use a c_v for any valve position less than 100%.

$$dP_{ratio} = \frac{P_{upstream} - P_{downstream}}{P_{upstream}}$$

$$Y = 1 - \frac{dP_{ratio}}{3 \cdot \left(\frac{k}{1.4}\right) \cdot \left[1 - \left(\frac{2}{k+1}\right)^{\frac{k}{k-1}}\right]} \tag{6.14}$$

$$q_{acf/hr} = c_v \cdot 1360 \cdot P_{upstream} \cdot Y \cdot \left(\frac{dP_{ratio}}{SG \cdot T \cdot Z}\right)^{0.5}$$

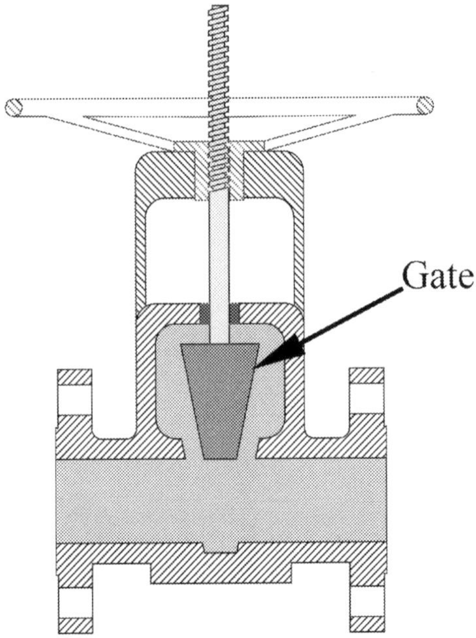

Figure 6.18 Gate valve.

6.3.6.1 Valve technologies

There are many types of valves to serve a wide variety of applications. Some technologies like swing check valves or small-diameter choke valves are widely used in the industry, but aren't often part of gas gathering systems. For that reason I won't discuss them in this section.

Other technologies can be described as "manufacturer specific" and lack a general means to describe them. Valves like the Fisher V-Ball or the Argus Pigging Valve are important to many gas gathering systems, but those applications are so specific that they need to be researched in light of the technology as it has evolved up to the time you are considering valve technology for a specific project.

In this section I'll limit the discussion to: (1) gate valves, (2) plug valves, (3) butterfly valves, (4) globe valves, (5) floating ball valves, and (6) trunnion ball valves.

Gate valve. Gate valves (Fig. 6.18) are the primary on/off block valve in water and steam service. These valves operate by sliding a "gate" between two seating surfaces. In most industrial service the gate is wedge shaped and the tighter you close the valve the more force is applied to seal the gate and the seats. In wellhead service the gate is flat and has a

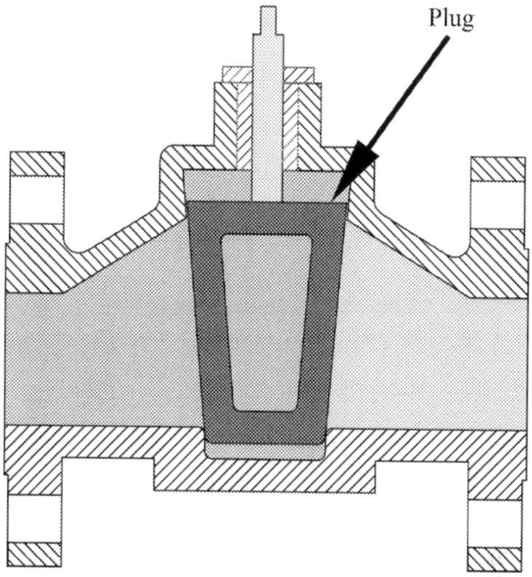

Figure 6.19 Plug valve.

hole that you align with the pipe to flow or offset to the pipe to stop flow. Wellhead gate valves work by the alignment of the port and applying more shutting force will not improve the seal.

Gate valves are mostly metal-to-metal seals and have proven to be reasonably poor at sealing natural gas lines and are primarily used in liquids or steam. Most of the wedge-gate valves are "rising stem" which means that as you turn the hand wheel the stem rises out of the valve and brings the gate along with it. These valves can be quite tedious to operate (e.g., one manufacturer's 12 in (300 DN) gate valve requires 100 full turns to go from fully open to fully shut). Gas field operators that are very familiar with ¼ turn valves like ball valves, butterfly valves, or plug valves will let you know their displeasure at having to operate rising stem gate valves, generally in very colorful language. Not effective for throttling.

Some gate valves are piggable, but the seats in some gate valves have a reduced flow area and are not piggable.

Plug valve. The valves on the Roman aqueduct system were "plug" or "stopcock" valves. Plug valves (Fig. 6.19) have a plug in the flow stream that is either positioned to stop flow or rotated 90 degrees to allow flow through the drilled hole in the plug. Modern plug valves have a mechanism to lift the plug while rotating it toward open to reduce the sliding friction and reduce the risk that the sliding plug will scratch the seating surfaces. Plug valves must be lubricated periodically, which generally makes them inappropriate for raw-gas gathering systems (since maintenance on those systems seems to be quite spotty), and certainly inappropriate for placing in valve cans (see later).

I once called for shutting a 4 in (100 DN) plug valve that had been in a valve can for 7 years without maintenance or operation; it wouldn't budge. The operator put a 4 ft (1.22 m) cheater bar on the valve and it still wouldn't budge. This progressed until the construction manager put a 20 ft (6.1 m) length of pipe on the valve and tried to close it using the bucket on a rubber-tired hoe; still wouldn't budge. Finally we moved the isolation back to an upstream ball valve and removed the plug valve for maintenance. When we tried to take it apart, the plug was fused to the seat and no combination (that we could find) of heat, cold, or force would free it. We finally just tossed it in the scrap metal bin.

My recommendation is to not use plug valves unless you can assure yourself that they will receive regular maintenance even with the next company reorganization and the one after that. Since no oil and gas company of my experience could make that promise, I don't recommend plug valves in gas gathering. Like gate valves, they have generally been metal-to-metal seats, and also like gate valves they are not effective for throttling. Plug valves are also not piggable.

Butterfly valve. These valves (Fig. 6.20) have a flat plate in the flow stream that rotates around a post near the center of the pipe. Recall from Chapter 5, Well-Site Equipment, that a bluff body in a flow stream will shed highly energetic vortices (called von Karman Streets). These von Karman Streets are quite energetic and can act on the bluff body (especially when the bluff body is attached to the plate that extends well into the disturbed area). It is common for the flat plates to break off in an energetic stream. These valves work marginally well in occasional flow (e.g., as a shutoff valve for tank filling where they are almost always shut, and when they are open they stay open for a brief period), but not so well in a normally open flow configuration.

Butterfly valves (Fig. 6.20) have the benefit of being available in a "wafer" style that allows them to sit between two pipe flanges without flanges of their own; this greatly reduces the space required for the valve. They don't work terribly well, but they fit in a small space. You can purchase variants of the inexpensive butterfly valves with names like "triple offset" and "double offset." The "offset" indicates that as the central shaft rotates, the center of the plate rotates about a different axis. This offset characteristic is intended to minimize wear on the elastomer seal, and

Figure 6.20 Butterfly valve.

some manufacturers claim that it improves throttling characteristics. In my experience these (more expensive) adaptations provide minimal improvement in throttling and I've never had a butterfly valve fail due to wear on the elastomer seat so I'm not sure if the reduced wear has real value.

The handle (or gearbox if it has one) must have a locking mechanism that will hold it in the position you desire, without that lock the forces on the plate will tend to slam it open and then shut on a very short cycle. You should not throttle with a butterfly valve (poor linearity and very high forces on the plate) and they are the least piggable valves available.

Globe valve. Globe valves (Fig. 6.21) include a change of direction and a disk that moves up and down relative to the centerline of the pipe, but in the same direction as the local flow. Globe valves include needle valves (very steep angle, the edge of the disk is close to 90 degrees from the face; so that a small movement results in a small change in flow area), dump valves (very shallow angle, the edge of the disk is closer to 45 degrees than 90 degrees; so that a small movement results in a larger change in flow area), and many "trim" combinations in between. The valve trim is a term that describes the effective size of the opening when the valve is fully open and the shape of the angle on the disk (e.g., in the valve in Fig. 6.17, when

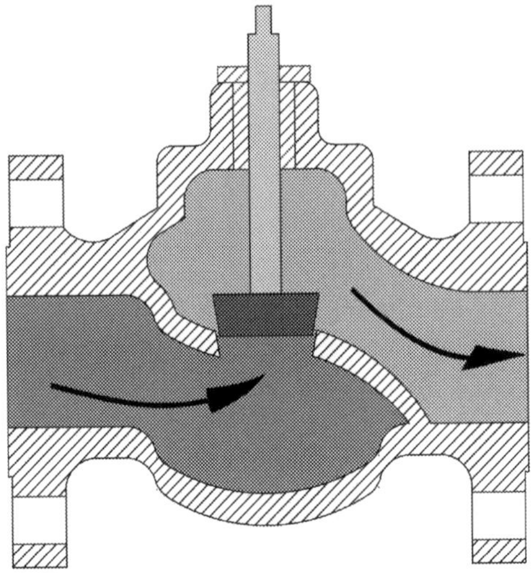

Figure 6.21 Globe valve.

the valve is fully open the flow area is equivalent to a 1 in hole, and the high degree of linearity indicates a fairly steep angle).

There are more globe valves in throttling service than all other valve configurations combined. There are non-globe valve configurations designated as "chokes" and "modified ball valves" that have a flow profile that is more consistently linear than a globe valve, but none of them has ever achieved the installed base that globe valves have. Globe valves are not piggable.

Floating ball valves. A floating ball valve (Fig. 6.22) has a drilled ball that floats (i.e., it is not mechanically connected to anything at all) between two seating surfaces. The actuator fits into a slot in the top of the ball and rotating the handle 90 degrees clockwise mates the smooth sides of the ball to the seating surfaces and shuts off flow. The hard-surface-ball seats against a resilient seat which is made of material consistent with expected pressures, temperatures, and fluids (e.g., high-pressure valves in CO_2 service cannot have Teflon seats because the CO_2 will fill the tiny void space in the Teflon and turn into dense phase which will rapidly expand when the pressure is eventually reduced; the expansion will damage the surface of the seal).

Rotating the handle 90 degrees counterclockwise aligns the drilled hole in the ball with the seating surfaces and allows flow. Floating ball valves seal via the differential pressure across the ball pushing the ball into the downstream seat. At the same time the ball is slightly off the upstream seat and the entire body cavity is exposed to upstream pressure. When dP

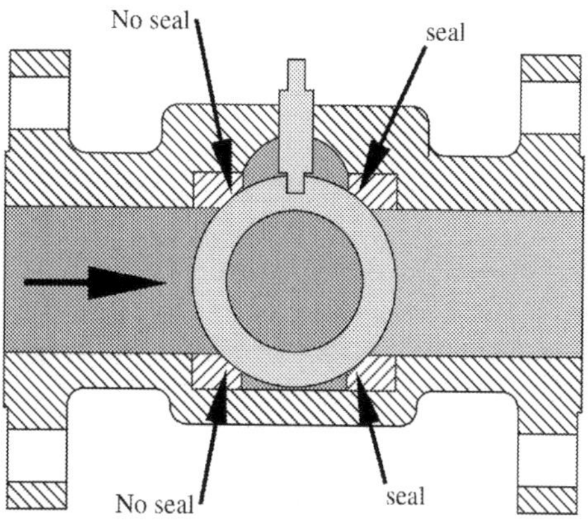

Figure 6.22 Floating ball valve.

is less than about 2–5 psi (14–35 kPa) differential pressure, the valve activation becomes questionable (i.e., it is not assured that the force from the dP will exceed the force of gravity trying to shift the ball off the seat), the larger the valve the larger the dP required to seat the ball.

Floating ball valves are available in "normal" and "full-port" configurations. Normal port valves are also called "reduced port" because the hole in the ball is one pipe size smaller than the valve size (i.e., a 10 in (250 DN) reduced-port valve will have an 8 in (200 DN) hole in the ball). Reduced-port valves came about because it is expensive to fabricate a large, smooth, shiny ball and the larger the ball, the more expensive. In the early days of ball valve deployment (the first patents for ball valves were in the 19th century, but the first successful commercial ball valve was marketed coincident with improvements in elastomers after World War II), the expense of fabricating the balls was very high and full-port ball valves were often twice as expensive as normal port valves. Over time improvements in materials science and manufacturing technics have greatly reduced the full-port premium nearly to the point that you are more likely to find full-port than reduced-port flanged valves on the shelves in supply houses for about the same price.

Floating ball valves are piggable, and they have poor throttling characteristics.

Trunnion ball valve. A "trunnion" is defined as (Webster) "a pin or pivot forming one of a pair on which something can be rotated or tilted." Early trunnions were used in cannons (Fig. 6.23). In that service the trunnion is used to change the elevation of the shot, support the weight of the cannon, and transfer the forces of firing a shot to the carriage. In a

Figure 6.23 Example of a trunnion.

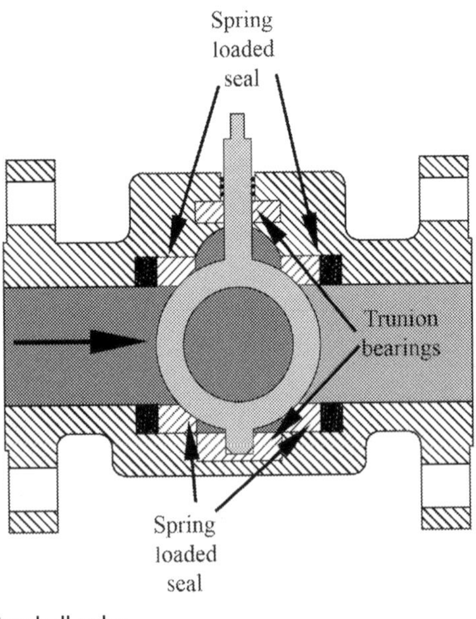

Figure 6.24 Trunnion ball valve.

trunnion ball valve (Fig. 6.24), the trunnion is used to hold the ball rigidly in position against the forces of flow (or no-flow), transfer the force of the fluid to the valve body (and ultimately to the ground), and allow for the positioning of the hole in the ball relative to the flow.

With the ball held rigidly and unable to shift along the pipe centerline to activate the seats, this necessary activation is provided by very heavy springs that push the seals against the ball with great force. Consequently, all trunnion ball valves come equipped with a body bleed that allows the

Figure 6.25 Pigging valve.

body cavity to be bled down to atmosphere while the valve is shut. Many people (including me) find this to be a superior positive isolation for maintenance to two floating ball valves with an evacuated length of pipe between them (remember that the ball in a floating ball valve requires some differential pressure to be activated, in this scenario the dP on the side of the piping being worked on has very low dP and seepage past the high-pressure side will find its way to the work site). People (including some regulators and some companies) that reject this concept point to the self-relieving characteristic of trunnion ball valves. These valves are designed such that if too much pressure builds up in the body cavity, the valves will relieve the pressure out the downstream seat. This is true and if the body cavity was not vented, the valves would not be effective for isolation for maintenance, but they do have a body bleed.

Pigging valves (Fig. 6.25) are always trunnion ball valves first. These valves have a port on the side of the valve that can be opened to allow you to load a pig into the cavity on the ball. If you were to add the same port to a floating ball valve the leakage across the upstream seat would prevent you from being able to safely open the body cavity. The "receiver" version of pigging valves has a perforated plate adjacent to the downstream side of the ball to capture the pig.

Trunnion ball valves should not be throttled (they have poor linearity) and are piggable.

6.3.6.2 Valve actuation

Most valves can be equipped to operate via pneumatic, electric, or mechanical actuators, but you always have to ask the question "should they be?" Valve actuators are expensive, and valves that are seldom operated are unlikely to save enough labor over their life to justify the cost. In general limit actuation to:

- Safety issues (e.g., an "emergency shutdown or ESD" valve must be actuated)
- Control issues (e.g., a valve that is part of a control algorithm must be actuated)
- Frequently operated big valves (e.g., the barrel-isolation valve and side valve on a 12 in (300 DN) launcher/receiver should be actuated, smaller equipment likely should not be).

Open/shut. For ¼ turn valves, we can actuate them with either pneumatic or electric rams attached to appropriate gearing. For vertical-movement valves (mostly globe valves), it is common to actuate them with diaphragm actuators using field gas or control air. Rising stem gate valves are rarely actuated outside of steam plants.

Throttling. Control valves that need to be held in an intermediate position are generally positioned with a pneumatic diaphragm, but increasingly we are able throttle with electric-motor actuators (not, solenoid-operated actuators, they are inappropriate for throttling).

Activation energy. Whether the energy to position valves comes from control air, process gas, or electricity, it is important to know that there is enough of it. Three examples illustrate how to fail at this:

- We installed a booster compressor station in the midst of a gathering system. When we lowered the line pressure on the upstream wells to under 5 psig (34.5 kPag), the pneumatic actuators on the separator dump valves would not operate and the separators flooded.
- Our first response to the low control-gas pressure was to set nitrogen bottles to provide the activation energy, but found that our careful calculations of required volume for weekly visits to the nitrogen bottles did not take into account that the seals on the level controllers were 12 years old and leaking badly—the nitrogen bottle lasted 3–4 hours. Our final solution was to simply bypass the separators and expect the compressor-suction scrubber to do the job.

- A device installed to get gas out of a water system used ¼ turn electric actuators on ball valves to let the gas out. These valves actuated every 6–20 seconds around the clock. We calculated that a 200 W solar panel with two 12 V batteries would provide adequate power. The system worked fine for three sunny days in a row, but the fourth day was cloudy and at 2:00 am the battery voltage got too low to operate the valve and refused to shut the gas outlet valve which allowed the downhole pump to pump directly into the gas gathering system until a small tank at the compressor-station inlet reached a high-level alarm and shut the site down. The failure in our calculation was that the service was dirty and the power required to operate the valve had almost doubled in 3 days of operation. We reevaluated both the size of the solar panels and the number of batteries and fixed the problem.

In other words, it pays to be very pessimistic about how much energy is available to operate actuated valves and how much is required.

6.3.6.3 Pipeline valve locations

Pipeline valves are located on the gathering system in several locations:

- At the start and end of each flow line (e.g., at the well-site and at the trunk, at both ends of the trunk, etc.)
- On pigging equipment
- On either side of any high-risk line section
  - On both sides of a river crossing
  - On both sides of a major highway crossing
  - At a classification break (i.e., a line designed for very low population density has to approach a housing development for a short distance, you can set block valves with actuators and process control before reaching the houses and another after you leave the houses; if your instrumentation and control are adequate then only the section between the block valves will need the lower design factors).

The next question is "should the valves be above ground or below ground?" If they are below ground, you have to further ask "do the valves need to be accessible for maintenance?" If the answer to these questions is "below ground due to freeze issues" and "no, we don't maintain valves" then you can possibly use an "extended-reach" ball valve which looks much like Fig. 6.22 or 6.24 except the valve bonnet (i.e., the area above the ball and below the stem seal) is extended 3–6 ft (1–2 m) above the ball so that the valve can be set with just the actuator above ground, without having to bury stem seals or having dirt acting on the stem.

If extended-reach ball valves are not an option for your project you are limited to the choice between: (1) valve cans and (2) doglegs.

Valve cans. Many producers-as-gathering-system-operators built water gathering systems in conjunction with gas gathering systems. While gas pipe is reasonable to bring to surface for tie-ins and block valves in any climate, water systems are less forgiving of freezing temperatures. The producer's response was often to put all block valves in "valve cans." Most commonly a valve can is a corrugated culvert, laid on its end, and a top cap is attached near ground level. Often, the producers specified that the top cap have remote operators for the valves in the can.

Putting valves in cans had the perceived benefits of providing freeze protection and protecting the valves from vehicular traffic. They do protect the valves from a vehicle sliding into them. Over time we found that the freeze protection was wishful thinking and pipes in cans would often freeze. A response to freezing pipes was to fill the can with organic materials like cottonseed hulls. This fill turned out to be a great bedding material for rodents so we started finding valve cans full of rats and mice. In the San Juan Basin one of the species of rodents that inhabited valve cans was the deer mouse, a known carrier of the hemorrhagic fever called Hantavirus. Now the valve cans are biohazard spaces. We also found that where mice and rats live, snakes will follow. You haven't lived until you've opened a valve can to see a 90 ft (20 m) long rattlesnake (I'm certain it was at least 90 ft long, I'm also certain that the guy that bravely went and removed it was just teasing when he showed me a pretty normal sized 3 ft (1 m) long snake). The rules for working in confined spaces also changed so that entering the can to maintain the valves required confined space precautions which include a lifting harness, winch, and someone to operate the winch (in addition to the biohazard suit and supplied air).

Other downsides to valve cans include:

- They fill with silt and mud (the can in Fig. 6.26 was opened to grease the valves, the water line was under 12 in (305 mm) of mud).
- The remote handles tend to bind with differential settling (and the u-bolts that people use to attach the remote actuator to the valve handwheel tend to rot off in a very few years).
- The cans themselves settle and have actually cut through nonsteel piping and they often damage the coating on steel pipe.
- Finally, the cans represent an attractive place for children to play "fort" and when valve cans are close to housing it is mandatory to

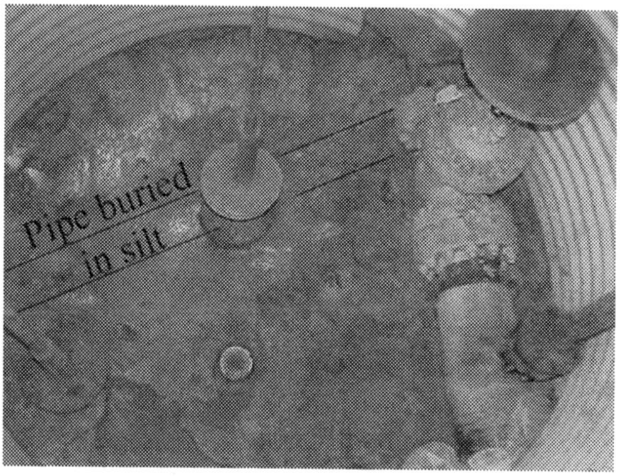

Figure 6.26 Valve can filled with silt.

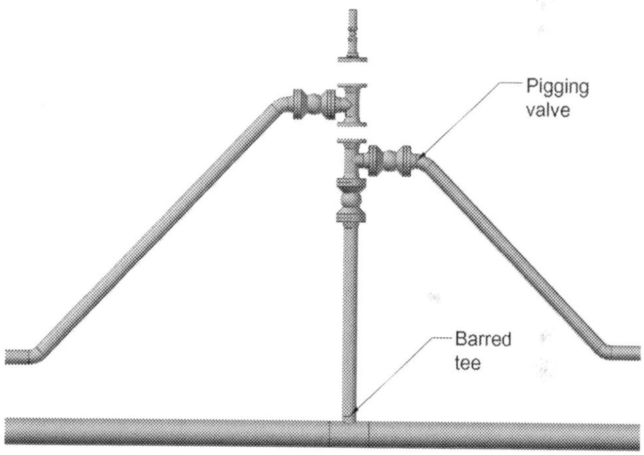

Figure 6.27 Valve dogleg.

lock the cans shut and to verify the locks are in place and intact on a rigid (and frequent) schedule. The lids on these cans tend to be quite heavy and there have been reported incidents of children getting into a can and not being able to get out.

I was working on a gathering-system design that included valve cans on the day that it was announced that in the future valve cans would be treated as biohazard confined space. I immediately changed the project valve-set philosophy and have never set another valve can.

Doglegs. A dogleg (Fig 6.27) is an alternative to a valve can for gas valves. There are as many variations of doglegs as there are engineers designing gathering systems. The functions that a dogleg should provide include:

- provide a purge point for the flow line;
- provide for isolation (including positive energy isolation) of the flow line;
- facilitate pigging flow lines.

The dogleg in Fig 6.27 has some unique features for a new gathering system:

- It is expandable. As wells are drilled, unbolt the pierced blind flange on the tee, bolt in another tee, and reattach the blind to the new tee.
- It allows tie in connections with minimal gas blowdown. One of my clients has a stack of tie-ins with 8 wells in one stack.
- Easy location for double block and bleed or insert blinds with minimal gas blowdown.
- The lines coming into the stack will normally be individual well-site/pad lines, but there is no reason that they can't be other trunks or line loops.

6.3.6.4 *Valve summary*

	Use	Actuate	Gas gathering applications
Plug	On/off	Piston	None
Gate	On/off	Pneumatic spinner	None
Globe	Throttling	Pneumatic diaphragm	Flow control
Butterfly	On/off	Piston (rare)	Low criticality intermittent service
Floating ball	On/off	Piston	Moderate to high pressures
Trunnion ball	On/off	Piston	All pressures

6.3.7 Positive energy isolation

For certain types of work, such as hot work in a confined space, there are significant hazards to life and health that must be managed carefully. One of the ways that we manage those hazards is with positive energy isolation. For someone standing in hip-deep brackish water working on an electrical circuit, just opening a breaker is probably inadequate isolation and she is going to want that breaker physically removed from the panel before she steps into the water. For pressure energy or explosive fluids, positive energy isolation is a bit more complicated, but still achievable.

Before getting into details on how to do positive energy isolation on pipelines, I need to point out that every pipeline welder of my

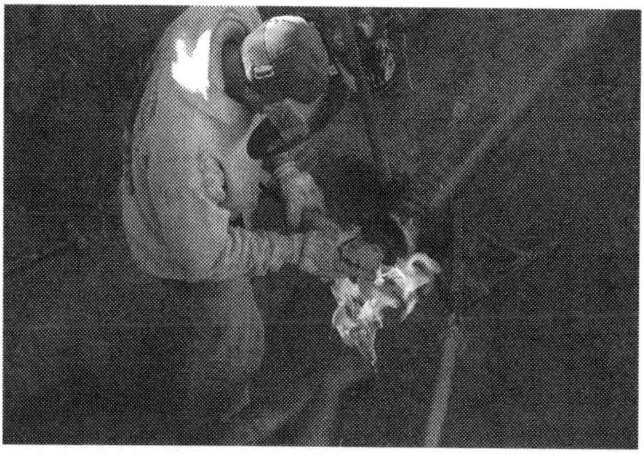

Figure 6.28 Preparing a pipe for welding.

acquaintance (which includes growing up in the home of a pipeline welder) prefers some fire at a weld site (Fig. 6.28, the lower pipe is belching fire) over no fire at a weld site. The reason for this is simple—as a blow-down pipeline sits, some amount of vapors will condense, lowering the pressure in the line to below atmospheric and welding on the line will "suck fire" and can result in an explosion deep in the pipe. If there is fire at the site then the welder is confident that the fire is in control rather than out of control. The 21st century engineers are often far too quick to require positive energy isolation for far too many jobs.

Another recent tendency is to make something "safe" by adding a nitrogen blanket on the pipeline. When doing this, we need to remember that workers breathe air, not nitrogen. A worker in a bell hole can easily be overcome with a lack of oxygen during a nitrogen purge and will often not notice he is in trouble until he collapses. If a job cannot be done without a nitrogen purge (rare), then the procedure must specify oxygen monitors for all workers in restricted spaces. This is a very visible and common example of where our "safety culture" is actually making things less safe.

In cases where positive energy isolation is actually necessary and useful, we most often talk about three techniques: (1) double block and bleed, (2) insert blind, or (3) remove/misalign piping.

Double block and bleed. In essence, double block and bleed is providing two independent sealing surfaces with the space between them open to atmosphere. Often there can be considerable distance (miles or kilometers) between the two block valves on a gathering system, which is perfectly acceptable to the isolation requirements. Any valves out of their normal position must be clearly labeled with the name of the project being worked

Figure 6.29 Spectacle blind.

on, the required position (i.e., "open" or "shut"), the date and time, and the words "Do Not Operate." As mentioned earlier, many of us consider a single trunnion ball valve to be better double-block-and-bleed than two floating ball valves (which are always acceptable to anyone's procedure) since the dP between the blowdown space and the work space is inadequate to activate the seals on the work side of the isolation. Some regulators and some companies disagree with this technical assessment.

Insert blind. The most common form of insert blind is the "spectacle blind" (Fig. 6.29) that is a manufactured unit made for the pressure rating of the flange. The left-hand photograph of Fig. 6.29 is set for no-flow, the center is set for flow.

Unfortunately for many projects these spectacle blinds are not always available where they are needed for a particular job. If they are not available then you might have to use an "insert blind" (also known as a "skillet blind" or "line blank"). An insert blind is a flat plate of steel sized to fit between flange faces. Blinds provide maximum positive isolation, are not prone to accidental or unauthorized operation, and are readily identifiable where installed. Proper blinding techniques, to be developed locally, should be followed. Blinding locations are sometimes difficult to access (i.e., buried pipe flanges) and may require portions of existing systems to be included in the isolation.

If an insert blind is used, it is typically fabricated from mild steel plate (such as ASTM A515-70 or A516-70). The width of the handle on the blind should be 1 in (25 mm), minimum and handle length should be 6 in (152 mm), minimum beyond the outer flange face. Minimum blind thickness is specified in Table 6.8 (thicker stock can be used if it is more readily available). For larger sizes use Eq. (6.15), but ensure that the deflection term is less than one-half of plate thickness (much more than that and removal of the plate can become very difficult).

Table 6.8 Insert blind thickness

	100 psid	285 psid	740 psid	1480 psid
2 in (50 DN)	0.125 in (3.2 mm)	0.125 in (3.2 mm)	0.125 in (3.2 mm)	0.125 in (3.2 mm)
3 in (75 DN)	0.125 in (3.2 mm)	0.125 in (3.2 mm)	0.125 in (3.2 mm)	0.139 in (3.5 mm)
4 in (100 DN)	0.125 in (3.2 mm)	0.125 in (3.2 mm)	0.156 in (4.0 mm)	0.185 in (4.7 mm)
6 in (150 DN)	0.125 in (3.2 mm)	0.184 in (4.7 mm)	0.233 in (5.9 mm)	0.278 in (7.1 mm)
8 in (200 DN)	0.189 in (4.8 mm)	0.245 in (6.2 mm)	0.311 in (7.9 mm)	0.370 in (9.4 mm)
10 in (250 DN)	0.236 in (6.0 mm)	0.306 in (7.8 mm)	0.388 in (9.9 mm)	0.462 in (11.7 mm)
12 in (300 DN)	0.283 in (7.2 mm)	0.368 in (9.3 mm)	0.467 in (11.9 mm)	0.555 in (14.1 mm)

This table is based on using ASTM A515-70/516-70 plate material.
Note: Blind thickness specified in this table is intended for insert blinds that will be used to isolate piping during a static test only, and is not appropriate for determining required thickness for nontest applications (i.e., permanent insert blinds).
Due to difficulty of fabrication, insert blinds thinner than 1/8 in (3.2 mm) should not be used.

$$
\begin{aligned}
f &= \left(\frac{3}{8}\right)\cdot(3+\nu)\cdot P_{psig}\cdot\left(\frac{r}{t}\right)^2 \\
y_{deflection} &= \frac{(1-\nu)\cdot(5+\nu)\cdot f\cdot r^2}{2\cdot(3+\nu)\cdot E\cdot t}
\end{aligned}
\tag{6.15}
$$

Misalign/remove piping. Finally, you can ensure that the work area is free of both pressure and explosive gases by removing a section of piping or unbolting a flange and shifting one side or the other the width of the flange (it is common to run a stud through a bolt hole on the top of one flange through a bolt hole on the bottom of the other flange).

In gathering systems this option is often the most difficult technique to achieve positive energy isolation. Field piping is generally subject to unpredictable forces due to uneven soil settling. Sometimes it is simply impossible to get two flanges to bolt back together after disconnecting them and the piping will have to be cut and reweleded. I've found that if field workers struggle with a makeup for more than an hour we will eventually have to involve welders (usually after several hours of nonproductive hard work) so I will proactively shut a struggle down after an hour and call in the welders.

6.4 DESIGN ISSUES (EPC)

Once you've selected pipe material and pipe size, there are a number of other things that you must be concerned about and decisions you must make prior to actually ordering material and acquiring ROW.

6.4.1 Acquiring ROW/route selection

Once you've picked the materials of construction, you have to pick the gathering-system route to get from the wells to the delivery point(s). Using tools like geographical information systems, digital topographical maps, and even Google maps, you can lay out a route that optimizes the system hydraulics and minimizes obstructions, but that is only the start. Someone owns the land you need to cross. Various obstructions need explicit government permits to cross. Sometimes there are cultural considerations that must be accounted for. Some land owners are easier to work with than others.

Once you have picked a route, you need to turn it over to someone who can acquire the ROW and permits. Sometimes company personnel can/will help with this, but more often than not you will have to find a land contractor. The land contractor will do some or all of the following list, it is often up to the project engineer to make the decision on how much the land contractor will do vs the engineer contracting for each task independently:

- Determine surface ownership on the land being crossed.
- Assess the land ownership for red-flag items (e.g., land owners who have refused to do business with your company in the past, known sites of historical interest, etc.). This red-flag assessment nearly always results in reroutes.
- Execute letters of intent with surface owners to lease/sell you the ROW you need contingent upon final routing.
- When the surface route is adequately secured, arrange for preliminary surveying and staking.
- Arrange for environmental and archeological surveys.
- Coordinate preparation of environmental assessment.
- Adjust the route to deal with archeological and environmental concerns and resurvey.
- Acquire permits and ROW.
- Arrange for construction surveying and staking.

- Sometimes arrange for construction drawings.
- Arrange for any required as-built surveying/drawings.

Four distinct reports/packages come together to make up the "environmental assessment" package: (1) the environmental assessment, (2) arch report, (3) threatened and endangered (T&E) species report, and (4) drawing package. All of these elements are packaged together into the environmental assessment.

6.4.2 Environmental assessment

Exact requirements vary widely from jurisdiction to jurisdiction (and project to project). Knowledge of a particular agency's hot buttons can be crucial to the acquisition of permits. Every jurisdiction has something that looks like an environmental assessment (although it often has other names). The environmental assessment document includes the arch report, T&E, and the drawing package, but it also has a number of additional sections:

- Climate at project site
- Site topography, soils, and geology
- Surface and subsurface water
- Safety standards
- Grazing impacts of the project
- Community health and safety
- Air quality impacts of construction activities and completed project
- Visual impacts of construction and the completed project
- Noise impacts of construction and the completed project
- Impacts to recreational activities
- Mode of transportation and primary routes of equipment and workers
- Waste management

Skipping any of these elements will often result in delays or outright rejection of the project. I failed to include a description of how we were going to transport workers to the job site once, and this happened to be a particular regulator's key issue. She called me and asked, and I responded flippantly that they were going to drive. She rejected the permit. We amended the environmental assessment to indicate that we were going to bus the workers from town and our permitting agent asked me to just not talk to regulators in the future—the regulator's issue was that more vehicles on the road would create more dust and road degradation, a reasonable concern that I handled poorly.

6.4.3 Arch report

A "cultural resources inventory" is generally called an "arch report" in the construction industry. It provides a detailed description of the results of literature searches and site surveys. The report must be written by someone recognized by the permitting body as an expert (often through a licensing process). The report is absolutely non-negotiable, once a site has been identified it will be placed in the arch report and you will be required to document how you will deal with it. Your choices for dealing with it are: (1) reroute to avoid the site, (2) remain adjacent to the site and provide arch monitors and site flagging to ensure that if you discover an extension of the site you will stop work and deal with it, and/or (3) protect the site (this is rare and the protection methods are quite technical).

At the end of the day, it doesn't matter that the vast predominance of the number of arch sites are midden piles (i.e., "a refuse heap" (Webster)) and the largest artifact will be a pea-sized pottery shard, if it is an arch site you have to treat it like it was the Colossus of Rhodes arisen from the sea. Never enter a marked arch site without the permission from and monitoring by an archeologist. We had a surveyor enter an arch site to search for arrowheads (which have actually been very rare on pipeline arch sites in our area). The search included scratching the surface with his boots (no digging tools). The archeologist saw the disturbance the next day, shut down the job, and called the state archeologist. The job was shut down for 12 days (with everyone on full pay) and the surveyor was fined by the state and fired by his employer. It was a very big deal and if the job had been 20 miles north of its actual location (another jurisdiction) he certainly would have ended up with jail time. Take archeology seriously.

6.4.4 T&E species

Some flora and fauna are protected by government agencies. Typically your project is not allowed to:

- disturb the nesting/calving/growing areas in any way;
- disturb the species while propagating;
- conduct any activity that has *any* risk of impacting reproductive activities.

This survey is conducted by a licensed biologist and often requires months of fieldwork to clear your site. If the biologist finds anything at all, you have the options of: (1) abandoning or relocating your project; or (2) spending the next decade in court prior to abandoning your project.

I have never heard of anyone winning one of these disagreements and the people who tried have all regretted the choice.

Sometimes the T&E report will set a deadline for ending activities. If you accept this challenge, make certain that all equipment has been moved out of the affected area prior to the deadline or it will have to stay until the end of the nesting/calving period, and your project has to pay rent on the equipment until the ban is lifted, it can get expensive.

6.4.5 Drawings

The workhorse drawing for gathering systems is the "alignment sheet." It shows land ownership, system route, "stations" (i.e., points of inflection or PI, crossings, tie-ins, etc.), elevation profile, and any environmental, archeological, or hydrological issues. Alignment sheets have a number of formats, with different approaches to providing the information, and most of the formats are useful.

Fabrication drawings like the example in Fig. 6.30 are used to show how the various bits and pieces of the project fit together. In this case the dark piping represents components and piping described on other drawings and only the light piping is related to this fabrication. Historically fab drawings were presented as single-line isometric drawings and were not shown to scale. Increasingly, they are shown like Fig. 6.30 which are

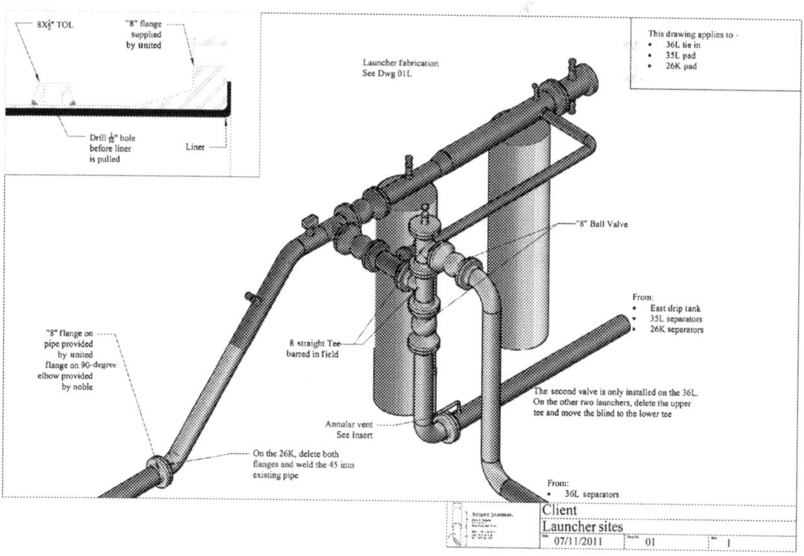

Figure 6.30 Fabrication drawing.

to scale and clearly show connection details. Many companies and construction managers are resisting this trend and still require the isometric drawings; the good news is that the same tools that are used to generate the 3D models can output isometric drawings with minimal effort.

Foundation drawings are sometimes required for gathering-system projects, but this is fairly rare.

The workhorse drawing for plant construction is the process and instrumentation drawing (P&ID). It shows all piping and instrumentation schematically, and displays relative hydraulic location of valves and piping, but it isn't to scale and provides no visual aids to the function of any particular line or component (and efforts to use colors and/or line weights have been met with outrage on the order of a medieval priest facing heresy). I worked in upstream for nearly 25 years before I ever saw a P&ID. I was quite happy without them. Since about 2005, you cannot get a project authorized without a P&ID covering every scrap of piping or control tubing in the entire project. Prior to the advent of P&ID requirements, it was rare to have more than a layout drawing for a well-site that showed the relative position of major equipment; I reviewed a project in 2010 that had 104 drawings for a single-well pad. This tendency has ballooned engineering costs while not adding any real increased value. I discuss this trend in Chapter 10, Integration of Concepts, so I'll end this discussion by saying that I know why we have gone in this direction but sooner or later we will have to go back toward significantly fewer drawings and my vote is to eliminate P&ID on well-sites (they make a lot of sense in plants, and some sense in compressor stations, not so much on well-sites).

6.5 PROJECT SAFETY PLAN

A gathering-system project involves many companies, each required to have their own safety standards, reporting, and processes. Your design document can simply refer to your safety manual, but you have to understand that not all of the contractors have access to your safety manual; your manual may not emphasize certain construction-related issues that the contractor considers important, and there will be a lot of material in your safety manual that does not apply to your project.

It is useful to do a job hazard analysis (JHA, the analysis, not necessarily a particular company's format) to break all of the risks into "ordinary hazards" and "extraordinary hazards." The ordinary hazards are things like driving safety, personal protective equipment (PPE) required for all workers, line-of-fire issues, etc. Things that a worker in Oil & Gas should be thinking about every day. Address ordinary hazards by listing required PPE, listing the required safety meetings, describing JHA format and requirements, isolation requirements, and lockout/tag-out procedures.

Extraordinary hazards are things that are unique to a specific project or unique to major projects in general. When we crossed the river in Fig. 6.5, we had detailed safety instructions and training on working at height and crane safety. Other projects have other emphasis. Every project should include detailed instructions and procedures dealing with purging air from lines and static testing.

6.5.1 Purging air from lines

Flow lines can represent a more complex problem than well-sites. Pipe volume is much bigger. There often is not a reasonable path for a clearing purge. There often is not a reliable source for purge gas. These complexities frequently lead to specifying (and designing) a dilution purge.

The detailed purge procedure including source of all gas being used, fill times, fill pressures, and soak times, with a description of atmospheric conditions or gas condition that should stop the procedure is required.

6.5.2 Static testing

An important step in a pipeline project is confirming that the piping and facilities have adequate strength to withstand the expected operating pressures. There is guidance in the primary pipeline construction codes (e.g., ASME B31.8) for many of the important considerations for reducing the risk associated with the tests required to confirm fitness for purpose.

When new piping is to be placed in service, various codes and company standards require that it be subjected to a leak test and/or a strength test. Leak tests are generally done at fairly low pressures and are only intended to prove that the pipe will in fact contain the fluids. Risks are generally reasonably low and leak tests are done without much risk of catastrophic failure.

The strength test is done with elevated pressure at some multiple greater than 1.0 of the system MAWP and held for a specified length of time. The pressure multiple and time duration vary considerably from one regulatory jurisdiction to another, from one code document to another, and from one company to another.

The primary kinds of tests are "hydrostatic" or "pneumatic static" (sometimes called "pneumostatic"). The "static" simply means that during a successful test the fluids under pressure have no net movement relative to a pipe end or the pipe centerline.

A hydrostatic test is done using a largely incompressible fluid like water (hence the prefix "hydro"), oil, glycol, or some mixture (e.g., glycol is sometimes added to hydrostatic-test water to prevent freezing during storage). In these tests, the line is filled with liquid, entrained gases are allowed to disperse to vents (often for days, generally at least 24 hours), and the pressure is raised within the system to the required test pressure and held there for the duration of the test. At the end of the test, the test medium is removed for disposal.

A pneumatic static test is done using a gas like compressed air, nitrogen, methane, or CO_2, (tests with CO_2 are rare and can be difficult because at elevated pressures the gas can change into a "dense phase" which behaves differently from either a gas or a liquid).

The industry standards do not show a clear preference for the selection of one particular test media over another. Some jurisdictions have written local regulations that do show a clear preference for hydrostatic testing over pneumatic testing. This preference manifests itself in several ways, but the primary representation is the requirement in statutes and regulations that a pneumatic test have an "exclusion zone" or "minimum approach distance" around the test to reduce the risk of injury during the test, but no similar exclusion zone for hydrostatic tests (e.g., see ASME PCC22, 2015, Mandatory Appendix III). The primary piping codes allow pneumatic tests under a list of reasonable conditions, and do not mention exclusion zones.

While on the surface, it does not seem to be unreasonable to have a bias in favor of hydrostatic testing over testing with compressed gases it is important to look below the surface. The risk being talked about here is that pressurized gas contains significantly more potential energy than a pressurized incompressible liquid. Rapidly converting this potential energy to kinetic energy can be a violent and destructive event. The question that is asked far too infrequently is "how much of the potential energy can be converted to kinetic energy in an explosive decompression?"

6.5.2.1 Energy involved in testing

The energy that needs to be considered in testing is the stored (or potential) energy of compression and the potential energy of stacking mass. The second one concerns the hydrostatic gradient. For gas the density is very low so the gravitational forces are not significant. For example, air at 900 psig would exert 0.034 psi/ft (0.758 kPa/m) (at the bottom of the pipe, decreasing as you move toward the top) which can be safely ignored. On the other hand, liquids have a significant mass. For vertical changes in the line, an elevation increase adds 0.433 psi/ft (9.81 kPa/m) to the pressure at the lowest point in the system. This means that in hilly country, it can be very difficult to design a hydrostatic test (e.g., if the elevation change is 1000 ft (305 m), then the pressure at the bottom will be 433 psi (2.99 MPa) higher than pressure at the top; for a 150% test on an ASME B16.5 Class 150 line, just filling this line would exceed test pressure at the bottom while leaving the top at atmospheric pressure). It is sometimes possible to segment the line to keep the elevation changes within a segment below some maximum, but that is not always possible (e.g., some lines have inaccessible segments in very rough terrain (Fig. 6.31), others do not have valves where needed to do the segmentation).

Compressive energy is more frequently discussed. To increase the pressure of either a gas or a liquid (at constant volume), you have to add mass. Gases are quite compressible so it is easy to visualize the mass going into the test volume. Liquids are generally considered incompressible, but during a test you do have to compress it. The bulk modulus (i.e., the amount of pressure required to reduce the liquid volume by 1%) of liquids is very large, so even in the most aggressive test the liquid will have very little compressive energy (e.g., the bulk modulus of water is on the order of 319,000 psi (2200 MPa), so a 900 psig (6.2 MPa) test would reduce the volume by about 0.003%).

It is misleading to talk about the energy involved in testing without a concrete example. I've seen comparisons where a 5 gal (19 L) propane bottle was tested to failure with water and a 500 gal (1900 L) vessel was tested with air, the results were dramatically different. For this comparison I will stick to the following example conditions:

- Length: 10 miles (161 km)
- Pipe size: 36 in (900 DN) Schedule 40
- Test pressure: 900 psig (6.2 MPag)
- Atmospheric pressure: 14.7 psia (101 kPa)
- System volume: 373×10^3 ft^3 (10.6×10^3 m^3)

Figure 6.31 Legoff.

- Hydrostatic test:
  - Mass of water at atmospheric pressure: 23.299×10^6 lbm (10.6×10^6 kg)
  - Mass of water at test pressure: 23.300×10^6 lbm (10.57×10^6 kg)
  - Added mass to reach test pressure: 304 lbm (668 kg)
  - Enthalpy at atmospheric pressure: 122.82 BTU/lbm (299.6 kJ/kg)
  - Enthalpy at test pressure: 128.82 BTU/lbm (285.7 kJ/kg)
  - $\Delta H = m_{final} \cdot \Delta h + \Delta m \cdot h_{final} = 140 \times 10^6$ BTU $(147 \times 10^6 \text{kJ})$
- Pneumatic test
  - Mass of air at atmospheric pressure: 0.285×10^6 lbm (0.129×10^6 kg)
  - Mass of air at test pressure: 17.729×10^6 lbm (8.042×10^6 kg)
  - Added mass to reach test pressure: 17.444×10^6 lbm (7.912×10^6 kg)
  - Enthalpy at atmospheric pressure: 118.09 BTU/lbm (275 kJ/kg)

- Enthalpy at test pressure: 124.31 BTU/lbm (289 kJ/kg)
- $\Delta H = m_{final} \cdot \Delta h + \Delta m \cdot h_{final} = 217 \times 10^6 \text{BTU} \quad (229 \times 10^6 \text{ kJ})$

In a test failure, the energy release is significant in either case, the real question is "what portion of the total energy is available to participate in the initial event vs how much energy is simply 'leaked out' after the explosive decompression is completed?"

Water at this test pressure is 99.997% incompressible. Archimedes principle says that a force applied (or removed) from any place within an enclosed mass of an incompressible fluid will be felt everywhere within that enclosed mass. That would imply that the conversion of enthalpy in the example would involve the entire volume simultaneously and 99.997% of the 140 MMBTU (147 GJ) of energy would be released in the initial event. This results in total energy release equivalent to 33 tonne of TNT (using the conversion 1,488,617 ft-lbf/lbm_{TNT} (ASME PCC22, 2015)).

The change in total system entropy in the pneumatic test is half-again as large, but Archimedes principle does not apply to a gas. We have to determine how much of the total system mass would participate in an explosive-decompression event.

NASA published a document in the 1990s which has come to be known as the "NASA Glenn Research Center Methodology." This document was really the first time that anyone had made an effort to quantify the risk of pressurized-gas static testing of pipelines (work on quantifying the explosive force resulting from a failed gas-filled pressure vessel was quite robust at the time of the "Methodology" paper (General Physics, 1988)). This paper was on NASA's website for several years but recent attempts to locate it have proven to be unsuccessful. Several regulations and many company policies were written based on the NASA document. Basically this two-page document said:

- A pipeline failure could properly be called an "adiabatic" process (i.e., there is no heat transfer).
- An adiabatic decompression results in a significant energy release.
- All of the material in the system will participate in the explosive decompression.

These are the same conclusions that had been reached for pressure vessels, where all of mass of the pressurized gas is physically in close proximity to any given point on the vessel. The adiabatic energy can be calculated by (ASME PCC22, 2015) Eq. (6.16).

$$W_{gas} = P_{psig} \cdot V_{system} \cdot \left(\frac{1}{k-1}\right) \cdot \left(1 - \left(\frac{P_{atm}}{P_{psig} + P_{atm}}\right)^{\frac{k-1}{k}}\right) \tag{6.16}$$
$$= 134.2 \times 10^6 \text{ BTU} \quad (141.6 \times 10^6 \text{kJ})$$

This calculation is 38% lower than the change in enthalpy shown for air in the example statement because it does not account for the energy in the mass that exits during the event while still considering the entire system mass as participating in the event. Using the Glenn Research Methodology the energy in Eq. (6.16) is equivalent to 32 tonne of TNT, 4% lower than the same volume of water, but very wrong.

The problem with the NASA Glenn Research Methodology is that an explosive-decompression event is very short duration which only allows time for a very limited amount of the mass to participate in the explosion. Experiments done at the University of Nebraska-Lincoln for the U.S. Department of Energy in 2012 (Nebraska, 2012) show that the gas temperature in an explosive decompression drops very rapidly to a minimum, and then increases to approximately initial temperature over the next few seconds. This minimum can be taken to be the end of explosive decompression and the start of depressurization. The University of Nebraska-Lincoln paper does not identify the duration of this nearly vertical temperature transient. Other, less formal sources indicate it occurs at 10–50 mS after an opening large enough to result in choked flow is created.

Natural events within a gas volume are limited to the speed of sound (Mach 1.0). As we've discussed elsewhere, this limitation is due to the creation of standing "shock waves" in the flow that inhibit communication from downstream to upstream. Prior to Mach 1.0, the existence of lower pressure downstream is communicated upstream through a failure to support the higher upstream pressure. At Mach 1.0 the shock wave is adequate to support the upstream pressure and only allow flow at the speed of sound.

So if we say that the near-vertical temperature transient is 50 mS and allow half of the available time for the notice of the event to communicate within the system and half of the time for the energy that now "knows" that there has been a failure to participate in the explosion then with the speed of sound (Eq. (6.17)):

$$v_{sonic} = \sqrt{k \cdot \frac{R_{air}}{SG} \cdot T} \tag{6.17}$$

For air at 60°F (15.6°C), the speed of sound is 1139 ft/s (347 m/s). That says that over the 25 mS available, the shock wave would travel 28 ft (8.5 m). Continuing our example, let's assume that the failure happened infinitely far (i.e., more than 28 ft (8.5 m)) from the end of the pipe so the amount of pipe involved is 56 ft (17 m) since stored energy from both sides of the failure participates. That is a volume in our example of 364 ft^3 (10.29 m^3) so using Eq. (6.16) (which is reasonable only because the mass participating is now small enough to be negligible), the energy is equivalent to 74 lbm of TNT—not a trivial event, but far from a tactical nuclear weapon. To put it in perspective, 74 lbm of TNT in a properly constructed and properly deployed "cratering charge" would result in a crater 6 ft (1.8m) deep and 30 ft (9.1 m) in diameter which is a volume of earth of about 50 yd^3 (38.1 m^3).

There are published equations to calculate a "restricted distance" (i.e., the closest safe point of approach while under a pneumatic test), the NASA version is Eq. (6.18):

$$d_{nasa} = \left[\frac{P_{psig} + P_{atm}}{125.208} + \frac{(P_{psig} + P_{atm})^2}{5.509 \times 10^5} - 20.287 \cdot \ln(P_{psig} + P_{atm}) + 10.812 \cdot \ln(P_{psig} + P_{atm})^2\right] \cdot \left(\frac{V_{system}^{\frac{1}{3}}}{10}\right) \cdot ft \quad (6.18)$$

Eq. (6.18) is an empirical equation (pressure must be in psi and volume in ft^3) yields 2687 ft (819 m) on either side of the pipe, over a mile for the example test. Changing the pipe length to the 56 ft (17.1 m) calculated in the previous equation changes the restricted distance to 274 ft (81.4 m)—still large, but considerably less than a mile. This calculation demonstrates the fallacy of this approach—if the 10 mile pipeline were operating at 300 psig (half of MAWP) the closest you could ever approach the line in operation using this approach the exclusion zone would be 1754 ft (535 m) and it would cover 2100 acres (860 hectares).

If the NASA Glenn Research Methodology had any validity at all, then every time a blowdown valve was opened, a PSV lifted, or a rupture disk failed we would have a kiloton-range explosion on location since a pipe failure is simply opening the line to atmosphere.

In this example a pneumatic test on pipeline has the potential to release 0.11% of the energy that could be released during a hydrostatic test. If we reduce the size of the test from pipeline scope toward vessel scope, once we get less than about 60 ft (18.2 m), the relative impact of the two tests

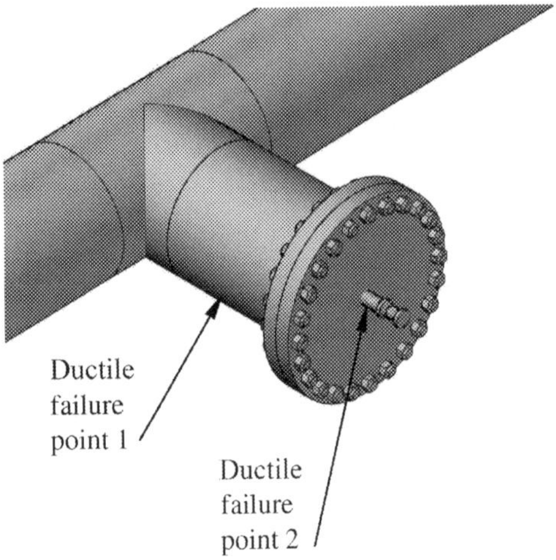

Figure 6.32 Pipe stub.

changes dramatically—a pneumatic test of 60 ft (18.2 m) of 36 in (900 DN) pipe would have more than twice as much energy participating in an explosive decompression than you would have in a hydrostatic test.

6.5.2.2 Alternate approach to evaluating stored energy impacts of pneumatic tests

If the total potential energy of all the fluids in the pipe is not a reasonable measure of the risks associated with a pneumatic test, what would be a better representation? That answer depends on whether the container exhibits ductile failure or brittle failure. Brittle failures will be discussed later. A ductile failure in the field of a pipe generally manifests as a tear and does not result in projectiles. Ductile failures do not represent a significant risk to personnel or equipment unless they occur on a dead-end like a blind flange or the piping leading up to a vent valve (Fig. 6.32). Dead-end failures do represent a risk and will be discussed here.

$$f = P \cdot A \tag{6.19}$$

$$f = m \cdot a \tag{6.20}$$

$$a = \frac{P \cdot A}{m} \tag{6.21}$$

The force of fluid pressure on any surface of a containment is given by Eq. (6.19). If a section of the containment were to fail and become a projectile, then the reference area is the area of the resulting opening.

Once a projectile can begin moving, the force on that projectile is given by Eq. (6.20). Combining these equations gives us Eq. (6.21) which lets us determine the acceleration of the projectile away from the pipe.

$$S = \left(\frac{a}{2}\right)\cdot t_{acc}^2 + v_0 \cdot t_{acc} + S_0 = \left(\frac{12{,}000\cdot\frac{\text{ft}}{\text{s}^2}}{2}\right)\cdot(0.05\cdot\text{s})^2 + 0 + 0 = 15.0\cdot\text{ft}\,(4.57\cdot m)$$

$$v = a\cdot t_{acc} + v_0 = 12{,}000\cdot\frac{\text{ft}}{\text{s}^2}\cdot 0.05\cdot\text{s} + 0 = 598\frac{\text{ft}}{\text{s}}\left[182\cdot\frac{\text{m}}{\text{s}}\right] \tag{6.22}$$

In our example, if we have a ductile failure of a stub of pipe on the branch of a 36 in (900 DN) tee (i.e., area of the opening is 6.5 ft^2 (0.603 m^2), mass of flange and blind = 2300 lbm (1043 kg)) then the acceleration is 12,000 ft/s^2 (3650 m/s^2). We determined above that it takes up to 0.050 seconds to reach the point where shock waves disrupt communication between upstream and downstream which would effectively remove the differential pressure from the acceleration stream and end the acceleration, then we can see from Kinematics equations (Eq. (6.22)) that the velocity at the end of acceleration will approach 600 ft/s (182 m/s).

If the flange is oriented horizontally, 3 ft (0.91 m) above the ground, then you can determine the time it takes to hit the ground by (select the reference plane so that initial displacement is zero) and the distance traveled (Eq. (6.23)) assuming air resistance is negligible.

$$t_{fall} = \sqrt{\frac{2\cdot S}{g}} = \sqrt{\frac{2\cdot(-3\cdot\text{ft})}{-32.176\cdot\frac{\text{ft}}{\text{s}^2}}} = 0.432\cdot\text{s}$$

$$S_x = t_{fall}\cdot v_{acc} = 0.432\cdot\text{s}\cdot 598\cdot\frac{\text{ft}}{\text{s}} = 258\cdot\text{ft}\,(78.7\cdot m) \tag{6.23}$$

If we allow a factor of 2 for the projectile bouncing, then a reasonable exclusion zone would be 500 ft (152 m) around a potential full-diameter projectile like a branch stub—considerably shorter than the 5621 ft that the NASA Glenn Research Methodology would suggest. A more reasonable approach would be to put sand bags in front of any potential projectile.

If the piping upstream of the 2 in (50 DN) blowdown valve on the blind flange (ductile failure point 2 in Fig. 6.32) were to fail instead of the 36 in (900 DN) pipe, the projectile would be much lighter, leading to higher acceleration. This high rate of acceleration would result in exceeding sonic velocity within the first 50 mS, so the force will stop when the projectile reaches Mach 1.0 or 35 mS after failure. This results in the projectile traveling 492 ft (150 m) before first impact with the ground. The sand bags that made sense for the 36 in (900 DN) pipe failing would also protect personnel and equipment from this projectile.

6.5.2.3 *Brittle failure*

Brittle failure is a different issue. Mild steel makes the transition from ductile-dominant to brittle-dominant at the ductile-to-brittle transition temperature (DBTT). Experiments show that this transition takes place over a range of temperatures (e.g., a sample may show simply ductile failure above the freezing point of water and simply brittle failure below −40°F (−40°C)). In that case the DBTT would be 32°F (0°C), realizing that a sample at 0°F (−17.7°C) would be more likely to exhibit ductile behavior than brittle behavior. DBTT is a function of the chemical makeup of a given steel and must be assessed for each specific project to determine minimum ambient temperature and minimum fluid temperature for a pneumatic test.

When a piece of steel fails in brittle fracture, the container walls do not thin out, the steel just breaks. These breaks are most common along grain boundaries. Places where the grains are not homogenous can cause the crack propagation mechanics to turn the crack, even back on itself. When a crack creates an unsupported island, that island then becomes a projectile. If we take a 1.5 in × 1.5 in (4 cm × 4 cm) rectangular island in our 36 in (900 DN) pipe, located 45 degrees from the horizontal, our projectile dynamics are very different. The mass of our projectile is 0.5 lbm (0.217 kg) and the effective hole diameter is 1.69 in (4.3 cm). The initial acceleration (disregard gravity for the acceleration period) shown in Eq. (6.24) is an order of magnitude higher than we saw for the full-size blind and flange.

$$a = \frac{P \cdot A \cdot g_c}{m} = \left(\frac{914.7 \cdot \frac{\text{lbf}}{\text{in}^2} \cdot 2.25 \cdot \text{in}^2}{0.477 \cdot \text{lbm}} \right) \cdot \left(32.176 \cdot \frac{\text{ft} \cdot \text{lbm}}{\text{s}^2 \cdot \text{lbf}} \right) \tag{6.24}$$

$$= 138,700 \cdot \frac{\text{ft}}{\text{s}^2} \left[42,270 \cdot \frac{\text{m}}{\text{s}^2} \right]$$

Gas escaping from a pressurized containment is limited to sonic velocity, so this very high acceleration limited to the duration required to reach sonic velocity is 0.8 mS (Eq. (6.25)):

$$v = a \cdot t_{acc} \therefore t_{acc} = \frac{v_{sonic}}{a} = \frac{1139 \cdot \frac{ft}{s}}{138,700 \cdot \frac{ft}{s^2}} = 0.0082 \cdot s \tag{6.25}$$

With the orientation of the projectile (i.e., 45 degrees from horizontal), the velocity has to be broken into an x and y component which gives us a flight time at sonic velocity of 25 seconds (Eq. (6.26)):

$$v_{x0} = \cos(45\text{degrees}) \cdot v_{sonic} = 806 \cdot \frac{ft}{s} \left[246 \cdot \frac{m}{s} \right]$$

$$v_{y0} = \sin(45\text{degrees}) \cdot v_{sonic} = 806 \cdot \frac{ft}{s} \left[246 \cdot \frac{m}{s} \right]$$

$$v_y = 0 = (-g) \cdot t_{up} + v_{y0} \quad \therefore t_{up} = \frac{-v_{y0}}{-g} = 25.038 \cdot s$$

$$S_{yRiseing} = \left(\frac{-g}{2} \right) \cdot t_{up}^2 + v_{y0} \cdot t_{up} + S_0 = 10090 \cdot ft\,(3075 \cdot m)$$

$$S_{yFalling} = \left(\frac{-g}{2} \right) \cdot t_{dwn}^2 + 0 \cdot t_{dwn} + 0 \quad \therefore t_{dwn} = \sqrt{\frac{-10090 \cdot ft \cdot 2}{-32.176 \cdot \frac{ft}{s^2}}} = 25.041 \cdot s \tag{6.26}$$

This says that the projectile will travel for 50.1 seconds prior to stopping. The distance from the pipe would be 40,000 ft (12.3 km) (Eq. (6.27)).

$$S_x = v_{x0} \cdot t_{total} = 806 \cdot \frac{ft}{s} \cdot 50.1 \cdot s = 40,030 \cdot ft\,(12,300 \cdot m) \tag{6.27}$$

In other words, if the pipe temperature is anywhere close to the DBTT then you don't want people to be within 8 miles (13 km) of this pipe under this test. Again very different from the 5621 ft (1713 m) you get from the NASA Glenn Research Methodology. In this scenario, if the projectile is launched from a position 4 degrees above the horizontal it could hit someone outside the NASA exclusion zone with enough velocity to be fatal.

If a given project determines the DBTT is 32°F (0°C), then if the test procedures specify that the test cannot be started if the predicted ambient temperature has a reasonable chance of falling below 50°F (10°C) while the pipe is under test will provide confidence that the chance of brittle failure is approaching zero. This approach provides a far superior result than specifying arbitrary "exclusion zones."

6.5.2.4 Static test design considerations

When designing a static test you need to give conscious thought to a number of issues and you should document your conclusions in the test procedure. As a minimum the test should cover:

- Hydrostatic tests:
  - Where will the test gauge and fill connections be? You need to make sure that the pipeline topography supports this location (i.e., no place in the test will have too high or too low a test pressure) and that water trucks can access the location.
  - Where are you going to get the water?
    - River? You need to make sure you have rights to take river water, often you don't.
    - Municipal water system? You need to make sure that your rate of withdrawal doesn't harm system capacity.
    - Oil & Gas produced water? You need a water analysis and a thorough understanding of the impact on your pipe of the water makeup.
  - Can any above-grade pipe supports carry the weight of the pipe when full of water?
  - What will you need to do to the water to use it
    - Biocides?
    - Filtration?
    - Degassification?
    - Freeze protection?
  - How will the system be filled?
    - Trucks: Do they have enough pumping power to overcome required topography?
    - Water gathering/distribution system: Does the system have enough pumping power and excess capacity to fill the system?
  - How will you degas the water?
    - What vents are available?
    - How long will you let the water sit after filling?
    - Do you have enough makeup water to replace the vented gas?

  - What is the minimum ambient temperature for the test?
  - What is the acceptable rate of pressurization?
  - How long will you allow the system to equilibrate temperature prior to starting the test?
  - How long will the test run? This is a really difficult question. The standards generally have a very short period (ASME B31.8 calls for 2 hours for offshore pipelines, but does not specify a duration for onshore, ASME B31.3 calls for a 10 minute test), but regulators and company policy can require longer test durations.
  - Can the operator add or remove water during the test? There are as many answers to this question as there are engineers writing test procedures. My answer is "yes you can remove water in response to thermal expansion, but you may not add water."
  - What constitutes a successful test? Again there are many answers to this question, my answer is usually "if the final pressure is above MAWP then the test was successful."
  - How will you remove the water from the pipe? The most common method is to chase pigs with air, but sometimes people just drain the system of as much water will come out and leave it—if you do that and leave it for many days you should put a 1–3 psig (7–21 kPag) nitrogen blanket on the system to try to do something about the inevitable corrosion issues.
  - How/where will you dispose of the water? The water that used to be potable is now a hazardous waste that must be disposed properly. The days of just letting it run down the bar ditch are far behind us.
- Pneumatic tests:
  - How much exclusion area is required?
  - What kind of gas are you planning to use
    - Air?
    - Nitrogen?
    - Methane?
  - How will it be delivered
    - On-site compressor?
    - Bulk truck?
    - Bottles?
  - What is the minimum ambient temperature for the test?
  - What is the minimum gas injection temperature? For air compressed on site this is not an issue, for a test from bottles it is just barely a problem, for bulk gases it is the biggest problem. The

boiling point of liquid nitrogen is −320°F (−195.8°C). Any temperature above that will deliver a gas. Flowing −40°F (−40°C) nitrogen into a test is certain to drop the pipe below the ductile-to-brittle transition and now you are applying high stresses to a glass jar. There are several people killed each year because engineers cannot be bothered to specify a gas injection temperature. If you don't specify the temperature, then the bulk-truck operator will inject gas at the lowest temperature that he gets an adequate flow rate (to save the cost of heating the gas). Pick a temperature and require a continuous monitoring device on it—if the temperature ever gets below 50°F (10°C) the injection should be shut down until the gas can be adequately heated.

- What is the acceptable rate of pressurization? This changes from job to job generally based on ambient temperature. One job I did called for 5 psig/min (35 kpag/min) up to 50 psig (344 kPag), soak for 30 minutes, 10 psig/min (700 kPag/min) to 150 psig (1034 kPag), soak for 30 minutes, 10 psig/min (700 kPag/min), etc. The slow start was because the change in stress with pressure is faster early and late than in the middle. You need to do your own analysis for the materials you are testing.
- How long will the test run? Same issues as for a hydrostatic test.
- Can the operator add or remove test gas during the test? No.
- What constitutes a successful test? Again there are many answers to this question, my answer is usually "if the final pressure is above MAWP then the test was successful."
- How will you remove the test gas from the pipe? You need to designate vent point(s) and control the blowdown to less than 25 psig/min (172 kPag/min) to prevent J−T cooling from causing brittle failure at the junction of the main pipe and the blow down pipe.

6.5.2.5 Static testing conclusion

The current tendency to assume that hydrostatic testing is inherently safe and that pneumatic static testing is high risk is based more on an incomplete understanding of the possible forces than on actual risks. Hydrostatic testing approached without proper respect for the mass and energy involved has a high potential for personal injury and property damage. Pneumatic testing approached with an adequate understanding of the risks and the important issues can reduce the potential for personal injury and property damage to an acceptably low level.

Regulations and standards that impose "exclusion zones" and "minimum approach distances" provide a false sense of security that can easily lead to dire consequences. As the examples in this section have tried to illuminate, a predefined exclusion zone assumes a failure mode that may be very different from another failure mode that may be more likely. There is no substitute for a competent analysis performed with a complete understanding of the issues followed by procedures that lead to a safe test.

6.6 BIDDING CONSTRUCTION CONTRACTS

The benefits of bidding the construction portion of projects are incredibly illusionary. If no one trusts anyone, and people within a company are not trusted by their management, bidding can prevent gross dishonesty—but why in the world a company would retain an employee that management doesn't trust is beyond me.

The alternative to bidding is "time and materials" (i.e., the contractor does the work and sends you a bill for the parts of the job at an agreed hourly or daily rate). Time and materials jobs used to be the norm in Upstream Oil & Gas projects, but increasingly the requirements of Supply Chain Management force bids. If you must bid a project, before preparing the bid package ensure:

- All design documents including pipe selection, pigging equipment design and location, drip design, drawings, environmental assessment, and maps are complete before sending the package out for bid.
- Detailed lists of things that will be included within the bid scope and a detailed list of things that will be excluded from the bid scope have been prepared. Remember that anything that you leave of the "included" list is excluded and anything you leave off the "excluded" list is included.
- Develop a list of qualified contractors (don't allow your EPC or Supply Chain Management group to specify the list or your low bidder could easily be a company you don't want to work with).

Bidding a job requires considerable effort by the project engineer, by the contractors bidding, and by the bid review team. I've found that creating a participatory environment and actually managing the project results in significantly lower project costs for time and materials jobs than a hard-dollar bid. It is common for contractors to pad a hard-dollar bid (one job that I had to audit had doubled their cost estimate in their bid and at the end of the job they found that they had been conservative in their estimate and made more than twice the profit they expected). On time and materials jobs that I've managed I've occasionally found contractors that padded their equipment list, left equipment on the clock after it was finished with my job, etc., but when that has come to light I applied for refunds and got them and then *never* used that contractor again. Over time I gravitated toward contractors that looked after my interests better than I did. Honest people are out there, you just have to create an environment where people see a benefit to not cheating you.

6.7 CONSTRUCTION ISSUES

The construction contractor's superintendent is the expert on construction issues. If you don't agree with that, then either: (1) you've chosen the wrong contractor; or (2) you have an inflated (and probably undeserved) view of your own competence. You can be confident that the contractor will look at every design decision you've made and translate it to number of workers of each type, number and type of pieces of equipment, and staging sequence for materials, equipment, and people. These are complex and narrow skills that few design engineers will ever master. I managed several construction projects and one of my first and most important tasks was to develop a rapport with the contractor's superintendent so that he was comfortable telling me where there were better ways to do things. If you ever find that your contractor's superintendent always agrees with you then you can be confident that he is keeping ideas to himself that could have saved the project hundreds of thousands of dollars or weeks on the schedule and you can be further confident that you have made a major error. I was always the boss, but I took great care to delegate the necessary authority to the people who needed it.

While the superintendent is the expert, even experts will have multiple nearly equivalent ways to accomplish any task. Imposing the processes of the last job on the next job almost never works. I built a job where the superintendent wanted a "parts runner" who was available to basically perform any short-duration task on the job from going to town to pick up a forgotten part or tool to running pipeline tape from the on-site storehouse to a lowering-in crew. I was skeptical that this person could be worth their salary, but the superintendent was adamant that they would add value and I needed to prove that I wasn't just an arrogant jerk that wouldn't listen to reason so I approved it. It turned out that the parts runner was the busiest person on the job and she saved the project hundreds of times the cost of her salary. On the next job I forced a parts runner on a superintendent who thought it was a dumb idea. That parts runner was as competent and enthusiastic as the first one, but had nothing to do because no one ever called. About half way through the project the parts runner job was eliminated and the person was shifted to another role. The difference did not lie in the person with the title "parts runner," it lay in the way that the individual superintendent managed their work. In the first job, the superintendent was very much "big picture" and the details that slipped through the cracks were picked up by the parts runner. In the second job the superintendent was very detail focused and nearly nothing slipped through the cracks. Both jobs were completed slightly ahead of schedule and well under budget so I don't know that I would ever say that one orientation is markedly better or worse than the other, just that different people accomplish the same task differently.

One area that the production-company representative to the project has to keep in mind is the equipment inventory. For a hard-dollar bid project you need to make sure that there is adequate equipment of adequate size, contractors will sometimes cut corners and try to use an underpowered machine that will often lower margin of safety below acceptable levels. For a time and materials job, you need to make sure that there is a reason that a particular piece of equipment is on your bill. For example, you may need the horsepower of a Cat D-9 bulldozer for one section of clearing the ROW. The D-9 should show up on your invoice a few days before that part of the job and should leave the invoicing a few days after it finishes the task. It is probably reasonable to have to pay rental on the D-9 for the entire month to make certain that it is available in case you find a section of ROW that the superintendent was wrong about being able to prepare with a smaller bulldozer. It is not reasonable for it to still be on the bill 2 months after the ROW is complete. Often these sorts of issues are oversights, but sometimes the contractor leaves the equipment on your bill because they don't have anywhere else to send it. I had that happen on a few jobs until word got out that contractors charging me for equipment storage became disqualified to participate in the next job.

6.7.1 Inspection

Various types of inspection are required by standards, regulations, and company policies. Most often we see requirements for: (1) welding inspectors, (2) ROW inspectors, (3) quality assurance/quality control (QA/QC) inspectors, and (4) arch monitors. Some of these may not be required for a specific job; other jobs may require multiples of each.

When employing inspectors it is important for the company representative to the project watch closely for: (1) unreasonable tension between the inspector and the construction superintendent; and (2) inspectors trying to "stretch" a job for personal financial reasons. There will always be tension between the inspector and the contractor, they often see their roles as being in conflict. When you start getting frequent "do you know what xxxx did?" kind of calls it is time to intervene and get everyone on the same page—inspecting and contracting should both have the goal of building a safe system good.

It is reasonably rare for inspectors to put their own financial interests ahead of their professional duties, but it happens. If you find that the welding crew is standing by too often "waiting for the inspector" only to find the welding inspector sitting on a rock in the shade "taking a break" you have to wonder if he is simply trying to delay completion of your job (and the end of his pay). If this happens too often welding (or ROW or arch) inspectors can be replaced.

Welding inspectors. ASME B31.8 requires that welds on steel pipe be inspected by someone "qualified by training and/or experience to implement the applicable requirements and recommendations of the Code" (ASME B31.8 Section 802.2.5). Section 826 of ASME B31.8 goes on to list the types of inspections required:

- Piping whose MAWP is less than 20% of SMYS
  - Visual inspection on a random selection of welds by each welder.
- Piping whose MAWP is greater than or equal to 20% of SMYS
  - Visual inspection of each weld
  - Nondestructive testing (some combination of x-ray, ultrasonic, magnetic particle, or other comparable and acceptable method of nondestructive testing)
    - 10% of each welder's welds in Location Class 1
    - 15% of each welder's welds in Location Class 2
    - 40% of each welder's welds in Location Class 3

- 75% of each welder's welds in Location Class 4
- 100% of welds within compressor stations, major river crossings, major highway crossings, and all railroad crossings
- All tie in welds not subject to static pressure test.

Criteria for acceptable welds, radiographic processes, and welder qualifications are given in API 1104: Standard for Welding Pipelines and Related Facilities.

ROW/ditch inspectors. This role is often performed by welding inspectors, but sometimes it is a separate function. The ROW/ditch inspector is responsible for:

- Ensuring that all work (including spoil piles) remain within the boundary of the ROW. In some projects this can be a critical function, in others it is merely important—it all depends on your project's relationship with the surface land owner.
- Inspect the surface of the pipe coating as it is lowered into the ditch to insure that "holidays" (i.e., places where the coating is not effective or is missing) are properly repaired.
- Inspect the condition of the ditch bottom just before the pipe is lowered in.
- Verify that the pipe fits the ditch prior to backfilling.
- Inspect backfill material prior to use and observe backfill procedure to ensure no damage occurs to the coating in the process of backfilling.

QA/QC inspector. Some jobs require a QA/QC inspector, but they rarely specify either the inspector's qualifications, company affiliation (do they come from the operating company, the construction contractor, or a third party?), or scope of control. This seems to be a "feel good" term that adds cost without adding value.

Arch monitor. Any arch monitoring required by the arch report and/or ROW stipulations must be done by a licensed archeologist whose only job on the project is to be present any time active work on the project is being done in the vicinity of documented arch sites. It has to be clear to the entire project team that no work can be done anywhere close to an arch site. If the monitor is not there, then the rest of the project shouldn't be there either.

6.7.2 Trenching equipment

The types of equipment used for making ditch (other than rock ditch) is shown in Fig. 6.33 and summarized in Table 6.9. For most large jobs,

Figure 6.33 Trenching equipment.

Table 6.9 Trenching equipment summary

	Speed	Ditch bottom	Foreign lines	Cost/mile	Cost/hour
Wheel ditcher	Very good	Very good	Worst	Low	Highest
Track hoe	Good	Acceptable	Very good	Acceptable	Low
Bulldozer	Good	Worst	Very bad	Very low	Low
Rubber-tire hoe	Worst	Poor	Poor	Very high	Low

some combination of track hoe and ditching machine will be used. For small jobs, most digging tasks will be done with a track hoe. For very small jobs (such as a line from a wellhead to a separator) a rubber-tire hoe is most often used. On large jobs, the only appropriate task for a rubber-tire hoe is material movement (not including lowering in).

Wheel ditcher. There is a saying around pipeline jobs that you need to budget 1 hour of operator time and 9 hours of mechanic time per day for a wheel ditcher. That is an exaggeration, but sometimes it seems true. When this equipment is working, the speed is impressive, but there are a lot of moving parts under very high stresses. Wheel ditchers can go through moderate rock and the spoil pile is generally able to be used for backfill without sifting. Their use on large jobs almost always makes economic sense, but on medium-sized and smaller jobs they tend to not pay for themselves.

Track hoe. Track hoes are the workhorse of any pipeline job. They are used for digging all bell holes (see below), materials movement, exposing all foreign lines, dealing with all obstructions, and a lot of open-country ditch. Track hoe operation requires considerable skill, and a good operator is a joy to watch (while watching a poor hoe operator will certainly raise your frustration level). Sometimes you will see cross-country ditch that was dug by a track hoe that is nearly indistinguishable (until you look at the spoil pile) from a wheel-ditcher dug ditch. Other times you will see ditch that looks like it was cut with a ripping blade on a bulldozer. It is all operator. Track hoes can (eventually) dig through any rock that a rock saw (see below) can cut through, just slower. That is an important thing to keep in mind—a track hoe can eventually do nearly anything, but the increased cost of specialized equipment can be offset by the time required to do it with a track hoe.

Bulldozer. Most of the time bulldozers are only used on pipeline jobs to clear ROW and push big things out of the way. They also have the ability to rip a rudimentary trench very quickly, but really rough. Occasionally really rough is good enough, but not often.

Rubber-tire hoe. Virtually every task on the pipeline that a rubber-tire hoe might be able to do can be done better by a track hoe. As I was transitioning from project-management/construction management to just project management, my on-site construction manager called me 3 times in one day to report that they had hit unmarked foreign lines. I went to the work site and saw that they were using a rubber-tire hoe to uncover the lines in rocky soil. The hoe did not have enough power to dig the rock without the bucket jumping all over the site. These "unmarked" lines were actually properly marked but the equipment lacked proper control because it was underpowered. We finished the job with a track hoe and no more "unmarked" line popped up. After that experience I banned rubber-tire hoes from ever digging on a job that included track hoes.

Side boom. Side booms (Fig. 6.34) don't fit into this category, they don't dig, but they have major role in materials movement and are especially important in lowering a pipeline into the ditch. Side booms can only boom-in or boom out, and raise and lower the hook using the cable winch. Significantly less capability than a track hoe with about the same horsepower (one contract I have charges 80% as much per hour for a side

Figure 6.34 Side boom.

boom as a track hoe, and a lower qualified operator can sometimes further lower the package price).

The side boom in Fig. 6.34 was the victim of several rookie mistakes concurrently: (1) the hook was retracted all the way to the top of the boom; (2) the boom was all the way in; (3) with a 30 ft (9.1 m) joint of 12 in (300 DN) pipe (about 1600 lbm (726 kg)) that high in the air the center of gravity of the equipment and pipe was very high; and (4) the 20-degree slope put the high center of gravity behind the back of the tracks. The operator was embarrassed, but unhurt. The boom was scratched, but unharmed. The operator eventually turned into an excellent equipment operator and was always aware of his center of gravity after that.

6.7.3 Trenching rock

"Rock" is always a definition item in construction contracts. Typically it is defined as "ditch that cannot be dug by the means described earlier under 'normal excavation' ". The "cannot be dug" phrase is important. A track hoe can (eventually) dig anything, but it is often not the most economical method.

The contractor and the company man need to agree on what is rock prior to digging it with a rock saw (Fig. 6.35). After the ditch is made, you really do have to pay the rock premium in a bid contract whether a wheeled ditcher or a track hoe could have made the ditch in about the same time.

Figure 6.35 Rock saw.

For a time and materials job, other than the cost of dulling teeth digging in dirt with a rock saw, it is often less expensive to dig short distances between rock outcroppings than to skip over the nonrock with the rock saw.

The spoil pile associated with a rock saw is the very best padding material that you can use. The saw grinds rock into rock dust that is quite homogenous and has nearly zero small, sharp rocks that can cut pipe coating.

6.7.4 Welding

Welding steel pipe is a very specialized activity and you need to specify that the pipeline welders on your job have passed an API 1104 certification within a reasonable time (some companies specify that the certification must be within 6 months of the start of a project, others have other criteria).

All welds must be done in compliance with a welding procedure. The company needs to specify the welding procedure. Some companies (all of the "majors") have their own procedures, but smaller companies often rely on the contractor to provide a welding procedure. A procedure must specify values for all "essential variables," and then the procedure must be tested to certify that welds done to that procedure meet the required parameters. The essential variables are:

- Base metal (or metals) to be joined. This variable addresses the published specification (such as API 5L) and the grade (e.g., Grade B or X60). If the job is joining two different base metals, then that would be a different procedure.
- Welding process. If the job is going to be shielded metal arc welding (SMAW) then a procedure that was developed using tungsten inert gas (TIG) welding would not be applicable.
- Filler metal group (API 1104 Table 1).
- Joint design. Changing from "V" to "U" joint is a different procedure.
- Position (i.e., can you roll the pipe to keep the weld progress on top or is the pipe fixed).
- Wall thickness group of pipe (i.e., Group 1 → <3/16 in (4.8 mm); Group 2 → 3/16 in (4.8 mm) ≤ t ≤ 3/4 in (19 mm); Group 3 → >3/4 in (19 mm)). A change within the group is not an essential change.
- OD group (i.e., Group 1 → <2-3/8 in (50 DN); Group 2 → 2-3/8 in (50 DN) ≤ OD ≤ 12-3/4 in (300 DN); Group 3 → >12-3/4 in (300 DN)), again a change within a group is not an essential change.
- Time allowed between passes.
- Direction of welding. It is a different weld if the joint is vertical or horizontal.
- Shielding-gas flow rate (if used).
- Speed of travel.
- Preheat/postweld heat-treat requirements.

The weld design will specify a root pass to fill the gap between the two pipes, some number of filler passes (ranging from zero upward) to fill in the pipe thickness, and a cap.

Welding activities take place on: (1) the firing line: (2) tie-in welds; and (3) fabrication welds.

Figure 6.36 Firing-line welding.

Firing line. The firing line (Fig. 6.36) represents the bulk of welds on a job. Often one welder does the root pass and moves on to the next joint. Behind the root pass a number of welders will be doing all the filler passes (and sometimes the cap weld, other times a specialist does the cap weld). The firing line works like a competition and the welder that delays the firing line moving to the next set of joints is often ridiculed by his peers.

The welding procedure specifies an acceptable maximum time between passes and preheat/postweld heat-treat requirements. It is important (and often difficult) to meet these requirements on a firing line and a significant component of the welding inspector's job is to monitor these parameters.

Tie-in welds. The firing line will stop at all road crossings, all foreign pipe crossings, many fence crossings, all fabrications, and sometimes at random-seeming places. The tie-in crew comes along after the firing line to connect the firing-line sections. This work is done in a "bell hole" (see later), and tie-in welds typically cost 5–10 times as much as a firing-line weld. Typically you will have to do 100% x-ray on tie-in welds.

Fig. 6.37 shows an especially deep bell hole because it had to accommodate the barrel on the drip, in this case the top of the ditch is about 4 ft (1.2 m) above the top of the drip riser and the maximum depth of the hole is 18 ft (5.5 m) below grade (Fig. 6.38).

Fabrications. Fabrication welds are typically done on the surface (often with a tie-in weld in the ditch, see Fig. 6.30) to the same welding procedure as the rest of the job. Fabrications require 100% x-ray.

Figure 6.37 Tie-in welds.

Figure 6.38 Bell hole in a hot ditch.

6.7.5 Bell-hole issues

As I've said, tie-in welds are made in bell holes. Bell holes are a civil engineering structure that:

- If greater than 4 ft (1.2 m) deep must be inspected by an "OSHA Competent Person" (i.e., the unfortunate designation by the US Occupational Safety and Health Administration, millions of actually competent people do not have the "OSHA Competent Person" certificate and often resent being told they are not competent by someone with the certificate who actually is not competent to do much). This inspection must be done every 8 hours while people are working in the ditch, documented, and confirmed to meet OSHA standards. In other countries there are similar inspection requirements that go by different (and less judgmental) names.

- If greater than 20 ft (6.1 m) deep must be designed by a licensed professional engineer and inspected to comply with the design every 8 hours while people are in the ditch. This design:
  - Evaluates soil conditions to see if it meets the requirements for straight walls, requires a slope (and defines a minimum slope angle), requires terracing, or requires reinforcement.
  - If reinforcement is required, specifies the location and materials of the shoring.
- Must have a means of egress on both sides of the pipe.
  - A walkout where a person could walk upright without using their hands.
  - A properly secured ladder can work.
- The spoil pile must be at least 3 ft (1 m) back from the edge of the bell hole (rocks falling from the spoil pile to the bottom of the ditch are painful and can be dangerous).

A "hot ditch" is any ditch that contains a pipe that has ever had hydrocarbons in it. Work in a hot ditch has all of the requirements of any bell hole, plus:

- You must minimize the number of people in the ditch (it is often useful to designate a certain color of fire-resistant coveralls (FRC) for people authorized to be in the ditch).
- Assign a fire watch (who should be in the required FRC for access to the ditch). The fire watch cannot have any other duties and must remain at the site for at least 30 minutes after the end of hot work.
- Energy isolation requires conscious decisions:
  - If fire is acceptable then there needs to be enough room in the hole for the worker to move away from the fire periodically.
  - If fire is not acceptable then the procedure needs to specify positive energy isolation and/or nitrogen blanketing.
  - If nitrogen blanketing is required then continuous oxygen monitoring (with an audible alarm) must be located in the ditch and must be on at all times that workers are in the ditch.

6.7.6 Taping and holiday checking

Before lowering in, you have to protect the pipe around the welds and make sure that the field of the coating is intact. Applying coating to a hot weld results in poor protection so the weld needs to be allowed to cool to ambient temperature prior to applying a coating. Before coating, you must apply a primer. Most welds are protected with either shrink sleeves or tape. Shrink sleeves are applied to the pipe and then a propane torch is used to shrink it to the pipe. Shrink sleeves are very tough and quite reliable if properly applied. Tape is more difficult to install properly and is not nearly as tough as shrink sleeves, but often tape is the only option available and it is an acceptable alternative.

Once the welds are coated, you need to run a "holiday detector" over the entire pipe to find where the insulation between the pipe and the outside of the coating is damaged or missing. The holiday detector is quite sensitive and when it sounds off, the pipe must be primed all the way around the pipe and it is good practice to tape three to four tape widths either side of the holiday.

6.7.7 Backfill and cleanup

The first layer of dirt on the pipeline is called "padding" and it needs to be consistent with the pipe material. Steel pipe is very tough, and the coatings used today are also very tough so steel requires minimal padding and the padding can be quite coarse. RTP and HDPE are also quite tough and probably do not require any more special handling than steel. GRP on the other hand is glass and needs to be treated like it. Padding for GRP needs to be largely free of rocks and well sifted.

Cleanup is the end of the project. You have to return the disturbed ground to the original contour, often you have to install water bars to help combat erosion and prevent the ROW from becoming a new road, and you have to reseed the ROW. After the project team demobilizes and moves on to the next job you have to inspect (and document the inspections) the ROW periodically (monthly is common) until the ground reaches 70% revegetated. If the seed doesn't take or erosion becomes excessive and/or widespread then you will have to install some sort of erosion control and possibly reseed.

6.8 GATHERING-SYSTEM OPERATION

Operation of the gathering system goes on for years and even decades after the project is done and paid for. As part of the project I prefer to hand over some "manufacturers suggested practices," sometimes they are followed, sometimes they are not. Most often they form the basis for procedures and schedules that work for the operating team.

It is good to provide: (1) procedures, (2) schedules, and (3) qualifications.

Procedures. Most of the operation a gas gathering system is pretty self-explanatory, but there are some things that work better with a bit of explanation. For example, if you have to shut off a lateral in the middle of a flow line to successfully run a pig, it would be good to know before the operators stick a pig and decide to never run another. The procedures should usually include:

- Pig launching and receiving which includes the list of system valves that must be operated, their normal position, and their position during pigging. I also include the valve sequence that works best for launching pigs, because people mess this up. That procedure is something like:
  - Reset the pig signal.
  - Blow the barrel down and load a pig (ensure that the pig is engaged in the throat of the barrel).
  - Close the barrel and purge it for 30 seconds through the kicker line.
  - Shut the vent and fully open the kicker valve to equalize across the barrel-isolation valve.
  - Open the barrel-isolation valve fully.
  - Slowly begin to shut the side valve until the pig leaves the barrel.
  - Fully open the side valve, shut the kicker and barrel-isolation valve.
- Drip operation is usually just what the valves are and where does the liquid go.
- Gas measurement procedures include calibration and inspection requirements. They also include reporting requirements. Many producers have gotten in trouble with state and federal agencies and with partners for not filing required production reports. When I took over the gathering systems in the coal, I found (via a strongly worded letter from the state) that the system had failed to report production for 84 consecutive months and if we did not remedy this within 30 days we would be fined. Our accounting department was unable to accomplish the reporting requirement in less than 12 months due to "high priority" programming requirements in other areas. We developed a reporting capability in the local office to serve as a stopgap until central accounting could get to us. We were able to meet the required schedule. Seven years later the stopgap was still being used.

- Corrosion control processes. These procedures usually revolve around cathodic protection, and should include the location of test leads on foreign pipes and identifying the owner of the foreign pipe. If the system has impressed cathodic protection then you need to describe where the groundbed is and expected current requirements. If the system has corrosion coupons then they need to be identified and indicate if anything needs to be done with them prior to running pigs.

Schedules. It is useful to turn over your expectation of how often things like pigging should be done. It is also a very good idea for the operator to have thought about how to change a schedule once it is set up and most importantly, who should be notified when a procedure is being run. Compressor-station operators really like to be informed if there is a pig coming at them at high velocity with several thousand barrels of water in front of it.

Qualifications. It is very useful to describe what level of training is expected for operators, and what (if any) certifications certain jobs require (e.g., operating cathodic protection systems requires a NACE certification).

6.9 CONCLUSION

This chapter tried to emphasize "what" needs to happen more than "how" to make it happen. The "how" is almost always project specific and it is certainly company/jurisdiction specific. Some of the things I've done on gathering systems would be shocking to regulators or management in other locations or other companies.

Every welder I've ever met has preferred to get fire from a hot pipe instead of having nitrogen in their face. Our safety culture really hates that idea. Hopefully after reading this chapter you know that for every decision you have to make there are at least two sides of the issue that each have strong points in their favor. You often have to ask yourself "is the appearance of safety more important than actual safety when you simply cannot do both?" You can't have both fire and nitrogen at a weld site, the welders say "fire equals control," the safety guys say "fire bad." Who do you listen to? How do you make the job safe when you are being pulled in two mutually exclusive directions?

There are many decisions that have to be made to take an idea through all the processes until gas is flowing. I tried to touch on as many of them as possible in this chapter, but there are many many more that I've left out.

REFERENCES

ASME PCC22 (2015). Repair of Pressure Equipment and Piping. New York, NY: The American Society of Mechanical Engineers (ASME).

Fisher (2005). Control Valve Handbook (Fourth Edition). Fisher Controls International, LLC.

General Physics. (1988). "A Review of Energy Release Processes From the Failure of Pneumatic Pressure Vessels", M. Coleman, M. Cain, R. Danna, C. Harley, D. Sharp, General Physics Corporation Cape Canaveral Air Force Station, Florida. http://www.dtic.mil/dtic/tr/fulltext/u2/a207549.pdf.

Nebraska (2012). "An Experimental and Theoretical Study of Heat and Mass Transfer During the Venting of Gas From Pressure Vessels", W.S. Winters, G.H. Evans, S.F. Rice, R. Greif, US Department of Energy Publications, 2012. http://digitalcommons.unl.edu/cgi/viewcontent.cgi?article = 1132&context = usdoepub.

Webster (n.d.). Retrieved April 21, 2016, from http://www.merriam-webster.com/dictionary/biogenic (other dictionary references replace "biogenic" with the referenced term).

FURTHER READING

Wikipedia, 1. ASME Boiler and Pressure Vessel Code (BPVC). 2015. Retrieved October 19, 2016, from https://en.wikipedia.org/wiki/ASME_Boiler_and_Pressure_Vessel_Code_(BPVC).

NOMENCLATURE

Symbol	Name	fps units	SI units
a	Generalized acceleration term	ft/s^2	m/s^2
A	Flow area	ft^2	m^2
c_v	Flow coefficient	decimal	decimal
$d_{holeCenterline}$	Linear (not curved around the pipe) distance between the center of the inlet holes on a two-line piggable drip	in	mm
d_{nasa}	Distance required from a pneumatic test using the NASA Glenn Research Methodology	ft	
dP_{ratio}	Differential pressure ratio	decimal	decimal
DR	Dimension ratio	decimal	decimal
E	Young's modulus	psi	MPa
E_f	Longitudinal joint factor	decimal	decimal

(*Continued*)

(Continued)

Symbol	Name	fps units	SI units
f	Force applied to a surface	lbf	N
F_{design}	Factor for fluid choice (0.32 for natural gas, 0.50 for water)	decimal	decimal
F_f	Pipe design factor	decimal	decimal
ID_{barrel}	Inside diameter of drip barrel	in	mm
ID_{pipe}	Pipe inside diameter	in	mm
k	Adiabatic constant (ratio of specific heat)	decimal	decimal
L_{avail}	Available length in a drip barrel for slots	in	mm
L_{slots}	Minimum required slot length in a drip	in	mm
m	Generalized mass term. Refers to mass in general, not a particular, quantifiable mass	lbm	kg
MW	Molecular weight	lbm/lb-mole	gm/gm-mole
n_{slot}	Number of slots	integer	integer
OD_{pipe}	Pipe outside diameter	in	mm
P	Pressure	psia	kPaa
P_{atm}	Local atmospheric pressure	psia	kPaa
$P_{downstream}$	Pressure downstream of a reference point	psig	kPag
P_{lths}	Long-term hydraulic strength	psig	kPag
P_{mawp}	Maximum allowable working pressure	psig	kPag
P_{psig}	Pressure in gauge	psig	kPag
P_{smys}	Specified minimum yield stress	psig	kPag
$P_{upstream}$	Pressure upstream of a reference point	psig	kPag
r	Radius of a circle	in	mm
$r_{lateral}$	Distance from the junction of the centerlines of the two pipes in a two-line piggable drip to the inside of the drip barrel	in	mm
$\overline{R}$	Universal gas constant	1545.3 ft × lbf/R/lb-mole	8.314 J/K/mole
R_{air}	Specific gas constant for air ($\overline{R}/MW_{air}$)	53.355 ft × lbf/R/lbm	287.068 J/K/kg
R_{gas}	Specific gas constant	R_{air}/SG_{gas}	R_{air}/SG_{gas}
S	Distance	ft	m

(Continued)

(Continued)

Symbol	Name	fps units	SI units
S_0	Initial height above a reference plane	ft	m
S_v	Distance in the vertical direction	ft	m
$S_{yFalling}$	Distance that a projectile travels from its maximum height to the ground	ft	m
$S_{yRising}$	Distance that a projectile travels from its starting point to its maximum height	ft	m
SG	Specific gravity	decimal	decimal
t	Thickness	in	mm
t_{barrel}	Thickness of the barrel in a drip	in	mm
t_{acc}	Time that a force is applied to a projectile resulting in acceleration	s	s
t_{corr}	Corrosion allowance	in	mm
t_{fall}	Time to fall distance S_0	s	s
t_{min}	Minimum wall thickness	in	mm
$t_{reinforced}$	Thickness of the inner flow layer of RTP	in	mm
T	Temperature	R	K
T_f	Temperature derate factor	decimal	decimal
v	Velocity	ft/s	m/s
v_0	Velocity at the start of an analysis	ft/s	m/s
v_{sonic}	Speed of sound in a referenced medium	ft/s	ft/s
V_{system}	Volume of the system	ft^3	m^3
W_{gas}	Work done on or by a gas	ft-lbf	N-m
W_{slot}	Width of slots in a drip	in	mm
$y_{deflection}$	Deflection of skillet blind	in	mm
Z	Compressibility	decimal	decimal
ν	Poisson's ratio	decimal	decimal
Subscripts			
1	Upstream value		
2	Downstream value		
avg	Average		
disch	Condition at the discharge of a pump or compressor		
gas	Gas		
liq	Liquid		
std	Standard condition		
suct	Condition at the suction of a pump or compressor		
tbg	Tubing		

EXERCISES

1. A well needs a gas gathering line to the compressor station. Using the data in Table 6.10 select a pipe schedule and grade. How does this compare to the schedule and grade selected in Exercise #1 in Chapter 5: Well-Site Equipment?
2. What is the minimum barrel diameter for a two-line piggable drip accommodating two 10 inch (250 DN) gathering lines?

Table 6.10 Exercise #1 data

	fps	Both	SI
MAWP	600 psig		4137 kPag
Length	13 miles		20.9 km
Population density		7 houses over entire length	
Design temperature	100°F		37.8°C
Nominal pipe	10 in		250 DN
Type test		Pneumatic	
Test pressure		150% of MAWP	
Corrosion allowance	3/16 in		4.8 mm
Normal op pressure	150 psig		1034 kPag

Pipe options	Sched 20	Sched 40 (std)	Sched 60 (X)	Sched 80
OD	10.750 in (273 mm)	10.750 in (273 mm)	10.750 in (273 mm)	10.750 in (273 mm)
ID	10.250 in (260 mm)	10.020 in (255 mm)	9.750 in (248 mm)	9.562 in (243 mm)
t	0.250 in (6.35 mm)	0.365 in (9.27 mm)	0.500 in (12.70 mm)	0.594 in (15.09 mm)

	SMYS (psig (MPag))	Maximum temperature (°F (°C)) before derate
API 5L Gr B	35,000 (241)	400 (204)
API 5L X42	42,000 (290)	400 (204)
API 5L X60	60,000 (414)	400 (204)
API 5L X70	70,000 (483)	400 (204)

3. According to the NASA Glenn Research Methodology, what is the closest allowed point of approach to the line in #1 while under a 150% test? What is the closest allowed point of approach for the line while operating at normal operating pressure?
4. A production company defines MAWP as "two-thirds of the lowest maximum pressure attained during a static test." Using the data in Table 6.11 and the profile in Fig. 6.39 find:
   - Maximum MAWP that can be achieved using a hydrostatic test within the company's limitations

Table 6.11 Exercise #4 data

	fps	SI
Pipe size	16 in std	400 DN std
Pipe ID	15.250 in	387.4 mm
Length	36,100 ft	11,003 m
Target MAWP	600 psig	4.14 MPag

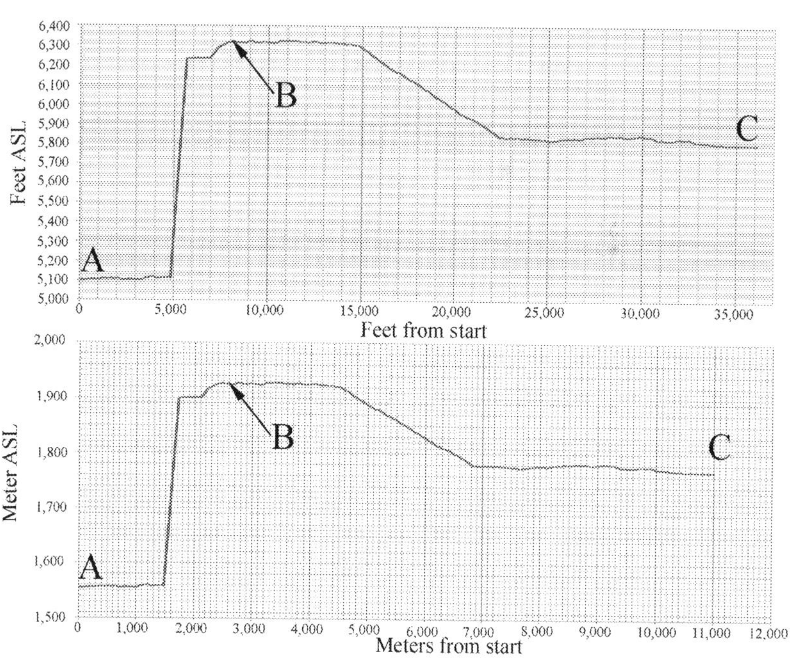

Figure 6.39 Exercise #4.

- Which of the three marked points would provide the monitoring station that resulted in the highest MAWP

5. What are the risks of doing hot work in a confined space? Which of these risks is reduced by applying a nitrogen purge to the line? What additional risks are added by applying a nitrogen purge?
6. What are the surface indications that a gathering line is past due for pigging?

CHAPTER SEVEN

Water Collection and Disposal

Contents

Managing produced water is the subject of thousands of pages of regulations and millions of pages of legal decisions. The intent of this chapter is to give you the feel for the magnitude of the subject, not prepare you to deal with its complexities—get help from environmental, regulatory/legal, and engineering professionals early in the process.

Regardless of product prices we are finding that wells remain economical with much higher lease operating expense than in the past. A big part of that increased operating expense is lifting and disposing of water. We talked about lifting costs in Chapter 3, Well Dynamics, but we stopped when the liquids got to the surface. Getting a handle on water quantity and actual water-handling costs is a real challenge because water is a waste product. We rarely, if ever, have any enthusiasm for spending money on waste management. This reluctance often leads to not measuring the water at the well-sites at all, not reconciling run tickets on water hauled, and building slip-shod water facilities. The regulatory

Practical Onshore Gas Field Engineering.
DOI: http://dx.doi.org/10.1016/B978-0-12-813022-3.00007-9

environment around the world is forcing a change in this mind set, but the existing water infrastructure inhibits the value of high-quality contemporary facilities.

According to the US Department of Energy (DOE) non-CBM (coalbed methane) onshore produced water production is 14 MMbbl/day (2230 ML/day). Independent estimates place CBM water production at 1 MMbbl/day (159 ML/day). Wild guesses put water from the shale plays at another 6 MMbbl/day (954 ML/day). As close as anyone can tell, industry explicit and implicit costs of lifting and disposing of produced water is on the order of $10 billion USD/year in the United States. This leads us to the conclusion that disposal costs average $0.80 USD/bbl ($5 USD/m^3). This is a misleading average. There are systems in the Antrim Shale in Michigan (Fig. 7.1) where water is pumped into a water gathering system and transported to water disposal wells that are on a vacuum—no pumps, no tanks, and piping that has been in the ground for 20 years lead to a disposal cost on the order of $0.005/bbl ($0.031 USD/m^3). On the other end of the spectrum, if you have to truck water from Pennsylvania to Ohio costs can be $15 USD/bbl ($94 USD/m^3).

All of these numbers are highly suspect. Look at the Antrim Shale example, most of the well-site measurement is turbine meters on the separator dump line (remember from Chapter 5: Well-Site Equipment

Figure 7.1 Antrim shale water disposal facility.

that this configuration results in the turbine meter spending most of its cycle time spinning faster than the device can measure) and the disposal wells (Fig. 7.1) do not have measurement at all. This all comes together to illuminate that any water numbers coming from this field are simply made up. This is a vivid example of the data quality, in systems with "high-quality" measurement at the source and "high-quality" measurement at the disposal site, we try hard not to ever do a system balance. I've done several of these balances and I am yet to be within $\pm 50\%$, and it is quite random which number will be higher—in one system I found that I was injecting 5 times as much water as I was producing, in another I was producing nearly twice as much water as I was disposing of. In both cases the numbers were simply nonsense.

7.1 WATER QUALITY

There are many measures of "water quality" and the actual quality of water is a market basket of parameters. Table 7.1 provides a reasonably typical water analysis for a mixture of CBM and shallow tight-gas wells.

It is important that the cations and anions balance on a milliequivalent (mEq/L) basis, but the ions will never balance on an mg/L basis (because of the items which have charges other than ± 1). The calculated total dissolved solids (TDS) number at the top of report is the most reliable number on the report. If you send 20 samples of the same water to 20 different laboratories, all of them will match the calculated TDS closely, but individual amounts of any specific ion will vary considerably due to the assumptions as to what a given peak on an instrument actually means (e.g., a bicarbonate spike could be $NaHCO_3$ or $KHCO_3$ or it could even be $Mg(HCO_3)_2$ each of which would give you a different TDS number for the cation and anion and the instrument cannot differentiate).

"Alkalinity" is the name given to the quantitative capacity of an aqueous solution to neutralize an acid (Wikipedia, 4). The parameters that constitute alkalinity in produced water include bicarbonate ions, carbonate ions, and hydroxide ions. Care must be taken in reviewing bicarbonate (HCO_3^-), carbonate (CO_3^{2-}), and hydroxide ion (OH^-) concentrations in water quality data as these parameters as sometimes reported as equivalent calcium carbonate ($CaCO_3$). To determine the TDS by summing the ions, the concentrations must be converted from

Table 7.1 Example water analysis

	Result	Units	Result	Units
pH	7.83	s.u.		
Conductivity at 25°C	34,200	μmhos/cm		
Total dissolved solids at 180°C	18,700	mg/L		
Total dissolved solids (calc.)	18,450	mg/L		
SAR	177.4	ratio		
Total alkalinity as $CaCO_3$	880	mg/L		
Total hardness as $CaCO_3$	280	mg/L		
Anions				
Bicarbonate as HCO_3^-	1,074.00	mg/L	17.60	mEq/L
Carbonate as CO_3^{-2}	<0.1	mg/L	0.00	mEq/L
Hydroxide as OH^-	<0.1	mg/L	0.00	mEq/L
Nitrate nitrogen as NO_3^-	1.20	mg/L	0.02	mEq/L
Nitrite nitrogen as NO_2^-	0.01	mg/L	0.00	mEq/L
Chloride (Cl^-)	10,560.00	mg/L	297.90	mEq/L
Fluoride (F^-)	1.55	mg/L	0.08	mEq/L
Phosphate (PO_4^{-3})	7.30	mg/L	0.23	mEq/L
Sulfate (SO_4^{-2})	<0.10	mg/L	0.00	mEq/L
Total anions	11,644.06	mg/L	315.83	mEq/L
Cations				
Iron (Fe^{+2})	0.16	mg/L	0.01	mEq/L
Calcium (Ca^{+2})	62.40	mg/L	3.11	mEq/L
Magnesium (Mg^{+2})	36.00	mg/L	2.96	mEq/L
Potassium (K^+)	18.80	mg/L	0.48	mEq/L
Sodium (Na^+)	7,110.00	mg/L	309.29	mEq/L
Total cations	7,227.36	mg/L	315.85	mEq/L
Total TDS	18,871.42	mg/L	631.68	mEq/L
Cation/anion difference			0.00%	

equivalent $CaCO_3$ to concentration as the ion. The conversions are as follows:

- Bicarbonate as HCO_3^- → Bicarbonate as $CaCO_3$ × 1.22
- Carbonate as CO_3^{2-} → Carbonate as $CaCO_3$ × 0.6
- Hydroxide as OH^- → Hydroxide as $CaCO_3$ × 0.34

Care must also be taken in interpreting the TDS at 180°C result for waters that are high in bicarbonate (HCO_3^-) ions. The TDS is determined by placing a measured sample into an oven at 180°C, boiling of the water, and weighing the residual salt. A temperature of 180°C is sufficient to cause the bicarbonate ion to decompose as per the reaction in Eq. (7.1):

$$(2HCO_3^-)_{aq} \rightarrow (CO_3^-)_{aq} + (CO_2)_{gas} + (H_2O)_{liq} \tag{7.1}$$

As the CO_2 and water produced as a result of Eq. (7.1) are removed from the sample as a result of the heating process, the TDS at 180°C test results in an artificially low measure of the TDS in the initial sample. For this reason, summing the ions is the preferred method for determining TDS in waters high in bicarbonate ions.

"Hardness" is the amount of dissolved calcium and magnesium in the water. "SAR" stands for "sodium absorption ratio" and is discussed in Section 7.3.3.7.

7.1.1 Water quality parameters

The four parameters that are nearly always relevant are: (1) total settle-able solids, (2) total suspended solids (TSS or "turbidity"), (3) TDS, and (4) pH.

Total settle-able solids: Many of the things that are found in water are there because of the velocity of the flowing water stream. Moving water has the ability to do work on its surroundings and that work is often reflected in the solids that it picks up from its surroundings and transports to another location. This is visible in floodwater where the settle-able solids can be as large as a house, a truck, or an entire tree. Once the flood subsides and the velocity of the water drops we see the houses and trees settle to the bottom of the river. Less dramatic is the mud and silt that are visible in many rivers flowing at less than flood stage. These solids will separate via gravity and settle to the bottom of the flow stream. A sample bottle filled with water containing settle-able solids would be expected to show a distinct solid strata in the bottom of the bottle within a few days. In Fig. 7.2 these solids would be in the $>100\ \mu m$ category.

TSS: From the Wikipedia definition of TDS, we can fall into the definition of TSS as "(TSS) cannot pass through a sieve of two micrometers and yet are indefinitely suspended in solution" (Wikipedia, 1). The units of TSS are "FTU" or formazin turbidity unit (the ISO refers to FTU as "formazin nephelometric units" they use the letters "FNU" to refer to it). Formazine is a heterocyclic polymer produced by reacting hexamethylenetetramine with hydraxine sulfate that is essentially not soluble in water (Wikipedia, 2). Mixing a known quantity of the chemicals into a known quantity of high-quality water at a known temperature will result in a "cloudiness" to the water that can be measured. Turbidity of a sample is a comparison of the light-refracting characteristics of the sample to the characteristics of a standard Formazine mixture. For example,

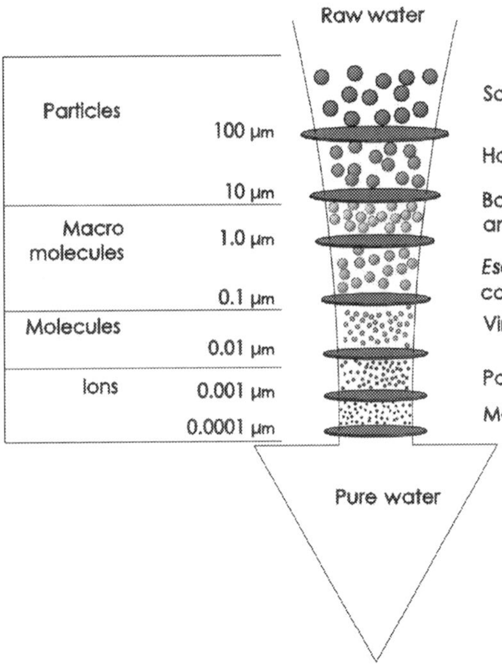

Figure 7.2 Contaminant size.

a suspension of 1.25 mg/L hydrazine sulfate and 12.5 mg/L hexamethylenetetramine in water has a turbidity of 1 FTU.

Unfortunately, since every possible dissolved solid in a water sample will refract light differently there is no way to get from FTU to a mass of suspended solids in a sample and since the suspended solids will stay suspended for an indefinite period of time (by definition) there is no good way to quantify the TSS of a sample.

Ions: When salts dissolve in water, the ionic bonds disassociate into "ions" or "charged particles." Positively charged ions are called "cations" (e.g., Na^+) and negatively charged ions are called "anions" (e.g., Cl^-). Ion exchangers are devices that replace undesirable ions with desirable ions, hold the undesirable ions until they can be dumped somewhere convenient during a "regeneration" step. Ion-exchange media is designed for specific cations and anions, and selection should be made in consultation with the manufacturer.

TDS: "TDS is a measure of the combined content of all inorganic and organic substances contained in a liquid in molecular, ionized, or micro-granular (colloidal sol) suspended form. Generally the operational definition is that the solids must be small enough to survive filtration through a filter with two-micrometer (nominal size, or smaller) pores" (Wikipedia, 1). In Fig. 7.2, bacteria, flour, blood, yeast, and clay would either be dissolved or suspended and everything smaller would be dissolved. Just like the discussion of the practical impossibility of mechanically separating two gases from each other, dissolved solids are an integral part of the water matrix and mechanically separating the TDS from the water, while much easier than separating two gases is still very difficult.

TDS is usually presented in units of "mg/L" which is equivalent to "parts per million by mass" since 1 L of pure water has a mass of 1 kg. Increasingly, TDS is presented as "electrical conductivity" in units of micro-Siemens/centimeter (μS/cm). The conversion is 1 μS/cm = 0.64 mg/L.

The two terms that tend to be the most widely used in talking about water are TDS and specific gravity (SG). You must know the makeup of the dissolved solids to convert one from the other, but you can approximate SG from TDS by assuming that all of the solids are NaCl (solid density is 135.157 lbm/ft^3 (2165 kg/m^3)) and using Eq. (7.2). This is just an approximation and if you (for example) replaced the NaCl with Fe_2O_3 (density 327.247 lbm/ft^3 (5242 kg/m^3)), the contribution of the dissolved solids would be twice as high (which would make seawater SG 1.028 instead of the 1.019 as given in Table 7.2). This would change the

Table 7.2 Example of TDS content

Water source	TDS (mg/L)	Approx. SG (using density of NaCl for TDS)
Rainfall	10	1.000005
Pristine freshwater lakes and rivers	10–200	1.000005–1.000108
Amazon River	40	1.000022
State water project deliveries	275	1.000148
Lakes impacted by road salt	400	1.000216
Agricultural impact on sensitive crops	500	1.000269
Colorado River	700	1.000377
Average seawater	35,000	1.018834
Oil & Gas brine production	>50,000	>1.026905
Groundwater	100 to >50,000	1.000054 to >1.026905

hydrostatic gradient at 10,000 ft (3048 m) from 4417 psig (30.5 MPa) for an NaCl approximation to 4458 psig (30.7 MPag) for an Fe_2O_3 approximation. Is that material? Sometimes. Not usually. If you feel that it is material then you can look at your actual water analysis and determine a density for the mix of components that exists in your water and replace it for the ρ_{nacl} in Eq. (7.2).

$$SG_{approx} = \frac{\left(1\,L - \frac{W_{TDS}}{\rho_{NaCl}}\right) \cdot \rho_{water} + W_{TDS}}{1\,L \cdot \rho_{water}} \tag{7.2}$$

Table 7.2 provides the water content of several common water sources.

The glass of water in Fig. 7.3 could be any from any of the sources in Table 7.2 since dissolved solids are simply not visible (it is a glass of seawater).

pH: In chemistry, pH is a logarithmic scale used to specify the acidity or basicity of an aqueous solution measured in units of moles per liter of hydrogen ions (Wikipedia, 3).

$$\begin{aligned} pH &= -\log_{10}[H^+] \\ pOH &= -\log_{10}[OH^-] \end{aligned} \tag{7.3}$$

Figure 7.3 Glass of water.

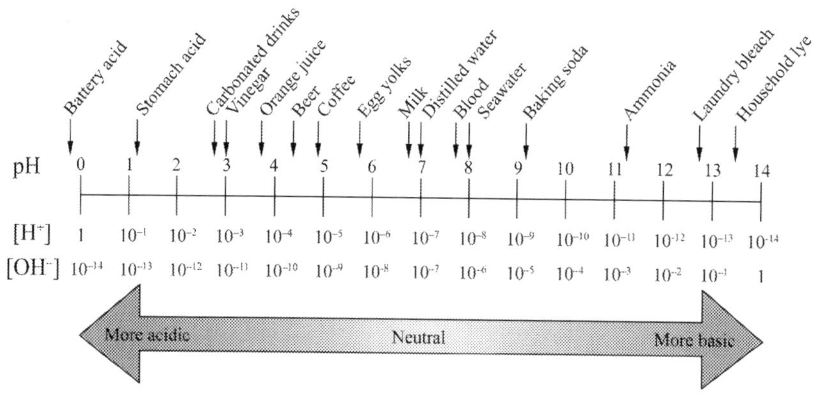

Figure 7.4 pH scale.

Each integer pH value is 10 times larger than the previous value (Fig. 7.4). Values less than 7.0 are acidic and values greater than 7.0 are basic. pH is a determining value for several kinds of corrosion and for the usefulness of a water stream in irrigation or livestock watering, but it is rarely reported in water analysis. It is useful to always make sure that field operators have pH test kits (a simple litmus paper test with a comparison scale) and they are trained to get a pH as an early step in any trouble-shooting exercise that involved water.

7.1.1.1 Summary

"Water quality" only has meaning in context. Without knowing where in the world the water is, regulatory environment at that location, and expected use, none of the parameters can have a subjective "good" or "bad" label affixed. Water with a pH of 8.8 would not be allowed into the Mississippi River, but would be fine in the Dawson River in Queensland (Table 7.3). On the other hand, water with 400 mg/L TDS would be fine for the Mississippi but not for the Dawson River.

7.1.2 Treatment

Much of the total volume of produced water needs and/or could benefit from some amount of treatment. Even something like an evaporation pond, the solids that can be removed by a filter will then not be available to form sludge that will shorten the effective live of the pond. In deep well injection, you are filtering the water whether you mean to or not—would you rather have the solids buildup (generally called "filter cake")

Table 7.3 Sample water quality

	US EPA Safe Drinking Water Act limits	Dawson River, Queensland, Australia, limits	San Juan River actual	San Juan Basin (SJB) coal typical
pH	6.8–8.5	6.5–9.0	8.5	7.8
Dissolved oxygen	No limit set	No limit set	11	0
Turbidity (FTU)	5	50 ppm TSS[a]	3.5	3
TDS (mg/L)	500	220	250	10,000
Oil & Grease (mg/L)	ND	0	ND	50

ND, none detectable.
[a]The Queensland regulations do not provide a methodology to demonstrate compliance with this limit.

collect on a disposable filter on the surface or would you prefer to have it build up on the entrance to the disposal formation?

7.1.2.1 Type of filtration

Filter media comes in many forms and configurations (Fig. 7.5). Sometimes the flow through the filter element is inside out, sometimes it is outside in. Filter media can be a bag over a strainer, a stand-alone filter element, and/or a combination. In general we think of filters as either "flow through" or "cross flow."

Flow-through filters (Fig. 7.6) are the thing that most people think of when they think of "filter." All of the water and contaminants go into the filter media and some stuff goes through, other stuff is rejected (Fig. 7.5). The rejected material builds up on the surface as filter cake and eventually requires that the filter be replaced or cleaned.

Cross-flow filters (Fig. 7.7) rely on an outlet restriction to hold pressure against a semipermeable membrane that will allow water and little else through. Cross-flow filters result in a waste stream that can be high in TDS and a permeate stream that is quite pure. Some portion of the (microscopic) rejected solids will accumulate on the surface of the membrane and they must be periodically removed through a "clean in place" process. While cross-flow filters will reject large particles, allowing large particles to arrive at the cross-flow filter is a poor operational strategy—you want to use lower pressure drop, lower cost filters to get rid of all of

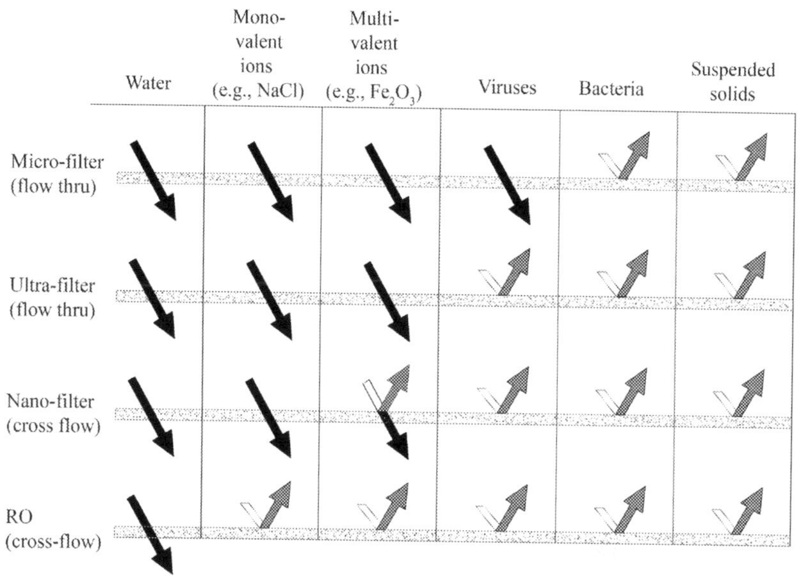

Figure 7.5 Removing water contaminants.

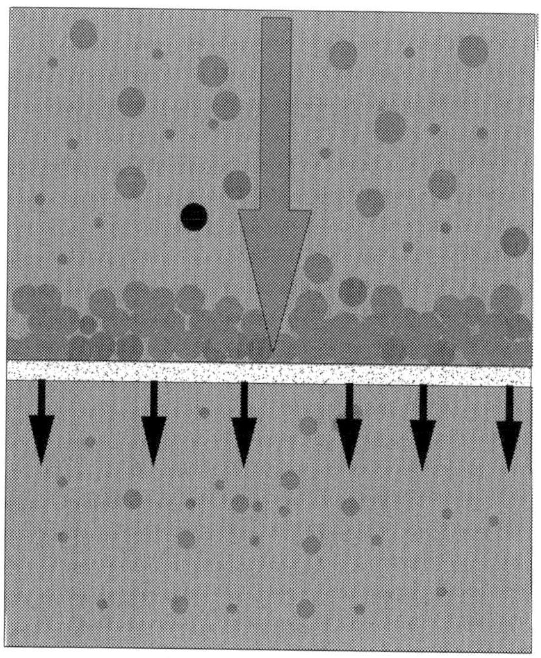

Figure 7.6 Flow-through filter.

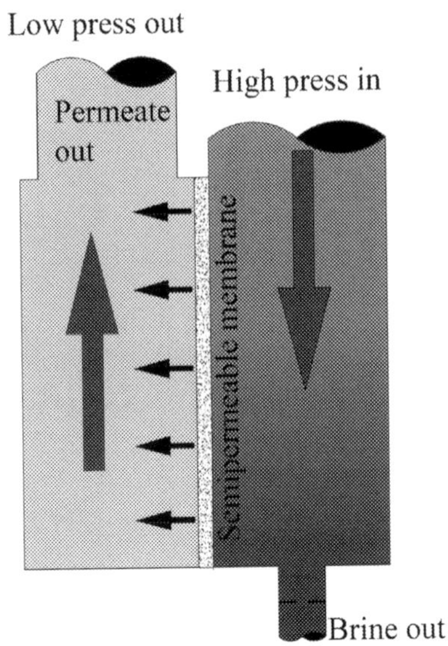

Figure 7.7 Cross-flow filter.

the larger particles that you can possibly remove prior to exposing the cross-flow media to the possibility of developing filter cake.

7.1.2.2 Filtration

We tend to classify filters by the size particle that they fail to reject. The classifications are generally: (1) micro-filter, (2) ultra-filter, (3) nano-filter, and (4) reverse osmosis (RO).

Micro-filter: Micro-filters (Fig. 7.6) are flow-through elements that come in sizes ranging from 100 μm down to about 10 μm (the term "μm" is pronounced "micrometer" and represents 10^{-6} m and is used in both fps and SI systems). Since the definition of TDS limits it to particles smaller than 2 μm, micro-filters won't lower TDS at all, but some of the tighter micro-filters can remove much of the TSS.

Ultra-filter: Ultra-filter units go down to 0.1 μm and are also flow-through elements. Putting even a 1.0 μm filter as the first element in a filtration strategy has a very short life since larger particles will quickly clog an ultra-filter. A 0.1 μm ultra-filter will remove virtually all biological elements like viruses and bacteria, but they are unable to reduce TDS in many cases.

Nano-filter: Nano-filter units are cross-flow filters (Fig. 7.7) that can start to make a dent a stream's TDS. They are able to filter down to about 0.001 μm (1 nm).

7.1.2.3 Reverse osmosis

"Osmosis" is defined as "movement of a solvent (as water) through a semipermeable membrane (as of a living cell) into a solution of higher solute concentration that tends to equalize the concentrations of solute on the two sides of the membrane" (Webster). In other words, nature will try to dilute a stronger concentration to drive both sides of the membrane toward the same energy. Freshwater will tend to flow through a semipermeable membrane toward saltwater. The driving force of this tendency to reduce purity is called "osmotic pressure."

The common term "RO" is the application of a dP across a semipermeable membrane that is greater than the osmotic pressure. An RO system will often reprocess the brine to concentrate it in stages (Fig. 7.8). In this example, the incoming stream has 100 L of 10 g/L water, indicating that the TDS is 1 kg. After the first stage, the permeate has 60 L of 900 mg/L water (54 g of the 1 kg of TDS). The brine stream consists of the other 40 L of 23.65 g/L water. The second stage permeate is 20 more L of 900 mg/L water (18 mg more of the TDS in the permeate). At this point the brine pressure is too low to overcome osmotic pressure and a pump has to be added. The final stage gets 10 L of 900 mg/L (9 g additional solids for a total of 81 g of solids in 90 L or 900 mg/L). The final brine stream is 10 L of 91,900 mg/L water containing the missing 919 g of solids, the entire 1.0 kg of TDS has been accounted for.

RO systems have been tried in Oil & Gas many times, with quite poor results overall. The failure in the five systems that I've tried to deploy has always been due to inadequate prefiltration. The most likely culprit has been trace hydrocarbons in the water which rapidly reduce the effectiveness of the RO unit and can actually damage the membranes.

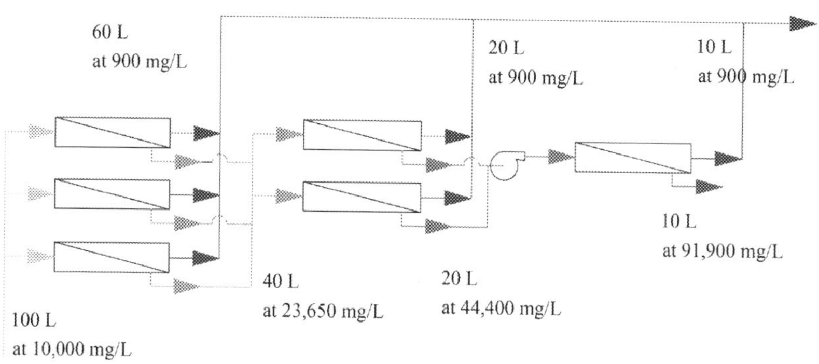

Figure 7.8 Recovery factors of RO unit.

When visiting a successful RO facility in Queensland, AU, I was able to see what it really takes to deploy a commercial RO system on produced water. The system was quite robust and contained:

- The water into the RO system was several days past its production date. The delay from the well-site to the treatment facility was spent in open ponds to facilitate any dissolved methane evolving out of the water. This field does not produce liquid hydrocarbons, but the ponds are still frequently inspected for any hydrocarbon sheen.
- The treatment started a simple basket strainer that would pick up leaves and bugs.
- Actual filtration started with three stages of flow-through filters ending with 1 μm flow-through filter element. All of the filters were duplex units that allowed the operator to shift to the backup filter when changing a clogged filter.
- The next step was the addition of a "coagulant" that added chemicals to attach to solids to form "micro-flocs" which were not visible to the naked eye.
- The water spent 20–45 minutes in a "coagulation/flocculation basin" where the micro-flocs were encouraged to aggregate into large clumps that are visible to the naked eye via paddles gently stirring the water.
- Out of the coagulation/flocculation basin the water entered a "sedimentation basin" where the water resided for 1–4 hours to allow the clumps to settle out of the water.
- As the water left the sedimentation basin it received a small dose of coagulant to clump with any iron or manganese that might have entered the water from the tank material.
- The next preprocess was a multimedia filter that had an anthracite layer to remove "large" particles, filter sand to remove smaller particles, and garnet sand and a gravel support layer to remove very fine particles. The multimedia filter was set up to allow back flushing dump solids out of the system.
- After the multimedia filter, the water entered into an ion-exchange softening process to exchange monovalent cations (e.g., Na^+) in the flow with a divalent cation (Ca^{2+}, Mg^{2+}, Sr^{2+}, or Ba^{2+}). This softening process was periodically regenerated with sequential doses of chemicals.
- Finally, the system had a two-stage nano-filter to ensure that nothing entered the RO unit that could possibly be filtered out prior to addressing the TDS in the RO (the primary concern at this point is silica).

This system processes upward of 100,000 bbl/day (18,000 m^3/day) of produced water into about 95,000 bbl/day (15,100 m^3/day) potable water for irrigation. For reference, to get 4 hours of residence in the sedimentation basin it must hold nearly 17,000 bbl (2700 m^3) or a cube 46 ft (14 m) on a side. The system is effective, but it is far from inexpensive in terms of capital, operating costs, or manpower.

Brine disposal. Any treatment process results in concentrating the solids into progressively smaller volumes. Disposing of this brine is a problem that has several solutions:

- Deep well injection. This is the most common method of brine disposal by a wide margin. The biggest concern is that solids content upward of 200,000 mg/L (approximate SG 1.11) tends to plug the entry to the disposal reservoir too quickly. This is often addressed by blending the brine with other produced water to attempt to flush the near-wellbore portion of the disposal reservoir, which works acceptably well.
- Solid salt to land farm. This is the common method used in most evaporation ponds. As long as the naturally occurring radioactive material (NORM) is in the acceptable range and there is minimal oil and grease the solids are simply "dirt" that do not require special processing.
- Salt recovery. The solids in the TDS can have commercial value as pure minerals. This opportunity is limited to those operations with high content of valuable minerals; if the primary component of the TDS is NaCl it is unlikely that the resulting minerals will be worth the cost of extraction. One of the major users of NaCl is the chlor-alkali industry that uses the NaCl to produce chlorine gas, sodium hydroxide, and hydrogen gas. The chlor-alkali industry places a high premium on the purity of NaCl, and attaining the purity levels required by the chlor-alkali industry may prove challenging when using produced water as the feed material to the crystallization process.
- Drilling fluid weight control. It is common for brine to have high SG (e.g., a brine with 200,000 mg/L would exhibit a pressure gradient of 0.48 psi/ft (10.9 kPa/m) which is about 12% higher than pure water). Sometimes there will be a market for the brine for pressure control in drilling operations. This use tends to be intermittent and requires immediate availability so it is generally not appropriate as a primary disposal method. Historically, drillers have used potassium-based salts

(primarily KCl) for weight control. These salts do not tend to cause hydrophilic clay's to swell. Sodium-based salts do tend to cause swelling clays to swell and this option is only appropriate in fields without hydrophilic clay.
- Send to marine waters. Brine can be sent to the ocean, permits are required, and pipeline capital cost limits this option to fields close to the coast. While technically feasible, this option may face considerable challenges so far as community acceptance is concerned.

7.1.2.4 Thermal processing (distilling)

You can achieve RO-quality water in a single step (with minimal prefiltering) by a distillation process. Distilling high-TDS water into potable water requires adding enough energy to raise the temperature of the water to the boiling point at the system pressure, and then boiling the water. Heating the water requires about 1 BTU/lbm/R (4.2 kJ/kg/K), but once you reach the boiling point you need to add about 970 BTU/lbm (2260 kJ/kg) to get it to boil. This is a lot of energy even if you are careful about using the waste heat for preheating, keeping the process at as low a pressure as possible, and recovering the latent heat in the condenser. The boiling water has a high scaling potential so getting the flow rate just a little bit wrong can result in significantly increasing the amount of energy required to distill a quantity of water.

The economics of lost revenue resulting from consumption of fuel gas (that could otherwise be sold as a product of the operations) that is required to provide the energy for the distillery then this option is rarely more economic than RO; however the use of thermal processing options allows recovery of more treated water from RO reject streams. The amount of treated water that can be recovered in RO systems is limited by the osmotic pressure of the concentrate stream. As such if a high degree of recovery is required it is preferable to recover as much as possible using RO and then further process the reject from the RO system using thermal treatment.

7.2 WATER GATHERING

There is no chapter of the ASME B31 series of standards that addresses water gathering. People sometimes use ASME B31.4: Pipeline

Transportation Systems for Liquids and Slurries because it is a liquid, but the exclusions section seems to exclude its use in water systems (the language is a bit ambiguous). Other people look at the amount of gas that gets transported with the water and apply ASME B31.8: Gas Transmission and Distribution Piping Systems, but again the exclusions section seems to exclude water systems, but it is not clear. Neither choice is wrong, but it is a good idea to document your choice and the reasons that you chose that standard.

Water gathering systems rarely (if ever) run full of water, so the pipe must be rated to withstand:

- the hydrostatic head of the sum of all uphill sections (no credit for downhill sections);
- added pressure to overcome friction losses;
- added pressure to overcome disposal site inlet equipment pressure drops.

Pressure ratings lower than ASME B16.5 Class 300 (600 psig (4100 kPag) MAWP) are almost never the best choice. Steel lines tend to have serious top-of-pipe corrosion issues (there is almost always condensation on the pipe walls that are not submerged) so any line running less than about 13,000 bbl/day (2000 m^3/day) (which should give you about 6 psi/mi (26 kPa/km) pressure drop and a velocity around 2.4 ft/s (0.74 m/s) in 8 in FlexSteel) should be built out of RTP. Trunks requiring a higher capacity should be steel with high-density polyethylene (HDPE) liners installed after the tank is welded together.

7.2.1 Degasifying

If a line is liquid-full then the only elevations that matter are the starting point and the ending point of the entire line.

Lines rarely run full as we saw in Eq. (4.6), so looking at Fig. 7.9, we need to consider the two lower cases. The middle case is fairly rare, but it happens when the pumping energy is inadequate to overtop a hill. At that point the pump stalls out and can't move liquid. Letting the pump sit idle for a time will not fix a gas lock since an idle line will tend to accumulate even more gas at the top of hills and lower the initial liquid level in the system, but not the height of the hill.

The bottom case in Fig. 7.9 is the most common. There is enough pumping power to overtop the hill, but not enough flow to fill the line and the flow runs along the bottom of the pipe over the crest which

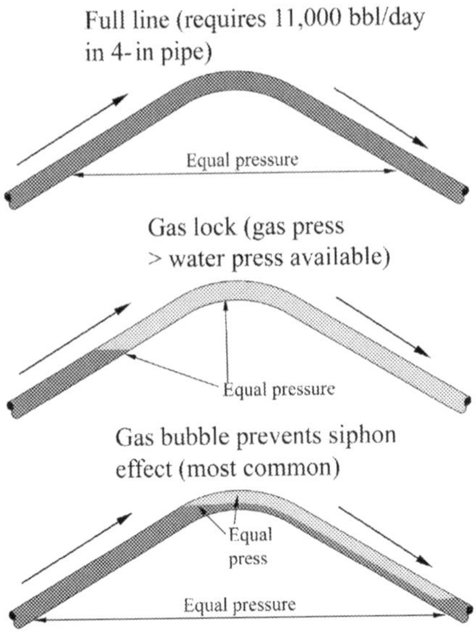

Figure 7.9 Gas in piping.

cannot create a siphon to pull the liquid up the hill. At some point in the line the hydrostatic gradient will sort itself out and a given elevation on the uphill will have the same pressure as the same elevation on the downhill, but that point is much lower than the top of the hill.

The stated purpose of high-point vents is generally "to prevent gas locking" and we can assume "to restore the siphon." It can't be done. If we were to ever completely remove all gases from a water system (not possible in my experience) then flowing less than the quantity in Eq. (4.6) would create a vacuum at various places in the system and the void space would be filled by water vapor with the same effect as it being gas filled. These vents add costs, create a potential leak point, and have no added value beyond the initial hydrostatic test of the system (having a high-point vent to degas test water is quite useful). It is much more effective to simply accept that there will be gas in the water system and remove it at the disposal site inlet facilities.

7.2.2 Infrastructure for aggregation and transport

For any well, the operator will have to decide if it is better economics to truck the water or build a gathering system. You need to look at capital costs, operating costs, and risks. One of my clients had a group of wells in the Barnett Shale whose water hauling costs were upward of $300k

Table 7.4 Infrastructure analysis

	Trucking	Pipeline
Capital cost	Very low	High
Operating cost	Very high	Very low
Main risk	Road accidents	Line failure

USD/day. Spending $10 million to build a water gathering system paid for itself in about a month and didn't take much analysis to reach the right conclusion once the wells are online and have proven that they are economic even with very high operating costs, but a decision this cut-and-dried is pretty rare. Table 7.4 tries to show some of these trade-offs.

It is often reasonable to have a hybrid system where some wells have tanks that are trucked to nearby aggregation points and then pumped to the disposal points. Maintaining relationships with landholders is critical to being granted access to build water gathering lines. Gaining the required access and having the agreement with the landholders legally ratified can be time consuming and as such addressing land access issues early in the project life cycle is recommended to mitigate the risk of delays due to land access issues.

7.2.3 Site entry facilities

Any method of disposal discussed later has to have expected water quality coming in. None of them can handle leaves, bugs, corrosion/scale chunks, construction debris, etc., so you will nearly always have duplex basket strainers with 1/64 in (397 μm) holes at the end of the pipeline. These strainers usually get cleaned based on a differential pressure across the filter and it is common after the first few months for operators to put 100–200 μm bag filters over the baskets. It can be difficult to find appropriate bags if there is any measurable oil or grease in the produced water.

Produced gas in the water gathering system is also often a problem. Historically we have used oil-removal equipment to let the gas evolve out of the water, but increasingly regulations are making this difficult (regulators are putting limits on vented methane at Oil & Gas production facilities) and operators are looking for separation equipment like the GasBuster (Fig. 5.8) used on well-sites for pumping wells. Conventional separation equipment has difficulty removing dozens of MSCF of gas from hundreds of bbl of water.

Oil and grease in produced water is both common and can be devastating for any of the common disposal methods. For deep well injection, the oil and grease will tend to plate out on the formation surface and restrict injection rate. For evaporation ponds, any oil on the pond surface becomes a hazard to wildlife and will reduce the evaporation rate. For beneficial use of "raw" water like using it for dust control, oil in the water becomes a hydrocarbon spill instead of a beneficial use. For beneficial use of treated water, filters that can deal with fine removal of water are rare, very expensive, and not terribly effective. It is far better to remove any oil and grease at an early stage and prior to any further processing. The usual way to remove oil from produced water is with a "gun barrel" (Fig. 7.10). The reason that these ubiquitous devices are called "gun barrels" is lost in time, but everyone calls them gun barrels. There are as many different designs as there are design engineers. The basic difference between the designs seems to be more style than substance. Virtually every gun barrel I've ever seen has been an adaptation of one of the two designs in Fig. 7.10. The example on the left has a weir plate welded to the inside of the tank that goes almost to the bottom. As fluid comes into the gun barrel, heavier water tends to fall and lighter oil tends to rise. The water-out nozzle is set lower than the oil-out nozzle to accommodate the different density between oil and gas. It is important to get the density of oil right enough to allow these nozzles to work together.

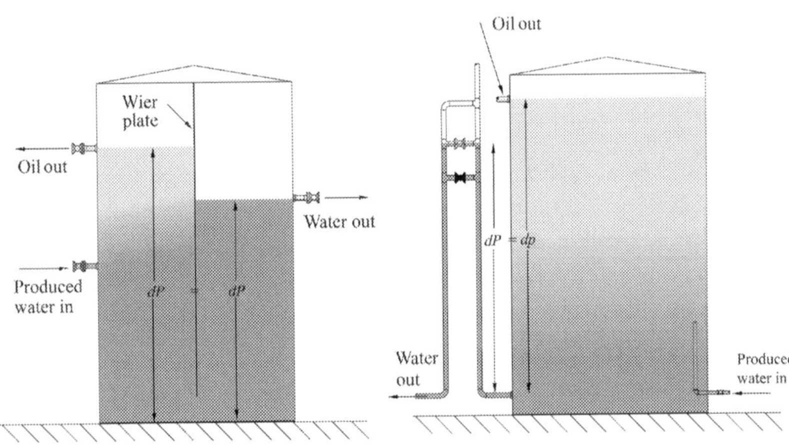

Figure 7.10 Gun barrel examples.

The gun barrel on the right in Fig. 7.10 can be applied to any tank. Again, the difference in height of the water overflow and the oil-out nozzle is a function of the difference in density of the oil and the water. In this example there are two possible overflow paths, therefore there are two different oil densities that can be accommodated. The third (non-valved) overflow is a fail-safe in case the oil density is too close to the water density. One frequent adaptation is the ability to micro-adjust the position of the water overflow, but this added complexity has tended to be poorly understood by station operators and have often been inappropriately adjusted.

All gun barrels will work better with heat. Oil tends to have a greater change in density with temperature than water has, so adding a bit of heat (something like 100°F (38°C)) can enhance the separation and help ensure that oil doesn't go into the water-treatment process.

A properly functioning gun barrel will have water right at the water outlet nozzle/overflow and oil right at the oil-out nozzle at all times. This means that if a quart (liter) of liquid enters the gun barrel then a quart of liquid must leave. This should make it clear that gun barrels are only effective in continuous processes, not in batch processes (e.g., truck unloading).

We frequently use gun barrels for truck unloading. In a real-world example:

- Design conditions
  - Temperature 160°F (71°C)
  - Water SG 0.96
  - Oil SG 0.75
- Truck conditions
  - Temperature 35°F (1.7°C)
  - 78 bbl (12.4 m^3) of water (SG 1.07)
  - 2 bbl (0.32 m^3) of oil (SG 0.98)
  - Empty truck in 15 minutes (7600 bbl/day (1208 m^3/day) flow rate)

The results of this event are shown in Fig. 7.11. The gun barrel had been sitting for several hours and all the fluids were at the design temperature. The oil on the truck was denser than the water that was in the tank so it fell like a stone. Its entry velocity was high enough that the incoming fluid could stay in a coherent stream until it had made it under the weir plate and nearly all of the truckload was on the water side of the weir plate when the truck was empty. The high flow rate raised the fluid levels on both sides of the weir plate to well above the nozzles. That

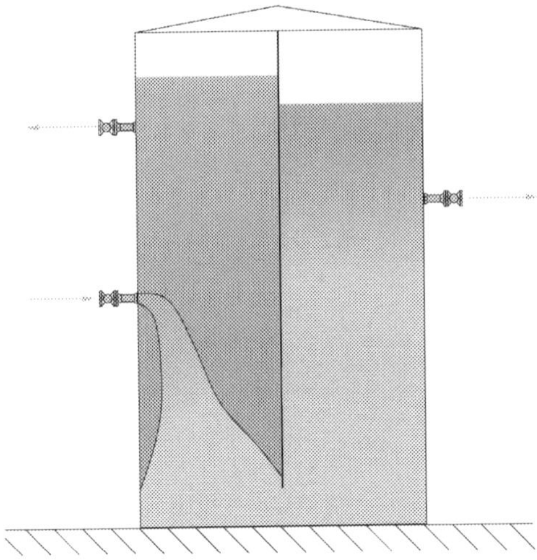

Figure 7.11 Truck unloading example.

much oil on the water side upset the balance of the fluids in the oil side and the end result was about 1.8 bbl (0.29 m^3) of oil in the water tank and 30 bbl (4.8 m^3) of water in the oil tank. This unhappy outcome happened 3–4 times/day at this facility. This facility was able to sell less than 60 bbl/year (9.5 m^3/year) of oil. The end result was that the disposal well had to be abandoned after only reaching 20% of its projected disposal volume because of high injection pressure.

When the problem became apparent, a project was initiated to convert the gun barrel from a continuous configuration to a batch configuration (Fig. 7.12). The 750 bbl (119 m^3) heated pretreat tank was designed with some unique characteristics:

- It had oversized heater elements to quickly raise incoming fluid temperatures to target values.
- The outlet nozzle was located such that the tank had enough capacity above the nozzle to hold three trucks worth of liquid. Each truck unloaded into the pretreat tank as fast as their equipment could unload and then moved on, the excess liquid was stored in the bottom of the pretreat tank until it could flow into the gun barrel.
- The inlet nozzle exit was in the lower one-third of the tank and the cold liquid was held against the burners by gravity.
- The tank stand was high enough (the bottom of the pretreat tank was about 8 ft (2.4 m) above the bottom of the gun barrel) that liquids could free flow into the gun barrel without a pump (pumps in

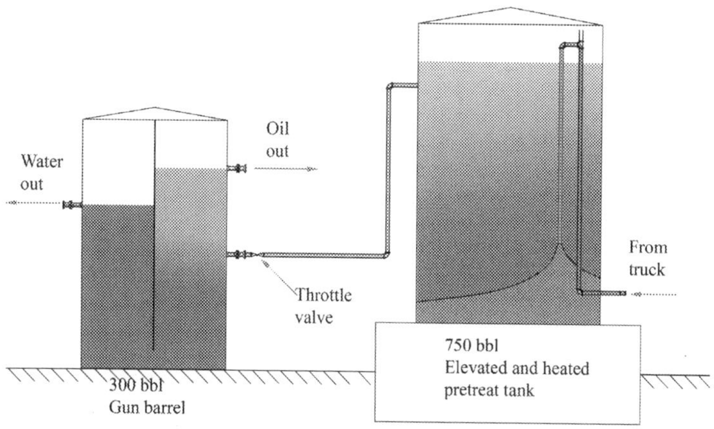

Figure 7.12 Gun barrel pretreat.

oil/water mixtures tend to create emulsions that can be difficult and expensive to break).

- The throttle valve between the pretreat tank and the gun barrel was set at 200 bbl/day (32 m^3/day) which was the design condition of the gun barrel.
- A new truck unloading area had to be built because the water trucks were unable to force water into the loop at the top of the elevated 750 bbl pretreat tank, so we moved the truck station to an adjacent hill.

This facility ran for 6 months before the regulators shut the site down because of high injection pressure in the disposal well (caused by many years of injecting oil into the disposal well while the gun barrel was not working properly). In that time the site sold an average of three truck-loads of oil per month (on track for 3000 bbl/year (477 m^3/year) of oil sales at $30 USD/bbl).

7.2.4 Water transfer facilities

There are a lot of reasons to break up a water gathering system, e.g., you may be making a transition from flat country to hilly country and have to change materials, or you may want to put a truck loading station in the middle of the system, or you may have to make up friction losses on a system that you cannot run at a high enough pressure to get from the well to the disposal facility. Including break tanks also reduces the inventory in any one section of a gathering network and hence it can be used as a means to manage risk associated with loss of containment. Flowing

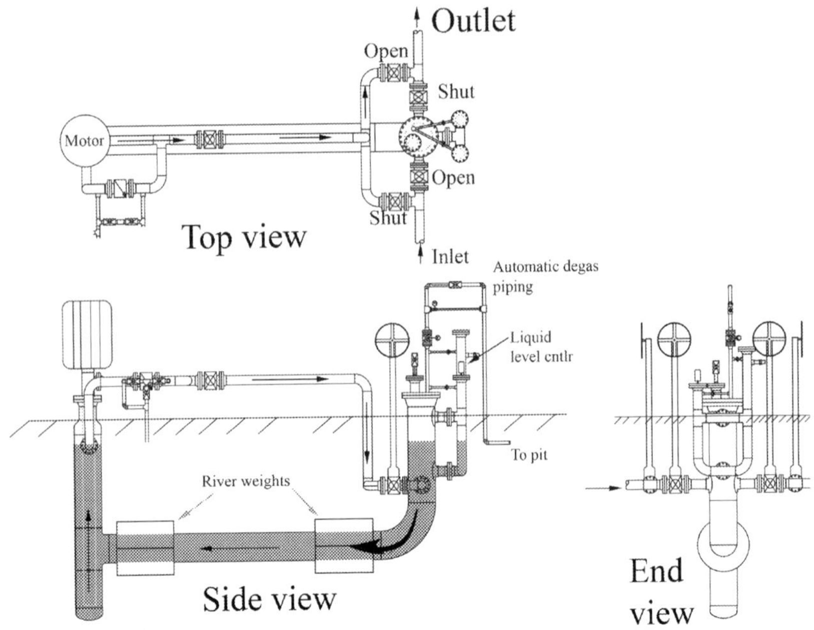

Figure 7.13 Water transfer station.

water from a wellhead separator to a collection tank means the water can be pumped which allows a reduction in the size of gathering lines in the gathering network as opposed to if wellhead pressure alone is used to push the water through the gathering lines.

It is common to think of "water transfer facilities" as a series of tanks, water transfer pumps, truck loading, fences, site security, office space, etc. I've seen transfer facilities that were as simple as a single tank and a small transfer pump to nearly as complex as a plant and costing millions of dollars.

Fig. 7.13 shows an alternative that costs less than $200k USD and has all the functionality of a full-blown water transfer station with the footprint of a few square meters. The valves are all extended handle to minimize water piping above ground, and the valving allows you to change direction on the system by repositioning four valves.

7.3 DISPOSAL METHODS

Once we've aggregated a field's produced water, gotten rid of oil, gas, and major boulders, we have to dispose of it. Disposal fits into one of three broad categories: (1) deep well injection, (2) evaporation, and (3) beneficial use. The first category removes the water from the biosphere and it won't be seen again for geologic time. The second category allows the H_2O portion of the produced water to enter the biosphere as clean water vapor and the non-H_2O portion to be treated as either solid waste or clean land-fill dirt depending on your politics. The last category is everything else. In a world with systemic water shortages it seems like the preference would be for either of the second two options, but the exact opposite seems to be true. Deep well injection accounts for over 80% of the world's produced water disposal and in the United States it is over 90%. Evaporation is heavily regulated and it has become difficult to successfully acquire permits for new evaporation ponds. The regulations and risks associated with all beneficial uses cause many companies to treat these options as unacceptable (there are exceptions, in Australia for example the regulators push hard for beneficial use while imposing some of the strictest limitations on beneficial use of produced water in the world).

It is important to understand that virtually any disposal option will (either immediately or eventually) run afoul of someone's "green agenda." Most of the obstacles to produced water disposal seem to lack sound logic, but nonetheless can be exceedingly difficult to overcome.

7.3.1 Deep well injection

The purpose of deep well injection is the permanent disposal of produced water into nonproductive formations. "Nonproductive" is a term of law (defined in the authorizing legislation for the US Underground Injection Control (UIC) program that regulates all wells used primarily for injecting fluids into the ground) that means that the formation is not: (1) a source or potential source for potable water; and (2) an economic source for hydrocarbon extraction.

Injection wells need to be permitted with state or federal regulators. Those permits can have almost any stipulation or performance requirement that is imaginable, some of the more common stipulations are (any given permit can have some of these, all of these, or none of these):

- Surface injection pressure plus the hydrostatic gradient must be less than the fracture gradient of the target formation. Often the permit will require some sort of injectivity test to confirm the fracture gradient.

- You will be required to protect groundwater with multiple barriers, and often with "annulus fluid" (discussed later) to provide a monitoring method for the continued performance of the barriers.
- There will be a list of required tests, monitoring points, and reporting requirements.
- Total injected volume will be restricted to a number you provide. This number is often just made up, but once you put it on the permit you have to live with it. If you make up a number that is too low you will be required to abandon the well with space remaining, if you make up a number that is too high your permit may not be approved.

Injection pressure is a function of the reservoir's permeability, the rate you are injecting, and any damage that you might have done to the interface between the wellbore and the reservoir (e.g., the accumulation of filter cake or the accumulation of oil and oil residue). Some wells (Fig. 7.1 for example) will accept water on a vacuum, meaning that the reservoir pressure is less than the hydraulic gradient of a column of water and the reservoir's ability to accept water is greater than the injection rate. Most wells will require some amount of injection pressure on the surface to transfer water at the required rates. This applied pressure is generally limited by the permit to drill and inject and must be closely monitored over the life of the well. This is a case where the regulatory requirement coincides with an engineering requirement—the rate of change in the injection pressure is an excellent indicator of actual or pending problems.

7.3.1.1 Typical wellbore

The wellbore in Fig. 7.14 is reasonably typical. For very deep injection wells there may be an intermediate casing string, but other than that most of them look like this example. The most significant difference between injection wells and gas wells is the production packer that isolates the tubing/casing annulus from the reservoir. This packer allows the annulus to be filled with annulus fluid. The main characteristic of this fluid is that it is not hazardous to groundwater, but it also needs to be quite noncorrosive. With the annulus filled with an incompressible fluid with a slight overpressure, you can monitor the annulus pressure for changes. If the annulus pressure begins to increase, then you have a good indication of a leak in the tubing into the annulus. If the annulus pressure begins to decrease, then you have a good indication of either a packer leak (on wells that take water on a vacuum) or (more likely) a leak in the production casing.

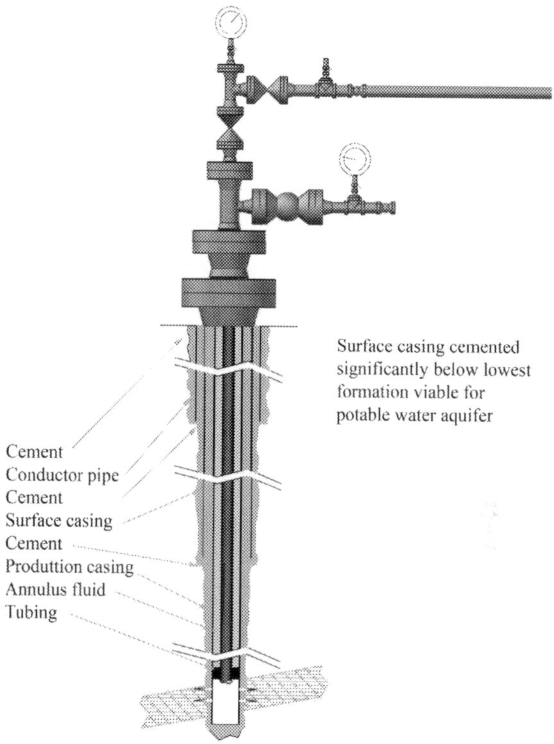

Figure 7.14 Injection wellbore.

7.3.1.2 Surface equipment required

Disposal equipment varies dramatically from operation to operation. At the low end of the spectrum is the disposal well in Fig. 7.1 where all of the incoming water is pumped (using downhole pumps) from wells directly into the disposal well with no surface equipment at either the source or the sink. At the other end of the spectrum is a disposal well drilled specifically for the brine from an RO facility where the disposal site looks more like a factory than a well-site. Most sites will be somewhere in between. The things to consider installing for the injection process include:

- Filtration. Every site is different, but it is common for sites to start with a basket strainer on the pump suction and a 5 μm filter on the discharge. Operators end up changing discharge filters daily and the costs for replacement filters and power become prohibitive. Over time many sites shift toward a 200 μm bag filter on the suction strainer and 25 μm pump-discharge filter (but some sites find their balance point with a 100 μm discharge filter). The trade-off is any solids that enter the injection well will tend to collect at the transition from the well bore to the injection formation and can shorten the life of the injection well vs the cost and effort required to change filters on the surface.

- Tanks. Any operation with injection pumps will need to have a buffer between inflow and delivery to account for variations in inflow, pump outages for maintenance and repairs, and outages for filter replacement. I am always nervous about the time that I have to order a pump repair part from Shanghai and I have to run at half capacity (or zero capacity) for a week, so I generally design sites with 5–7 days of storage capacity. Most other engineers I've worked with are more optimistic than I am and a content with 1–2 days of storage—it is a risk/return balance.
- Pumps. Historically the lion's share of downhole injection pumps have been plunger pumps with 3, 4, or 5 plungers. Positive displacement plunger pumps cannot actually suck on the tanks and will rapidly heat up and fail if they are not supplied with adequate net positive suction head (NPSH, see Chapter 3: Well Dynamics). I've found that supplying plunger pumps with centrifugal charge pumps significantly improves their long-term reliability. Increasingly, injection sites are using progressing cavity pumps (PCP, see Chapter 3: Well Dynamics) for injection. PCP have a much lower NPSH-r than plunger pumps and generally do not require a charge pump. Whatever technology is used, it needs to be able to pump the required volume (with a safety factor) into the permitted injection pressure. I prefer to put my safety factor in a second pump and design stations with two (or more) parallel injection pumps each with at least 75% of required volume flow rate.
- Automation. For sites with multiple parallel pumps, I use an automation strategy to designate one as the primary pump that runs continuously, and then use tank levels to cycle the second pump (i.e., when the tank levels reach a designated high level the second pump comes on and pumps until the tank levels reach a designated low level and the second pump turns off). I like for the automation system to automatically swap the operating and standby pumps daily. Automation also needs to be able to shut the site down if the injection pressure approaches the permit limit, the tanks reach a low-low status, the filter *dP* becomes too high, or if annulus pressure changes rapidly.

7.3.1.3 Monitoring requirements

Different permits set different information that must be captured and maintained, but at a minimum, you will need to be able to supply an auditor with:

- continuous monitoring of pump injection pressure;
- continuous monitoring of tubing/casing annulus pressure;
- volume injected;
- pump run time.

7.3.1.4 Testing requirements

The testing requirements vary widely from jurisdiction to jurisdiction and even from permit to permit. Most will require an initial hydrostatic test of the casing, but some will require periodic retests. Some will require periodic injectivity tests (also called "step-rate tests"), and some of those will require that a regulator supervise the test. This is an area where understanding the permit stipulations and designating someone with the responsibility of meeting those obligations will pay significant dividends since failing to adhere to the permit requirements will nearly always result in severe penalties up to and including having to plug your well prematurely.

7.3.2 Evaporation

At the surface of any coherent gas/liquid interface the gas that touches the surface of the liquid will be at 100% relative humidity (RH) (Fig. 7.15). As you move a very short distance away from that interface,

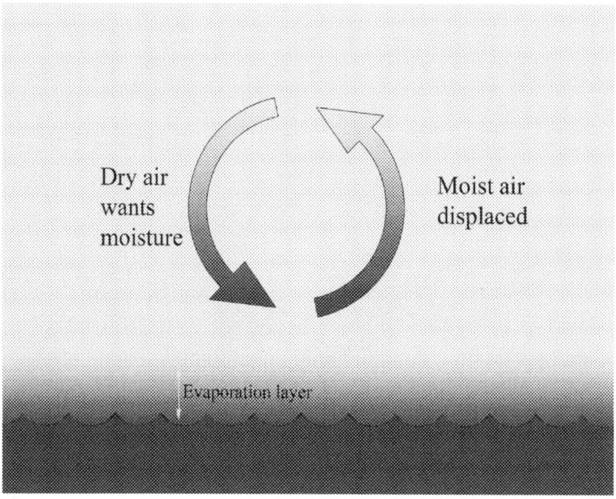

Figure 7.15 Evaporation cycle.

the humidity rapidly moves toward ambient values. The thickness of this "evaporation layer" is a function of ambient RH and wind speed. As "dry" air moves toward the evaporation layer (via osmotic pressure, the dry air wants to be contaminated with water vapor), it displaces moister air within the evaporation layer that moves into the bulk humidity of the region. This exchange is accelerated by increasing temperatures (because for the same water vapor mass in the ambient air, a higher temperature will lower the RH because at higher temperatures the air can hold more water vapor).

This natural water cycle seems like an obvious improvement over deep well injection—the water stays in the biosphere to fall as wholesome rain somewhere and the solids become soil. Environmentalist spend more effort fighting against evaporation ponds than most issues in Oil & Gas. I am certainly not qualified to speak for environmentalists, but I think I can paraphrase their concerns into "overspray is toxic." Frequently operators of evaporation ponds spray water from the ponds to either aerate the water (to control odors) or to increase the evaporation surface (to accelerate evaporation). As these airborne droplets lose H_2O to evaporation, the dissolved solids in the droplets are concentrated and the TDS increases dramatically. If the droplets migrate outside of the pond boundaries, then this high-TDS water will kill local plant life and future rainfall can easily create high-TDS puddles where animals expect rainwater. Rather than controlling overspray, environmentalists and regulators make it exceedingly difficult to get a permit for an evaporation pond at all.

7.3.2.1 Calculations

The nominal average daily solar energy reaching the earth's surface is something like 1360 W/m^2 (431 BTU/ft^2/h) (sorry for all of the weasel words, but this discussion needs a number, your mileage may vary). The latent heat of vaporization of pure water is 2250 W-s/g (967 BTU/lbm). Putting these numbers together and you can get a nominal evaporation rate shown in Eq. (7.4). This simplified factor significantly understates required pond size as you move further from the equator:

$$q_{evap} = \frac{R_s}{Q_{latent} \cdot \rho_{water}} = \frac{1360 \cdot \dfrac{\mathrm{W}}{\mathrm{m}^2}}{2260 \cdot \dfrac{\mathrm{W \cdot s}}{\mathrm{g}} \cdot 1000 \cdot \dfrac{\mathrm{kg}}{\mathrm{m}^3}} = 0.0520 \cdot \frac{\mathrm{m}^3}{\mathrm{m}^2 \cdot \mathrm{day}} = 0.0304 \cdot \frac{\mathrm{bbl}}{\mathrm{ft}^2 \cdot \mathrm{day}} \tag{7.4}$$

7.3.2.2 Pond size

For a first cut feasibility analysis, you can get a quick rough cut pond size by taking the inverse of Eq. (7.4) (Eq. (7.5)). Outside of about ± 40-degree latitude I tend to double the Eq. (7.5) factor.

$$\frac{1}{q_{evap}} = 1.92 \cdot \frac{\text{hectare}}{\text{ML/day}} = 0.756 \times 10^{-3} \cdot \frac{\text{acre}}{\text{bbl/day}} \tag{7.5}$$

Eq. (7.5) says that in most of North America, if you want to evaporate 1500 bbl/day (119 m^3/day) then you need something like 1.13 acre (0.46 hectare). To account for factors like elevation above sea level, distance from the equator, and changes in solar irradiance with mean temperature, it is more accurate to use the PenPan equation (Eq. (7.6)). Eq. (7.6) is an empirical equation and only works for the units shown in the nomenclature section.

$$E_0 = (0.015 - 0.00042 \cdot T_m \cdot h_{asl} \cdot 10^{-6}) \cdot (0.8 \cdot R_s - 40) + 2.5 \cdot F \cdot u_m \cdot (T_m - T_d) \tag{7.6}$$

A couple of the terms are quite obscure, but they can be represented by other empirical relationships shown in Eq. (7.7).

$$\begin{aligned} F &= 1.0 - 1.7 \times 10^{-5} \cdot h_{asl} \\ R_s &= 10.8 \cdot T_m + 153 \end{aligned} \tag{7.7}$$

Eq. (7.6) wants a single value for mean temperature and dew point. As is obvious these values change moment to moment, day to night, summer to winter. The US National Oceanic and Atmospheric Administration (NOAA) has an extensive compilation of weather data from around the world (more concentrated in the United States, but I've had pretty good luck using this resource in Nigeria, Botswana, and Australia). The NOAA data has normal values, mean values, and extreme values for temperatures, dew points, rainfall, and wind by month. It also allows you to calculate a worst case (i.e., what happens if mean temperature is lower and rainfall is higher for a couple of years in a row?) This allows you to calculate evaporation rates minus average rainfall for each month in the year in both a pessimistic, optimistic, and expected case, and then set your pond size such that the surface of the pond is appropriate for the average case, the freeboard is appropriate for the pessimistic case, and the pond doesn't run dry in the optimistic case.

This discussion is crying out for an example, so here it is:

A pond is to be built in Farmington NM, conditions are:

- Location: 36.75°N latitude, 108.19°W longitude (about the same as Mosul, Iraq; Nagano, Japan; Auckland, NZ; or Kosciuszko National Park, NSW, Australia)
- Elevation: 5200 ft ASL (1580 m ASL)
- Detailed design called for 1.16 acres (0.47 hectares). The quick and dirty calculation in Eq. (7.5) called for 1.13 acres (0.46 hectares), which is much closer to Eq. (7.6) than you normally see.

7.3.2.3 Pond permitting and construction

Evaporation ponds are actually quite complex systems. They must be impervious to percolating down to groundwater, you must be able to verify that they remain impervious to percolation, you need to be able to combat odors, the walls must be sloped enough, but not too much, and you have to have a plan to remove accumulated solids.

Regulations are becoming more complex and requirements for protecting and monitoring groundwater vary widely from jurisdiction to jurisdiction. In New Mexico, the regulation requires (at a minimum) that you submit:

- Operating and maintenance procedures with monitoring and inspection plans
- Pond closure plan
- Hydrologic report which includes sufficient information on the site's topography, soils, geology, surface hydrology, and groundwater hydrology (this requirement results in drilling at least four test wells to more than 50 ft (15 m) below the projected bottom of the disturbed area, any water at all in the wells will get your permit rejected)
- Dike protection and structural integrity
- Leak detection
- Liner inspection procedures and compatibility report
- Freeboard and overtopping prevention
- Nuisance and hazardous odor prevention
- Emergency response plan
- Type of waste stream (including chemical analysis)
- Climatological factors including freeze/thaw cycles

The regulation goes on to say that each pond must be designed, constructed, and operated so as to contain liquids and solids in a manner that will prevent contamination of freshwater and protect public health and the environment. It must have a foundation of firm, unyielding base free of rocks, debris, sharp edges, or irregularities. The dike walls are limited to 2H:1V maximum slope inside pond and 3H:1V maximum slope outside pond; and it must be wide enough to include anchor trench, room for inspection, and maintenance. The pond may not be larger than 10 acre-ft (1.23 hectare-m) (larger than that puts under "Dams and Dikes" regulations which are much more stringent). The pond must be netted or have other approved means of protecting migratory birds.

The rules for liners are quit specific as well:

- Primary (upper) liner made of synthetic material
- Secondary (lower) liner can be synthetic or other material approved by regulators
- Upper and lower liners must be separated by at least 2 ft (0.61 m) of compacted soil with leak detection equipment between the liners and under lower liner
- Shall meet:
  - At least 30 mils (0.030 in (0.762 mm)) thick
  - Impervious to hydrocarbons, salts, acidic and alkaline solutions
  - Resistant to UV light
- At any point where fluid enters pit, liner must be protected from fluid force and mechanical damage

These New Mexico requirements are more explicit than the requirements in many jurisdictions, but they are not stricter. When you anticipate building an evaporation pond, get a copy of the applicable regulations first—read and understand them before you even put a budget together.

7.3.2.4 Aeration requirements

The best information about the effect of spraying on evaporation comes from the irrigation industry—any water that evaporates in the air doesn't benefit the irrigated crops so they want to stop it. They find that with a droplet size greater than 100 μm, the evaporation is a surface function and below about 50 μm it becomes a body function—an order of magnitude greater evaporation rate. Fig. 7.16 shows the effect of this observation. The picture on the left is a water cannon that is sometimes used in irrigation to reach inaccessible parts of a field. The water cannon results in droplets that are on the order of 200 μm, they don't evaporate much. You can see this by the amount that the droplets disturb the surface of the water and by how little spray remains in the air. The picture on the

Figure 7.16 Effect of droplet size on evaporation pond.

right of Fig. 7.16 is a device designed to facilitate evaporation. It is spraying about twice as much water as the water cannon, but it has a fan blade in the flow stream that literally chops the droplets into approximately 25 μm droplets. These droplets remain in the air for a long time (you can see that the flow does not disturb the surface of the pond). The droplets remain in the air for so long that overspray is a serious problem and the wind fences were installed to catch the drift. The company that makes the evaporator on the right-hand picture also has a line of floating units that reduce the height of cloud to reduce overspray.

Standing water in an evaporation pond can accumulate microbes that can have a rank odor. The worst of these odors are from anaerobic microbes, so you frequently have a need to introduce aeration equipment to control the odors. These units will increase evaporation slightly, but their main purpose is to cause droplets to absorb oxygen and fall back into the pond—aerators don't throw the water very far (see Fig. 7.21). Odors can sometimes be controlled with chemicals, but this tends to be expensive and seldom provide a long-term solution.

7.3.2.5 Solids accumulation

The TDS and TSS in produced water represents mass. A produced water stream with 10,000 mg/L TDS has about 3.5 lbm/bbl (10 kg/m^3) of solids that will remain behind after evaporation. Evaporating a million barrels (159 ML) of 10,000 mg/L water will leave 1750 ton (1590 tonne) of solids. If the solids were all NaCl (which they are not), they would occupy a volume of 26,000 ft^3 (734 m^3), which is not available for water. When the pond fills up with solids, the solids have to be removed manually. Typically if they pass the test for NORM, then the stuff in the pond is just dirt that can be used anywhere that dirt can be used. Mixed with organic material like processed cattle manure it often makes excellent soil. It is common to dump these solids in facilities land farming oil-contaminated soil so that the outlet is a light, mineral-rich soil that has a high value.

Because solids accumulation is reasonably rapid, I always design evaporation facilities with two ponds of equal size. Generally you inflow into one pond, suck on that pond with evaporators, and spray them over the other pond. Solids will tend to accumulate in the second pond. When you reach your limit on solids on the second pond you turn the sprayer over the first pond and remove the solids from the second. Removing the solids is a reasonably violent activity and it is generally better to remove the top of the pile with powered equipment (like a bobcat with a bucket) and as you get close to the liner, roll up the inner liner with the solids inside and lift the package out with a crane to be finally processed outside the pond. Then you can inspect the leak detection equipment and verify that it is still functioning and install a new liner. Without using the two-pond approach you have to shut the facility down for weeks while the solids are removed.

7.3.2.6 Wildlife

Birds and beasts like access to water. You really don't want your pond to become a watering hole. A pond designed for Botswana had to be "hyena proof" because their claws are so sharp that they rip a liner to shreds in short order, and hyenas are quite tolerant of marginal water. A pond designed for Australia has the same concern with kangaroos and wallabys. Around the world, it is more common to be concerned about birds and cattle. Cattle (and deer, elk, moose, etc.) can be controlled pretty well with perimeter fencing. Birds are another problem. For years we pretended that putting wooden or plastic owls (see the left side of the right-hand picture in Fig. 7.16) to "scare them off." It is too small to see, but the owl in Fig. 7.16 is covered in bird droppings, it is not doing a very effective job as a "scare crow." Another site used plastic alligators to the same effect. A site in Australia that put real crocodiles in their pond had success until the crocodiles left for better hunting and tore up the liner as they walked out. Companies have also spent considerable money on putting flagging across the pond. Birds are often seen sitting on the wire holding the flagging.

Bird netting is more common and more effective. The problem with bird netting is that the pond is big. The 1.16 acre pond in the example above was 90 ft × 400 ft (27.4 m × 122 m) at the bank in each of two ponds where the anchor posts for the netting would be. Getting a net to stand above the pond over a 300 ft (91 m) expanse is a difficult undertaking. The best solution that I've found to this problem is a grid of cables located about every 10 ft (3 m) over the entire pond. The net is then pulled tight over the grid and a second cable grid is located above the net. Then ties are used to pull the two cables together sandwiching the net. The net will rot out in about 5 years, and they are a pain to remove (all of the ties have to be cut from a man-basket hanging off a crane or from a boat), but every 5 years is better than every thunderstorm. This grid system is also able to carry a snow load that no other system can carry. Typical grid size for the netting is 2 in (51 mm), and each strand will carry a few grams of snow per unit length. With miles of strands the total weight of snow can be significant.

An interesting alternative to netting is pond balls (Fig. 7.17). These black plastic hollow balls are designed to limit evaporation. Wait. Why are we talking about a device to limit evaporation in an evaporation pond discussion? The thermodynamics of the balls is interesting. They are air filled and sunshine on the balls heats up the black top surface, but the heat-transfer characteristics of the air in the balls are so poor that little of the heat makes it to the water. Also, the balls prevent free air movement from ambient humidity to the evaporation zone. On the other hand, the balls get quite hot. When you use an evaporator over the pond balls you can hear the droplets sizzle as they immediately evaporate on contact.

Figure 7.17 Pond balls.

This process leaves solids on the surface which unbalances the balls and causes the hot side to rotate into the water which brings up a new wet surface to evaporate. The pond in Fig. 7.17 has the problem of outrunning the inflow and it often sits empty of water.

7.3.2.7 Freeze/thaw purification

The water in icebergs in the ocean is quite pure and drinkable. This is because water with lower TDS will freeze before water with higher TDS and osmotic pressure will tend to replace high-TDS water with lower TDS rain/snow within the ice.

The rub is that when the ice melts while floating in high-TDS water it will remix with the high-TDS water and return to unusability. If you can remove the (nearly pure) ice from the high-TDS source then it can be used as any other pure water.

The company that makes the evaporator in Fig. 7.16 (Snow Machines Inc.) got their start in providing snow-making equipment to ski resorts to allow the sky resort to have a limited opening early in the season. The company noticed that a large portion of the source water did not return when the snow melted. Their conclusion was that the latent heat of fusion resulted in significant evaporation which then condensed and became snow, and they went looking to expand their target client base to people who needed to accelerate evaporation.

Between the natural purification of freezing that we see in icebergs and the accelerated evaporation in freezing temperatures, it seems that spraying water in freezing temperatures and then removing the resulting ice can purify large quantities of water.

Amoco Production Company did an (unpublished) study on freeze/thaw purification in the San Juan Basin in the winter of 1996–97 (Fig. 7.18). The results were interesting:

- Started with 8000 bbl (1.27 ML) of 12,800 mg/L TDS water
- Operated for 60 days (spraying water over a grate above the pond)
- Ended with:
  - 6400 bbl (1.01 ML) of 1010 mg/L TDS
  - 1600 bbl (0.025 ML) of 44,900 mg/L TDS
- They removed the ice from the process by picking up the grate with two rubber-tire hoes and shaking it off over another basin.

Where freezing temperatures are common (at least seasonally) this process can be useful, and at the least it shows that turning spray equipment off during the winter is counterproductive.

Figure 7.18 Freeze/thaw purification.

7.3.3 Beneficial use

In many locations water rights law can be extremely complicated and contentious. Operators may be reluctant to pursue beneficial uses because once they have made the investment to clean and use the water, their rights may be challenged. Even if the challenge is unsuccessful, the cost and uncertainty associated with litigation may make the pursuit of beneficial produced water use unattractive. Another legal concern is the potential for unknown future liability. While there are no known problems with using treated produced water, the specter of liability issues arising in the future still looms. Other industries have faced huge liabilities (e.g., asbestos and tobacco) from products once thought to be benign. In addition, the possibility exists for lawsuits to be filed alleging problems where none exist. Whether these fears are founded or not, these are real concerns that can limit the beneficial uses of produced water.

A vivid example of the contentiousness of water management can be seen in the lawsuit *Vance v. State of Colorado.* The plaintiff sued the state claiming:

- In situ water must be removed before CBM can be produced.
- Therefore, all CBM produced water is "beneficial use" instead of a waste product.
- The state had no real incentive to defend the case (and the industry was not allowed to join the suit) so the state agreed:
  - CBM wells in Colorado must be permitted as both gas and water wells (two different applications to two different departments).

- CBM operators are required to purchase water rights prior to producing any water.
- The decree did not explicitly require that producers must pay mineral owners royalties on the water, but left the door open to that interpretation.

The point here is that you shouldn't assume that everyone is going to embrace your idea for beneficial use of produced water. No matter what you decide to do with the water will cause someone concern.

7.3.3.1 Reuse raw

Sometimes there are opportunities to use produced water without any treating at all. Some of the applications that raw produced water include drilling fluids, frac water, hydrotest water, and dust control on roads. All of these applications seem reasonable, but you still must take care. Some examples are as follows:

- Spent hydrostatic test water must be treated as an industrial waste (not an Oil & Gas waste) because of picking up grease, oil, and mill scale from the pipe. Water that could have gone into an Oil & Gas disposal well UIC Class II) now must go into a UIC Class I well generally at a much higher cost.
- Transportation departments are responsible for roads and road maintenance. A phone call to the transportation department may get you permission to spray produced water on dirt roads for dust control. That permission does not preclude environmental departments from citing you for a produced water "spill" of the water you thought you had permission to spray on roads (and the fines can be significant).
- Some jurisdictions limit how many times water can be reused for hydraulic fracture stimulation. These limitations seem like an odd subject for legislatures to spend their time on, but the chemical lobby can be powerful in some places. The offshoot of this is that in those places you often cannot use produced water in frac jobs.

7.3.3.2 Manmade wetlands

Nature's way of converting marginal water to wholesome water is to evolve plants and organisms to use the contaminants in it. If you look at a salt marsh on the verge of oceans, you'll see water that could not possibly be used for irrigation that is full of plant life. The right mix of plants, microbes, and fauna can pull the contaminants from any water source and result in clean water and useful soil. There have been cases of industrial waste water being used to feed a wetland and the effluent of the wetland being high-quality water.

Environmentalists have a soft spot in their hearts for wetlands. Consequently, disturbing an established wetland is among the highest crimes to the planet. Creating a wetland will sometimes be met with enthusiasm by environmental regulators. Once it is established and thriving, it becomes an "established wetland" and anything you do to disturb it will not be met with nearly as much enthusiasm and will generally be prohibited. It is common for new wetlands permits to be marked that the wetlands must be maintained in perpetuity. "In perpetuity" is a long time. It basically says that if the water source for the wetlands *ever* dries up, then you have to find an alternate source of water forever.

While creating a wetland has some real benefits to the planet and to your operation, they can be a trap that has no escape.

7.3.3.3 Managed aquifer recharge

In many of the locations that have Oil & Gas operations, availability of groundwater for residential, agricultural, and industrial uses is limited. It is possible for treated produced water to be injected into these groundwater aquifers to be used by whoever is tapped into the groundwater. The consequences of an upset in the treatment facility are staggeringly large, consequently regulators (in places where this has been proposed) have a long list of strict limits on water quality and a longer list of points that must be monitored. They set strict limits on both injection rate and total injected volume. Regulators further limit their exposure by being slow to approve permits. Managed aquifer recharge has enthusiastic support from the environmental community (in spite of their reservations about the attention span of Oil & Gas to make sure that it always works) but as of this writing no permits have been issued.

7.3.3.4 Surface discharge to rivers

There was a time that surface discharge of untreated produced water to rivers was the go-to option all over the world in all industries. Industry abused this option to the point that the US Congress passed the Clean Water Act (and governments around the world quickly followed suit) that makes it difficult to get approval to put either treated or untreated produced water into rivers at all.

Permits can be acquired, but at a minimum the water must:

- have the same (or better) turbidity as the river;

- have about the same TDS as the river, with about the same mix of contaminants (i.e., if the river has 300 mg/L TDS that is nearly all MgCl, introducing 200 mg/L water that all of the TDS is NaCl will not be allowed);
- be at the same temperature as the river;
- have at least the same dissolved oxygen as the river;

In addition to meeting all of these requirements, you may be subject to additional tests. One of the favorite tests is the "fish kill" test. In this test, a particular species of fish (it is rumored that the particular species has an expected life span of 45 days, but that can't be confirmed) is put in your water and must survive for at least 60 days. It is rare for a water stream to pass this test.

Water quality standards for surface discharge to rivers are the highest of any of the beneficial use options. All of the (failed) RO projects I've worked on had the ultimate goal of discharging to a river. The discharge standards have proven too difficult (and expensive) to accomplish.

Temperature is a major hurdle. Two of my clients have solved both the temperature and aeration problems by dumping their water into surface structures (one uses a canal, the other uses a historical dry wash or gully), that flow several miles to the river. The treated water flowing over rocks and up against the bank acquires oxygen and reaches ambient temperature. When it reaches the river it is just water, indistinguishable from the river water. Transporting the water by pipeline from the treating site to the river would have had the water entering the river at ground temperature instead of surface temperature, and balancing the dissolved oxygen would have been done through added chemicals which is difficult to get right over time.

Using canals to transport produced water from well-sites to treating facilities, especially as we move toward multiwell pads and have multiple wells' water production at one place, has a significant pretreatment benefit, but it can be difficult to sell a regulator on using canals instead of pipelines—they tend to want to permit the canal as an evaporation pond and installing a double liner over miles of canal is prohibitively expensive.

7.3.3.5 Stock/wildlife watering

Nonlactating cows need 1 gal (4 L) of water per 100 lb (45 kg) of body weight in warm weather. In cold weather that number doubles. Lactating cows need to double those numbers again. With domestic cattle weight on the order of 2000 lb (900 kg), that is a lot of water. The problem is that they need pretty good water. Generally accepted TDS limitations are given in Table 7.5. It can be difficult to convince regulators that a catchment designed for stock/wildlife watering is not required to meet the design standards of an evaporation pond (which standards would make accessing the water by animals very difficult).

7.3.3.6 Irrigation

According to the University of Texas A&M Extension Service (A&M) irrigation water should meet the limits in Table 7.6.

The TDS is not the end of the story. Just like aquatic life needed oxygen in the water, plants cannot tolerate too much sodium in the soil. Too much Na^+ in the water will replace Ca_2^+ and Mg_2^+ in the clay fraction of the soil which will convert a granular structure into a hard/compact structure that will inhibit root development. To determine if a water stream is suitable for irrigation, you need to look at the "SAR." There are many forms of the SAR equation, all expect the input to be in "milliequivalents/liter" (mEq/L), which is not a unit that is common in Oil & Gas so Eq. (7.8) has a conversion that allows the use of values that we actually have. Eq. (7.8) includes this conversion.

Table 7.5 Produced water use for animals

TDS (mg/L)	Comments
<1,000	Excellent for all stock and wildlife
1,000–2,999	Very satisfactory, may cause mild diarrhea in animals until they become accustomed to it
3,000–4,999	Satisfactory, may be refused by animals not used to it
5,000–6,999	Avoid use for pregnant or lactating animals
7,000–10,000	Avoid use with very young or very old animals
>10,000	Unsatisfactory for all classes of animal

Table 7.6 TDS limits for irrigation

TDS		Comments
mg/L	μS/cm	
<175	<273	Excellent
175–525	273–820	Good
525–1,400	820–2,187	Permissible
1,400–2,100	2,187–3,281	Doubtful
>2,100	>3,281	Unsuitable

$$SAR = \frac{\frac{PPM_{Na}}{23.0}}{\sqrt{\left(\frac{1}{2}\right)\cdot\left(\left(\frac{PPM_{Ca}}{20.02}+\frac{PPM_{Mg}}{12.16}\right)\right)}} \quad (7.8)$$

The relationship between TDS and SAR is shown in Fig. 7.19. The section marked "excellent TDS" is completely within the section marked "marginal SAR," this is because at low TDS, the ability for the very low magnesium and calcium to counter the sodium becomes less effective.

To actually use high-quality produced water for irrigation requires (in addition to permission from regulators) a willing farmer to take the water. One significant success story has been under way in the San Joaquin Valley of California. In addition to being one of the premier agricultural regions of the world, this valley is home to significant Oil & Gas operations and is in an arid region subject to frequent water shortages. Since late in the 20th century the industry has been selling about 450,000 bbl/day (72 ML) of water to the local water district and putting it into irrigation canals. The sales price is under market price for groundwater and revenue from water sales helps to offset treatment costs (Waldron).

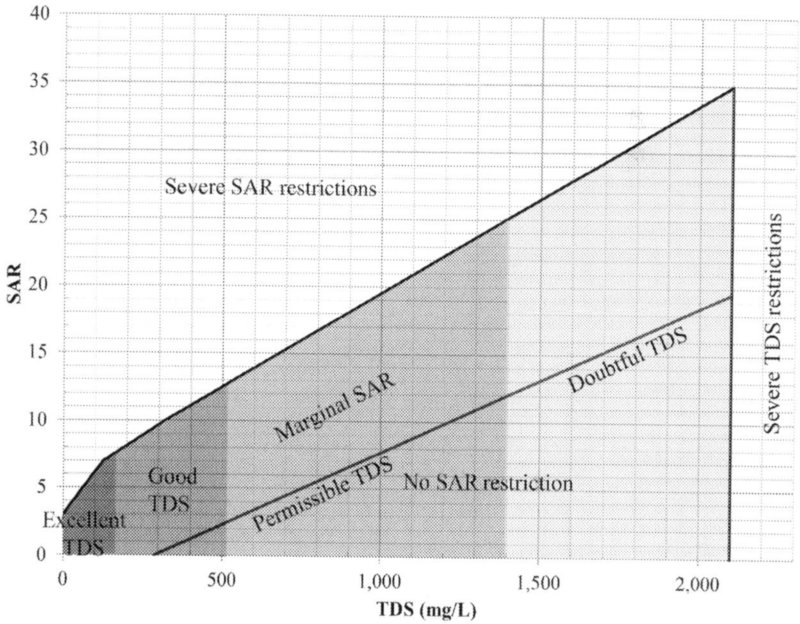

Figure 7.19 TDS vs SAR.

If you can't find farmers willing to take your produced water for irrigation, you have the possibility of purchasing land and hiring farmers. As CBM development has advanced in the Bowen Basin of Queensland, Australia operators have done just that. Two success stories have been: (1) Luecaena and (2) the Chinchilla White Gum.

Luecaena is a fast-growing crop that has been used as feedstock for cattle in Australia since it was introduced in 1921 and has been a factor in increasing beef production. One operator has planted many acres of luecaena under pivot irrigation of treated water and has been using a significant amount of water during the growing season.

Chinchilla white gum is a white bark eucalyptus with an attractive redish hard wood that was harvested nearly to extinction primarily to support the flooring industry. This tree thrives on the water from coalseam gas (CSG, the Australian term for CBM) produced water in the Bowen Basin of Queensland. The trees grow to maturity (about 1 m (3.3 ft) in diameter and 40 m (130 ft) tall) in about 12 years and can be harvested for lumber production. One operator is irrigating 1.2 million trees using drip irrigation. Each tree gets about 3 gal/day (11.4 L/day) and the tree farm uses 80,000 bbl/day (12.7 ML/day). As the field has developed, additional plots have been planted.

The Chinchilla white gum is an endangered species in its native Tasmania. There is a possibility that using this lumber could violate international treaties on endangered species (using the same argument that raising rhinoceros on a farm to harvest their horns violates the treaties), but this is unclear at this writing.

7.3.3.7 Large-scale industrial cooling

Power plants use a lot of water. A 1800 MW (gross) conventional power plant needs to evaporate something on the order of 500,000 bbl/day (79 ML/day) which is 15,000 gpm (946 L/s). Power plants that are not close to the ocean get this water from rivers. Water evaporated in the power plant is not available to irrigation and recreational uses downstream.

We conducted a feasibility study at one power plant of replacing 10% of the river water with produced water (the limitation is due to TDS limitations in the cooling towers) and in this case the leasing of the plant's water rights to downstream users would have brought in enough revenue to cover the capital cost of the project in 5 years. The area was under a severe drought when the study started and the state legislature was enthusiastic about it and offered significant tax incentives to start the work. Shortly before the tax incentives came up for a vote, it started raining and the drought broke. The bill never left the committee and the owner of the power plant lost interest.

While this project did not come to fruition, it is a good idea to try to look for unconventional uses for produced water.

Injected water into the combustion chamber on gas turbines is occasionally done to control the flame temperature (and thereby the NO_x in the exhaust), but the tolerance of this application for dissolved solids is not assured, and salting up a multimillion dollar turbine might not be a career enhancer.

7.3.3.8 Local-scale equipment cooling

Both water-cooled processes and evaporative coolers are common in industrial processes around the world. There must be a way to apply produced water to these applications

Evaporative coolers (also known as "swamp coolers") pull the latent heat of vaporization from the air flowing through them and cool the air in the process. A properly designed swamp cooler in a relatively arid location can lower air temperature by about 20°F (11°C) will add significant mass to the air in the form of water vapor which improves the heat-transfer characteristics of the air. These characteristics should get us thinking about where we have air-cooled processes. The first one that comes to mind is the cooler on a compressor. We frequently have to reduce the throughput on compressors because we can't get enough ambient air through the coolers to maintain the process temperature in an acceptable range. Swamp coolers can be an effective solution to this issue.

Fig. 7.20 shows a couple of examples of places where swamp coolers were used on large compressors. The left-hand compressor is 1000 bhp

Figure 7.20 Evaporative coolers.

(nominal) and the evaporative cooler was built by the field foreman for just under $40,000 USD without any engineering input at all. Before the cooler was installed, there were 20–30 days per summer where the operation of the compressor had to be curtailed due to inadequate cooling. After the swamp cooler was installed the site was never curtailed again. This unit tolerates produced water very well and the mist pads are replaced in the spring of each year and operate without plugging for a year (and when pulled they are generally in good shape).

The right-hand compressor in Fig. 7.20 is a 3600 bhp (nominal) unit and the swamp cooler was designed by a team of engineers for significantly higher cost. It was intended to use produced water, but it was found that the mist pads plugged quickly in produced water service and the site converted to deionized water.

There is no real point to Fig. 7.20 beyond trying to point out that dealing with beneficial uses of produced water, you need to remain aware of limitations of the ultimate use and design the process such that it is the appropriate level of complexity.

In Oil & Gas we use air-to-air heat exchangers for compressor cooling because air is what is always available. In offshore, the ocean is pretty readily available and using seawater for cooling compressors is common. If we are careful where we put our produced water catchments (Fig. 7.21), then it is quite reasonable to use that produced water as cooling water, transferring the waste heat from the compressors into the evaporation process—the definition of a win–win. The design in Fig. 7.21 was a retrofit of compressors that were already installed and had air-to-air heat exchangers that were failing to allow 100% production for nearly 3 months per year due to equipment overheating. After completing this design it became clear that having both air and water cooling was a good thing. In this site the fans on the coolers were electric-motor driven

Figure 7.21 Water cooling equipment.

(with variable speed controls) and the retrofit planned to take the required cooling streams as they entered the air cooler and send them to the water cooler, returning to the entrance to the air cooler. Most of the time the motors on the cooler fans would be off-line, but during problems with the cooling water supply (e.g., while mucking the pond out or field start-up) the air-to-air heat exchangers could carry the load. This project was caught up in a reorganization of the client company and it was never built.

7.4 PLANNING FOR OPERATION

Produced water management is required for gas production in virtually all gas-field developments. We often do a good-enough job of planning for gas well deliquification and understand how it fits into the overall field-development strategy, but it is rare for the accumulated-water management to fit into any plan. We keep getting surprised by how hard it is to transport and dispose of. One of my clients started their field development with a list of "overriding principles" for produced water management that was both forward-looking and integrated into the field-development strategy. These principles were practical and achievable.

- Business requirements
  - Water management must not constrain gas production 95% of the time
  - When water management does constrain gas production
    - Water production must not be constrained by more than 20%
    - Gas production must not be constrained by more than 2%
- Environmental/social requirements
  - Maintain license to operate
  - Be a valued member of the community
  - Absolutely avoid a legacy of enduring environmental damage

The business requirements dictated that the water-treatment/disposal system needed to be both dispersed and interconnected so that an outage in a facility could allow shifting water to another facility during the duration of the outage. It also mandated that all catchments and storage facilities be significantly oversized with useful storage capacity. Without thinking about the business impact of water management on gas production, many operators would build one or two large facilities with a total capacity less than 150% of projected water production. This business plan led to six facilities that each have the capacity to process nearly 40% of projected water production. The cost of the overdesign was made up the first time an equipment failure did not cause gas production to be curtailed.

The planning for the water-system design included a staged response to upsets.

- First response: Increase use of produced water for irrigation and dust suppression
- Second response: Shift water toward more flexible options (like surface discharge and managed aquifer recharge) without increasing flow in the flexible options (i.e., increase stored inventory to those options)
- Third response: Increase flow to flexible options
- Fourth response: Increase flow of operating options into contingency region (use design safety factors)
- Fifth response: Reduce flow from specific wells where water/gas ratio (WGR) is high
- Sixth response: Shut in specific wells with high WGR based on a schedule
- Seventh response: Shut in the field

Thinking through various credible scenarios at the design stage is an excellent start toward having a water disposal system that will enhance gas production rather than curtailing gas production.

7.5 CONCLUSION

Produced water is a large and growing problem. All solutions are expensive and all have drawbacks:

- Deep well injection requires considerable manpower and wells do not have a predictable life, and it can be difficult to find a suitable disposal reservoir.
- Evaporation ponds require considerable real estate and overspray of concentrated solids is harmful to plants and animals.
- Beneficial use options can have unintended consequences and can present significant seasonality (e.g., you can't put water into a river that is at flood stage and plants don't need irrigation water in the winter).

Any option should be reviewed by environmental, regulatory, legal, land, production operations specialists early in the design process. The rules, laws, and regulations are quite complex and often contradictory.

REFERENCES

A&M. Irrigation Water Quality Standards and Salinity Management Strategies, April, 2003, Texas A&M Agrilife Extension Service, Retrieved January 11, 2017, from http://soiltesting.tamu.edu/publications/B-1667.pdf.

Waldron, R.G. C.Hg, Chevron U.S.A., Inc., Southwest Hydrology, November/December 2005.

Webster. (n.d.). Retrieved April 21, 2016, from http://www.merriam-webster.com/dictionary/biogenic (other dictionary references replace "biogenic" with the referenced term).

Wikipedia, 1. Total dissolved solids, retrieved January 3, 2017, from https://en.wikipedia.org/wiki/Total_dissolved_solids.

Wikipedia, 2. Formazine, retrieved January 3, 2017, from https://en.wikipedia.org/wiki/Formazine.

Wikipedia, 3. pH, retrieved January 3, 2017, from https://en.wikipedia.org/wiki/PH.

Wikipedia, 4. Alkalinity, retrieved January 13, 2017, from https://en.wikipedia.org/wiki/Alkalinity.

NOMENCLATURE

Symbol	Name	fps units	SI units
E_0	Evaporation rate for Eq. (7.6)		mm/day
h_{asl}	Height above sea level		m
F	Density correction factor for Eq. (7.6)		decimal
H^+	Molar concentration of hydrogen ion	moles/L	moles/L
OH^-	Molar concentration of hydroxide ion	moles/L	moles/L
PPM_{ca}	Calcium content a water sample	mg/L	mg/L
PPM_{mg}	Magnesium content in a water sample	mg/L	mg/L
PPM_{na}	Sodium content in a water sample	mg/L	mg/L
SG_{approx}	Approximate specific gravity of water with nonzero TDS	decimal	decimal
q_{evap}	Rate of evaporation	$bbl/ft^2/day$	$m_{water}^3/m_{surface}^2/day$
Q_{latent}	Latent heat of vaporization	BTU/lbm	kJ/kg
R_s	Solar irradiance	$BTU/ft^2/h$	W/m^2
SAR	Sodium absorption ratio	decimal	decimal

(Continued)

(Continued)

Symbol	Name	fps units	SI units
T_d	Mean dew point temp for Eq. (7.6)		C
T_m	Mean daily temperature for Eq. (7.6)		C
u	Wind velocity 2 m above ground level		m/s
W_{TDS}	Total mass of dissolved solids in 1 L of water	mg	mg
ρ_{nacl}	Density of sodium chloride as a solid	135.157 lbm/ft^3	2165 kg/m^3
ρ_{water}	Density of pure water	62.4 lbm/ft^3	1000 kg/m^3

EXERCISES

1. Describe the three possible scenarios that can occur at the high point of a water gathering system and discuss the role of high-point vents in each scenario.
2. Which of the following is not a generally used parameter for deciding if produced water can be released to rivers? Choose an item.
   a. TDS less than 5000 mg/L
   b. Turbidity less than 5 FTU (ISO refers to the same unit as FNU)
   c. Oil & Grease less than 35 mg/L
   d. Dissolved oxygen less than 25 mg/L
3. Of the various treatment options which would produce the most concentrated (e.g., highest TDS) brine for a given feed stream? Choose an item.
   a. RO
   b. Freeze/thaw evaporation
   c. Industrial-scale cooling towers
   d. Distillation
4. An evaporation pond is 250 ft × 600 ft (76.2 m × 182.9 m) (at the vertical center of useable space) and 12 ft (3.7) deep to the minimum freeboard height. The water coming into the pond contains 100,000 mg/L TDS. The TDS is primarily NaCl. Inflow is expected to be 2000 bbl/day (318 m^3/day). How long will the pond operate before 80% of the available volume is filled with solids?

5. Water with 1200 mg/L TDS can be used for:
   a. Home drinking water
   b. Irrigation without restriction
   c. Stock watering with minimal restriction
   d. None of the above
6. If the primary concern of a water supply is bacteria, what style of filter would be able to accomplish the task of removing the bacteria?

CHAPTER EIGHT

Gas Compression

Contents

Practical Onshore Gas Field Engineering.
DOI: http://dx.doi.org/10.1016/B978-0-12-813022-3.00008-0

Reservoir gas is rarely suitable for delivery to end users. First, it is not located where the end user is located, so it must be transported. Reservoir gas is also rarely of a suitable quality for end-user equipment and must be processed to a required quality standard with the proper energy content. Every step in the transportation and processing of reservoir gas reduces the pressure of the gas. To boost the gas pressure to the values required for the next step, we compress the gas. This compression-and-use-of-energy cycle is repeated several times in the journey of a molecule of gas from the reservoir to the burner tip. As the gas moves through the systems, it evolves toward becoming a commodity that is universally consistent for all end users. The most appropriate choice for compression technology evolves as well.

Raw well-head gas is quite variable in terms of contaminants, pressures, and temperatures. The best choice from a fluid-mechanics viewpoint might not be the best choice from a space-management or equipment-availability viewpoint. We always have to modify preferences to accommodate the realities of operations and logistics. As we prioritize the various constraints, we need to make the requirements of the reservoir the top priority. For example, the movement of fluids within a reservoir is chaotic and constantly changing; no reservoir is going to provide gas at a constant flow rate at a constant pressure. We can "overcome" this variability by putting an aggressive control valve on the suction of the compressor to take a large pressure drop during maximum flow while still taking a significant pressure drop at minimum flow to allow the compressor to see a constant suction pressure. We've seen in earlier chapters that this kind of variability in flowing bottom-hole pressure does not result in optimum recovery of reservoir fluids. A better solution for the reservoir would be to use a compressor that can adapt to maximum flow and unload to accommodate minimum flow.

The work that a compressor does is the combination of how much stuff it is lifting (mass-flow rate) and how high it is lifting that stuff (compression ratios). The same amount of work is done at a low mass-flow rate and high compression ratios as is done at a much higher flow rate and lower compression ratios. The simplest (and most accurate) representation of compressor work is the change in specific enthalpy times mass-flow rate (Eq. (8.1)). The accuracy of this equation is dependent on the quality of the equation of state that is used to determine the specific enthalpy and the quality of the gas analysis used to describe the gas.

$$W = \dot{m} \cdot \Delta h \tag{8.1}$$

People throw the word "efficiency" around like it has exactly one mathematical meaning and that meaning is an immutable constant. Poppycock. You can define efficiency as "the ratio of work applied to the process divided by the power input to the process" Eq. (8.2).

$$\eta = \frac{\text{Work applied to the process}}{\text{Power input to the process}} \tag{8.2}$$

Here is where it gets messy. Where do you define the "power input?" For a compressor driven by an internal combustion engine, do you consider the energy inherent in the fuel times the mass-flow rate of the fuel? Or do you consider the power that is actually transferred to the input shaft on the compressor, which is generally around 1/5 of the energy in the fuel? On the output side, should you look at conditions at the points you take suction/discharge pressures or should you look at the skid edge? There are no general answers to any of these questions. We have conventions that start to move us toward having a common understanding, but they are far from universal. At the end of the day, you need to pick a compression technology before you are able to assess compression efficiency, so we will develop this topic further in sections talking about specific technology.

8.1 TYPES OF COMPRESSION

Fig. 8.1 shows the families of compressors. The vast majority of well-site compressors and booster compressors on gas gathering systems are positive displacement. Of those, reciprocating compressors ("recips" in the lingo) significantly outnumber all other types of compressors, both in terms of installed units and installed horsepower. Whether that mix of technologies represents an optimum or not is open for debate, but that mix is the reality as of this writing. Dynamic compressors are quite rare on onshore well-sites for the good reason that they don't handle varying conditions well. However, we will introduce a family of dynamic compressors (thermocompressors) in this chapter that could possibly have a significant role in well-site equipment.

There are a number of compressor technologies shown on Fig. 8.1, which we will not discuss in this chapter. In general, these technologies simply lack the reliability and/or operating range to stand up to 24/7 unmanned operations. A machine that needs to be fiddled with seldom fits with our operating strategies.

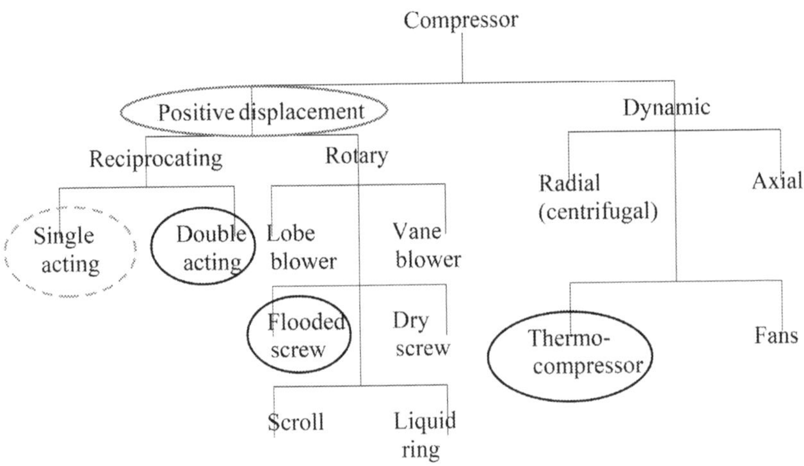

Figure 8.1 Types of compression.

8.1.1 Dynamic

A dynamic compressor is a device that uses added kinetic energy to accelerate a gas at approximately constant pressure. A common type of dynamic compressor is the "centrifugal," which is shown in Fig. 8.2 that the ribs on the impeller are farther apart at the rim of the impellor than at the eye in the center, consequently the flow in the impeller is not precisely isobaric, but it is reasonably close. Velocity of the gas is rapidly increased in the impeller. Once the gas enters the volute, Bernoulli's Law describes how pressure and velocity change as a function of the cross-sectional area of the volute.

For well-defined flows, dynamic compressors can move a lot of gas. Joining a gas turbine to a centrifugal compressor can compress many times as much gas than a positive displacement machine for the same footprint. This characteristic is vitally important on offshore platforms where available real estate is in short supply. A centrifugal compressor can do fewer compression ratios per stage than a positive displacement (PD) compressor, but it is possible (at skid design time) to add stages to the compressor like stages are added to electrical submersible pump (ESP) downhole.

As discussed in Chapter 0, Introduction, Bernoulli's Law assumes that flow is incompressible. At a fairly small fraction of the speed of sound, that assumption is no longer valid. Looking at the performance envelope of a centrifugal compressor (Fig. 8.3), you will see a "choke line" toward the right-hand side. This line represents a real physical limit and can be

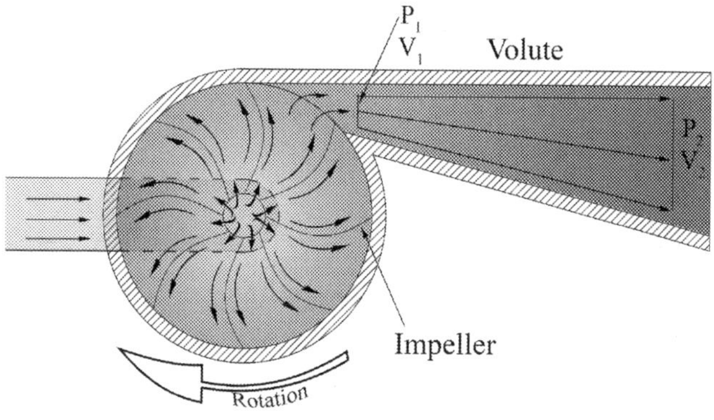

Figure 8.2 Centrifugal compressor.

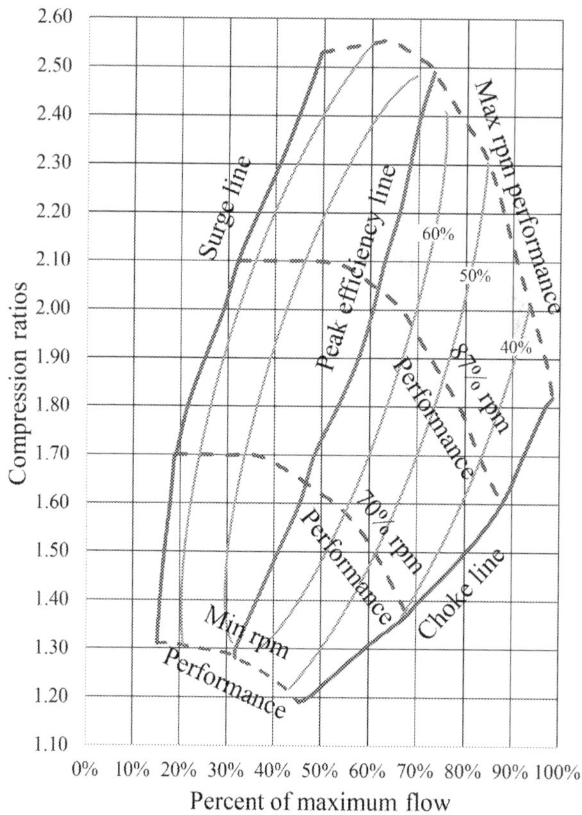

Figure 8.3 Centrifugal compressor performance envelope.

thought of as the point where Bernoulli stops "working" (i.e., where the math stops representing physical reality). One of the major results of Bernoulli stopping is that during compressible flow, an increase in cross-sectional area results in an *increase* in velocity (rather than a decrease as you would expect during incompressible flow) and no change in pressure (rather than the increase you expect during compressible flow).

The line on the left-hand side of Fig. 8.3 marked "surge line" is another physical limit. When downstream pressure approaches the pressure at the outlet of the volute, flow reverses from the discharge line back into the volute, which creates actual surges that can physically damage equipment and must be prevented. We prevent surges with complex recirculation systems that prevent discharge-header pressure from approaching the outlet pressure of the volute. These systems have an impact on skid efficiency, which already isn't very good even before you start sending gas back from the discharge to the suction.

The "peak efficiency" line represents the machine's peak efficiency. In the example in Fig. 8.3, at the minimum rpm, you get best performance if you are trying to move 32% of rated flow with about 1.28 compression ratios (we will talk about compression ratios below, for now treat it as discharge pressure in absolute terms divided by suction pressure also in absolute terms). At maximum rpm in this example, you get the best efficiency at 72% of rated flow and 2.48 compression ratios. The only way to reach 100% of rated flow is to drop compression ratios to about 1.82 at maximum rpm and to accept a fuel efficiency that is significantly worse than on the peak-efficiency line.

The other significant dynamic compressor technology is called "axial" because the compression takes place in a direction parallel to the axis of rotation (as opposed to centrifugal compressors where the flow direction is perpendicular to the axis of rotation). These compressors are very similar to impulse turbines used in steam plants (which simply extract work from steam by uncompressing the steam in a particular configuration). The rotating blades are analogous to the impeller in a centrifugal compressor and the fixed blades have a function similar to the volute. The air-end of a gas-turbine is an axial compressor.

8.1.1.1 Dynamic compressor efficiency

It is common for compressor maps (Fig. 8.3) to have contour lines of constant efficiency. To get an efficiency value you find your flow rate and compression ratios and plot the point on the map. You can do a visual interpellation of the contour lines to find the efficiency.

The overall skid efficiency of a dynamic compressor is less than we typically see in positive displacement compressors (so they can require more fuel than a positive displacement machine for the same throughput). Numbers in the range of 60%–70% are common for the "peak efficiency" line and getting off that line can drop the compression efficiency to under 40%.

8.1.1.2 Polytropic heat of compression

Dynamic compression is a polytropic process, meaning that the polytropic index (n) in Eq. (8.3) is not equal to ratio of specific heats (k) or 1.0 ($n = k$ would be adiabatic, $n = 1$ would be isothermal). Entropy need not be constant, the process need not be reversible (but it generally is), and heat transfer need not be reasonably close to zero. Determining the polytropic index (n) is usually done experimentally.

$$P \cdot V^n = \text{constant} \tag{8.3}$$

If you have a value for n, then you can determine the heat of compression by Eq. (8.4). It is common to look at the measured suction and discharge temperatures and compression ratios and determine the polytropic index that satisfies observed conditions (and then using that value as time moves forward).

$$T_{\text{disch}} = T_{\text{suct}} \cdot (R_c)^{\frac{n-1}{n}} \tag{8.4}$$

8.1.2 Positive displacement compressors

Positive displacement compressors are devices that physically reduce the space available for the gas to occupy as the gas moves from suction conditions to discharge conditions. This is quite visible in a recip (Fig. 8.5) where you can easily visualize the piston moving up in the cylinder, closing the suction valve, and pushing against the discharge valve to make a fixed mass of gas occupy a progressively smaller space.

Positive displacement compressors are able to lift the load much higher than dynamic machines, but they require physically larger components to move the quantity of gas that a dynamic compressor moves in smaller components.

8.1.2.1 Positive displacement compressor work

Eq. (8.1) contains all of the information required to determine the work that has been successfully applied to a gas, but specific enthalpy is not a term that is always readily available. Using terms that are more accessible, we can calculate the work using Eq. (8.5). The enthalpy in Eq. (8.1) contains all of the adjustments for adiabatic changes and for compressibility, and if you are rigorous in determining efficiency, density, and compressibility Eq. (8.5) is amazingly close to the results you get from Eq. (8.1).

$$W = \frac{C}{\eta_{\text{total}}} \cdot P_{\text{suct}} \cdot q_{\text{mmscf}/d} \cdot \left(\frac{\rho_{\text{std}}}{\rho_{\text{suct}}}\right) \cdot \left(\frac{k}{k-1}\right) \cdot \left((R_c)^{\frac{k-1}{k}} - 1\right) \cdot \left(\frac{Z_{\text{avg}}}{Z_{\text{suct}}}\right) \tag{8.5}$$

"Compressor efficiency" is a term with considerable gray area and space for (mis)-interpretation. "Compression ratios" on machines without suction/discharge valves is the outlet pressure divided by inlet pressure (in absolute pressure units). We will discuss compression ratios on compressors with valves in Section 8.2.1.3.

8.1.2.2 Positive displacement compressor efficiency

In simplest possible terms, "efficiency" is given in Eq. (8.6). Driver output is the power delivered to the shaft of the compressor. The delivered power could be the reading on a power meter times a motor-efficiency value. It could be a value calculated from an engine manifold pressure. It could be the reading from a dynamometer.

$$\eta = \frac{W_{\text{out}}}{W_{\text{in}}} = \frac{\dot{m} \cdot \Delta h}{\text{DriverOutput}} \tag{8.6}$$

Compressor efficiency can be calculated without knowing driver input at all. Those techniques are quite different for rotary and reciprocating equipment and will be discussed below.

8.1.2.3 Impact of changing suction pressure on a positive displacement compressor

Determining what will happen inside a compressor when suction pressure changes is crucial to the application of positive displacement compression technology. If you take Eq. (8.5) and rearrange it to group all the terms that remain essentially constant with a change in suction pressure (Eq. (8.7)), it is clear that a change in suction pressure per unit volume would change the work required in the opposite direction (suction pressure is on the bottom of the R_c term so a decrease in suction pressure would raise R_c and therefore work). At the same time, positive displacement compressors move the same physical volume per revolution of the shaft regardless of suction pressure and at a lower pressure the volume flow rate at standard conditions would be lower than it would be at higher pressures. Understanding what happens when you change suction pressure requires an example.

$$W = \left(\frac{C \cdot P_{std} \cdot k \cdot \cancel{T_{suct}} \cdot \cancel{Z_{avg}}}{\cancel{T_{std}} \cdot Z_{std} \cdot (k-1) \cdot \eta_{total}}\right) \cdot \left(q_{mmscf/d}\right) \cdot \left(\left(\frac{\cancel{P_{disch}}}{P_{suct}}\right)^{\cancel{\frac{k-1}{k}}} - \cancel{1}\right) \tag{8.7}$$

Example of changing suction pressure. A screw compressor is installed on a well-site. The conditions at the start of the example are in Table 8.1.

We will discuss screw compressor efficiency later, but for now, the efficiencies are: (1) case 0 is 73%, (2) case 1 is 61%, and (3) case 2 is 69%.

Table 8.2 shows that energy input requirements change in the same direction as the change in suction pressure because the specific work change is smaller than the change in flow rate. It is very common to look at the specific work per MMSCF and draw conclusions (Lea, Table 6.1) that approach generally leads to poor decisions and even worse results.

8.1.2.4 Adiabatic heat of compression for positive displacement compressors

Gas compression using positive displacement equipment is generally taken as an "adiabatic" process. As we've discussed before, "adiabatic" means

Table 8.1 Suction pressure change example initial conditions

	fps	Both	SI
VI		4.6	
k		1.306	
Speed		1200 rpm	
q_0	1 MMSCF/day		28.3 kSCm/day
$V_{revolution}$	0.053 ft^3		0.001503 m^3
P_{atm}	12 psia		82.7 kPaa
P_{suct0}	14.5 psig		100 kPag
P_{disch}	84 psig		579 kPag
T_{suct}	80°F		26.7°C
T_{disch}	205°F		96.1°C

Case 1 → Suction pressure changed to 29 psig (200 kPag)
Case 2 → Suction pressure changed to 0 psig (0 kPag)

Table 8.2 Change suction pressure example conclusion

	Case 0	Case 1	Case 2
Suction pressure	26.5 psia (187.7 kPaa)	41 psia (283 kPaa) (+55%)	12 psia (82.7 kPaa) (−55%)
Flow rate	1.0 MMSCF/day (28.3 kSCm/day)	1.55 MMSCF/day (43.9 kSCm/day)	0.452 MMSCF/day (12.8 kSCm/day)
Compression ratios	3.62	2.34	8.00
Efficiency (%)	73	61	69
Specific work/MMSCF	95.2 hp (71 kW)	71.3 hp (53.2 kW)	179.8 hp (134.1 kW)
Total work	95.2 hp (71 kW)	110 hp (82 kW) (+16%)	81.3 hp (60.6 kW) (−14%)

that the flow exhibits no appreciable heat transfer from or to the environment (which tends to imply is both isentropic and "reversible"). If you've ever laid your hand on the discharge pipe from a reciprocating compressor, you know with confidence that there is heat lost to the environment, but in terms of the total thermal energy in the compressed gas, the heat that burned your hand is an insignificant portion of the total and the adiabatic assumption is acceptably valid for practical purposes.

For any adiabatic compression process, the discharge temperature is the same as it was in Eq. (3.18) and Eq. (4.31) and repeated here as Eq. (8.8). The temperatures are in absolute units.

$$T_{\text{disch}} = T_{\text{suct}} \cdot (R_c)^{\frac{k-1}{k}} \tag{8.8}$$

8.2 RECIPROCATING COMPRESSORS

Reciprocating compressors have been used since the 1800s, originally in steam driven air service (Fig. 8.4). Pistons moving inside of cylinders draw gas in through suction valves, then the pressure is increased by reducing the physical volume inside the cylinder and the higher pressure gas is exhausted out of the discharge valves.

Figure 8.4 Steam driven air compressor (Thanks to San Juan County Historical Society Mining Heritage Center, Silverton, CO for allowing photography inside).

8.2.1 Operating principles

In its simplest configuration, a reciprocating compressor has a cylinder, piston, connecting rod, and suction/discharge valves. You can see from Fig. 8.5 the connecting rod cannot be directly connected to a crank shaft in a double-acting cylinder as the rod must physically enter the compression chamber, so it must be sealed. We make the transition from the distinctive motion of a rod that accommodates a crank shaft to a rod that accommodates entering a compression chamber in a "crosshead" which converts a two-dimensional motion to a linear motion.

The "throw" in Fig. 8.5 is a fairly rare configuration. It is much more common for the cylinder diameter to be the same on the crank-end (first stage in Fig. 8.5) as on the head-end (second stage in Fig. 8.5).

We categorize recips by (1) number of "throws," (2) "action," (3) number of stages, (4) separable or integral, and (5) high speed or low speed.

Compressor throw. An industry term for a cylinder/piston pair (Fig. 8.5 is one throw).

Action. An industry term indicating if a cylinder has one compression chamber ("single acting") or if it has two compression chambers ("double acting") (Fig. 8.5 is double acting).

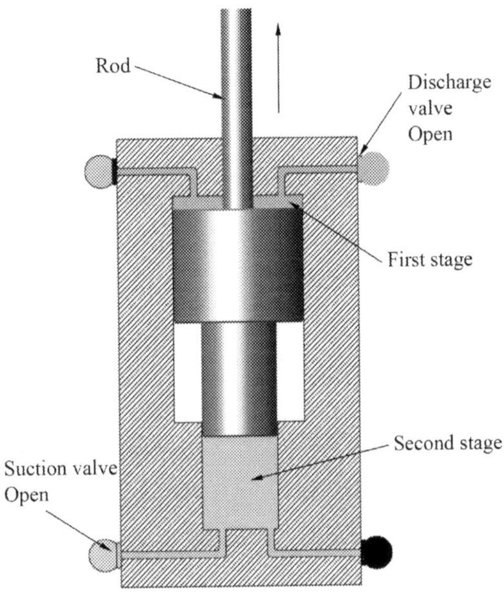

Figure 8.5 Double acting recip cylinder mock up.

Stages. The gas leaving a compressor cylinder can go to many places. On some skids, the gas goes directly to an end-use, on other skids the gas goes to an after-cooler before going on to an end use, other skids send the gas to an interstage cooler and to another cylinder on the same skid. In Fig. 8.5 the gas goes from the upper right-hand valve (first-stage discharge) to an interstage cooler and back to the lower right-hand valve (second-stage suction), so this double-acting throw would be two stages, meaning that the gas is compressed twice. It is important (and often difficult) to balance the load between stages of compression. The main tool that we use to allow us to balance stages is to adjust "clearance" to have the early stages to do more (or less) work to transfer the work toward (or away from) later stages. As discussed below, "clearance" is the portion of the cylinder volume that is in the compression cylinder when the discharge valve closes.

Integral vs Separable. In an integral compressor there is a single crank-shaft shared by the driver and the compressor. A separable compressor is designed for the compressor to be matched up with a driver by the packager and each component stands alone. Integral compressors were quite common in the early days of internal-combustion-engine driven compressors (i.e., the end of the dominance of steam power) but have fallen out of favor as separable machines have come into their own.

High speed vs low speed. Originally recip compressors were all designated as low speed and typically had a maximum speed less than 300 rpm. Machines able to operate well above 1000 rpm were developed to better utilize the load curve of electric motors and industrial engines. Today, people tend to equate "integral" with "low speed" and "separable" with "high speed" and the high-speed/low-speed designation is largely meaningless.

8.2.1.1 Compressor valves

Rotary compressors transfer gas from one section of the compression chamber to the next section in a manner that the discharge gas is physically removed from the suction chamber as the pressure is increased. In a recip, the same space is used for both suction and discharge, so a mechanical barrier is required between the inlet/outlet plenums and the compression chamber. In most recips this mechanical barrier is provided by spring-loaded check valves. When the differential pressure across the valve is high enough to overcome spring tension and is in the right direction, the valve opens. As the piston descends in the cylinder, the bypassed gas (i.e., the gas in the clearance) must be expanded. As the pressure drops in the expanding gas trapped in the clearance, eventually the differential pressure gets high enough to open the suction valve and allow suction gas into the cylinder. When the piston reaches the lower limit of travel ("bottom dead center") and starts back up, the gas in the cylinder begins to compress and the dP between the compression chamber and the suction plenum falls below the spring tension and the suction valve closes. Further up the cylinder, the pressure within the cylinder gets high enough for the dP to exceed the spring tension on the discharge valve, and the discharge check valve opens to allow the gas to exit into the outlet plenum. This process repeats every revolution of the crank (Fig. 8.6).

The "stiffness" of the valve springs (i.e., the amount of dP required for them to open) is selectable from a list of valve options. The stiffer the spring, the greater the dP required to open them. People tend to use stiffer springs on the discharge valves than the suction valves because the volume flow rate out of the cylinder is so much greater than the volume flow rate into the cylinder and on the discharge side you can tolerate a shorter open time.

A recip compressor will generally be intolerant of varying suction pressure. If you lower suction pressure, the suction valve opens later (for

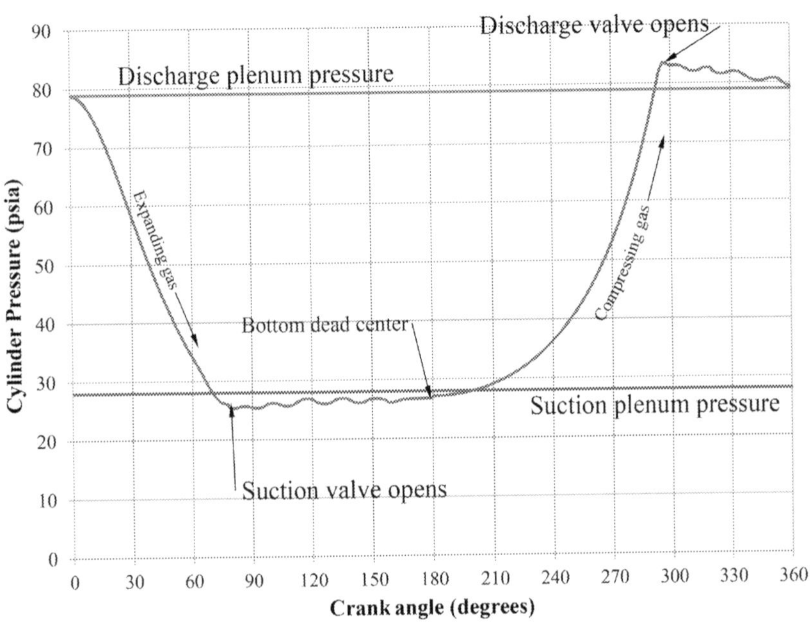

Figure 8.6 Recip compressor crank angle.

the same clearance and discharge pressure) which means that it doesn't have as much time for gas to flow into the cylinder, but more importantly, the lower pressure inside the cylinder when the suction valve closes requires more compression ratios which causes the temperature of the discharge gas to increase. It is very common for a seemingly small decrease in suction pressure to result in a large increase in discharge temperature.

In general terms, you should try to maintain the suction pressure on a recip within ±5% of design conditions. This means that if you have designed the compressor for 30 psia (207 kPaa) suction pressure, you want the suction pressure to remain in the range of 28.5 psia (197 kPaa) to 31.5 psia (217 kPaa), which represents less than 7 ft (2.13 m) of liquid level change in a well-bore. If the design suction pressure were 12 psia (83 kPaa) then the allowable liquid level change drops to 1.2 ft (366 mm)). For suction pressures much over 145 psig (i.e., the "high pressure" region of Fig. 0.4), the impact of this limitation is significantly reduced. Pressures that are high are common in second and third stage of multistage compressors and most midstream compressors.

People get around the ±5% limitation by installing a pressure-control valve on the compressor suction that is set to take a pressure drop that is larger than expected variation in flowing pressure. In general, this practice has a detrimental effect on reservoir performance, but some people see the benefits of recips to be greater than the harm of throttling the reservoir.

8.2.1.2 Compression ratios

For compressors with suction/discharge valves, you have to account for the pressure drop (cracking pressure) of the suction/discharge valves. As it is rare to have a pressure transducer on the piston-side of the valves, you have to estimate the cracking pressure of the valves and correct the measured pressures for this cracking pressure (Eq. (8.9)). If you don't have any manufacturer's data to help you estimate the valve cracking pressure, using 5 psi (34.5 kPa) for the suction and 10 psi (69.0 kPa) for the discharge provides reasonable answers in most well-site and gathering applications.

$$R_c = \frac{P_{\text{disch}} + dP_{\text{disch}}}{P_{\text{suct}} - dP_{\text{suct}}} \tag{8.9}$$

Example. For a specific compressor, the conditions are as follows:

- Atmospheric pressure: 12 psia (82.7 kPaa)
- Suction: 14.5 psig (100 kPag((26.5 psia (182.7 kPaa)) at 70°F (21.1°C) (529.7 R (294.3 K))

- Discharge: 74.5 psig (514 kPag) (86.5 psia (596 kPaa))
- Techniques used:
  - Ratio of gauge pressures: 74.5/14.5 = 5.14
  - Ratio of absolute pressures: 86.5/26.5 = 3.26
  - Ratio of cylinder pressures: (86.5 + 10)/(26.5 − 5) = 4.49
- Discharge Temperatures
  - Ratio of gauge pressures: 314°F (157°C)
  - Ratio of absolute pressures: 237°F (114°C)
  - Ratio of cylinder pressures: 290°F (143°C)

It is common for a discharge temperature over 300°F (149°C) to result in an automatic compressor shut down, so if the correct compression ratio was "gauge pressure" then the compressor would be down on high discharge temperature. If the ratio of absolute pressures was right then it would be reasonable to run the compressor harder to try to move more gas (the normal design limit for recips is 4.5 ratios per stage), but that would almost certainly shorten the life of the compressor. Using the ratio of cylinder pressures in this case, allowed the calculated value to exactly match the measured value in the field.

For design conditions, engineers try to keep compression ratios above 3.5 and below 4.5. Below 3.5, the efficiency decreases (see below) and above 4.5 the temperatures and mechanical stresses become too great.

8.2.1.3 Compressor efficiency

For recips, you need to account for: (1) unswept volume (also called clearance); (2) leakage around valves, piston rings, and rod packing; (3) friction losses; and (4) number of stages.

Volumetric efficiency. The manufacturer of the compressor will provide a clearance number as a percentage of the cylinder volume. If your compressor has the ability to adjust clearance in the field, then the manufacture will have a conversion from number of turns on the variable volume pocket to total clearance. Eq. (8.10) provides the volumetric efficiency. In the compression ratio example above, with 5% clearance the volumetric efficiency would be 0.891.

$$\eta_{\text{volumetric}} = 1 - \left((R_c)^{\frac{1}{k}} - 1\right) \cdot \left(\frac{V_{\text{clearance}\%}}{100\%}\right) \tag{8.10}$$

Compression efficiency. At low compression ratios, time that the valves are open become too large a proportion of the total time and the adiabatic assumption becomes less valid. Consequently, compression efficiency is

- Below 1.5 compression ratios → 0.50
- Above 3.5 compression ratios → 0.92
- Between 1.5 and 3.5 compression ratios → $0.21 \times R_c + 0.185$

For the compression ratio example above you would use 0.92.

Mechanical efficiency. Friction losses are very difficult to assess and if there is adequate mass-flow rate to carry off the heat generated by friction, then its impact is typically small. Mechanical efficiency is generally taken to be 0.95.

Stage efficiency. These three factors are combined in Eq. (8.11) to get a total efficiency for each stage. Each stage will have a different stage efficiency, but for a properly balanced multistage compressor (rare), the values will be similar and it is common to assume that all stages are equal to the first stage conditions.

$$\eta_{\text{stage}} = \frac{\eta_{\text{volumetric}} \cdot \eta_{\text{compression}}}{\eta_{\text{mechanical}}} \tag{8.11}$$

Total efficiency. The efficiency of the compressor is shown in Eq. (8.12). If the example in the compression ratio discussion is a single stage, then it would have 86.2% efficiency. If there are two additional stages with similar compression ratios, then the compressor would have 64% efficiency.

$$\eta_{\text{total}} = \eta_{\text{stage1}} \cdot \eta_{stage2} \cdot \eta_{stage3} \cdot \ldots \approx \eta^{n}_{\text{FirstStage}} \tag{8.12}$$

Meaning of efficiency. In general terms, we hope that compressor efficiency is a clear indicator of fuel consumption and that a compressor with higher efficiency would use less fuel (with the same driver, same gas, and the same conditions) than a compressor with lower efficiency. On the contrary, that is almost never the case.

The compressor skid in Fig. 8.7 is reasonably representative. The operator was having a problem loading the second stage (high first stage temperature), so he installed a backpressure valve.

Flowing conditions are as follows:

- q_{gas} → 1 MMSCF/day (28.3 kSCm/day)
- q_{water} → 10 bbl/MMSCF (56.1 m^3/MSCm)

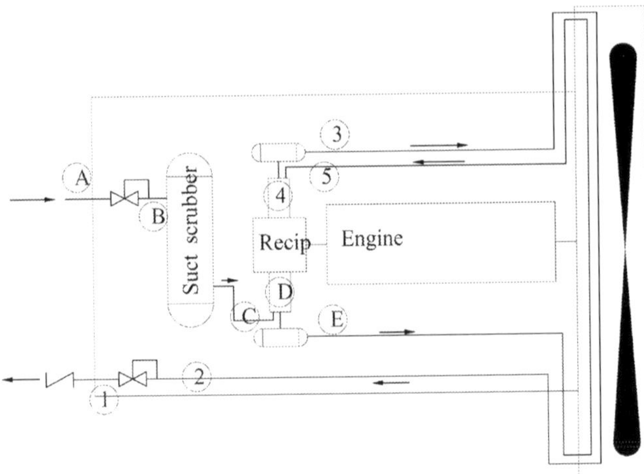

Figure 8.7 Recip efficiency example.

- c_p → 0.1925 BTU/lbm/R (0.8060 kJ/kg/K)
- c_v → 0.1473 BTU/lbm/R (0.6167 kJ/kg/K)
- k → 1.306

The conditions at each point in Fig. 8.7 are tabulated in Table 8.3.

The change in energy from one point to the next will either be $W = m \times \Delta h$, $W = m \times c_p \times \Delta T$, or $W = m \times c_v \times \Delta T$ depending on what is going on in the step.

The energy traverse is shown in Table 8.4. The normal way to calculate compression hp is to ignore the heat lost to the atmosphere in the inner and after coolers as that is energy that is not usable.

Using the excellent Performance.EXE program from Ariel Corporation (Ariel), this compressor should have used 74 hp in the first stage and 81.4 hp in the second stage for a total of 155.4 hp (about ±5% difference in each cylinder from the energy traverse, probably due to a slight difference in the unswept volume of the two cylinders, total energy was the same as the energy traverse).

The compressor skid in this example had a Waukesha F-18 LE driver. Output horsepower calculated from manifold pressure was 220 hp. Fuel usage was 2.54 MMBTU/h (744 kW h/h) (73 MSCF/day (2.1 kSCm/day)). This would make skid efficiency one of the following:

- Theoretical compressor efficiency → 81%
- Compressor net (E → 3) vs engine output → 70%
- Compressor net (E → 3) vs energy in fuel → 22%
- Compressor skid (A → 1) vs engine output → 7%

Table 8.3 Efficiency example data points

	Map point	Press (psig)	Temp (°F)	Specific enthalpy (BTU/lbm) (REFPROP)
Well-head	A	60.0	70	353.13
After suction controller	B	20.0	68	353.31
Upstream of first stage	C	15.0	68	353.45
In first stage cylinder when suction valve opens	D_s	13.3	77	357.72
In first stage cylinder at end of discharge stroke	D_d	71.4	251	443.45
Downstream of discharge valve	E	66.9	251	443.72
Upstream of second stage suction valve	5	65.5	120	376.98
In second stage cylinder when suction valve opens	4_s	61.4	127	380.49
In second stage cylinder at end of discharge stroke	4_d	264.8	332	485.31
Downstream of discharge valve	3	255.2	332	485.43
Before backpressure valve	2	250.0	120	372.59
Line pressure	1	160.0	115	372.27

Table 8.4 Efficiency example energy traverse

	Equation	ΔT or Δh	BTU/h	hp
A→C	$m \times c_v \times \Delta T$	−2 R	−593	−0.2
C→E	$m \times \Delta h$	90.3 BTU/lbm	181,612	71.4
E→5	$m \times c_p \times \Delta T$	−118 R	−45,700	−18.0
5→3	$m \times \Delta h$	108.5 BTU/lbm	218,187	85.8
3→2	$m \times c_p \times \Delta T$	−192 R	−74,359	−29.0
2→1	$m \times c_v \times \Delta T$	−5 R	= 1482	−1.0
Traverse total (including heat transferred to ambient in cooler)				108.4
Traverse total (without heat lost to ambient in cooler)				155.4
A→1	$m \times \Delta h$	19.14 BTU/lbm	3697	15.1

In other words, due to the limitations of the technology and the design choices, this unit is only able to apply 7% of the energy from the engine to the problem of getting the gas up to line pressure, so it would use twice as much fuel as a skid that could transfer 14% of the engine output into the gas. We will revisit this example in our discussion of oil flooded screws.

8.2.1.4 Limiting capacity

A recip will move about the same volume every revolution of the crank. At a constant suction pressure it will move the same mass. If the compressor is not keeping up with inflow then suction pressure will increase along with the mass-flow rate per revolution; this means that the work required compress the gas will increase and can cause the driver to stall. We generally address inadequate compressor capacity using suction pressure controllers that keep the pressure at the compressor inlet plenum at an acceptable level.

If the compressor is trying to move more mass than is being supplied, then suction pressure will drop, which will lower the mass-flow rate per revolution. If the compressor keeps out-running the inflow, the suction pressure will continue to drop until either the mechanical forces get so high that you exceed the allowable load on the rods or the discharge temperature gets so high that the compressor controls stop the machine. On a recip, we try to prevent overrunning supply by progressively:

- Reducing compressor speed
- Increasing clearance
- Removing some of the suction valves (generally called "crippling the cylinder")

These actions are progressively more intrusive. Changing compressor speed (on an engine driven compressor) is a minor adjustment to a governor that can be done with the engine running. For an electric-motor driven compressor, if the motor has a variable speed drive then changing speed is a simple adjustment. For very small compressors that are belt driven, you can adjust the size of the sheaves to change the compressor speed.

Increasing clearance is sometimes just a matter of turning the compressor off and turning an adjustment wheel. Other times, you have to remove the valve(s) and add a spacer (called a "chair") under the valve(s).

Completely removing suction valves turns the cylinder into a place in the line that does no work. Suction gas flows into the cylinder as it descends and then flows back into the inlet plenum as the cylinder ascends.

8.2.2 Other recips

The most common compressors in Oil & Gas are high-speed, separable, and double-acting recips. There are a couple of manufacturers that still make integral machines (still double acting), but they are fairly rare and tend to be used in locations where the decision maker "has always used" integral machines from a specific manufacturer. They tend to be massive machines that require more extensive foundation and other site work than separable machines do.

The other frequently-discussed recip is a modified industrial internal combustion engine. Generally, these will be a V-8 configuration with one bank of four cylinders converted from internal combustion to compression by changing the heads. Obviously, these compressors are integral (single crank shaft) and single acting (traditional connecting rod below the piston can't be sealed). These compressor/engine combinations have a water-cooled block, so the lion's share of the heat of compression is carried away by the coolant. These hybrid machines have reported long-term performance at over 10 compression ratios without excessive gas-discharge temperatures.

8.3 OIL-FLOODED SCREW COMPRESSORS

Work began on developing a positive-displacement rotary compressor to overcome surge problems with dynamic compressors in the 1930s and proceeded through the 1960s (see reference (SRM) for an interesting discussion). In the reference there is a list of companies that have licensed Svenska Rotor Maskiner (SRM) technology and it reads like the Who's Who of the screw-compressor industry. Initial development was for a dry screw where the male rotor was driven by a motor or engine and the female rotor that was driven by a timing chain to minimize wear on the lobes. Dry screws are available today, but they tend to have a very poor reputation in Oil & Gas because of inconsistent fluids, poor lubrication, and problems with rotor timing.

In the late 1940s, basic patents were issued to SRM for an "oil injected screw" that is functionally identical to the units commonly called "oil flooded screw compressors" or "oil injected screw compressors." Oil flooded screws still have the male rotor driven by the driver, but the female rotor is now driven by the male rotor. The oil injection was intended to: (1) prevent metal-to-metal contact between the rotors, (2) seal the area around the rotors to minimize leak-back, (3) lubricate the rotors, and (4) cool the process.

Initially, oil-flooded screw compressors were used within plants either for refrigeration (generally called "process derivative") or for compressed air (generally called "air derivative"). Plant compressors usually don't have to be very flexible:

- The oil only has to be compatible with one gas.
- The gas generally has a very low water vapor content (even air is rarely more than 20% RH when compressed to useful values).

- The process is consistent enough to allow the differential pressure across the skid to pump the oil.
- Compressors are generally set on a rigid foundation, so the skids do not have to provide all of the structural support.

In the 1990s, we started moving oil-flooded screw compressors to well-sites where the gas was subject to change significantly day to day let alone year to year. We did not have much control over either suction or discharge pressure. To top that off, we nearly always had a very large water vapor content.

Unmodified plant packages turned out to be an unmodified disaster on well sites. We immediately found the following:

- Oil selection had to be compatible with a range of condensable hydrocarbons, not just one.
- The oil-temperature control resulted in the oil being too cool (more about this below) and the control method was too inflexible.
- Differential pressure across the skid was too inconsistent to allow safe use without an oil pump.
- Accumulated solids were quite common and the package did not anticipate having to open pressure vessels to shovel the solids out.
- The plant skids were too light to allow rough field handling (things like tail-rolling the skid off a truck or total lack of a prepared foundation).

Over time, packagers have gotten progressively better at designing packages that are suitable for field use, but even today you occasionally see a plant package being deployed to a wellsite. The second most-important lesson we learned on our very steep learning curve with screw compressors was that the packager was far more important than the compressor manufacturer or the derivative industry. We found that a well-packaged process-derivative machine from one manufacturer would have about the same life expectancy and field performance as a well-packaged process-derivative machine from another manufacturer. We found that a well-packaged air-derivative compressor from a third manufacturer would perform just as well. We also found that poorly-packaged machines broke frequently and horribly regardless of the industry or manufacturer.

8.3.1 Configuration

The basic shape and relative size of the rotors (Fig. 8.8) was developed mostly through trial and error and has been refined by many companies based on computer modeling to both make the rotor more efficient and less expensive to manufacture.

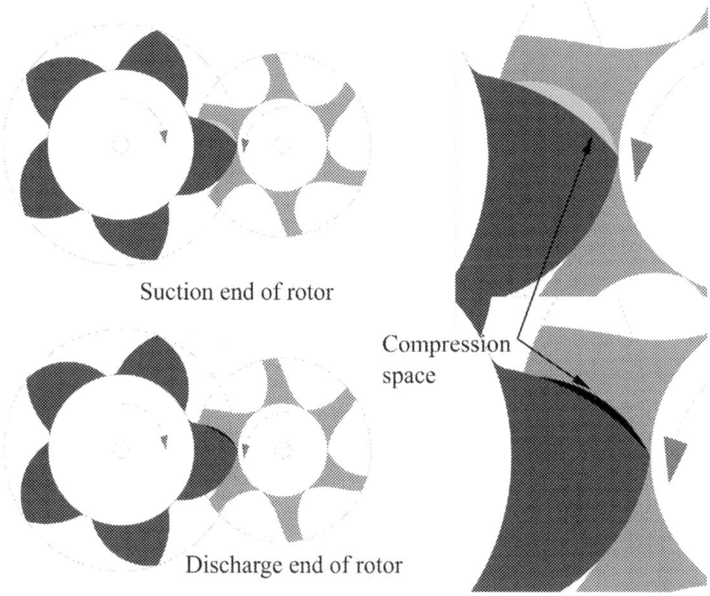

Figure 8.8 Flooded screw rotors.

Notice in Fig. 8.8 that the compression space is much larger at the suction than at the discharge. This relationship is called the "Volume Index" or "VI." A useful representation of the VI relationship is shown in Eq. (8.13). This says that you will reach maximum efficiency if discharge pressure (in absolute units) is equal to the suction pressure (also in absolute units) times VI raised to the adiabatic constant. Other values for discharge pressure are possible (see below) at a lower efficiency.

$$\mathrm{VI} = \left(\frac{P_{\mathrm{disch}}}{P_{\mathrm{suct}}}\right)^{1/k}$$
$$P_{\mathrm{disch}} = P_{\mathrm{suct}} \cdot \mathrm{VI}^{k} \tag{8.13}$$

Most compressors have VI established in the factory and it is not field-adjustable (some manufactures sell a replacement "VI-plate" that can be changed by mechanics in the field in about a day's work). Some compressors have adjustable VI configurations. These "adjustable VI" or "variable VI" machines often allow the VI to be changed by field-automation on the fly. My preference is to never allow the field-automation to adjust the VI at all, I have operated compressors where changing the VI was the first step in an unloading scheme, and the results were universally poor (increased failures and machines operating at a bad place in their load curves at all times).

In addition to the VI, we also talk about the "unloader." This is a variable port to take suction gas from early in the rotor and dump it back to the suction, only doing minimal work on the gas. In process-derivative machines, this unloader is nearly always a "slide valve." In air-derivative machines, it may be a turn valve, a poppet valve, or missing altogether. I have found no operational difference between turn valves and slide valves. Poppet valves have poor throttling characteristics so an unloading scheme with poppet valves can be too abrupt.

If you include unloaders (or variable VI for that matter) in your automation design, drive the operators with oil instead of gas. Pneumatic actuators on these devices tend to overshoot and often move independent of the control signal.

Some compressor manufacturers include reduction gears in some of their frames that have either step-up or step-down gearing. These integral gears can be a very effective way to maximize driver/compressor performance. Trying to accomplish the same result with 3rd party gear sets has not been nearly as effective since the lubrication requirements of general-purpose gears are not compatible with the characteristics of the oil available on the skid.

Finally, we talk about "rotor diameter" and "length to diameter ratio (L/D)." The rotor diameter refers to the maximum diameter of the male rotor, and all other things being equal, a larger rotor will move more gas. The larger the L/D ratio is, the larger the compressor capacity, at the cost of lower maximum differential pressure. Shorter L/D ratios can move less gas, but at a higher differential pressure.

8.3.2 Efficiency

As the gas moves from the suction plate (100% point in on *x*-axis in Fig. 8.1) to the discharge plate (which is located at about 28% in Fig. 8.9), the pressure changes in a predictable manner. A compressor configured to satisfy Eq. (8.13) will smoothly compress the gas up to the discharge plenum pressure and the gas will flow out the end of the rotor with maximum efficiency.

If Eq. (8.13) predicts a higher pressure than the actual discharge plenum pressure then the compressor will actually satisfy Eq. (8.13), and then dump pressure to get to actual outlet pressure (wasting energy). This is called "over compression."

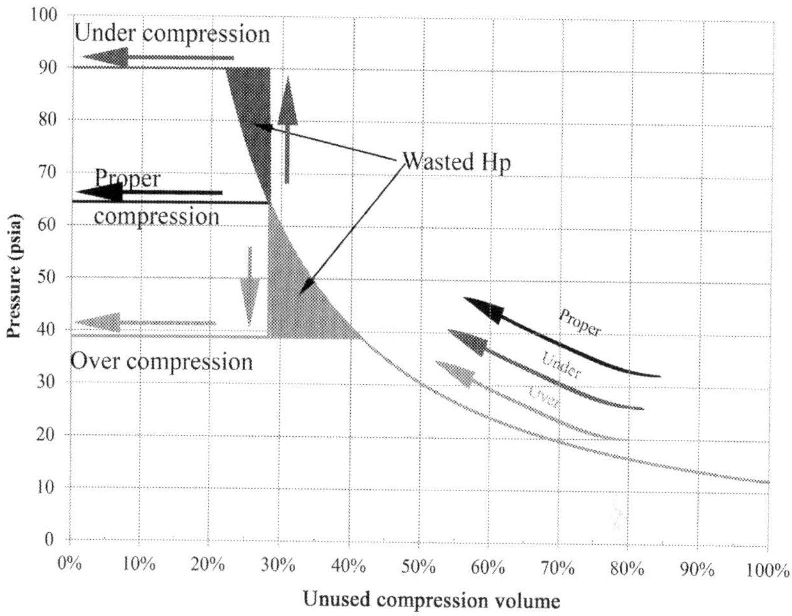

Figure 8.9 Screw compressor PV.

If the actual discharge pressure is higher than Eq. (8.13) then the compressor has to "stuff gas" into the outlet plenum which results in gas compressing gas (the vertical up-arrow in Fig. 8.9) instead of steel compressing gas (the smooth line), again resulting in wasted hp. This is called "under compression."

The two wedges marked "wasted hp" on Fig. 8.9 are different size. "Over compression" wastes more hp than the same amount of "under compression." Fig. 8.10 shows why. Notice that all of the curves are much steeper to the left of the peak and then taper off slowly to the right. This is an indication that a slight under compression has a smaller impact on unit efficiency than the same magnitude of over compression. The curves in Fig. 8.10 are based on Eq. (8.14). The peak enthalpy of the gas is calculated at target discharge temperature and theoretical discharge pressure (using the second equation in Eq. (8.13) to calculate discharge pressure). You can see from Eq. (8.14) that peak efficiency is determined by the work done on the oil. If the mass-flow rate of oil is high and the change in enthalpy of the oil is high then maximum efficiency is very low.

$$\eta_{\text{screw}} = \frac{\dot{m}_{\text{gas}} \cdot \left(\Delta h_{\text{gasPeak}} - \left|\Delta h_{\text{gasPeak}} - \Delta h_{\text{gasActual}}\right|\right)}{\dot{m}_{\text{gas}} \cdot \Delta h_{\text{gasPeak}} + \dot{m}_{\text{oil}} \Delta h_{\text{oil}}} \tag{8.14}$$

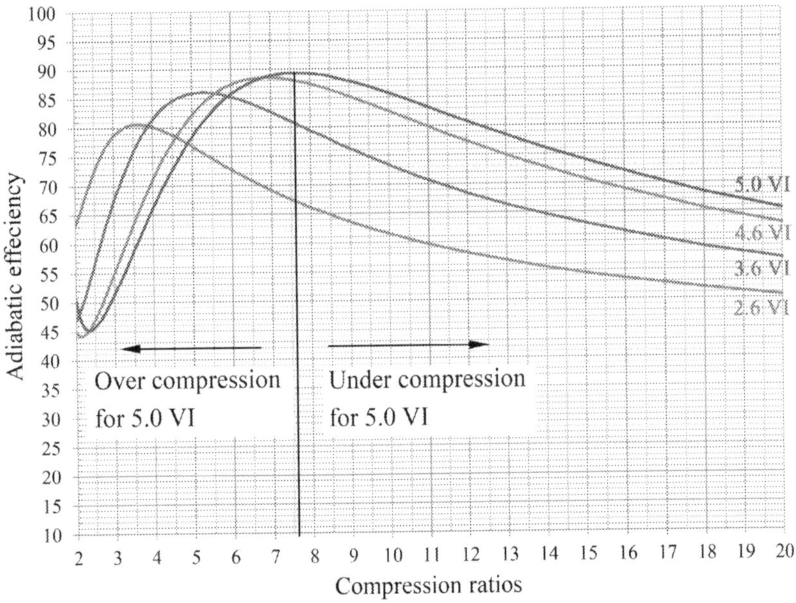

Figure 8.10 Screw compressor efficiency.

Example. Let's revisit the example in Fig. 8.7, but with the screw compressor in Fig. 8.13. Replicating Tables 8.3 and 8.6.

This energy traverse matches the formal computer models which also say that this is a 74 hp application. Calculated input horsepower to the shaft is 102 hp. Notice that the work done on the oil ($D \rightarrow 4_o$) is 31% of the total power out of the compressor, which means that in this case the compressor efficiency can't be better than 69%.

The results in Table 8.6 will be compared to Table 8.4 in Section 8.8.

8.3.3 Compressor oil

The compressor oil is key successful screw compressor operation. The kinds of oil that we talk about are (1) mineral oil, (2) synthetic oil, and (3) semisynthetic oil. Each has strengths and weaknesses and each has a place where it excels.

Mineral oil. Least expensive, but it tends to be an excellent solvent for heavier hydrocarbons and when it takes on butanes and propanes (etc.) the characteristics of the oil changes until it stops performing per specifications. Mineral oil also tends to begin breaking down above 210°F (99°C), and when it has been exposed to a high-temperature transient, the oil does not return to original performance when cooled.

Table 8.5 Screw Compressor efficiency example data points

		Press (psig)	Temp (°F)	Enthalpy (BTU/lbm)
Well-head	1	60.0	70	353.13
After suction controller	2	50.0	70	353.41
Compressor inlet	3	48.0	70	353.47
Out of compressor (gas)	4_g	170.0	205	417.58
Out of coalescer (gas)	5	168.0	205	417.62
Out of cooler (gas)	6	160.0	120	374.74
After backpressure valve (gas)	7	160.0	120	374.74
Suction end of compressor (oil)	C	48	192	7.309
Out of compressor (oil)	4_o	170.0	205	13.528
Out of coalescer (oil)	A	165.0	205	13,519
Out of oil cooler (oil)	B	164.0	192	7.511
Out of oil pump (oil)	D	200.0	192	7.576

Table 8.6 Screw compressor efficiency example energy traverse

	Equation	ΔT or Δh	BTU/h	hp
$1 \rightarrow 3$	$m \times \Delta h$	−0.34 BTU/lbm	−684	−0.3
$3 \rightarrow 5$	$m \times \Delta h$	64.1 BTU/lbm	128,981	50.7
$5 \rightarrow 7$	$m \times c_p \times \Delta T$	−118 R	−32,919	−12.9
$D \rightarrow 4_o$	$m \times \Delta h$	6.2 BTU/lbm	59,352	23.3
$4_o \rightarrow B$	$m \times c_p \times \Delta T$	−13 R	−31,222	−22.6
$B \rightarrow C$	$m \times \Delta h$	0.065 BTU/lbm	630	0.03
Traverse total (including heat transferred to ambient in cooler)				38.5
Traverse total (without heat lost to ambient in cooler)				74.0
$1 \rightarrow 7$	$m \times \Delta h$	19.14 BTU/lbm	3697	15.1

Synthetic oil. Most expensive (usually by a good margin), but it tends to not act as a solvent for heavier hydrocarbons. Synthetic oils are very stable at high temperatures and values above 350°F (177°C) have been reported without permanent degradation. As always, you need to follow manufacturer's recommendations for maximum temperature.

Semi-Synthetic oil. Mineral oil and synthetic oil can be blended to achieve specific intermediate properties. These blends are often very cost-effective.

Compatibility with water vapor. All of the oil types are significantly hydrophilic and have a prodigious capacity for absorbing water vapor. When the oil absorbs water vapor, it: (1) becomes more viscous (i.e., it is harder to pump), (2) loses lubricity (i.e., so you need to pump more), (3) increases surface tension (i.e., this allows bigger droplets to fail to coalesce in the outlet separator vessel), and (4) raises the oil level in the reservoir (increasing foaming and carryover).

There is simply no way that you can prevent the oil from absorbing the water vapor that is present in the gas, and removing the water vapor from the suction stream is very expensive. The only way to successfully operate an oil-flooded screw compressor in raw-gas service is to manage the temperature in the outlet separator vessel to cook the water vapor out of the gas. Fig. 8.11 shows an example of how a small change in discharge temperature can change the process. The inlet gas is saturated with water vapor and contains 5061 lbm/MMSCF (81 g/SCm). Inside the compressor, the gas is heated from 110°F (43.3°C) at 12 psia (82.7 kPaa) to 192°F (88.9°C) at 112 psia (772 kPaa).

At these discharge conditions, the gas can only hold 4225 lbm/MMSCF (70.9 g/SCm), so 826 lbm/MMSCF (13.2 g/SCm) stays in the oil, which represents a volume of nearly 100 gal (375 L) of water that remains in the oil for every MMSCF (28.3 kSCm) of gas. If the temperature out of the compressor is raised just 13°F (7.2°C), then the gas is able to carry more water vapor than what is present in the system, so it completes the cycle with no water left in the oil and the gas at 89% relative humidity.

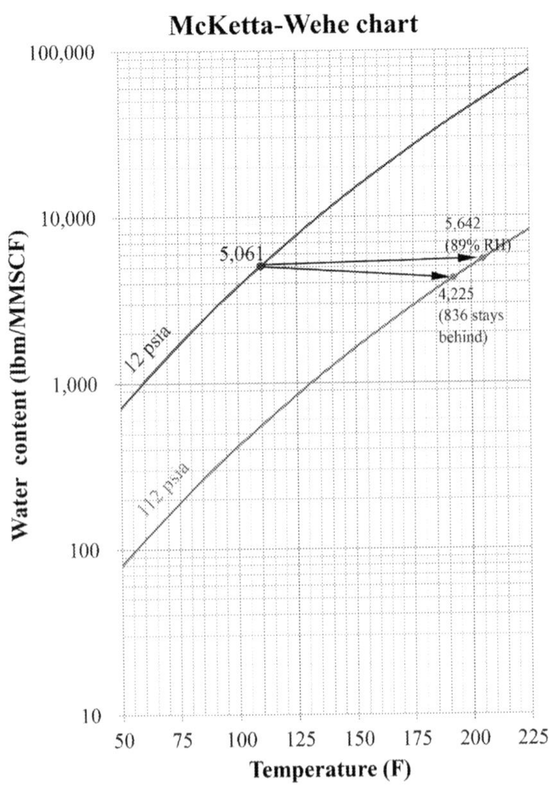

Figure 8.11 Screw compressor outlet water content.

8.3.3.1 *Outlet temperature*

A screw compressor compresses gas. The process can be assumed to be adiabatic. That means that Eq. (8.8) should have some role in predicting the discharge temperature. If we have a compressor that has a suction pressure of 12 psia (82.7 kPaa) at 80°F (26.7°C) and a discharge pressure of 112 psia (772 kPaa), Eq. (8.7) would predict the outlet temperature would be 405°F (207°C) with $k = 1.28$. That temperature would be a very difficult problem for metallurgy and for operations in general. The injected oil must have a role to play in that temperature. First you should use Eq. (8.8) to calculate the heat of compression. Then you have to convert the flows to mass-flow rate. Eq. (8.15) is used to determine the thermal energy that the act of adiabatic compression added to the process.

$$Q_{gas} = \dot{m}_{gas} \cdot c_{pGas} \cdot (T_{dischTheo} - T_{suct}) \tag{8.15}$$

The discharge temperature can now be determined using Eq. (8.16).

$$T_{disch} = \frac{Q_{gas} + T_{suctGas} \cdot \dot{m}_{gas} \cdot c_{pGas} + T_{oilIn} \cdot \dot{m}_{oil} \cdot c_{pOil}}{\dot{m}_{gas} \cdot c_{pGas} + \dot{m}_{oil} \cdot c_{pOil}} \tag{8.16}$$

Example. A screw compressor has the conditions in Table 8.7.

This data is used to calculate the temperature of the oil out of the compressor (Eq. (8.17)).

$$\dot{m}_{gas} = q_{gas} \cdot \rho_{gasStd} = 300{,}000\,\text{SCF/day} \cdot 0.046\,\text{lbm/SCF} = 13{,}770\,\text{lbm/day}$$

$$\dot{m}_{oil} = q_{oil} \cdot \rho_{water} \cdot SG_{oil} = 40\,\text{gal/min} \cdot 8.34\,\text{lbm/gal} \cdot 0.81 \cdot \frac{1440\,\text{min}}{\text{day}}$$

$$= 389{,}000\,\text{lbm/day}$$

$$T_{disch-gas} = (110 + 460)\left(\tfrac{100+12}{0+12}\right)^{(1.28-1)/1.28} = 929\text{R}$$

$$Q_{gas} = 13{,}770\,\text{lbm/day} \cdot \left(0.52669 \frac{\text{BTU}}{\text{lbm} \cdot \text{R}}\right) \cdot (929\text{R} - 570\text{R})$$

$$= 2.60 \times 10^6\,\text{BTU/day}$$

$$T_{disch} = \frac{2.60 \times 10^6\,\text{BTU/day} + 570R \cdot 13{,}770\,\text{lbm/day} \cdot 0.52669 \frac{\text{BTU}}{\text{lbm} \cdot \text{R}} + 640\text{R} \cdot 389111\,\text{lbm/day} \cdot 0.45 \frac{\text{BTU}}{\text{lbm} \cdot \text{R}}}{13{,}770\,\text{lbm/day} \cdot 0.52669 \frac{\text{BTU}}{\text{lbm} \cdot \text{R}} + 389{,}000\,\text{lbm/day} \cdot 0.45 \frac{\text{BTU}}{\text{lbm} \cdot \text{R}}}$$

$$= 652\text{R} = 192°\text{F} \tag{8.17}$$

Table 8.7 Screw compressor temperature example

	Gas	Both	Oil
Atmospheric pressure		12 psia (82.7 kPaa)	
Suction pressure	0 psig (0 kPag)		40 psig (276 kPag)
Discharge pressure	100 psig (689 kPag)		100 psig (689 kPag)
Inlet temperature	110°F (43.3°C)		180°F (82.2°C)
Flow rate	300 MSCF/day (8.5 kSCm/day)		40 g pm (151 L pm)
Specific gravity	0.6		0.81
c_p	0.527 BTU/lbm/R (2205 J/kg/K)		0.450 BTU/lbm/R (1884 J/kg/K)
k	1.28		

This example is plotted on an extract of the McKetta–Wehe Chart in Fig. 8.11. In this example, the ability of the gas to hold water vapor at outlet conditions is less than its ability at inlet conditions, so 836 lbm/MMSCF (13 g/SCM) remain in the oil—increasing the volume of the oil reservoir by 30 gal/day (113 L/day).

8.3.3.2 Oil temperature control

The "standard" way to control oil temperature is shown in Fig. 8.12. In this common scheme, the 3-way "constant temperature valve" looks at the temperature into the screw and bypasses the cooler to try to maintain a constant input temperature. In the example in Table 8.7 the 3-way temperature control valve is set to maintain 180°F (82.2°C) into the process. You can see from the example, this setting (and the oil flow rate) result in a 12°F (6.6°C) ΔT, and 192°F (89°C) outlet temperature which causes considerable water to collect. Changing the set point would help for these exact conditions, but if gas-flow rate increases, or discharge pressure decreases, etc., it will no longer be correct. The long and the short of the story is that controlling the inlet temperature is a ineffective practice that has come to us from plant compressors (where, if you'll recall, water-vapor in the gas is much lower) and really should have stayed there.

The packagers of the first screw compressors that I deployed failed to understand field use to the extent that the 3-way valve was set for 140°F (60°C) into the screw and the oil flow rate was expecting about three compression ratios (we had 12 compression ratios) so the oil was flowing

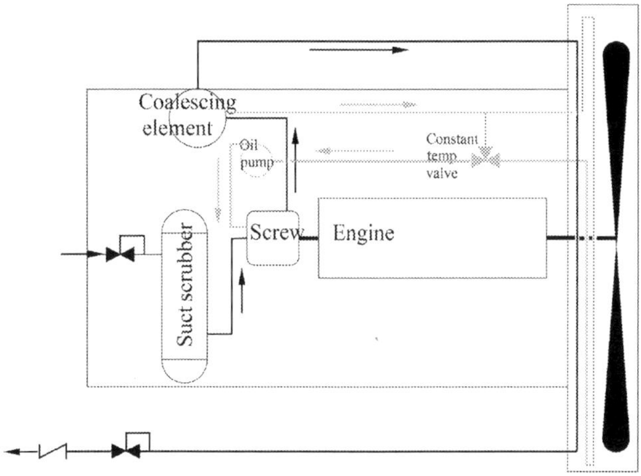

Figure 8.12 Screw compressor "standard" temperature control.

so quickly that temperature out of the compressor was 142°F (61.1°C) and we had a considerable number of serious issues and very high oil-replacement/replenishment costs.

These skids were so basic that they didn't even have a temperature gauge anywhere on the compressor, let alone on the compressor outlet.

After many months of fighting with controlling the wrong thing, I realized that life gets easier if you control the right thing (I've always been a slow learner). The "right thing" is to control the temperature out of the compressor in the face of frequently changing conditions. In Fig. 8.13, there is a temperature sensor looking at the temperature out of the screw. Based on that temperature, the PLC controls (it goes through the following steps in sequence with a 2-minutes pause before going to the next step)

1. Adjust the speed of the glycol pump (can be adjusted to a minimum, and then on the next time the PLC looks at the glycol pump it can be turned off).
2. Adjust the setting on the thermostatic valve on the oil pump discharge (this valve can go fully shut, the orifice in the bypass is sized for required lubrication and minimum oil-injection).
3. Adjust the speed of the fan (can go to a minimum, but not zero).
4. Adjust the speed of the oil pump (can go to a minimum, but not zero).
5. Repeat.

A group of skids designed to this temperature control scheme was able to run for 3 years with zero unscheduled downtime and nearly zero added

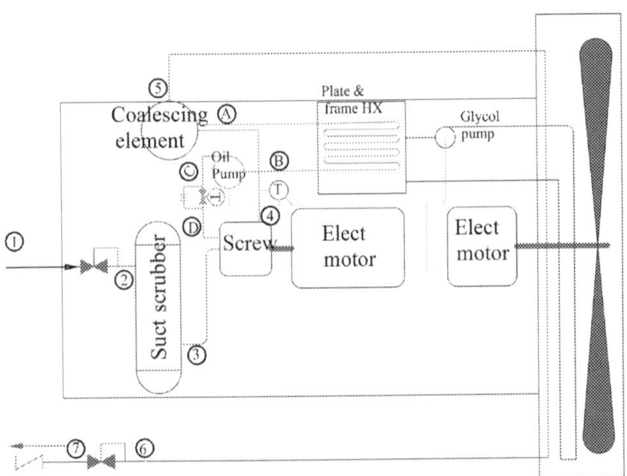

Figure 8.13 Flooded screw compressor effective temperature control.

oil. This skid had variable speed drives on all electric motors. Before this skid was designed, I was reluctant to set an electric-motor driven compressor on a well site because that puts the weakest link under someone else's control—a hard concept for a well-site engineer. After reviewing the results of this compressor, I am prepared to set a genset to run site facilities with electric power. Genset's combined with variable speed electric motors has a high capital cost, but low-operating cost and the crossover happens in the first couple of years.

8.3.3.3 Oil pressure management

As mentioned above, plant machines historically do not have oil pumps, but rely on the differential pressure across the skid to move oil. This can work in steady-state operations, but well-site and gathering operations are rarely steady state. Experience has shown that field compressors without oil pumps will tend to be on the top of down-time lists and failure reports. On engine-driven compressors oil pumps can be run off the pony shaft. On electric skids the oil pump should have its own variable speed drive.

8.3.3.4 Coalescing element

We often call the big lump of steel on the backside of the skid a "coalescing filter." It is not a filter. The coalescing elements look somewhat like filter elements, but they serve a different function. The coalescing element is intended to force small droplets (that are buoyant in the gas stream) to crash into other droplets and coalesce into larger drops that are not buoyant in the gas. As is normal with any piece of equipment, there is a range where they are more effective. It is recommended to try to get the same magnitude of velocity in a coalescing element as the target velocity for in a separator mist pad (see Chapter 5: Well Site Equipment).

8.4 THERMOCOMPRESSORS

"Thermocompressors" are devices that exploit the laws of thermodynamics to allow a high-velocity fluid to transfer flow energy to another fluid in a manner that leaves the second fluid at an intermediate pressure between the pressure of the power fluid and the pressure of the suction fluid. The terminology is a bit confusing:

- Venturi: a constricted area of a pipe, not a thermocompressor.
- Eductor: a thermocompressor designed for a liquid power fluid.
- Ejector: a thermocompressor designed for a gaseous power fluid.

Mostly, terminology doesn't matter, but in this case using an eductor with a gaseous power fluid or vice versa has poor results. As you can see from the notation on Fig. 8.14 an ejector can reach supersonic velocities, but an eductor cannot. The ejector has a choke point near the end of the nozzle where a shock wave is created and gas velocity reaches sonic velocity, that choke point is followed by a divergent section that increases velocity at a constant pressure (see Table 4.4). This high-speed fluid has significant dynamic pressure and is quite dense. The combination of high speed and enhanced density provides significant momentum to transfer to the suction fluid. An eductor does not provide these enhancements to flow because incompressible flow is not amenable to very high velocities.

8.4.1 Eductor vs ejector

Both eductors and ejectors are in the family of equipment that includes air ejectors, evacuators, sand blasters, certain kinds of paint sprayers,

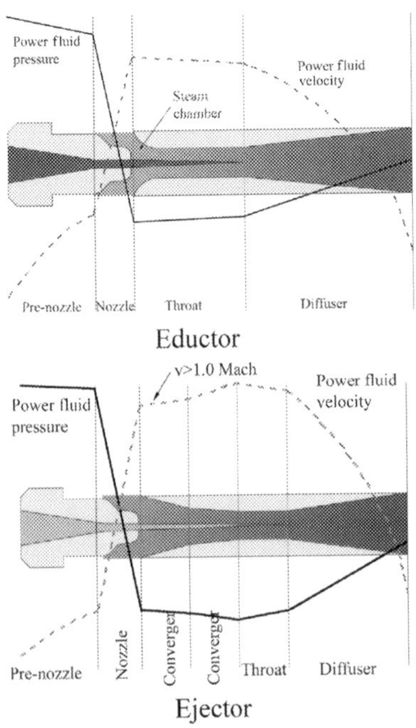

Figure 8.14 Thermocompressor pressure/velocity map.

hose-end sprayers, and jet pumps. In both cases the high pressure/high velocity power fluid entrains the suction fluid at the no-flow boundary between the two fluids, which causes energy to transfer from the power fluid to the suction fluid. The combined stream is left at an intermediate pressure.

The exhaust pressure for an eductor is limited to about 1.5–3 times the suction pressure (in absolute units). The exhaust pressure for an ejector can be as much as 10 times suction pressure, but there is a strong relationship between compression ratios and mass-flow rate of power fluid—more ratios requires significantly more power fluid. Except in very specific cases, greater than three compression ratios require uneconomic quantities of power fluid. Oddly, the more power fluid you use, the higher the overall efficiency of the unit will be. Low power-fluid usage is associated with efficiencies as low as 30%. High power fluid usage is associated with efficiencies as high as 70%.

8.4.2 Cases

It can be difficult to visualize where it makes sense to take 100 psig (690 kPag) gas and pump it through a device that ends up with 40 psig (276 kPag) gas that you now have to do something else with. We will present four cases where eductors and ejectors have been used to provide a real benefit.

8.4.2.1 Critical flow

Fluid flow in a well-bore is a complicated system, especially at low pressures. The target pressure for optimizing gas flow may not be close to the optimum pressure for flowing liquid. A common way to solve this problem is to restrict casing flow to force an adequate amount of flow up the tubing (this is the theory behind the tubing flow controller in Chapter 3: Well Dynamics, Section 5.5.2.3). However, when target pressure gets very low, managing that differential becomes impossible in real time. One solution is to install a second compressor to manage flowing tubing pressure independent from flowing casing pressure to optimize each independently. Fig. 8.15 shows an application of this concept. The ejector takes gas from the oil-flooded screw that is drawing from the production separator at 10 psig (68.9 kPag) suction. At the well's flow rate, this separator pressure results in a flowing bottom-hole pressure of 12 psig (82.7 kPag), which corresponds to a flow rate up the tubing of zero (because of the cracking pressure of the check valve on the tubing). The ejector is just

Figure 8.15 Critical flow ejector.

enough compression horsepower to pull the tubing to 4 psig (27.6 kPag) with 250 MSCF/day (7.1 kSCm/day) flowing up the tubing. Using the technique from Fig. 3.21 we determined that the actual critical flow rate at these conditions was 220 MSCF/day (6.2 kSCm/day), so this configuration was able to keep the well unloaded for over 4 years.

The ejector in Fig. 8.15 used 28 hp (20.9 kW) of the 500 hp (372 kW) oil-flooded screw compressor output. Boosting 250 MSCF/day (7.1 kSCm/day) from 4 psig (27.6 kPag) to 10 psig (68.9 kPag) is a 10 hp (7.5 kW) job, so the ejector was 36% efficient while avoiding the cost of a downhole pump.

In the first 4 years of this project, critical flow ejectors were installed on 32 CBM wells, and all of them showed flatter declines and reduced variability. The installed cost (for sites that already had well-site compression) was $4.5 k/site, and the contribution to net income of this $144 k investment was $16.2 million over 4 years. There was a change in staff in 2003 and all of the ejectors were replaced with nodding donkey sucker rod pumps, so long-term performance was not assessed.

8.4.2.2 Tubing flow control

One of the techniques we discussed for deliquifying gas wells using reservoir presser we discussed in Chapter 3, Well Dynamics (Section 5.5.2.3) was Tubing Flow Control. Under this method, we put a control valve on the casing and a flow-measurement device on the tubing. As long as the tubing is above critical, the casing is allowed to flow. At very low pressure, the amount of throttling required on the casing can cause the gas flow to become unstable and even log the well off. The well in Fig. 8.16 was experiencing this problem. There would be periods of high flow rates (on the order of 3–4 MMSCF/day (85–113 kSCm/day)) followed by several days of no flow or slugging flow. The ejector tee in Fig. 8.17 was intended to use the pressure drop up the tubing to allow

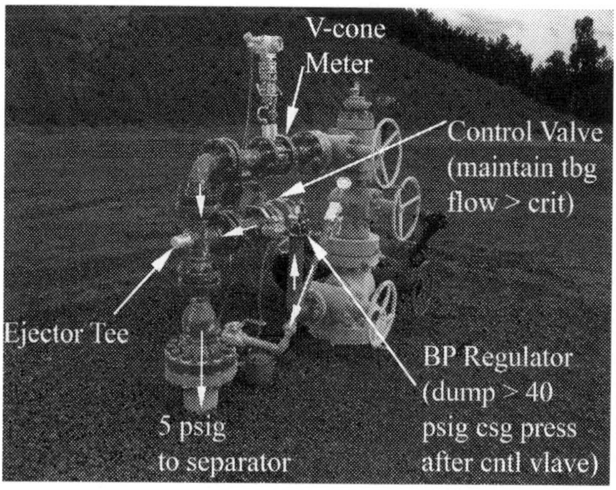

Figure 8.16 Tubing flow control ejector.

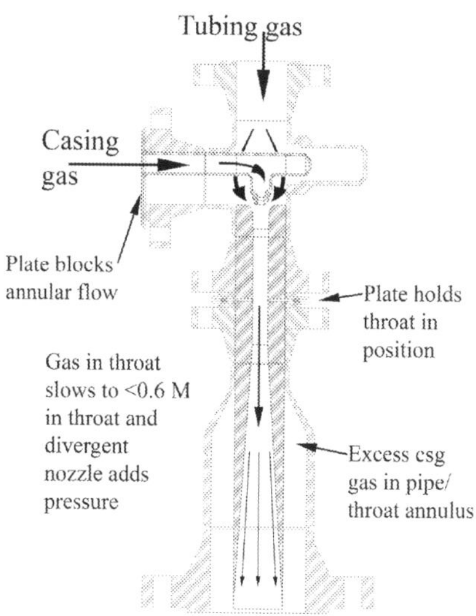

Figure 8.17 Ejector tee internals.

the casing valve to be open further (while still maintaining critical flow in the tubing). We didn't modify the tubing-flow control equipment or software, everything we did was after the casing control valve.

The ejector tee (Fig. 8.17) was designed to have very high efficiency with 40–45 psig (276–310 kPag) power gas (upward of 65%) at the cost of poor efficiency outside that range.

After the installation:

- If tubing flow was greater than the preset value, then the program bumped the casing control valve toward open.
- If the pressure downstream of the control valve is less than 40 psig (276 kPag), the casing gas just goes through the ejector nozzle without doing much work.
- If the pressure downstream of the control valve is in the design range, then the ejector sucks on the tubing.
- If the pressure downstream of the control valve is greater than 45 psig (310 kPag), then the backpressure valve allows some of the excess gas to bypass the ejector.

The compressor maintains the exhaust pressure at 5 psig (34.5 kPag).

The initial daily cumulative production was just over 25% higher after installing the ejector tee and the project paid out in 8 days. Everyone was so happy with these results that they got greedy. They surmised that if 45 psig (310 kPag) would give them a 25% uplift, then 85 psig would have to give them a 50% uplift (see below for a discussion of the effects of changing power gas pressure, they are not good). When that caused tubing pressure to increase dramatically, they decided that the tubing-flow control ejector was a failed idea and removed it. I chalked that one up to the importance of making sure that field staff understands the technology you are deploying.

8.4.2.3 Add-a-stage compressor

A CBM well was making 600 MSCF/day (17 kSCM/day) into 9 psig (62.1 kPag) suction pressure and was experiencing significant slugging of liquid. Efforts to lower that well-head pressure failed, and we were unable to lower suction pressure in spite of the fact that the compressor driver was running at only 35% of rated power. This was a two-stage integral recip with a poured foundation that would cost upward of a million dollars to replace. Reconfiguring a two-stage integral recip for three stages is nearly as expensive.

Figure 8.18 Add-a-stage thermocompressor.

Our solution was to remove the tubing from the well-bore for vacuum operations and install the ejector in Fig. 8.18 between the well-head and the separator. What the compressor now saw was as follows:

- Recip compressor suction pressure: remained the same at 9 psig (62.1 kPag).
- Recycled power gas: 1700 MSCF/day (48.2 kSCm/day).
- Well-head gas: 900 MSCF/day (25.4 kSCm/day).
- Engine load: 65% of rated power (engine-fuel increased 10 MSCF/day (288 SCm/day)).
- Ejector efficiency 48%.

At the same time, the flowing well-head pressure was −5 psig (6.75 psia (46.5 kPaa)) and the well stopped slugging. This unit that cost $20 k USD has delayed the requirement to replace the well-site compressor for 7 years so far.

8.4.2.4 Add a compressor

There are few things that will turn a cooperative neighbor into an adversary faster than installing a compressor in their front yard. In the example in Fig. 8.19, "Well 2" had been free flowing for nearly 10 years when the adjacent plot of land sold to someone who built a house sharing both a million dollar view and a fence line with Well 2. Pressures inevitably

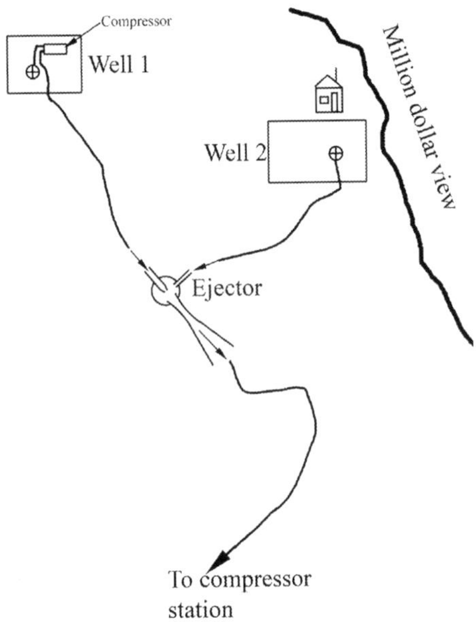

Figure 8.19 Add-a-compressor.

declined and the well needed a compressor, but putting one in would spoil a good relationship. The solution was to modify the compressor on Well 1 (it was a single-stage recip, but it was designed to allow converting it to a 2-stage in a couple of hours site work) to get high enough pressure to supply an ejector that could pull on Well 2 and exhaust into the preexisting line pressure. The necessary ejector was of a similar size to the Add-a-Stage ejector with a different nozzle and throat.

This project had passed the design stage when I stopped working in that basin and I never heard if it was ever installed or not, but there is still no compressor on the well with the million dollar view.

The most famous add-a-compressor ejector application (Caltec) was in the North Sea. The BP Indefatigable 23A platform handled its own gas plus 11 satellite platforms using 120,000 hp (89.4 MW) of centrifugal compression. This equipment plus necessary separators, dehydrators, valves, and piping took most of the available space on the platform. Shell proposed bringing the gas from two of their platforms on to the 23A platform, but these wells needed 50 psig (345 kPag) instead of the 80 psig (552 kPag) that the compression was designed for. Caltec Ltd designed an ejector able to boost the gas from the Shell platforms from 50 to 80 psig using the installed compression on the platform. Everything on a North Sea platform is very expensive, but even at those prices, this project paid out very quickly.

8.4.2.5 Case studies conclusion

These four cases were included to provide a feeling for why someone should care about thermocompressors. Every time I get a new problem that has resisted solution (because no one ever calls me first, everyone looks for cheaper alternatives before agreeing to my exorbitant hourly rate) I ask myself "is there any application for eductors or ejectors as part of the solution set?" The answer is usually a resounding "NO," but occasionally I can solve a problem with an ejector that cannot be solved economically any other way.

8.4.3 Rule of twos

The design of an ejector or eductor involves some complicated arithmetic (the printout of the FORTRAN program for sizing is close to 100 pages each for eductors and ejectors), but for scoping purposes you can use the "Rule of Twos." If you can satisfy (all pressures are in absolute terms and all volumes are mass-flow rate or volume flow rate at standard conditions) these bullet points, then there is a reasonable chance that a thermocompressor could be part of the solution:

- Achievable suction pressure is more than half of maximum exhaust pressure (i.e., if the exhaust pressure is expected to be less than 50 psig (345 kPaa) then the lowest suction pressure you can reach is 25 psia (170 kPaa)).
- Minimum power gas pressure is twice exhaust pressure.
- It takes twice as much power fluid as you are planning on pulling through the suction.
- You are going to get something like 50% of the work out that you put in.

All of these rules are useful for answering the question "can an eductor/ejector solve part of this problem?", but specific designs can be significantly different. The add-a-stage compressor does three compression ratios with a 1.89:1 mass flow ratio and is 48% efficient, but requires the power-gas to exhaust ratio to be 3.4:1. The tubing-flow-control ejector tee is designed for 8:1 power gas to suction gas ratio to ramp up the efficiency, but the design trade-offs made the operating range to the power gas very narrow.

In general terms we need to remember the following:

- A limited amount of mass can flow through the throat.
- Increasing compression ratios requires increasing the mass-flow rate of the power gas (you are trying to do more work so you need more power input).

- Adding power fluid decreases the amount of suction mass that can flow through the throat.
- To increase compression ratios, generally requires a design change, not an operational change.

8.4.4 Ejector response to changing conditions

There are four parameters that can be adjusted: (1) power fluid pressure, (2) exhaust pressure, (3) suction pressure, and (4) power fluid flow rate (which changes with power fluid pressure and is not independent). Table 8.8 summarizes these changes. The two highlighted cells are quite counter intuitive. First, if you increase power gas pressure, you are putting more power-fluid into the throat and leaving less space available for suction gas so you decrease your suction flow rate. Second, the fluid dynamics within an ejector are quite complex and the transition to and from compressible flow creates shock waves that limit communication upstream and downstream—in short the steam chamber and the throat do not "know" what the exhaust pressure is as long as it is low enough to allow the compressible/incompressible transitions to occur, which means that decreasing exhaust pressure will not change suction flow rate at all.

8.4.5 Thermocompressor conclusion

Thermocompressors are not magic or any reasonable facsimile thereof. They are a tool that should be in every facilities engineer's tool box. There are thousands of potential applications across a world with millions of wells. Being the engineer that finds the next *Indefatigable* application may turn you into a rock star for a minute, but dang, a minute is far better than never.

Table 8.8 Ejector response to changes

		Power fluid pressure	Exhaust pressure	Suction pressure	Suction flow rate
Power fluid pressure	Decrease		Constant	Increase	Decrease
	Increase		Constant	Constant	Decrease
Exhaust pressure	Decrease	Constant		Increase	Unchanged
	Increase	Constant		Constant	Decrease
Suction pressure	Decrease	Constant	Constant		Decrease
	Increase	Constant	Constant		Increase

Look for wasted horsepower:

- A compressor that can't be loaded to a good place on the driver load curve
- Well-head chokes
- A pump that is too big for the application.

Then look for someplace that needs a little horsepower:

- Well-site too close to housing for a compressor
- VRU requirements
- Focused power required
- A need for a sump pump.

Finally, use the rule of twos to see if you can match the wasted power to the power requirement. It is surprising how often there is a good fit, and if you are wasting power anyway, the costs can be scandalously low. In fact, they can be so inexpensive that it can be economic to take advantage of short-term conditions (e.g., you expect to have to choke the tubing flow for 1–2 years before the reservoir pressure decreases to the point you can remove the choke, if an eductor project has an 8 day payout you get free money for 722 days by using that wasted power for two years).

8.5 VACUUM OPERATIONS

As we saw in Section 3.5.3.3, the amount of water that a gas can carry as water vapor in vacuum conditions can be enough to deliquify a reservoir. In addition, requirements to remove the vapors from atmospheric tanks forces vacuum operations in many cases. On the other hand, there is significant resistance from gathering companies to operating wells in a vacuum due to (largely irrational) fears of oxygen corrosion. Sometimes the gathering companies add a concern about ingesting enough air to support combustion as we discussed in Chapter 4, Surface Engineering Concepts, but this is a threat without any basis in the physical world—there has never been a compressor fire or explosion that could be traced to vacuum operations, because to support combustion you would have to ingest six times as much air as the gas you are aggregating, not possible. Regardless of resistance, the economic benefit of operating in vacuum conditions will increasingly overrule the resistance.

When we talk about vacuum operations, the first technology that comes to mind is liquid ring compressors. These machines are similar to a fan or dynamic blower operating in an asymmetrical housing mostly filled with liquid. The blades tend to sling the seal liquid to the outside of the casing and as a chamber in the blades moves from the near casing-wall to the far casing-wall the liquid moves outward and creates a low pressure area at the suction. As that chamber moves toward the discharge side, the casing-wall gets closer and raises the pressure. The choice of seal liquid is important (e.g., a water seal will tend to evaporate, and most glycols and oils will have compatibility issues with heavier hydrocarbons). With a maximum of 4.5 compression ratios, if the target pressure is 2 psia (13.8 kPaa) the maximum discharge is only 9 psia (82 kPaa), so you need an additional stage just to get above atmospheric pressure. The casings tend to be limited to about 20 psig (138 kPag) operating pressure. On the face of it, liquid ring compressors seem to be a reasonable choice for vacuum operations, but maintenance and make-up seal liquid make them a poor choice for well-site use.

The biggest strength of centrifugal and axial compressors is the ability to move large volumes of gas at very constant conditions. There is no technical reasons why you could not use this technology for vacuum operations, but the lack of mass in the flow stream would be an issue for maintaining internal temperatures (high rotational speeds and low mass-flow rate is a recipe for high friction losses).

Thermocompressors can have a role in vacuum operations, but subsequently removing the power fluid from the combined stream can be a problem. There are a couple of vapor-recovery-unit manufacturers who use an ejector to pull on the tank and then use recips to boost the ejector-exhaust (positive suction pressure on the recip) up to line pressure while providing power gas to the ejector.

Recip compressors can certainly work in vacuum service, but that $\pm 5\%$ of design conditions limitation creates a real problem as suction pressure moves into a vacuum. If design suction is 2 psia (13.8 kPaa) then we want to control the suction pressure ± 0.1 psi (689 Pa), which is quite unlikely to be successful and temperature control will be very difficult.

The best choice I've found for well-site vacuum operations is oil-flooded-screws. With the ability to provide 20 compression ratios, it is a single step from a deep vacuum to a positive pressure that is high enough to be useful. Also the injected-oil mass-flow is adequate to control temperatures and friction losses.

There is an apocryphal story (circulated by salesmen of process-derivative screws) that air derivative screws are inappropriate for vacuum operations. This story goes something like: air-derivative screws have the low-pressure end of the screw at the shaft-penetration end of the casing and the shaft seals are inadequate to keep air out. They go on to say that process-derivative screws have the high-pressure end at the shaft penetration and will not suck air. The only true part of this story is that the typical air-derivative screws have the suction on the end of the casing with the shaft penetration. In fact, both the shaft seal and the journal bearings are pressurized with oil and every air-derivative screw manufacturer will provide you with an affidavit that certifies that their compressors are rated for vacuum service without modification.

In Chapter 3, Well Dynamics, we covered the issues that we had to resolve to successfully implement vacuum operations for deliquification. It seems appropriate to repeat those findings here. We were required by state regulations to put oxygen sensors and slam valves on every site with vacuum operations. What we found was:

- Setting the O_2 sensors at a spike to 10 ppm resulted in most wells being shut in most of the time during the start-up phase of the project (we reset them to >25 ppm for 30 seconds which helped a lot).
- All pressure safety valves (PSV) are DESIGNED to leak when downstream pressure is greater than upstream pressure. We fixed the problem of PSV sucking air by putting rupture disks (best) or check valves (not as effective, but nearly) under the PSV.
- Sight glass packing nearly always leaks and the leaks can be very hard to find. We replaced sight glass packing, but eventually just isolated the sight glasses out of service.
- Finger-tight plugs in open-ended tubing are far worse than worthless (a visual inspection shows the line as plugged, but the plug doesn't really do anything).
- Threaded connections often leak, and leaks can be difficult to find.

8.6 FUEL GAS

Gas compression needs a reliable and economic power source. Most well-site compressors are engine driven or electric-driven with an on-site engine-driven genset. The engines want clean, dry gas, but well-site gas tends to be dirty and wet. The typical solution is to use something called a "fuel gas dryer" and a great deal of wishful thinking. These units are characterized by the following:

- Smaller than 6.0 in. (152 mm) ID to circumvent the requirements of the BPVC

- Fairly short (usually less than 18 in. (457 mm) long)
- No mist pad
- No level control
- No pressure safety valve
- Manual water drain.

These units are far worse than worthless. They do collect some water in spite of themselves, and a person must manually drain them often. If the water content of the gas is 4000 lbm/MMSCF (64 g/SCm), and the engine burns 75 MSCF/day (2 kSCm/day), then there is 36 gal/day (136 L/day) of water going through the dryer, but the unit only holds 2.1 gal (8.2 L) so it doesn't take long to fill the dryer.

I asked several field techs how often they drained their fuel scrubbers and they all say "about once a week" which likely meant "monthly." I asked them all to drain all of their dryers into a bucket daily for a week. They grumbled, but did it. Every one of them reported that on wells with compressors or pump jacks, the scrubber was full of liquid every day on every unit. That means that the dryers were probably full within a few hours of being drained. Collecting water in a place that requires manual intervention to drain was a bad idea in gathering systems and it is an even worse idea on fuel-gas systems. Water in fuel gas is always the number-one cause of compressor, genset, and pump downtime on well sites. I recommend that if a device does not have an automated dump-valve then remove the device. A straight piece of 1-in. (25 DN) pipe is far better than blowing the gas through a water bath. If the dryer does have an automated dump, then frequently confirm that it is still working (these little dump valves have a terrible track record for plugging off).

For very cold-weather operations, it can be economical to use a deliquescent (salt) dryer (discussed in Chapter 9: Interface to Plants) that can actually remove water vapor, but this is a serious commitment to maintaining the salt level and keeping the brine drained off.

8.7 COMPRESSOR CONTROL

When we talk about "compressor control," we are really saying, "how do we match inflow at our target pressure to the capacity of the compressor." If the capacity of the compressor is less than the well is able to produce, then pressure will increase until the two match. Unless you are able to install additional compression, there is not much you can do about the compressor having less capacity than the supply except install a suction-control valve to keep the flow into the compressor in line with the compressor capacity.

The other end of the spectrum is more interesting. When the compressor has more capacity than the well, we have to "shed rate," or reduce the flow capacity of the compressor to match the inflow. Compressor control schemes are almost exclusively rate shedding programs.

There are three common compressor "control valves" that should not be part of program logic to control the compressor since all three of them are external to the compressor. The valves are: (1) suction pressure control valve, (2) discharge back-pressure control valve, and (3) recycle valve.

Suction pressure control valve. The ability of a compressor to protect itself from too much suction pressure is quite limited. Consequently, we typically put a suction pressure control valve in front of the compressor to establish the maximum pressure we will allow to come into the compressor suction. Ideally, these valves should have a near-zero dP in "steady state" operations so that we don't waste reservoir energy to protect our artifacts. When a suction controller is locally controlled it is called a "Max-inlet control." When a suction controller is controlled within a range of values (called "range control") you must use a PLC. Range control is sometimes used to keep multistage machines in balance and will always have a nonzero dP and the need for this type of logic is a positive indication that you have the wrong compressor on the job.

Discharge back-pressure control valve. It is occasionally necessary or desirable to increase the dP across a compressor skid. This can be done to increase the compression ratios on a screw to heat up the oil. Discharge back-pressure control valves can be used on recips to balance stages. There is almost always a better way to accomplish your goals than to install a discharge back-pressure control valve, but people do it as a stop-gap or because they don't know any better. If you are going to use a discharge back-pressure control valve it should always be local control, PLC control does not add value and can risk having the valve nonresponsive at key times.

Recycle valve. A compressor going out of service due to low suction-pressure is a pain. To prevent this, recycle valves are often used to send discharge gas back to the suction to reduce down time. This is a common practice on machines without speed control or unloaders, but recycle valves are occasionally seen on machines with speed control. On one skid, I was distressed to see that the PLC controlled the recycle valve and the priority in the PLC was to operate the recycle valve before it activated speed control.

One of the instructors at Ariel's Advanced Service School said "Gas you never compress is infinitely efficient, compressed gas you throw away is zero percent efficient." This sentiment has stuck with me over the years. There are times when you have no option but to install a recycle valve, but it needs to be locally controlled and never part of the program logic.

8.7.1 Local vs PLC options

Programmed logic control (PLC) devices get more powerful and more capable every year. We often REALLY want to use the latest Gee-Whiz device, but sometimes that is not the best-ever idea. I recently reviewed a compressor-skid design that had taken the Gee-Whiz to amazing lengths. In addition to every external valve being PLC controlled, this skid had a blow-case with an electric level switch on the suction scrubber that went to the PLC and the PLC told the dump cycle to begin (if you'll recall from Chapter 5: Well Site Equipment, this involves opening the dump valve and opening the power gas valve). The problem that they asked me to look at was that at least once a day the compressor went down on high suction scrubber level. I'd never seen the compressor or the design, but the downtime report told all—the time stamp on every level transient was within 90 seconds of the top of the hour. The PLC was reporting to the central database at the top of the hour and when it got a notice to drain the blow-case it had to wait until the data transfer was completed (in a queue with every other field device trying to update the database at the same time). While waiting for the data transfer to end, the scrubber filled with water. This was an easy fix because they had used electric level switches that were compatible with the electric dump and power valves, and it was reasonable to terminate (and power) the level switches at the valves, it is not always that easy. Some of the other functions on that skid were harder to correct and we installed a second PLC to handle data transfer while the functions were being reassessed.

Well-site automation has historically been one-controller-one-end-device, and that is an appropriate technology mix for nearly everything on a compressor skid. A PLC is likely required for capacity control, but that function needs to limit the levers it can pull to speed and unloader valves that need more complex logic than can be economically provided with a local device.

8.7.2 Capacity control

It is common in rate-shedding situations to include early use of unloaders (on screw compressors) and this rarely works as well as people hope it will. Rate-shedding schemes have a variety of costs:

- Reducing speed within a design range that honors the driver torque curve is the most efficient (i.e., it costs zero dollars to not compress gas).
- Adjusting engine manifold pressure to keep rpm constant is next most efficient.
- Unloader valves are less efficient, but still viable.
- Recycle valves are least efficient and should be a last resort.

The three most common control sequences that in general use are (1) suction control, (2) fuel-manifold pressure control, and (3) suction control with manifold pressure override.

Suction control. This is different from a suction controller. In this scheme, the PLC looks at suction pressure and adjusts driver speed to keep it within a range. This control needs to be able to adjust engine fuel-manifold pressure as the load changes in response to the speed change.

Fuel-manifold pressure control. This only applies to engine-driven compressors. An internal combustion engine gains or loses speed in response to changing load. Controlling fuel-manifold pressure allows the compressor to maintain a constant speed under changing load. If suction pressure is constant (usually accomplished by taking a large dP across a suction controller) then constant rpm is a good thing, but a large dP across a suction controller is not a good thing.

Suction control with manifold pressure override. This scheme is a combination of the other two. As long as manifold pressure is less than the maximum, the PLC controls on suction pressure. If the manifold pressure approaches the maximum, then the priority in the PLC shifts to the fuel-manifold pressure and begins ignoring suction pressure.

The most effective load shedding sequence I've ever seen is for screw compressors:

- Adjust compressor speed between a minimum and a maximum with the unloader closed.

- If the speed reaches the minimum and suction pressure is still dropping, operate the unloader (suction pressure control).
- If the speed reaches the maximum and suction pressure is still increasing, operate the unloader to keep the driver from exceeding maximum power.

This scheme is called "Suction pressure control with power override" and it works well for screw compressors. For recips, the only lever you can pull is speed control.

Where to put sensors? Typically we sense suction pressure after the suction controller and sometimes after the suction scrubber. When the dP across both the suction controller and the suction scrubber is near zero then this is just fine, but what about start-up conditions? Say a well has been shut in for a week and near well-bore pressure has built up considerably above both normal operating pressure and the design suction pressure of the compressor. You know that you will be able to draw this pressure off eventually, but how long will it take? With the usual placement of the suction-pressure transducer, the compressor does not have any indication that there is a huge dP across the suction controller and the program logic does not place a priority on pulling the head off the well. If you move the sensor to the well side of the suction controller, the PLC sees the head pressure and shifts into "DRAWDOWN" mode to lower it.

	Conventional transducer placement	**Transducer placed on well side of suction controller**
Well-head pressure	125 psig (862 kPag)	125 psig (862 kPag)
Suction controller setting	30 psig (207 kPag)	30 psig (207 kPag)
Input to PLC	30 psig (207 kPag)	125 psig (862 kPag)
Initial Flow rate	2.8 MMSCF/day (79.3 kSCm/day)	4.9 MMSCF/day (138.9 kSCm/day)
Day 4 flow rate	3.7 MMSCF/day (104.8 kSCm/day)	3.7 MMSCF/day (104.8 kSCm/day)
Time to reach zero dP on suction cntlr	4 days	8 h
4 days sales total	12.92 MMSCF	15.04 MMSCF

In a competitive reservoir and $3/MSCF gas this difference is worth $6400. It didn't take many of these transients to pay for sending an automation tech to the site to move the transducer.

8.8 COMPRESSOR COMPARISON

The examples of recip and screw compressor efficiencies in Tables 8.4 and 8.6 can be combined into Table 8.9. As the only real reason for caring about "efficiency" is fuel cost, this table shows a clear winner in this case. Raising the suction pressure on the recip and getting rid of the backpressure valve on the outlet would bring the numbers closer together, but it is rare for the fuel use on a two-stage recip to be close to the fuel use on an oil-flooded screw for the same volume of gas and the same pressure change requirement.

In more general terms, Table 8.10 shows a qualitative comparison. Again, the flooded screw shows up better.

It would be easy to assume from Table 8.10 that recips are junk and oil-flooded screws are wonderful. That would be a mistake. It is appropriate for the first compression that a well sees to be an oil-flooded screw because of the flexibility on the suction side, but after first compression it makes a lot of sense for the remainder of raw-gas compression to be recips and there are good reasons for commodity gas being moved with dynamic compressors (very high throughput in a single frame).

8.8.1 Lots of ratios

If you need a lot of ratios, you can combine technologies into an interesting configuration. Oil flooded screws deal well with varying suction pressure but seem to work better with a fairly constant discharge pressure. Recips want a very constant suction pressure and deal well with varying discharge pressure. This seems to say that letting a screw see varying suction pressure and a recip see varying discharge pressure looks to be the best of both worlds. Oil flooded screw compressors can do 20 compression ratios. You can go from an atmospheric pressure of 14.7 psia (101.4 kPaa) to

Table 8.9 Comparison of screw and recip efficiency example cases

	Recip	Screw
Theoretical compressor efficiency (%)	81	64
Compressor net vs engine output (%)	70	73
Well-head to line vs engine output (%)	7	15
Annual fuel cost at $4/MMBTU (assuming on-site genset for screw)	$89 k	$42 k

Table 8.10 Comparison of recips and screws

Recip		Flooded screw	
Strengths	**Weaknesses**	**Strengths**	**Weaknesses**
When everything is perfect, best use of hp	Narrow suction range	Wide suction range	Moving oil requires energy
Operating staff is comfortable with them	Not tolerant of changing conditions	Very tolerant of changing conditions	Technology unfamiliar to operating staff
Few consumables	Valves are high maintenance	No valves, No rods	Oil is expensive
Some packagers do field machines well	Difficult to balance stages	No stages to balance	Few packagers understand field requirements
High operating pressure	Low compression ratios per stage	High compression ratios per stage	Limited operating pressure
Rugged and reliable	High temps	Very low temps	
	High maintenance	Low maintenance	
	High capital cost	Lower capital cost	

294 psia (2030 kPaa) in a single machine. To do that with a recip it would have to be three stages (about 2.7 ratios/stage) which would be less fuel-efficient and would be very difficult to properly balance the stages. On the other hand, you can do up to 4.5 ratios/stage in a recip, and if you put a screw in front of a two-stage recip you can do 405 ratios—from atmospheric pressure to 5900 psia (40.7 MPaa) with two skids.

8.8.2 Technology summary

Key parameters of the various compression technologies are summarized in Table 8.11.

8.9 LEASE VS BUY

The discussion concerning owning equipment vs leasing it seems to come up more often in compressor decisions than in any other upstream capital decision. The economists say that the only differences are (1) cost of acquiring capital; and (2) salvage value of equipment.

Table 8.11 Compressor technology comparison

	Eff	Limiting parameter	Max ratio/ stage	Typical use	Well site use?
Liquid ring	40%–50%	Discharge press	5	Deep vacuum into lp	No
Dry screw	50%–65%	Discharge temp	4	Control air	No
Eductor/ Ejector	40%–70%	Power fluid flow rate	10	Focus hp	Yes
Axial	60%–70%	Discharge press	1.1	Large volume, low head	No
Centrifugal	65%–75%	Discharge temp	2.5	Large volume	No
Oil-flooded screw	70%–88%	Differential pressure	20	Varying suction	Yes
Recip	78%–92%	Rod load or discharge temp	4.5	Varying discharge	Yes

Large production companies typically have a very low cost of capital because they tend to have a very good credit rating (not nearly always, but often enough to be significant). Compressor leasing companies tend to have a higher cost of capital than large production companies. Based on just cost of capital, it is rare for a decision to be "lease" instead of "buy." But what about salvage value? If you can redeploy a piece of gear immediately, then the "salvage value" is very good. If you can't find a place within your operation to redeploy it then the salvage value could easily be less than zero (i.e., you have to pay rental on the space the machine is sitting and rotting away). Some of the supposedly second-order effects that should go into the lease vs buy decision are as follows:

1. How long will the compressor be on the original site? The longer it is there, the more the analysis should favor purchase.
2. What will you do with it when you are done with that site? Does that disposition have a reasonable chance of still being viable in 2 years? 10 years? 30 years?
3. What value will it transfer at? This can be a major stumbling block when a machine that has been running for several years needs an overhaul prior to being redeployed. Does it leave the lease at salvage value or "used" value? If it leaves at salvage value do you have a place to charge the overhaul cost before you "sell" it to the next lease?
4. What partner permissions do you need in order to authorize disposal and/or replacement? A leased machine does not become part of the lease equipment, so the monthly cost of the lease and the removal

cost generally do not require an Authorization for Expenditure (AFE). Moving a purchased compressor certainly requires an AFE, and your partners may have the right to take possession of the compressor rather than paying for the overhaul.

Having managed a fleet leased from a third party, a fleet owned by the company and leased to the wells, a fleet jointly owned by another production company and leased to the wells, and a fleet owned by the wells, the worst of all worlds was the fleet owned by the individual wells. From the time we decided to move a compressor from one well to another until the move was accomplished was often over six months. On too many occasions, by the time we were authorized to do the move the machine was no longer right for the target location.

The other configurations had similar lead times to each other and were all better for the reservoir than for a well to own a compressor. For fields where you lack certainty on the duration of the decline period, leasing from a third party, while generally higher cost, has the positive characteristic of being able to just let the third party take the compressors back after a (fairly short) initial contract period.

Whatever convoluted capitalization scheme the economists come up with, you need to make very certain that you completely understand the terms of the "lease agreement," and that you know what you have to do to remove, replace, or reconfigure any or all the machines in the fleet. Before agreeing to the lease agreement run some scenarios. Ask "if we need to take a "Size A" compressor off joint-interest Lease XYZ, send that compressor to Lease MNO that we own 100% and replace the "Size A" compressor with a "Size B" compressor that came from the yard, what reporting and accounting processes must be triggered?" Then walk through the contract and see if you can live with what you find. I did this on one contract and found that the fleet we were setting up (it was jointly owned 50/50 with a working interest owner who was in most of our operated wells 50/50) could only sit on wells that were owned 50/50 with that owner, not a limitation that was in the best interests of the field as a whole.

8.10 COMPRESSOR CONCLUSION

Compressors are everywhere in Oil & Gas, and far too often technology and operating decisions are based on "what the last guy did" rather than "what the reservoir needs." I have spent many hours in conversations with people who were certain that the only viable technology was "recips" or "Ajax recips" or "Gas Jacks" or "flooded screws" trying to filter the fear and superstition from actual facts. Sadly, the discussions have always been very short on facts. I've had a PhD mechanical engineer say "reciprocating compressor technology is the highest efficiency and the only rational choice is based on fuel consumption related to efficiency," and then when I showed them the efficiency example in this chapter their response was "you manufactured a case that showed recips in a bad light, the real world is not like that." I admittedly selected a real-world example that illustrated my point, but I could have selected any one of several hundred similar cases.

There is one rational basis for selecting compressor technology—what optimizes ultimate profitability of the reservoir? If that answer is to install a recip for a few years and then install a flooded screw (and you've included reasonable capital costs for the swap out and reasonable redeployment assumptions) then that is what you should do. The idea that the first compressor you set on a site will be the last compressor you set is rarely consistent with the best interests of the reservoir. When I designed a compressor fleet for a group of CBM wells, the one position that I never deviated from is that the machines had to be interchangeable. The relative position of every suction flange relative to the position of the discharge flange and the edge of the skid was the same, and every one had the same size flanges from suction to discharge and from skid to skid. This allowed me to move compressors from site to site with very low costs, and we made 362 compressor moves in seven years. The design allowed for optimizing the impact of compression on the reservoir while keeping operating costs very low (about \$0.04/MSCF (\$1.41/kSCm)). When this attitude was abandoned (that fleet had about 10 compressor moves in the next 14 years); the operating costs started increasing while production dropped rapidly. Few things are more expensive than a "cost avoidance" policy.

Choice of technology is key, but close behind it is the choice of packager. There are examples of every manufacturer's compressor working well. There are also examples of the same models of machine that work poorly. The difference is almost always the packager. One of the best examples of packager differences is seen in Fig. 8.20. In both pictures, the package is the same: Both skids had the same 5000 hp (3730 kW) engine driving the same 4-stage, 6 throw separable compressor (which ended up with the same size cylinders in the same places on the frame). Both packagers were working from the same company specifications (including a

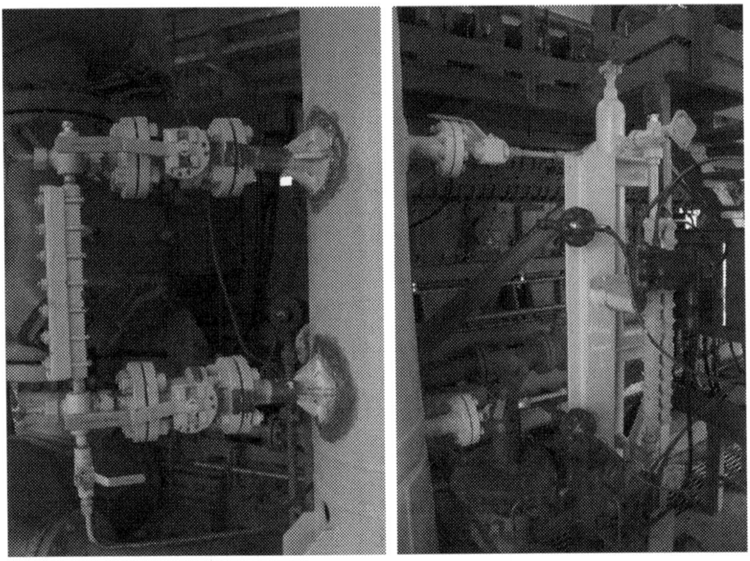

Figure 8.20 Example of compressor packaging.

requirement that there be no threaded connections on the skids). The requirements included a sight glass on the suction scrubber.

The company on the left-hand picture took all of the specifications literally. Without critical analysis of those specifications they hung a sight glass on the vessel with flanged valves. The packager on the left-hand side did not consider what vibration would do to a rigidly mounted cantilevered load with two points of support that could move independently—and put enough force on the structure to rip the sight glass off the vessel, which it did several times.

The packager on the right-hand side saw a potential problem and got an exception to the "no threaded connections" mandate. This packager did several other things properly as well. The sight glass weight is carried by a beam tied into the skid base. The piping from the vessel is heavy wall 2-in. (50 DN) that swedged down to 1-in. (25 DN) after the flange and then turned 90° so that vibration would translate to independent torque in the two pipes that could not set up harmonic vibrations. Finally, if the repeated cyclical torque did result in a failure, it would be in piping, not in the BPVC certified part of the vessel structure.

The same producer engineering team reviewed the design for the two skids and never questioned the layout of the sight glass. There were simply too many details on the drawings for anyone who had never had a nozzle fall off a pressurized vessel to question a sight glass. Obviously, the packager on the right-hand side in Fig. 8.20 had a design engineer who had seen an unsupported nozzle fall off a vessel and designed the sight glass to prevent that. While there are also many examples of the packager in the right-hand picture getting it wrong, they tend to do a workman-like job, but so does the packager on the left-hand side. I prefer the packager on the right-hand side and have used them often, but I still spend considerable time reviewing design documents and make sure that I visit the fab shop several times during the fabrication process. Further, if I find an error that I approved, I accept my mistake and pay for the fix.

REFERENCES

Ariel Performance Program. (n.d.). Retrieved January 30, 2017, from https://www.arielcorp.com/Performance-Program/.

Lea J.F., Nickens H.V. and Wells M.R., Gas well deliquification, second Edition, 2008, Gulf Professional Pub/Elsevier; Amsterdam, ISBN 978-0-7506-8280-0.

Peeran, S. M., Beg, D. N., & Sarshar, S. (2013, March 10). Novel Examples Of The Use Of Surface Jet Pumps (SJPs) To Enhance Production & Processing. Case Studies and Lessons Learned. Society of Petroleum Engineers. doi:10.2118/164256-MS.

REFPROP. (2017, January 17). Retrieved January 30, 2017, from https://www.nist.gov/srd/refprop.

Svenska Rotor Maskiner AB, "The Screw Compressor Development at SRM", November 9, 2010, retrieved January 17, 2017 from http://opconenergysystem.com/wp-content/uploads/2015/11/The-History-of-SRM.pdf.

NOMENCLATURE

Symbol	Name	fps units	SI units
C	Constant in Eq. (8.04)	3.03	1.0
c_p	Specific heat at constant pressure	BTU/lbm/R	kJ/kg/K
c_{pGas}	Specific heat at constant pressure of a gas	BTU/lbm/R	kJ/kg/K
c_{pOil}	Specific heat at constant pressure of an oil	BTU/lbm/R	kJ/kg/K
c_v	Specific heat at constant volume	BTU/lbm/R	kJ/kg/K
dP_{disch}	Diff press across discharge valves	psi	kPa
dP_{suct}	Diff press across suction valves	psi	kPa
k	Adiabatic constant (c_p/c_v)	decimal	decimal

(*Continued*)

(Continued)

Symbol	Name	fps units	SI units
Δh	Change in specific enthalpy	BTU/lbm	kJ/kg
$\Delta h_{gasActual}$	Change in specific enthalpy for actual conditions	BTU/lbm	kJ/kg
$\Delta h_{gasPeak}$	Change in specific enthalpy for maximum efficiency	BTU/lbm	kJ/kg
$\dot{m}$	Mass-flow rate	lbm/h	kg/h
n	Number of stages of compression	integer	integer
n	Polytropic exponent	decimal	decimal
P	Pressure	psia	kPaa
P_{suct}	Suction pressure	psia	kPaa
P_{disch}	Discharge pressure	psia	kPaa
$q_{mmscf/d}$	Gas flow rate at standard conditions	MMSCF/day	kSCm/day
Q_{gas}	Rate of heat transfer to the gas	BTU/h	kJ/h
R_c	Compression ratios	decimal	decimal
T_{disch}	Discharge temperature	R	K
$T_{dischTheo}$	Theoretical discharge temperature	R	K
T_{oilIn}	Inlet oil temperature	R	K
T_{suct}	Suction temperature	R	K
$V_{clearance\%}$	Recip unswept volume	%	%
VI	Volume index	decimal	decimal
W	Work	hp	kW
W_{in}	Work input to a fluid process	hp	kW
W_{out}	Work done on a fluid by a process	hp	kW
Z_{avg}	Average compressibility $(Z_{suct} + Z_{disch})/2$	decimal	decimal
Z_{std}	Compressibility at standard conditions	decimal	decimal
Z_{suct}	Compressibility at suction conditions	decimal	decimal
ρ_{std}	Gas density at standard conditions	lbm/ft^3	kg/m^3
ρ_{suct}	Gas density at suction conditions	lbm/ft^3	kg/m^3
η	Efficiency	decimal	decimal
$\eta_{compression}$	Compression efficiency	decimal	decimal
$\eta_{FirstStage}$	Stage efficiency, first stage	decimal	decimal
$\eta_{mechanical}$	Mechanical efficiency	decimal	decimal
η_{stage}	Stage efficiency	decimal	decimal
$\eta_{volumetric}$	Volumetric efficiency	decimal	decimal
η_{total}	Total efficiency	decimal	decimal

EXERCISES

Scenario: A four stage recip compressor with 4% clearance in the first stage is taking 6 MMSCF/day (170 kSCm/day) of CBM gas (use the example analysis in Chapter 0: Introduction) from 50 psig (345 kPag) at 5400 ft (1645 m) elevation and 85°F (29.4°C) to 10,000 psig (68.9 MPag). Ambient temperature is 95°F (35°C) and all four coolers have the ability to cool the gas to within 20°F (11.1°C) of ambient.

1. What is the efficiency of the compressor?
2. What is the expected work required to accomplish this task? What assumptions do you have to make to calculate this?
3. What is the fourth stage discharge temperature before the after cooler?
4. What engine output is required to provide the necessary work?
5. If the Carnot efficiency of the engine is 22%, how much of the gas being compressed must be diverted to fuel the compressor?
6. How would the total work and fuel consumption change if this unit were replaced with an oil flooded screw (with a 4.6 VI) in front of a 2-stage recip?

CHAPTER NINE

Interface to Plants

Contents

We have followed our squashed dinosaur through the decomposition process, through migration within rock strata, to aggregation within a reservoir, to accessing the reservoir with a well, selecting and installing wellsite equipment, aggregating the hydrocarbons in a gathering system, and adding energy in compression, with a side trip to getting rid of waste products that can be economically removed via mechanical means. But we still do not have a commodity product. To remove gaseous contaminants and hydrocarbon products that are too energetic for most industrial and residential burners, we need a plant. A "plant" is simply a collection of that process equipment needed to remove specific components from the intermediate gas stream to develop a commodity gas that is made up of

Practical Onshore Gas Field Engineering.
DOI: http://dx.doi.org/10.1016/B978-0-12-813022-3.00009-2

precisely defined components in precisely defined ratios. Every plant is different, because the predominant components of reservoir fluids vary widely from reservoir to reservoir.

This chapter is intended to provide very broad-brush information on what you might see if you look over a plant fence, not in nearly the detail that a plant operator or process engineer would be expected to know.

9.1 TERMINOLOGY

Plants have their own terminology (like any field of study), some of the more common esoteric terms are as follows:

Adiabatic: a process that occurs without heat transfer (adiabatic processes also tend to be isentropic and reversible, but these characteristics are not necessary for a process to be adiabatic)

Contactor: a vessel designed to facilitate the contact of one fluid with another (e.g., an amine contactor to remove acid gas or a triethylene glycol (TEG) contactor in a dehydrator tower)

Cryogenics: the study of "very low" temperatures. National Institute of Standards and Technology (NIST) defines "very low" as less than −300°F (−184°C), GPSA uses −51°F (−46°C) based on application limits for low temperature carbon steel

Isobaric: constant pressure

Isenthalpic: constant enthalpy

Isentropic: constant entropy

Isothermal: constant temperature

Lean oil: a term used for an absorption media used in a contactor to extract natural gas liquids.

9.2 RISK

As you move inside the plant fence, the first thing you notice is that there is a lot of densely packed equipment, complex piping, and the appearance of confusion in layout. With all this stuff in such close proximity, you have to ask yourself "what happens if this stuff breaks?"

That is a crucial question for plant designers and plant operators and it all comes under the broad category of "risk."

"Risk" has been defined as "a measure of economic loss or human injury in terms of both the likelihood and magnitude of the impact" (CCPS). Risk management is an attempt to affect both the likelihood of an event and/or the magnitude of the event's impact. Risk management strategies are a combination of:

- Inherent (or intrinsic)—using materials that are nonhazardous
- Passive—equipment design features
- Active—using controls, interlocks, and emergency shutdowns
- Procedural—using operating procedures, administrative controls, emergency response, or other management approach.

  From the perspective of the ultimate consequence(s) of an undesirable event (e.g., release of toxic material), safeguards can be preventive and mitigative. A preventive safeguard stops the occurrence of a particular loss event after an initiating cause has occurred (i.e., a safeguard that intervenes between an initiating cause and a loss event in an incident sequence). A mitigative system is designed to reduce the consequences of an incident in an effort to maintain a safe and operable plant. Mitigative systems provide a layer of protection after there has been loss of containment or the incident has progressed to a point that the preventive safeguards will not be of value. Some examples of preventive and mitigative safeguards are

  - Ignition control—generally captured by electrical area classification areas (API RP 500, IP 15, IEC 60079) and ventilation/purging, static electricity control, etc.
  - Safety instruments systems
  - Pressure/Vacuum relief systems
  - Equipment isolation/sectionalizing
  - Fire and gas detection
  - Blowdown
  - Emergency evacuation and rescue
  - Firefighting.

Table 9.1 shows some examples of each kind of risk-mitigation strategy.

Table 9.1 Risk management examples

Category	Example	Comments
Inherent	Replace an ammonia refrigeration system with a propane system	This is replacing a toxic and flammable substance with a flammable nontoxic substance
Passive	For a vessel with 550 psig (3.8 MPag), normal operating pressure uses ASME 16.5 Class 600 instead of Class 300	This increases the safety factor from 9% (i.e., MAWP of Class 300 is 600 psig and Class 600 is 1440 psig) to 62%
Active	Plant emergency shut down (ESD) system	If any one of a number of possible conditions happens, then the ESD drives the plant toward a stable condition
Procedural	Lock out/Tag out	Energy isolation procedures are intended to minimize risk likelihood

9.2.1 Process safety management

US Occupational Safety and Health Administration (OSHA) regulation 1910.119 defines the 14 elements of process safety management (PSM) (OSHA) as:

1. Process safety information (PSI): Includes information pertaining to hazardous substances (i.e., toxicity, exposure limits, physical data, corrosivity data, stability data, mixing information, etc.), processes (e.g., block flow diagrams, process chemistry, safe upper and lower limits on pressures and temperatures, and an evaluation of consequences of deviations), and equipment (e.g., materials of construction, P&ID, electrical area classification, codes and standards applied, etc.).
2. Process hazard analysis: Focuses on the consequences of a failure/release of the items listed in the PSI.
3. Operating pProcedures: Detailed step-by-step instructions for operating the equipment.
4. Training: Required initial and refresher training for operators and maintenance staff.
5. Contractors: Deal with the safety aspects of contractor management.
6. Mechanical integrity: This section deals primarily with inspections.
7. Hot work: Focus is on what constitutes hot work, what work needs a hot-work permit, and what the hot work permit should address.

8. Management of change: Requires written procedures for documenting and approving changes to the process, equipment, or chemicals.
9. Incident investigation: Required employers to investigate and document findings for incidents.
10. Compliance audits: Establish a frequency that the facility must review the operation and procedures to ensure compliance with OSHA 1910.
11. Trade secrets: This prohibits trade secrets from being withheld from an incident investigation team (but allows a requirement for a non-disclosure agreement).
12. Employee participation: Requires employee participation.
13. Prestartup safety review: Requires that the start-up procedures include a review of the plant status.
14. Emergency planning and response: Requires development of a response plan.

PSM is the law in the United States for any facility that contains an inventory of any chemical in quantities greater than the thresholds listed in the act (OSHA, Appendix A). Very few wellsites or upstream facilities exceed those limits (e.g., 5000 lbm (2200 kg) of isopropylamine which is 830 gallons (3140 L); or 1500 lbm (630 kg) of hydrogen sulfide which is 16.8 MSCF (475 SCm)), but it is common for a plant to exceed the PSM limit on multiple chemicals and PSM is a key element in the functioning of most plants in oil and gas.

9.2.2 Hazard and operability

HAZOP has become the universal name for a "Hazard and Operability" which is intended to accomplish the first two elements of PSM but has grown to be used extensively in oil and gas for every project, regardless of size. The process of performing a HAZOP tends to be quite regimented and is supposed to look at every step in a process, every piece of equipment, and every chemical and determine a likely outcome of a failure and a mitigation method for that outcome that has a high likelihood of success.

9.2.3 Layers of protection analysis

One way that hazards are identified and mitigated is a layers of protection analysis (see Fig. 9.1). The labels above the center are the layers that can provide protection, and the labels below the center are examples

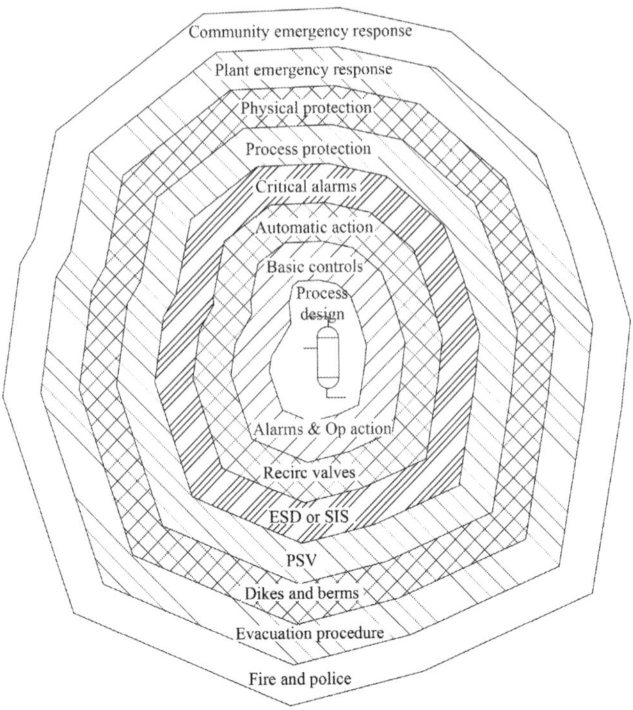

Figure 9.1 Layers of protection analysis.

(e.g., you can provide "physical protection" by installing "dikes and berms"). This analysis takes the list of layers and determines the actual element of a safety plan that satisfies that layer.

9.2.4 Safety integrity level

Safety integrity level (SIL) looks at failures of protection systems and the consequences of those failures. A high-likelihood failure that results in a high-consequence failure would warrant increasing the SIL. A basic SIL level would have a single input to a safety decision. The next level might have two inputs that trigger the event if either input exceeds a set point. The next level might have three inputs going into a logic controller that initiates the safety protocol if two out of the three exceed the limit (this type of SIL generally requires the safety protocol to be activated if the logic controller fails). Each SIL implies a greater level of redundancy and the highest level requires different sensor technology for the various inputs.

SIL is focused on preventing a failure much more than reducing the consequences of that failure.

9.3 GAS QUALITY

Well-head gas is what it is. It likely has a high liquid-water content, considerable water vapor, varying gas composition with both time and from well-to-well, and it can have solids. Sales gas is a commodity. Every lump of gas that someone purchases must be quite similar to any other lump of gas of the same size that they could purchase.

9.3.1 Wobbe Index

The difference in heating value between natural gas and liquefied petroleum gas (LPG) is significant. A combustion process designed for one gas will perform differently if you provide the other gas. In 1927, Goffredo Wobbe, a physicist in Bologna, Italy observed that (Wobbe):

- The heat output of a burner is proportional to the volume flow rate per unit time (at a constant pressure and temperature).
- The flow velocity through a given orifice size at constant pressure is proportional to the specific gravity of the gas.
- The calorific value, or heating value, of a gas is proportional to its specific gravity.

These observations indicate that just knowing the heating value of a gas is not enough to predict how that gas will perform in your burner. This observation has led to the use of a "Wobbe Index" (Eq. (9.1)) to allow comparison of a given mix of gases to a facility's fuel requirements.

$$WI = \frac{GHV}{\sqrt{SG}} \tag{9.1}$$

In previous tabular representations of gas properties we've use Net Heating Value (also known as "Lower Heating Value"), but that value is not appropriate for calculating a Wobbe Index. Table 9.2 includes the Net Heating value for reference, and adds the data for calculating a Wobee Index.

Wobbe Index has not been an issue in getting gas into a plant, but it is often the crucial parameter for a successful commercial transaction ranging in size from many public utilities around the world pricing home delivery of gas based on Wobbe Index to transactions the size of a supertanker full of liquefied natural gas.

The Wobbe Index addresses the issue of burner design for a given gas being inappropriate for another gas. If you have a burner on your

Table 9.2 Wobbe Index

	Net heating value	Gross heating value	SG	Wobbe Index
Methane	909 BTU/SCF (33.88 MJ/SCm)	1010 BTU/SCF (37.6 MJ/SCm)	0.5539	1357 BTU/SCF (50.6 MJ/SCm)
Ethane	1618 BTU/SCF (60.28 MJ/SCm)	1770 BTU/SCF (65.9 MJ/SCm)	1.0382	1737 BTU/SCF (64.7 MJ/SCm)
Propane	2315 BTU/SCF (86.25 MJ/SCm)	2516 BTU/SCF (93.7 MJ/SCm)	1.5226	2039 BTU/SCF (76.0 MJ/SCm)
n-Butane	3010 BTU/SCF (112.18 MJ/SCm)	3262 BTU/SCF (121.5 MJ/SCm)	2.0068	2303 BTU/SCF (85.8 MJ/SCm)
i-Butane	3000 BTU/SCF (111.79 MJ/SCm)	3252 BTU/SCF (121.2 MJ/SCm)	2.0068	2296 BTU/SCF (85.5 MJ/SCm)

hot-water heater designed for natural gas (Wobbe Index from 1400 BTU/SCF (52 MJ/SCm) to 1490 BTU/SCF (55.5 MJ/SCm)) will likely burn very hot with propane (Wobbe Index 2039 BTU/SCF (76 MJ/SCm)) and could fail catastrophically if used for extended periods. There have been many reports of house fires caused by installing natural gas equipment in a house supplied with propane, industrial processes are even higher risk. Matching Wobbe Number to equipment ratings is more effective than using a raw heating value.

9.3.2 Pipeline tariff

Every pipeline has a "tariff" that defines the limits of what they will take, and while the exact limits vary from pipeline to pipeline, they are quite close to each other. Looking at 24 pipeline tariffs in North America, Europe, and Australia, the most common values and the ranges are shown in Table 9.3. Various tariffs can have other components like a limit on Naturally Occurring Radioactive Material or Net Heating Value instead of Wobbe Index, it is necessary to review your specific tariff prior to making decisions that will affect the plant gas quality.

The fps units for sulfur in Table 9.3 are interesting. "Grains per 100 SCF" seems to be an awkward unit. There is no clear reason that it hasn't been updated to less obscure units (that limit is equivalent to

Table 9.3 Sales gas quality

		Min (if specified)	Average (if specified)	Max (if specified)
Wobbe Index[a]	Min	1366 BTU/SCF (50.9 MJ/SCm)	1402 BTU/SCF (52.2 MJ/SCm)	1408 BTU/SCF (52.5 MJ/SCm)
	Max	1450 BTU/SCF (54.0 MJ/SCm)	1490 BTU/SCF (55.5 MJ/SCm)	1624 BTU/SCF (60.5 MJ/SCm)
Oxygen	Max	0.2%	0.3%	1.0%
H_2S	Max	0.25 g/100 SCF (5.7 mg/SCm)	0.51 g/100 SCF (11.6 mg/SCm)	1 g/100 SCF (23 mg/SCm)
Total Sulfur	Max	0.75 g/100 SCF (17 mg/SCm)	8.8 g/100 SCF (202 mg/SCm)	20 g/100 SCF (460 mg/SCm)
Water content	Max	4 lbm/MMSCF (65 mg/SCm)	5 lbm/MMSCF (83 mg/SCm)	7 lbm/MMSCF (112 mg/SCm)
Hydrocarbon dew point	Max	−5°F (−21°C)	14°F (−10°C)	20°F (−7°C)
Total inert (incl CO_2 and O_2)	Max	3%	4%	7%
CO_2	Max	2%	2%	3%

[a]For tariffs that have heating value instead of Wobbe Index, a composition was assumed to match heating value provided within the limits of the tariff.

0.357 lbm/MMSCF which would put it in the same units as water vapor), but cut-and-paste seems to be as ubiquitous in contracts and regulations as it is in the rest of engineering.

The oxygen limits are especially interesting. Remember from Chapter 4, Surface Engineering Concepts that gathering companies are pushing for limits on oxygen in gas-gathering systems on the order of 25 ppm (0.025%), but the tariffs are still limited to 0.2%−1.0% (2000−10,000 ppm). This disconnect also does not have a clear explanation.

Table 9.4 shows some examples of blends that could reach specific Wobbe Index targets within the limits of the tariff if you wanted the maximum ethane and CO_2 allowable within the tariff—there are approximately an infinite number of other combinations that could reach the same targets, but this is one.

Table 9.4 Sample mixtures to reach Wobbe Index targets maximizing ethane and CO_2

	1366 BTU/SCF (50.9 MJ/SCm)	1446 BTU/SCF (53.9 MJ/SCm)	1624 BTU/SCF (60.5 MJ/SCm)
Methane (%)	84.7	74.2	56.4
Ethane (%)	13.3	15.0	15.0
Propane (%)	0.0	8.8	15.0
n-Butane (%)	0.0	0.0	11.6
Total inerts (%)	2.0	2.0	2.0

9.4 MASS AND ENERGY BALANCE

Well-head gas is priced on an energy basis, but measured on a volume basis. This can lead to confusion:

- CO_2 removal reduces volume, but it doesn't change energy.
- H_2S removal reduces both volume and energy.
- Natural Gas Liquids (NGL) removal represents a small reduction in volume, but a very large reduction in energy.

To make any sense at all of this, you have to work in mass and energy units, volume units are just too confusing. "Plant shrink" is usually charged, but it doesn't always mean the same thing:

- Sometimes it is a percentage of the total inlet volume.
- Other times it is a percentage of the total inlet energy.
- Plant shrinkage is generally set by contract, not by performance.
- Contractual shrinkage is deducted from field gas to get to a financial transaction.
- Physical shrinkage is a function of the plant processes and is the responsibility of the plant.

For example if a sweetening plant (i.e., a plant set up to remove acid gases, but not extract NGL) has a contractual shrinkage of 3% (of energy) receives 1000 MMBTU (1055 GJ) then they are obligated to deliver 970 MMBTU (1023 GJ) to the people that the owners of the gas have sold it to. If they find that there is only 950 MMBTU (1002 GJ) available at the plant outlet then the plant owners must purchase the missing 20 MMBTU (21 GJ) to make up the difference. On the other hand, if the plant does an exceptional job of controlling fuel, flare and leakage and they are able to deliver 990 MMBTU (1045 GJ) then they can add to the plant profitability by selling the excess for their account.

Sometimes processing agreements are done on a volumetric shrinkage. For the sweetening plant above, the 1000 MMBTU (1055 GJ) that was delivered to the plant might have contained 30% CO_2, so the volumetric shrinkage cannot be less than 27% (assuming that the maximum CO_2 allowed in the sales gas is 3%) and shrinkage based on volume would be a disaster. On a plant designed to extract NGL, the difference between volumetric shrink and energy shrink is not as marked, since NGL extraction includes both volume reduction (small) and energy reduction (large). For NGL plants, the shrinkage is generally based on a combined energy of all value streams. Sometimes, a large portion of the NGL is retained by the plant as a processing fee, which makes the process of balancing the plant a bit easier.

9.5 PLANT PROCESSES

Plant processes can include any number of components in a wide range of configurations, and depending on the nature of the gas some or all of the components in Fig. 9.2 will be present or some may be absent.

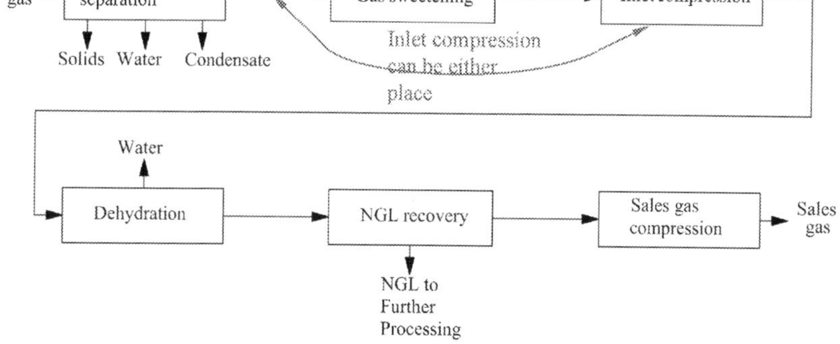

Figure 9.2 Process plant block diagram.

9.5.1 Inlet facilities

The inlet of the plant is raw field gas. As we've seen through this entire book, it is not a safe idea to assume that raw field gas is anything but "raw" (often it has been dehydrated at the intermediate compressor station, but these lines are rarely pigged and many plants have been shocked by how much water can condense out of a 7-lbm/MMSCF (112 mg/SCm) line over a few months). Slugs of liquid, solids, varying gas mixtures, and other variations will occur from minute to minute. Consequently the plant inlet facilities must be both robust and rugged. They must be large enough to process the liquid from a pig run without taking the plant down. They must be able to remove solids and emulsions without those components getting into sensitive and delicate processes further into the plant.

Liquid hydrocarbons are a special problem in many of the plant processes and need to be positively removed before getting all the way into the plant. Some of the more common inlet facilities include:

- Slug catchers and drips
- Scrubbers (generally without mist extractors)
- Oversized three-phase separators
  - Horizontal are slightly more common.
  - Vertical are less common because determining the water/oil interface in raw well-streams is difficult and their liquid capacity tends to be limited.
- Coalescing elements.

This equipment tends to have wide operating range and needs minimal maintenance or operations oversight.

9.5.2 Inlet compression

Inlet compression may or may not be needed. If it is absent, then the plant is relying on field compression to operate the plant. If it is present it may be before or after Acid Gas removal. The trade-off is interesting, if you put it before the acid–gas process then you have to compress the gas you are about to throw away (can be as much as 30% of the inlet volume), but if you put it after the acid–gas process then you might be limited in the amount of pressure drop you can afford, through the acid gas treatment process. Additionally all processing plant (including compression) upstream of the acid–gas process is subject to a more corrosive environment and hence corrosion resistant materials are more likely to be required. Both positions have their proponents and both are common.

If inlet compression is present it is nearly always recip compressors, but it is rarely present on large facilities.

9.5.3 Acid gas removal

The primary acid gases that are a concern in natural gas production are CO_2 and H_2S. Both can be removed by absorption, adsorption, or membranes.

9.5.3.1 Absorption processes

Absorption using an aqueous solution of an "amine" chemical is the most common by far. "Amines" are "any of a class of basic organic compounds derived from ammonia by replacement of hydrogen with one or more monovalent hydrocarbon radicals" (Webster). Amines used for acid gas removal are in molecules with organic alcohols. Common amines are:

- Monoethanol amine
- Diethanol amine (DEA)
- Methyl DEA (MDEA)
- Activated MDEA.

The specific amine formulation in a given plant is based on the expected mix of acid gases and the target sweet-gas quality.

The process is laid out in Fig. 9.3. Notice that the process gas is only on the left-hand side of this schematic. The sour gas is passed through a separator to remove any liquids, then it is bubbled up through a contactor tower while an amine solution flows from the top down (this allows the most contaminated gas to see the most contaminated amine and vice versa). The rest of the schematic is all trying to regenerate the amine for reuse and recover as much of the added heat as possible.

The process in the contactor is exothermic and considerable heat is generated in the tower. Often times this heat is the limiting factor in system capacity since the reversible reaction between the amine and the acid gas is temperature sensitive and to get the amine to release the acid gas you heat the amine.

The sweet gas can approach zero acid gas, but there are economic incentives to leave as much of the acid gas as the tariff will allow. Sometimes, this is done by processing part of the stream down to approximately zero acid gas and then blending in unprocessed gas to approach the target composition. Gas into the process has often been dehydrated in

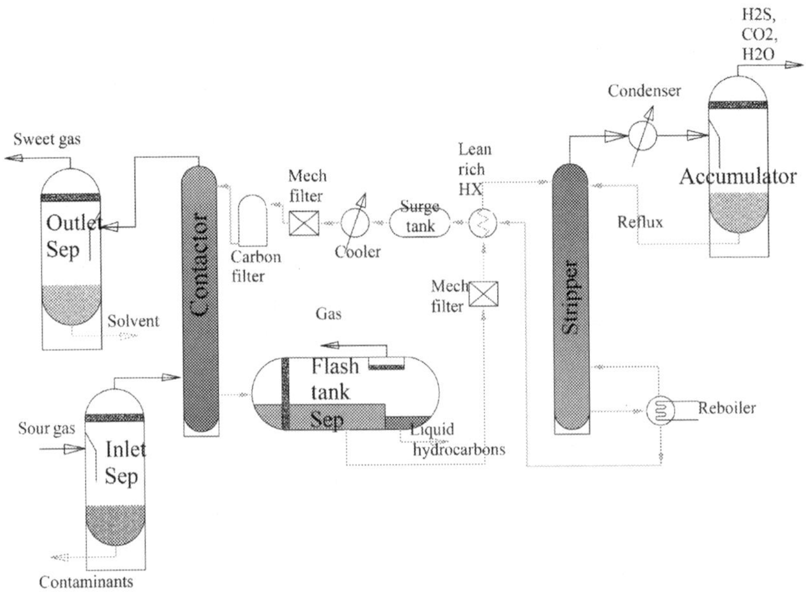

Figure 9.3 Acid gas removal.

mid-stream compressor stations, and gas out of the amine outlet separator is at 100% relative humidity, blending dry gas back in reduces the dehydration load.

The saturated amine, leaving the bottom of the high-pressure contactor, has the pressure reduced in the flash tank. Gas produced from the flash tank can contain sufficient hydrocarbon to allow its use as fuel gas or to warrant delivering it flare to ensure hydrocarbon destruction. The vent gas off of the accumulator stack will be >98% acid gas (dry basis) saturated with water vapor. For CO_2 removal that gas is either vented to atmosphere or goes into some sort of CO_2 sequestration (or reuse) process. If the acid gas of concern is H_2S or some other sulfur compound then it cannot simply be vented. The typical limit for venting H_2S is 10 mg/L. As elemental sulfur has an economic value, using something like the Claus Process can recover the sulfur for sale. There will be some amount of H_2S in the vent stack off the Claus Process, and that stream (or the entire stream off the accumulator if the plant doesn't do sulfur recovery) must be flared in a forced air combustor.

The water in the aqueous amine is evaporated to some extent by the dry gas going through the system. It is necessary to closely monitor the condition of the amine and add makeup fluids as necessary.

9.5.3.2 *Molecular sieve*

A molecular sieve or "mol sieve" removes target contaminants by adsorption instead of absorption. The media in the mol sieve is a granular solid with the ability to preferentially accept the target substance at adsorption sites and allow hydrocarbon molecules to pass through. This process is similar to the gas storage mechanism in coalbed methane (CBM) wells.

When the bed begins losing effectiveness it must be regenerated to remove the target contaminants and prepare the bed for the next cycle. Adsorption is temperature dependent and at elevated temperatures the adsorption sites lose their ability to hold the target molecules. Typically the contaminant is removed by passing very high temperature (on the order of 550°F (288°C)) gas over the bed and exhausting it. Occasionally the gas can be exhausted to atmosphere, but this is increasingly prohibited under "clean air" regulations. Other options include sending it to a flare or in some cases the waste stream can be used for fuel gas.

When you are treating natural gas you have a dilemma—if you regenerate with air, then the bed is air-filled and introducing natural gas requires a purge, if you regenerate with natural gas you avoid having to purge the system but you are discarding a considerable volume of valuable gas.

You can select adsorption media to preferentially target any contaminant that you are interested in removing. Common target contaminants are CO_2, H_2S, and water vapor. The adsorbent that the media prefers is a statistical preference, not an absolute. Just like CBM "prefers" to adsorb CO_2 if there isn't enough CO_2 to saturate the media, then it will adsorb methane and water. Mol sieve media works the same way.

The mol sieve in Figs. 9.4 and 9.5 was a wellsite application that was indistinguishable from a mol sieve you would see in a plant. The problem it was trying to address was that that the CO_2 content in the well gas had increased to the point that it exceeded the engine manufacturer's specification for the carburetor's installed on the engines and downtime was increasing. The idea was to pull fuel gas off the discharge of the compressors and process it in the mol sieve to lower the CO_2 from 19 mole% to 6 mole% while lowering the water content. The media was charcoal made from coconut hulls, so of course the project was called "Piña Coloda." When the project was initially started up (Fig. 9.4) it was assumed that there was plenty of heat in the engine exhaust and there was no reason to insulate the heat exchanger—regeneration gas at the dryer was never warmer than 300°F (150°C) instead of the 550°F (288°C) that

Figure 9.4 Wellsite molecular sieve.

Figure 9.5 Mole sieve regeneration circuit.

was actually required to release the CO_2. When the jacket heat exchanger (i.e., the inside of the heat exchanger is the full diameter of the exhaust pipe and there is a jacket around it where the gas to be heated was flowed) and the piping were insulated the regen gas was consistently 575°F (302°C) and the device worked well. As the project was winding down with a recommendation to purchase a large number of these very expensive units and accept a large regeneration exhaust stream, one of the engine-manufacturer's reps asked "why don't you just install a carburetor designed for 19% CO_2, it will cost a few hundred dollars and will return the engines to full power." Piña Coloda was abandoned in favor of a trivial modification to the engines.

9.5.3.3 Membrane

Membrane units use a nonporous media to diffuse specific molecules from the flow stream into a waste stream. Since the membrane in nonporous (unlike the semipermeable element in a reverse osmosis system for example) molecule size is irrelevant. The media will diffuse different molecules at different rates.

- CO_2, H_2, He_2, H_2S, and water vapor are "fast" molecules.
- Methane, ethane, and larger hydrocarbons are "slow" molecules.

The membrane takes fast molecules as they arrive, but if it still has capacity it will accept some slow molecules. There are always significant quantities of hydrocarbon gases in the permeate: (1) 10% hydrocarbons in the waste stream is best in class; (2) 50% is about normal; and (3) 90% hydrocarbons have been reported. Diffusion rate is very dependent on operating pressure (higher pressures work better) and the amount of product in the waste stream is reduced for all concentrations of contaminants at higher pressures.

Membrane technology is the only acid–gas technology that doesn't require regeneration, but it is also results in the highest quantity of wasted hydrocarbons when the gas is processed to very low levels of acid gas. It is fairly common for plants to use membranes to reduce the amount of acid gas upstream of an amine process to reduce the required regeneration load on the amine.

Membranes are also used for dehydration, and this application has the benefit of not needing regeneration.

9.5.3.4 *Other technologies*

As we discussed in Chapter 8, Gas Compression, the oil used in oil-flooded screw compressors is hydrophilic and is an excellent (albeit very expensive) dehydration media. This should make someone think that there are other technologies to remove acid gases, and they would be correct. One of the more common processes is called the "Benfield Process" where hot potassium carbonate is used as an absorption media. Another is the UOP Selexol Technology that uses proprietary solvents.

You can expect new processes to come (and most to go) every few years. Most of it will not be able to compete with the other technologies and will quietly fade away. Some will actually be game changing and could easily be the next generation's "amine" (which was also a new process once). It is useful to mention these technologies simply to point out that no science is ever "settled" and everyone needs to keep an open mind about technology proposals and give them a fair trial if you are in the market for new technology at the same time as some bit of new technology becomes known to you.

9.5.4 Dehydration

When third party gathering was the only option for gas-well production, it was common to remove water vapor at the wellsite. As we will see in a moment, with decreasing gathering pressure wellsite dehydration economics suffered dramatically. Today it is fairly rare to see wellsite dehydration and when you do see the equipment installed it is bypassed more often than not. Generally we see wet gas up until the first compressor station where the gas will be dehydrated after compression. Largely water-free gas comes into the plant, then if there is acid–gas removal it will be resaturated with water. Even without acid–gas removal, the 7 lbm/MMSCF (112 mg/SCm) specification that is typically used in compressor stations is too much water vapor for cryogenic processes.

Liquid water is removed in mechanical separators, water vapor is a tougher problem. Common methods to remove water vapor are: (1) glycol absorption; (2) molecular sieve; (3) deliquescent dryers; and (4) membrane systems. All of these systems have different ways to reject extracted water. Glycol absorption systems add heat in a reboiler to boil the liquid and since water has a lower boiling point than glycol, the water will boil off and leave regenerated glycol behind. Molecular sieves use a heated gas to evaporate the water into the regeneration gas that is then rejected to atmosphere.

Deliquescent dryers use a consumable media and the water vapor actually melts the salt in the dryer and creates a brine that must be drained off. Membranes preferentially pass water through the nonporous media and it is rejected to atmosphere along with some portion of the process gas.

9.5.4.1 Glycol absorption

A lean (i.e., low water content) hydrophilic liquid is circulated into the top of a contactor tower with the wet gas flowing into the bottom of the tower. The gas and liquid contact each other and the water vapor is absorbed into the hydrophilic liquid. The rich (i.e., high water content) liquid is circulated into the reboiler for regeneration. The most common hydrophilic liquid is TEG, but other liquids are used. This process is largely independent of system pressure.

If the process is largely independent of pressure, why did we stop putting dehydrators on wellsites? The answer is that cooking the water out of the glycol requires 970 BTU/lb (2.26 MJ/kg) of heat with about 10% of that heat unrecoverable.

As gathering pressures have decreased across the board, the required size of reboiler burners and glycol pumps became prohibitive, and equipment that was fine 100 psig (689 kPag) was one-fourth the required size for 10 psig (68.9 kPag) (Table 9.5).

9.5.4.2 Deliquescent dryer

These dryers use a "hygroscopic" (i.e., a hydrophilic substance that is changed by acquiring water) salt like calcium chloride ($CaCl_2$) which has

Table 9.5 Reboiler capacity on dehydrator

	1000 psig (6.89 MPag)	100 psig (689 kPag)	10 psig (68.9 kPag)	Atmospheric
Water content of feed gas[a]	60.4 lbm/MMSCF (968 mg/SCm)	466.3 lbm/MMSCF (7470 mg/SCm)	1819 lbm/MMSCF (29,300 mg/SCM)	3077 lbm/MMSCF (49,00 mg/SCM)
Reboiler size[a]	2400 BTU/h (2.53 MJ/h)	21,000 BTU/h (22.2 MJ/h)	81,000 BTU/h (85.5 MJ/h)	137,000 BTU/h (144.5 MJ/h)
Glycol flow rate[a]	3.8 gal/h (14.4 L/h)	29.1 gal/h (110 L/h)	282.1 gal/h (1068 L/h)	2811 gal/h (10,640 L/h)

[a]Reboiler size to process 1 MMSCF/day (28.3 kSCm/day) at 100°F (37.8°C) down to 7 lbm/MMSCF (110 mg/SCm)

an affinity for water vapor. As the salt takes in the water vapor, the body of the salt dissolves into a brine that must be drained. This media can remove up to three mass units of water vapor for every mass unit of salt. That would say that to remove 3077 lbm (1396 kg) from 1 MMSCF (28.3 kSCm) of gas you would have to add 1025 lbm (465 kg) of salt. For a vessel that holds say 100 lb (45.4 kg) of salt that you expect to replenish when it is 50% consumed, you would have to replenish the salt and drain the brine 21 times a day, about every hour. By the smaller size of these vessels it should be clear that they reach an upper limit of throughput at a very low level. There are some larger vessels on the market, but the goal with them is more providing a longer interval between replenishment than processing a larger volume.

Deliquescent dryers can achieve a 15–20 R (8–11 K) dew point depression which may or may not be adequate to reduce the freeze potential on a small natural gas fired engine to an acceptable level.

9.5.5 NGL removal

Natural gas liquids are hydrocarbons that are gaseous at ambient pressure and temperature, but can be liquefied at reasonably achievable pressures and temperatures. We normally think of NGLs as ethane, propane, normal butane, iso-butane, and pentane combined with all heavier hydrocarbons (usually called "C5 + "). Since the boiling point of normal pentane is 82°F (28°C) (see Table 9.6) and the other components of C5 + are all higher boiling points it will be rare for C5 + to survive mechanical separation so the predominant products are ethane, propane, and the two butane products.

NGL products sometimes have their own markets (e.g., you put propane in your residential tank, butane is used in cigarette lighters, and ethane is a key feedstock used in the manufacture of many plastic products), at times a price premium may be realized by supplying NGL products into these markets while at other times they have more value blended with methane for commodity gas. Modern NGL plants have the ability to make the recover/reject decision for each product on an hourly basis to accommodate immediate price swings.

The primary methods of extracting and separating NGL from natural gas are "Lean Oil Absorption" and "Cryogenic Expansion."

Table 9.6 Boiling points and freezing points of selected components

	Formula	SG (relative to air)	Boiling Point at atmospheric pressure	Freezing point at atmospheric pressure
C1—Methane	CH_4	0.5539	−259°F (−162°C)	−296°F (−182°C)
C2—Ethane	C_2H_6	1.0382	−128°F (−89°C)	−297°F (−183°C)
C3—Propane	C_3H_8	1.5226	−44°F (−42°C)	−306°F (−188°C)
C4—i-Butane	C_4H_{10}	2.0068	11°F (−12°C)	−255°F (−159°C)
C4—n-Butane	C_4H_{10}	2.0068	31°F (−1°C)	−217°F (−138°C)
C5—i-Pentane	C_5H_{12}	2.4912	82°F (28°C)	−256°F (−160°C)
Carbon dioxide	CO_2	1.5196	−109°F (−78°C)	Not defined[a]
Hydrogen sulfide	H_2S	1.1767	−76°F (−60°C)	−122°F (−86°C)
Water vapor	H_2O	0.6220	212°F (100°C)	32°F (0°C)

[a]Not defined below 75 psia (517 kPaa) above 75 psia CO_2 freezing point is −69.8°F (−56.6°C)

9.5.5.1 Lean oil absorption

Lean oil absorption plants operate on exactly the same principle as amine treating plants, however instead of using an amine solution to selectively absorb CO_2, a lean oil is used to selectively absorb heavier hydrocarbons. As is the case with amine treating, the working fluid becomes saturated and must be regenerated.

A Lean Oil Absorption plant starts with mechanical separation followed by a contactor tower and a second mechanical separator. The raw stream is brought into the bottom of a contactor and a chemical with an affinity for heavier hydrocarbons (called "lean oil") is pumped into the top of the contactor. The pentane, butane, some of the propane, and a bit of the ethane will absorb into the lean oil and the methane, most of the ethane, and the rest of the propane will leave the top of the tower to the outlet separator. The absorption chemical (now called "rich oil") is processed in a series of distilleries ("'stills") that are maintained at temperatures that allow the products to selectively boil off and be removed.

Looking at Table 9.6 you can see that the 'stills must be kept at fairly cold temperatures. A 'still that is warmer than −44°F (−42°C) would have methane, ethane, and propane in the overhead. Transporting the liquid to a still that is maintained above 11°F (−12°C) but below 31°F (−1°C) would only have iso-butane in the overhead. A third 'still that is maintained between 31°F (−1°C) and 82°F (28°C) would have normal butane in the overhead and C5 + as a liquid.

The value of ethane for plastics manufacture and propane for LPG deliveries has significantly reduced the popularity of lean oil plants.

9.5.5.2 *Cryogenic expansion*

Cooling a rich gas to a value between −217°F (−138°C) (the freezing point of iso-butane) and −128°F (−89°C) (the boiling point of ethane) will leave methane in the overhead and all other hydrocarbon components will be liquid (Table 9.7). Any solids-formation will inhibit the process at best and can completely clog it up. Temperatures colder than −217°F (−138°C) will result in any iso-butane freezing into a solid. If there is any water present, then temperatures below 32°F (0°C) will cause water-ice. If there is any CO_2 in the stream then temperatures below −69.8°F (−56.6°C) will result in "dry ice" (Fig. 9.6). CO_2 is an especially troublesome compound for any cold-temperature processing. The direct transition from a gas to a solid and back occurs at a relatively high temperature and pressure and can easily plug up piping, pumps, heat-transfer equipment, and vessels even at a very low CO_2 concentration (the common limit for CO_2 into a cryogenic process is 50 mg/L).

Table 9.7 Fractionation vessels

	Overhead	Max. temp. for recovery	Liquid	Next step for liquid
Demethanizer	CH_4 −	−128°F (−89°C)	Everything heavier	Deethanizer
Deethanizer	C_2H_6 −	−44°F (−42°C)	Everything heavier	Depropanizer
Depropanizer	C_3H_8 −	11°F (−12°C)	Everything heavier	Debutanizer
Debutanizer	C_4H_{10} −	82°F (28°C)	Natural gasoline (C5 +)	Condensate stabilization

The " − " symbols in the "overhead" column indicate that any lighter component that wasn't extracted in earlier steps would be in the overhead in the next step.

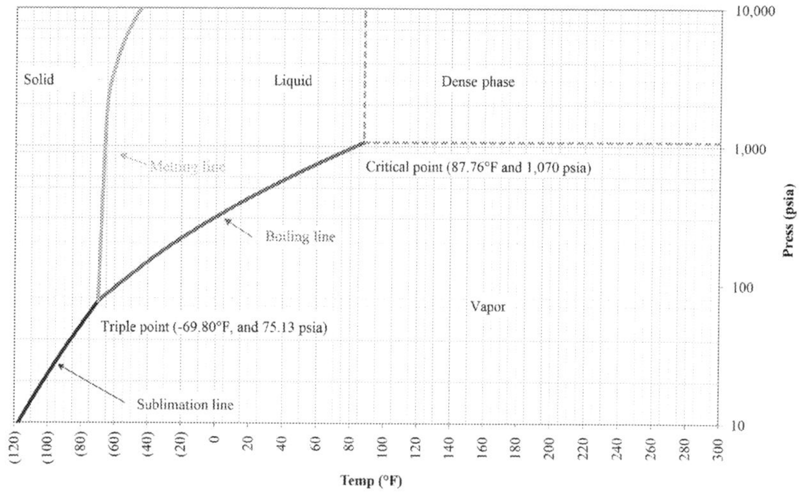

Figure 9.6 CO_2 phase diagram.

After the methane has been extracted, the liquid goes to the next vessel which is maintained at a temperature that will allow ethane to boil, but keep the other hydrocarbons in liquid form.

The gas is usually chilled in several steps. You can use a conventional refrigeration cycle (usually with propane or ammonia refrigerant) to chill the gas to around −40°F (−40°C) in a single step. Using two or more refrigeration cycles in series can allow the gas to reach −90°F (−68°C). Refrigeration cycles are nearly an isobaric process.

The Joule–Thomson Effect says that an isenthalpic pressure drop across a restriction will result in lower temperatures. The μ_{JT} constant in Eq. (9.2) is very much dependent on upstream pressure and temperature.

$$T_{\text{final}} = T_{\text{initial}} - \mu_{JT} \cdot \Delta P$$
$$\mu_{JT} = \frac{V}{c_p} \cdot (\alpha \cdot T - 1) \tag{9.2}$$

For example, if you drop the pressure of a volume of methane from 1400 psig (9.65 MPag) at −40°F (−40°C) to 400 psig (2.76 MPag), the temperature would drop to −110°F (−79°C). A useful change in temperature, but a big drop in pressure. To get to really significant temperature drops you have to move away from isenthalpic processes.

To reach the necessary very low temperatures for NGL processing, plants generally pre-cool the gas with refrigeration and then use a "turbo-expander" to make the gas do work. A turbo-expander uses a pressure drop in the gas to spin a turbine wheel. The work that the gas does on the wheel results in a large change in enthalpy that is represented in both dropping pressure and dropping temperature. These turbines are usually driving a compressor that will then (in a polytropic process) replace some of the pressure used in the turbine. Turbo expanders represent the highest capital cost, but the lowest operating cost.

When we talk about NGL recovery we use terms like "overhead" (i.e., the gas volume above the liquid) and "fractionation tower" (i.e., a pressure vessel used to remove a specific compound or "fraction" from the stream).

The reject/recover decision is implemented by managing the temperature of the demethanizer, anything in the overhead of that tower is going to be used as natural gas so if you raise the demethanizer temperature to above −44°F (−42°C) then all of the ethane will be sold as "natural gas" instead of "NGL." Each plant has procedures for the reject/recover, and all of them have some combination of by passing the turbo-expander and adjusting the outlet temperature of the refrigeration cycles.

9.5.6 Sales-gas compression

The sales gas is a commodity. No contaminants. No water vapor. Constant density. Constant heating value. Constant suction pressure and temperature. Constant discharge pressure with a wide range of acceptable discharge temperatures. Reasonably constant throughput. Extreme penalties for adding compressor oil to the stream. This is a perfect application for a gas-turbine driven centrifugal compressor, and that technology dominates the population of sales-gas compressors.

9.5.7 Process flares

One of the most visible characteristics of many plants is the flare stack. It would be good for community relations for the flare (either a stack or a ground flare) to never be ignited, but that rarely proves to be possible. Since equipment is so densely packed inside a plant, local releases of flammable or toxic gases have too high a chance of finding an ignition source and/or a receptor to be allowed. Consequently all vents, drains, unloaders, and PSV exhausts must go into closed systems. For gases the systems come together into a flare header that may or may not have forced air to

Table 9.8 Exercise #6 data

	fps	Both	SI
Plant inlet flow rate	200 MMSCF/day	Sweet gas	5663 kSCm/day
Inlet pressure	27 psia		186.2 kPaa
Inlet temperature	80°F		26.7°C
Shrinkage		2% of inlet volume	
Minimum outlet gas energy	950 BTU/SCF		35.4 MJ/SCm
Maximum outlet gas energy	1050 BTU/SCF		39.1 MJ/SCm
Max CO_2		3%	
Water vapor	<7 lbm/MMSCF		<112 mg/SCm
Gas sales price	\$5/MMBTU		\$4.74/GJ
NGL sales price	\$0.75/gal		\$0.198/L
		Liquid SG	
Ethane (C2)		0.35619	
Propane (C3)		0.50698	
iso-Butane (iC4)		0.56286	
n-Butane (*n*C4)		0.58402	
iso-Pentane (iC5)		0.62441	
n-Pentane (*n*C5)		0.63108	
Hexane plus (C6 +)		0.66404	

ensure complete combustion, and either has a pilot light or a high-reliability ignitor.

The flare lighting-up the roadway in front of a plant, may mean that a piece of equipment has just been shut down and everything is normal, or it may mean that something dangerous is underway. When working around a plant it is always a good idea to be aware of the flare, what its status is, and notice when that status changes.

9.6 PLANT INTERFACE CONCLUSION

Design and operation of gas processing plants is a specialized skill that requires considerable focused training and experience. The skill set that makes a good plant engineer may not contribute to a person being a good field engineer.

Inside plants:

- The equipment density is very high.
- Many processes operate very close to real physical limits on equipment.

Outside plants:

- Equipment density is very low.
- Few processes operate at more than 20% of physical limits so failure is less likely.

Many engineers have successfully gone between plants and fields, but that is more a statement of those individual's ability than an implication of inherent compatibility of the two environments.

REFERENCES

[API RP 500] Recommended Practice for Classification of Locations for Electrical Installations at Petroleum Facilities Classified as Class 1, Division 1 and Division 2, 3rd edition, December 2012, Washington, D.C.: American Petroleum Institute.

[CCPS] Center for Chemical Process Safety (CCPS) of the American Institute of Chemical Engineers (AIChE), *Guidelines for Engineering, Design for Process Safety*, AIChE, New York, 10993, 2012, ISBN 0-8169-0565-7.

[IP 15] Area Classification Code for Installations Handling Flammable Fluids, Model Code of Safe Practice Part 15, 3rd edition, July, 2005, Energy Institute, London, ISBN 0 85293 418 1.

[IEC 60079] Electrical Apparatus for Explosive Gas Atmospheres – Part 0, General Requirements, IEC 60079-0, 6th edition, June, 2014, London: IEC.

[OSHA] United States Department of Labor. (n.d.). Retrieved February 01, 2017, from https://www.osha.gov/pls/oshaweb/owadisp.show_document?p_table = STANDARDS&p_id = 9760.

[Webster] (n.d.). Retrieved April 21, 2016, from http://www.merriam-webster.com/dictionary/amine, (other dictionary references replace "amine" with the referenced term.

[Wobbe], The Wobbe Index and Natural Gas Interchangeability, Bristol Babcock, LTD a division of Emerson Process Management, July 30, 2007, retrieved February 8, 2017 from http://documentation.emersonprocess.com/groups/public/documents/application_notes-tech._briefs/d301334x012.pdf.

NOMENCLATURE

Symbol	Name	fps units	SI units
c_p	Specific heat at constant pressure	BTU/lbm/R	kJ/kg/K
GHV	Gross heating value (also known as "Higher heating value," see Chapter 0: Introduction)	BTU/SCF	MJ/SCm
SG	Specific gravity relative to air	Decimal	decimal
T_{final}	Final temperature	°F	°C
T_{initial}	Starting temperature	°F	°C
V	Volume of gas	ft^3	m^3
WI	Wobbe Index	BTU/SCF	MJ/SCm
α	Coefficient of thermal expansion	R^{-1}	K^{-1}
μ_{JT}	Joule–Thomson coefficient	R/psi	K/psi
ΔP	Change in pressure	Psi	kPa

EXERCISES

1. A facility chooses to install a vessel rated at ANSI 600 (MAWP is 1440 psig or 7.860 MPa) on a system who's normal operating pressure is 500 psig (3.45 MPa). What category of risk management would that be? Choose an item.
   a. Inherent
   b. Passive
   c. Active
   d. Procedural
2. In a cryogenic plant, which of the following should be kept out of a Demethanizer tower? Choose an item.
   a. Carbon dioxide
   b. Propane
   c. *n*-Butane
   d. All of the above

3. A gathering system has a 3% (by volume) shrink in their gathering contract. The sweetening plant has a 5% (by energy) shrink in their processing contract. A producer puts 500 MMSCF of 84% CH_4, 16% CO_2 gas onto the system, what is their wellsite total energy? How much energy should they expect to sell at the tailgate of the plant?
4. A fuel supply at 45 psig (310 kPag) and 75°F (23.9°C) is supplying 50 MSCF/d (1.4 kSCm/d) of 100% RH gas to a compressor through a deliquescent dryer that holds 100 lbm (45.4 kg) of salt. If the dryer is able to lower the dew point by 20°F (11°C), and the operator wants to add salt when 50% of the inventory is depleted, how often would he have to add salt?
5. If the Joule–Thomson coefficient for methane at 400 psia (2.76 MPaa) is 0.087°F/psi (7.01°C/MPa) and the temperature on the high-pressure side of a permanent pressure drop is −40°F (−40°C), what is the temperature of the gas on the low pressure side if the choke drops the pressure to 150 psia (1034 kPaa)? What assumptions did you make to reach this number?
6. For the conditions in Table 9.8, find:
   a. Gross dollars received at the plant for minimum energy in gas
   b. Gross dollars received at the plant for maximum energy in gas

CHAPTER TEN

Integration of Concepts

Contents

This book has covered a very wide range of material from how hydrocarbons are created through final processing of sales gas and produced water. Few people will ever have responsibilities that span this range over an entire career, but every individual's job will be touched by the entire range of specialties. It should be obvious that the better you are at communicating with other specialties, the better your chances of a favorable outcome. Saying to the driller "I've got a schedule to keep, when will your rig thingy get off my location so I can do the important work?" will most likely be less effective than "How long does the cement have to set up before you can run your CBL?" Both statements are asking the same thing, but the second one shows that there is a chance that the two of you can communicate while the first one looks like you are searching for someone to blame for your (inevitable) failure.

Mastering that communications difficulty is the intention of the rest of this book. The intention of this chapter is to help understand other's successes and mistakes to try to facilitate copying the successes and avoiding the mistakes.

10.1 PROCESS OWNERSHIP

When you look at people who frequently succeed versus people who usually fail in any field of endeavor, you get the sense that there is

Practical Onshore Gas Field Engineering.
DOI: http://dx.doi.org/10.1016/B978-0-12-813022-3.00010-9

some unifying principle that can facilitate (certainly not "ensure") success by its presence and facilitate (again, certainly not "ensure") failure by its absence. Mike Rowe calls this unifying principle "passion" and says "don't follow your passion, but do bring it with you" (Rowe). This statement can be taken to mean that a personal drive toward a goal ("passion") influences the outcome of your efforts toward that goal. I call this characteristic "ownership." It means that

- You question things being done to the things you own.
- You understand "steady-state" conditions and can recognize the onset of deviations.
- You learn to recognize "anomalies" before they become "problems."

Ownership is the key element to becoming the best that you can be.

10.1.1 How to achieve process ownership?

Most of us have purchased a car. When we get the keys, we find that the seat is in the wrong position, the mirror looks the wrong way, all of the knobs and switches are in the wrong position, and there are a couple of sounds that you are not sure are right. You then

- Spend time with the owner's manual.
- Spend time adjusting all of the ergonomic things.
- Pay attention to every squeak, rattle, and roar until your subconscious knows what is "right".
- You drive it slow, you drive it fast, you begin to understand how it will perform in marginal situations before you enter marginal situations.

A gas well costs a lot more than a car, but the same principles apply. You have to learn what "normal" looks like. The "user manual" on a well is the culmination of thousands of observations, opinions, and decisions. Every formation is different, and every well is different, but there are things that will always move you toward ownership.

10.1.1.1 Owning a reservoir

If you can't reliably say what the reservoir pressure is today, how do you set target pressures, forecast future performance, or predict the outcome of an intervention? Determining reservoir pressure can be difficult, but it is worth it. For conventional and tight gas, you will rarely have reliable mass balance or pressure build up data so you can get your best estimates from reservoir models. For coalbed methane (CBM) and shale gas we can do a material balance:

- Find the underlying data that went into the isotherm generation (Fig. 10.1). If it isn't possible to find the underlying data, then get

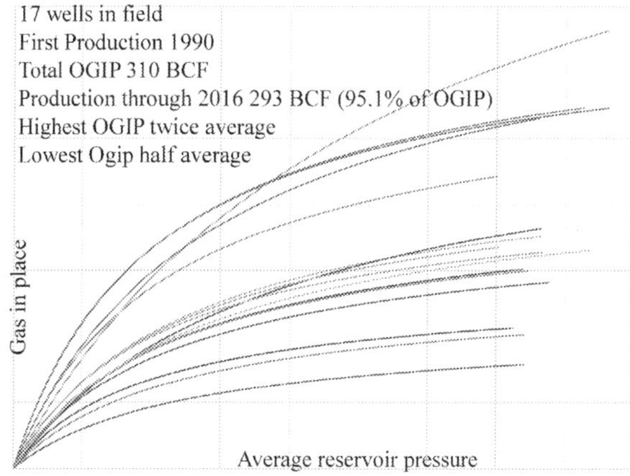

Fig. 10.1 Isotherms for a field.

a copy of the paper isotherm—it exists somewhere and the owner of the data will likely be delighted that someone is using her hard work.

- Compare the last pressure build up data to the cumulative production at that time.
- If the data doesn't match the isotherm, then adjust your least confident input parameter (it is usually drainage area) to make the line fit the data, all of the necessary arithmetic is in Chapter 1, Gas Reservoirs.
- Look at the isotherm again the next time there is a shut in. Most wells will match the isotherm data after 1–2 iterations, now you have an idea of how pressure is going to change with production and can use that to develop reliable forecasts, plan the timing on future interventions, and set target wellsite pressure that support the needs of the reservoir.

10.1.1.2 Owning a well

A well's "owners" include the production engineer, the production foreman, the lease operator, and the facilities engineer. Each and all of these people need to feel free to look at the other team member's stuff while honoring the prerogatives of the others. One division of responsibilities for the well-management team that has worked well is as follows: (1) the production engineer has primary responsibility for everything from the wing valves to the reservoir, (2) the facilities engineer has primary responsibility for everything from the wing valve to the plant, and (3) the field foreman has responsibility for field manpower.

The things that a facilities engineer needs to look at to take ownership include: (1) well-bore configuration, (2) surface drawings, (3) contractual obligations, (4) automation configuration and data, (5) gas-flow measurement, and (6) liquid-flow measurement.

Well-bore configuration. It seems odd for a facilities engineer to care about well-bore sketches, but we are all working together, right? Once you get a sketch,

- Verify that it is current (they often aren't).
- Look at every piece of downhole jewelry and ask the questions:
  - Why is this device in the well?
  - Is it the right device for this task?
  - Is it located in the best place?
  - What constraints will this piece of kit impose on production?

It is amazing how often things are in wells because "that is the way we do it" instead of "that is what the reservoir needs." You probably cannot change policy the day you walk in the door, but this conversation will help you understand what can and cannot be changed. Two attitudes are worth the effort to try to change: (1) you can only produce a gas well up the tubing because of the risk of corrosion in the casing (the casing must be protected via a packer) and (2) our policy is to use xxx for artificial lift ("xxx" could be progressing cavity pump, electric submersible pump, Gas Lift, etc.). You can start to chip away at the second one by always asking, "I didn't think these wells made oil, why aren't you doing deliquification instead of artificial lift?" That trivial (almost pedantic) question begins to shift the conversation away from "policy" toward the needs of the reservoir.

The first attitude is very difficult and very important to change. You cannot successfully produce late-life gas wells with a packer in the well. It simply cannot be done and the packer will always significantly increase the gas abandoned at the end of life.

Surface drawings. You need to get a copy of the pipeline map (if someone else operates the take-away pipeline, you still need to know the size of the line, what other volumes are on it, where pigging facilities are located, etc.) and study it. You need to develop the confidence that any efforts you take to maintain or increase production won't just shift a bottleneck.

If wellsite drawings exist, then it is a good idea to verify that the drawing matches the equipment on the ground (they often don't) so that when you pick up a drawing, you have confidence that it matches reality. If there is a P&ID (increasingly there is), you need to go to the field and understand why any valve is not in the position (i.e., open, closed, or throttled) indicated on the design drawings. You may find that the reason for problems is the field ignoring the design intent. You may also find that the design intent was ill conceived. Both situations are good to know.

Contractual obligations. You may have partners in a well, you may have a gathering agreement (even if you operate the gathering system), and you do have gas-sales contracts. It is important to understand what you are expected to do under the terms of these contracts.

For example, for any joint-interest well, there are spending limits before you have to get the partner's permission to spend more money. If you decide to set a tank for a new function or to extend the capacity of an existing function, the cost of materials and labor will probably exceed spending limits and you will be required to send an authorization for expenditure (AFE) to partners and get their approval to spend their money prior to spending it. Every operating agreement has the right to set those AFE limits at a point that works for those partners. As someone who will be doing projects on well sites, the new facilities engineer needs to know those limits on every well (the process for submitting an AFE will nearly always be someone else's responsibility, but you need to know when the process needs to be initiated).

Same with gathering shrink. What is the magnitude? What unit is it measured in (energy or volume)?

Same with gas quality. What are the pipeline limits on things like free water, H_2S, total inert gases, etc.?

Finally, what does "standard temperature and pressure" mean? It is sad how often data will be stored in one set of standards and required to be reported in another set of standards. For example, your company standard may be 14.73 psia (101.56 kPaa) and 60°F (15.6°C) while the state requires 15.025 psia (103.59 kPaa) at 32°F (0°C), for CBM gas (say 8% CO_2 and 92% CH_4), using the wrong standard temperature and pressure on the report will understate volume (and attendant severance taxes, etc.) by 7.9% which can be enough to trigger fines, court actions, and even criminal prosecutions in some places. The standard temperature and pressure can be different for each: (1) joint interest agreement, (2) gathering agreement, (3) sales contract, and (4) reporting entity. As a facilities engineer, you will often be the only one in a position to understand the impact of this Tower of Babel.

Automation. Too often, the story that automation tells is a fairy tale. The most valuable data on a well site for trouble-shooting well-bore problems are tubing and casing pressure. Unfortunately, the most neglected instruments on a well site are often tubing and casing pressure—this disconnect can only be corrected by being the squeakiest wheel on the train, and insisting the very day that a tubing transmitter cannot be reconciled with line pressure (an instrument that gets the most attention in an automation system) that an automation tech investigates. After a few times down this road, the automation techs will begin to calibrate tubing/casing transmitters when they calibrate flow meters and the data will get more reliable and useful.

It is common to have a disconnect between calibrated range on the transducer and the calibrated range in the Remote Terminal Unit (RTU). Most transmitters send an amperage signal between 4 and 20 mA to indicate the measured value. If the transmitter is calibrated to send 4 mA at 0 psig (0 kPag) and 20 mA at 100 psig (690 kPag), but the RTU is expecting 20 mA to be 500 psig (3.4 MPag) then a 10-mA signal indicating 37.5 psig (259 kPag) would be entered into the database as 188 psig (1.3 MPag). There are many incidences of this sort of disconnect going on for years without correction, everyone "just ignores" that instrument and never sees developing problems that the instrument could help solve.

The goal of automation is to facilitate field operations. When our first reaction to an anomaly in the data is "that instrument is probably messed up," we are rendering the value of field automation to the category of "expensive noise." If we demand that known problems are corrected promptly and that we have reliable data, then our first reaction becomes "why did the tubing/casing differential change that much since yesterday, are we liquid-loading?" (i.e., we are "using" the data instead of "questioning" the data.)

Gas-flow measurement. The gas-flow meter is the cash register of the operation. The ability and attention to detail of measurement techs is probably the best of any discipline in field operations, but just like any field, quality varies widely. You really cannot own a well until you understand:

- Do you have a reasonable material balance between the sum of the wells and the off-system delivery? Do you have processes in place to confirm the system material balance on a regular, ongoing basis?
- What is the technology, calibrated ranges, and β ratio if applicable? Are they all consistent with industry and/or manufacturer's specifications?
- Is the gas analysis in the RTU exactly the same as the latest available gas analysis? Do you have a gas analysis for each meter station?

Sometimes, companies try to save money by putting a field-wide gas analysis into the RTU, this is likely illegal and usually in violation of one or more contracts, but it happens.

- Is the RTU doing the right calculations? They usually are, but there have been occasions where the wrong compressibility calculation is selected or even the wrong meter type.
- Are all the meter parameters associated with the well you are looking at? Mostly RTU configurations are created via cut and paste, and too often a tube ID, or gas analysis, or meter type, etc. doesn't get changed after the paste. These errors can stand for years and just contribute to the system imbalance.
- How often are meter tubes and orifice plates inspected? The answer is too often "never."

Liquid-flow measurement. It can be difficult to get much traction with water measurement. People see water as a waste product that doesn't warrant much effort. This can be true, until you end up with a large quantity of it on the ground or in a river. You can confirm that you don't have turbine meters downstream of dump valves. You can take a look at the piping and verify that the meter has a reasonable chance of remaining full of liquid. You can confirm input parameters. It is unlikely that you will be able to spend much money fixing water measurement.

10.1.2 What do you do with a well you "own"?

In spite of "owning" the well, you probably can't sell it. If you've done the steps in the previous section, you likely now have data that can be used instead of questioned. This results in

- Anomalies and changes from day to day become vivid.
- You can see potentially expensive issues while they are still small enough to fix inexpensively.
- You can participate (and contribute) in discussions with the other well owners.

When this approach was implemented in the San Juan Basin:

- Lease-operating expense dropped from $0.26/MSCF ($9.18/kSCm) to $0.04/MSCF ($1.41/kSCm) even after adding $250k USD/month in compressor rental and increasing field staff so that instead of 60 wells/operator, they had 22 wells each.
- Identified "water in gas gathering" as the biggest opportunity and increase pigging from 1-2 pig-runs/month to 2–3 pig-runs/day.

10.2 CASE STUDIES

Case studies are all the rage in technical writing, and this seems to be a fad that actually adds value. The four case studies that are presented here include two that have been published elsewhere and two that have never been published.

10.2.1 CBM plan of depletion (POD)

This book has mentioned Amoco's San Juan Basin Plan of Depletion several times in various contexts, and you can find a detailed review of this project in the literature of the Society of Petroleum Engineers (SPE) (Simpson, 1). Before discussing the project, it is useful to set the stage. As discussed in Chapter 6, Gas Gathering Systems, the Section 29 Tax Credit provisions of the 1980 "Windfall Profits Tax" created an incentive to try to learn if coalbed methane production could be made economic. In the late 1980s, this tax credit was worth more than the sales price of the gas and CBM started to look interesting. As you can see in Fig. 10.2, the dominant CBM play in the United States was the San Juan Basin of New Mexico and Colorado. What was it about the San Juan Basin that created this dominant position?

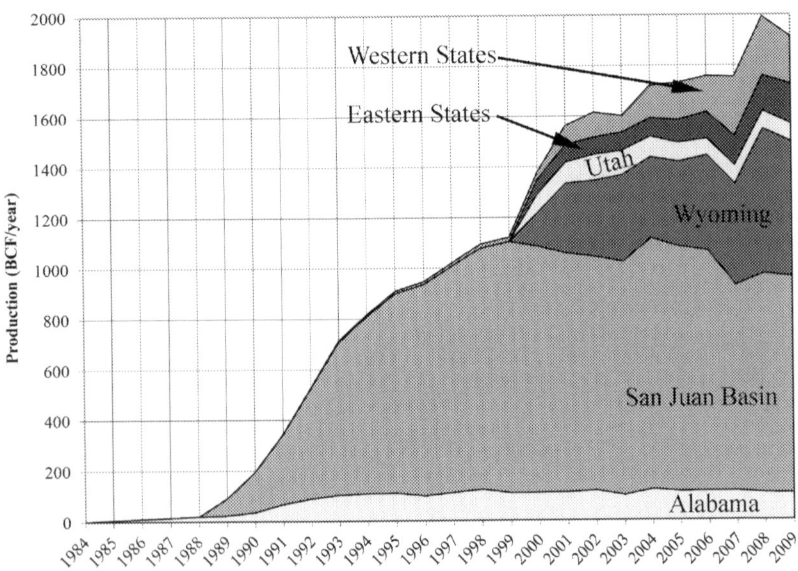

Fig. 10.2 CBM Production in the United States.

The Fruitland Coal seam had been encountered in over 20,000 well bores in San Juan between 1927 and 1988. The drillers knew it was there and they saw it as an impediment to progress because they often got a significant kick as they went through it, but completing a well in the seam hadn't added measurable production. When the Fruitland Coal became a viable target, the anticipated hurdles were

- More water production than the basin had infrastructure to handle.
- Significant CO_2 that was incompatible with existing cryo plants.

The "discovery" well (i.e., the first well that was drilled with a target formation in the Fruitland) produced 16,000 bbl/day (2540 m^3/day) with 9% CO_2 in the small amount of gas produced. That well was a learning experience and in short order CBM development was underway in earnest. In 1986, "CBM production" (which was primarily tight gas from an adjacent formation that had been mischaracterized) was 2% of San Juan Basin production (0.024 BSCF/day (680 kSCm/day) out of total basin production of 1.3 BSCF/day (36.8 MSCm/day)). By 1996, the CBM production had grown to 64% of basin production (2.4 BSCF/day (68.0 MSCm/day) out of 3.6 BSCF/day (101.9 MSCm/day)). The factors in the initial success included:

- The central "fairway" portion of the basin had rock mechanics that made these wells suitable for cavitation and that resulted in large production rates that encouraged management enthusiasm.
- Existing commodity gas take-away pipelines provided excellent access to nearly insatiable markets.
- Gathering companies were unwilling to pick up gas at well sites (which forced producer-operated gathering systems as discussed in Chapter 6: Gas Gathering Systems).
- Producers were large enough companies to "encourage" the development of necessary sweetening capacity.

By 1996, it was becoming clear that:

- Production decline was mostly over for the early wells.
- There were no available new drilling sites in the Fairway.
- All the wells that could be recavitated, had been.
- The early wells were seeing a 60%–70% annual decline.
- None of the traditional reservoir performance models adequately described either the incline or the rapid decline.
- The next stage called for an unconventional reservoir, well bore, and infrastructure model that could properly evaluate proposed interventions.

Amoco management commissioned an analysis (the term "study" implies more resources than they put on this project, one facilities engineer was detached from other duties and turned loose for 4 months to see what he could come up with) to develop a 10-year plan to determine:

- Required wellsite, gathering system, well bore, and deliquification modifications to stem the decline.
- Predicted reservoir performance with the modifications.
- Project size and timing required.
- Staffing levels required.

The starting point for the study was:

- 1989 predictions from the reservoir model had failed to match the performance of any of the 62 high rate Amoco-operated San Juan Basin CBM wells.
- Decline rates significantly higher than expectations.
- Original downhole equipment, surface facilities, and gathering systems were inadequate for current needs.
- Excellent data were available on reservoir characteristics, but it was not widely disseminated.

The traditional approach to this sort of analysis had some significant hurdles:

- Reservoir engineers were certain that the drillers ruined the reservoir by poking holes in it.
- Production engineers saw themselves as victims of both reservoir uncertainty and inflexible surface facilities.
- Facilities engineers wanted the wellhead to be an invariant plant feed.
- These three groups each tried very diligently to never talk to the other group.

Amoco management was able to identify a facilities engineer who had worked with reservoir engineers, drillers, and production engineers and who had never worked inside a plant to try to avoid these internecine conflicts. The overarching philosophy of the POD was

- The single goal is to maximize reservoir long-term profit.
- Interventions on surface facilities are every bit as valid a reservoir-management tool as rig work.
- Rig work only makes sense if the surface facilities are adequate to accept the additional production from a successful intervention.

In short, the combination of reservoir, well bore, and surface facilities all had to work together if any of it was to work at all. Like any analysis done in the information age, the POD started with a model. This model was an integration of

- Reservoir model
  - Calibrate the input parameters to the Langmuir isotherm (these input parameters are held constant with regard to time) and the empirical "skin" equation discussed in Chapter 3, Well Dynamics and develop isotherm curves for each well.
  - Compute pressure drops from average reservoir pressure to the well bore and develop a "characteristic length" of a 3-in. pipe that would have the same pressure drop, hold that length constant with time.
  - Model system from first production to abandonment.
- Well-bore model—use standard vertical flow correlations.
- Wellsite model—convert existing pressure drops to another equivalent length of a 3-in. pipe.
- Gathering model—use a commercial pipeline model.
- Integrating model—custom program and database allowed iterating the various pressure drops within a step and then captured output from one time step and fed it into the input of the next step.

This integrated model allowed predicting the short- and long-term impacts of various interventions that were available. The interventions that were fed into the model included

- Cleanouts and/or recavitations.
- Adding/removing liners.
- Impact of tubing size and position.
- Deliquification equipment.
- Wellsite debottlenecking.
- Gathering system debottlenecking.
- Well site and lateral compression.

The model was not really sensitive enough to properly evaluate the ramifications of the tubing set depth; so when the project actually got under way, there was considerable trial and error around where to set the tubing.

Wellsite compression analysis was an interesting academic exercise. This part of the evaluation looked at a "no compression" case, an "immediate compression" case, and a "staged compression" case. The model predicted that 23 of the 62 wells needed wellsite compression in 1997. The model also predicted that 21 of the 62 wells would never need wellsite compression. Of course, the first three wellsite compressors were installed on "no compression" wells to confirm this category, all three showed significant uplift and flattening of decline so we determined that the "no compression" category should have been named "undiagnosed well-bore damage." All of the "no compression" wells had well work and wellsite compression in 1997 with good results.

For the other 21 wells, I got really fancy. The analysis started with the "no compression" case and developed empirical equations for gas-flow rate (q), flow constant (c_p), and decline rate. Then it took the first and second derivatives with regard to time. Compressor installations were scheduled for the date of a local minimum (i.e., upward inflection point) in any of the six derivative curves. This was an interesting theoretical exercise, and it worked in that we were able to install compression as the wells were entering distress instead of waiting for a well to log off before starting the compressor-acquisition process—but the amount of work involved was not justified in the results.

The model was optimized into a series of interventions spread over 10 years, with production profiles and necessary expenditures by well. Details of the interventions are available in the SPE paper on the project. Staff changes in late 2003 resulted in the 10-year plan being abandoned in early 2004. The results as of the plan prior to abandonment are shown in Table 10.1. The project delivered 114% of projected cumulative production over the 10 years of the POD projections (including the 3 years where the company was no longer following the 10-year plan). Excluding the wells in the "no compression" case, the production rate

Table 10.1 POD results

	Target 2/1/2004 gas rate (MMSCF/day)	Actual 2/1/2004 gas rate (MMSCF/day)	Well count
Actual <50% of target	2.3	1.3	5
50% of target < actual <90% of target	10.7	6.9	10
Actual within ± 10% of target	9.4	9.4	10
110% target < actual <150% target	12.5	15.8	16
Sub total	35.8	33.6	41
Actual >150% target[a]	10.8	19.2	21
Total	46.6	52.8	62

[a]This group is predominantly those wells that the model said "no compression" and compressors were set anyway. The category was actually a reflection of wells needing undiagnosed well work.

was 94% of projected production when the plan was abandoned. If you include those wells, the ending rate was 113% of projected rate.

The POD was very successful. In 1989, when the development of the field was authorized, the projected estimated ultimate recovery was 176 BSCF (4980 MSCm). Some highlights of the POD results were:

- In January, 2004
  - Original predicted rate without the POD—13 MMSCF/day (368 kSCm/day).
  - POD predicted rate—52 MMSCF/day (1473 kSCm/day).
  - Actual rate—67 MMSCF/day (1897 kSCm/day).
- Project confirmed that very high production rates were achievable at very low reservoir pressure (the 17 POD wells in Fig. 10.1 were making 5 MMSCF/day (142 kSCm/day) in January 2017, from a rate-weighted average reservoir pressure of 24 psia (166 kPaa)).
- Project confirmed that recovery rates over 95% of original gas in place (OGIP) were achievable (the 17 wells in Fig. 10.1 have recovered an average of 95.1% of OGIP as of December 31, 2016).
- As of May 2009, the 62 POD wells had recovered 820 BSCF (23.2 MSCm).

The philosophical conclusions that can be drawn from this project are

- Unconventional analysis is required for unconventional reservoirs:
- Late-life CBM operations require a significantly different mindset from early-life operations:
- It is profitable to manage a reservoir from the burner tip (see the next case for more on this).

10.2.2 Managing the reservoir from the burner tip

Every decision that you make in Oil & Gas operations can either look back toward the reservoir or look forward toward the end use (Simpson, 2). Looking back asks the question "how will this decision affect reservoir performance, ultimate recovery, and profitability?" Looking forward asks the question "how will this decision affect installed facilities?" The first decision that installs equipment incompatible with full reservoir pressure has shifted the focus forward.

The usual reason for changing the focus is a decreased risk tolerance. We want to avoid safety risks, environmental risks, and performance/profitability risks. The tools we use to eliminate risk are process safety management (PSM), supply chain management, and processes/procedures.

10.2.2.1 Process safety management

As we discussed in Chapter 9, Interface to Plants, PSM is a set of processes and procedures designed to

- Ensure that system design contains appropriate risk mitigation.
- Ensure that system modifications meet the standards of the original design as applied to current operating parameters.
- Ensure that procedures used will minimize the risk to the environment, the public, workers, and equipment.

The basic tenet of PSM is to balance "risk mitigation" with "risk density." Risk mitigation is simply the steps that you've taken to reduce the damage (i.e., consequences) of a failure. Risk density is a term that is not used explicitly in PSM, but it can be thought of as the intersection of likelihood and consequences of an event to harm employees, damage equipment, harm the environment, and/or harm the public.

When you look at a plant, you see that they tend to have "round-the-clock staffing" and a significant number of workers. There are numerous components. The plant contains fluids and pressures that can do real harm. Plants are often located near population centers. This all combines to result in very high-risk density.

When you think of a wellsite, you see that people are rarely on the site, there is not much equipment, very small quantities of stored liquids (and those are mostly benign), more often remote rather than within population centers. This all combines to result in very low-risk density.

Successful risk-mitigation strategies will always consider risk density in the establishment of processes and procedures.

When you ignore risk density, then it is reasonable to apply processes appropriate for a refinery to well sites and

- Require management of change (MOC) and hazard and operations (HazOp) review to change orifice plates in a meter run.
- Require shutting in a well and applying full lock-out/tag-out protocols to spray a location for weeds.
- Develop extensive drawing packages for well sites and require that drawings be updated before work can be started and again after work is complete.
- Develop rigid procedures that can only be deviated from with management approval.

When you consider risk density for well sites, the work environment looks quite different:

- MOC and HazOp are not required for routine activities (e.g., you can define a fleet of compressors that are all interchangeable within the fleet, changing orifice plates, replacing a plunger, gauging a tank).
- Lock-out/tag-out is only required when multiple, unrelated activities are done concurrently or the well is left in an unstable configuration.
- Procedures are guidelines and if they do not add value, they can be ignored at the discretion of the field operator.

10.2.2.2 Supply chain management

In the 1970s, automakers and large retailers began using computers to work toward the goal of "just-in-time" inventory control. This concept is based on the idea that a part or subassembly will arrive on the loading dock just as needed to allow the warehousemen to take it off the truck (or train or boat or airplane) and directly to the assembly floor without the requirement to warehouse it between delivery and use. For a retailer, this inventory control approach required the analysis of consumer trends and adjusting both the magnitude and mix of products on the shelf to satisfy those trends. Both of these fields were ripe for improvement and the advent of widespread computing power facilitated that improvement.

By the 1980s, PhD and Masters candidates were writing thesis on "just-in-time" inventory control and had coined the term "supply chain management" to speak to the entire supply chain from raw materials through to retail—customer purchase. The focus and intent of supply chain management (I cannot bring myself to shorten that to SCM) is to

- Manage **units of production** to provide commodities as required with minimal on-site warehousing.
- Manage **tools of production** to minimize the amount that they constrain the production process.

These new terms are crucial to understanding supply chain management and why I contend that its implementation in Oil & Gas has curtailed production while adding costs.

Units of production are the things that go into the final product.

Tools of production are the things that stay in the factory or store when the final product leaves (like robots, assembly lines, factory lighting, compressed air systems, shelving, etc.).

It is easy to imagine that if an assembly facility that relied heavily on pneumatic tools was to lose its compressed air source, production would grind to a halt which encourages a facility that relies on pneumatic tools to have redundant air compressors with extensive spare parts warehousing for the compressed-air system components.

Supply chain management moved to Oil & Gas in the late 1980s and a careful analysis of a proper implementation of these management concepts would show

- Our **units of production** are hydrocarbon molecules and there is no really good method to manage the supply of those molecules.
- Our **tools of production** are valves, valve repair kits, pipe, tanks, pumps, compressors, gensets, etc.
- A proper implementation of SCM to Oil & Gas would take extraordinary measures to ensure that
  - Repair/replacement equipment was immediately available.
  - Field workers are adequate in number and extensively trained in repairing and diagnosing failures in all of the **tools of production**.
  - Cost control would focus on unit costs not total costs.

Unfortunately, proper implementation was simply not the direction that the Oil & Gas industry took. Our industry has decided to apply techniques appropriate to **units of production** to **tools of production.** Before this process was imposed on the industry, field workers decided for themselves what tools and spare parts they carried on their vehicles. Supply chain management (as implemented in Oil & Gas) vilified the idea of carrying spare parts, calling them "squirrel stores," and making it a dismissal offence to have them. Prior to this practice, if a field tech found a dump valve diaphragm leaking (a common occurrence), he replaced it from his truck and the well was shut-in for somewhere between zero time and an hour depending upon the specific field tech. After supply chain management got involved, the field tech was required to shut in the well and request a work order to fix the leaking diaphragm, acquire authorization, generate the paperwork to charge the repair kit to the well, pick up the parts, install the parts, and try to get the well back on line—3 to 5 days of lost production.

Stories like this are repeated across the industry a hundred times a day and result in cost per dollar revenue increasing by a multiple of several hundred times the unit costs that were common prior to implementing these processes. There are many published papers showing that supply chain management has reduced total costs, but these reports are always on a total-cost basis, and not related to changes in production. When you look at unit costs or total profit, it is easy to see that the "benefit" is a huge negative.

10.2.2.3 Processes/Procedures

A "process" is a description of something that must be done. A "procedure" is a description of how to do something. Both are intended to minimize the risk of an error and to ensure that everyone does the same task in the same way on every location every time. The actual outcome of proscriptive processes and procedures is to

- Force workers to lie about having followed procedures that are inappropriate for a given location.
- Stifle innovation.
- Provide an easy excuse for failure (instead of providing a reasonable path to success).

10.2.2.4 Examples of implementations

Now that we've discussed the things that cause us to change our focus from the reservoir to the artifacts that we've chosen to install, it should be useful to discuss how this change in focus has manifested itself. We'll talk about: (1) completion techniques, (2) managing bottom-hole pressure, (3) wellsite equipment design, (4) gathering systems, (5) administrative processes, and (6) some general conclusions.

Completion techniques. Experience shows that CBM wells that can be completed with open-hole cavitation significantly outperform any other completion technique (often by a factor of 20–40 times). Cavitations only work in a limited number of wells (and it can be difficult to identify candidates prior to drilling the well). Cavitations are messy and have an unpredictable required duration which impacts the rig schedule.

- Looking toward the reservoir, any well with any reasonable chance of a successful cavitation must be cavitated.
- Looking toward the expense (or capital) budget and the rig schedule, it is more reasonable to case and frac the coal even if the result has a high probability of providing a small fraction of the production rate and ultimate recovery that the well would provide if cavitated. Evaluating drilling staff based on meeting a budget and a rig schedule ensures that cavitations will not be considered.

Experience shows that coal is self-healing and frac proppant quality/quantity/type is largely irrelevant in the coal.

- Looking toward the reservoir would not use hydraulic-fracture stimulation or would have frac's with large carrier volume and only enough sand to enhance abrasive action.

- Looking toward supply chain management, you farm out the decision to the service company and get a huge sand load.

Managing bottom-hole pressure. Steady pressure improves the affected reservoir area and results in higher flow rates and more ultimate recovery.

- Looking toward the reservoir, you would make an effort to determine the most effective pressure relationship between reservoir pressure and flowing bottom-hole pressure and try to stay as close as possible to that value over time—if there is a wellhead choke, it is a "backpressure" choke that holds flowing tubing pressure constant.
- Looking toward the lease equipment and gathering system, you put a "pressure regulating" choke that ignores upstream pressure and keeps downstream pressure constant.

Wellsite equipment design. Typically, each pressure class will result in costs about 10%–20% higher than the next lower pressure class.

- Looking toward the reservoir would have you select a maximum allowable working pressure (MAWP) based on reservoir pressure and will typically be something like ASME B16.5 Class 600 (1440 psig (10 MPag)) and not require wellhead chokes to protect the artifacts.
- Looking toward a low pressure gathering system would select ASME B16.5 Class 150 (280 psig (1.9 MPag)) or less and would require wellhead chokes and wellsite emergency shutdowns (ESD's) to protect the artifacts.

Gathering systems. Gathering systems can either be a tool of reservoir management or a sales tool.

- Looking toward the reservoir
  - The system MAWP is consistent with reservoir pressures.
  - The system anticipates difficult reservoir fluids (large quantities of condensed water, significant potential for corrosive fluids).
- Looking toward the sales line
  - Pressure rating is largely irrelevant (you can build compressor stations to maintain whatever MAWP you select).
  - Cost is king, and assumptions about installation costs are often naïve.
  - Assumptions about the long-term reliability of remote, automated equipment can be very naïve.

Administrative processes. We use administrative processes and procedures to relieve individuals of the risk of making the wrong decision.

- Looking toward the reservoir
  - Individuals have the authority to make changes that are required to optimize reservoir performance.

  - A meter change, changing pump speed, or running a pig requires budget money, not MOC.
  - Local control of maintenance resources and local ability to change priorities.
- Looking toward process-driven activities
  - Every decision refers to a process document.
  - Every change requires MOC.
  - Maintenance resources are centrally controlled and the work-order system has goals like "all work will be scheduled 30 days in advance," "no spare parts will be issued without a work order," and "no squirrel stores."

Conclusions. The POD discussed above provided a unique opportunity—it demonstrated that the CBM was different and allowed the management team the latitude to remain outside of the supply chain management/PSM/process-control environment as long as the team was willing to fight for deviations. When reassignments/retirements broke up the core management team, their replacements did not see this as a good fight to continue and immediately implemented a work-order system and inventory control. Within 6 months, the production (that had been inclining at about 3%/year for the 6 months previous) had begun declining at 7%/year. Cost per unit volume went from $0.04/MSCF (which was published) to much higher (proprietary) values. All of these differences can be traced directly to changing the team focus from the reservoir to the processes and artifacts.

The reason for the very existence of our industry is to exploit Oil & Gas reservoirs for profit. Any activity that loses that fact will make less profit than it could have. Any statement that contains the phrase "reservoir _______ is irrelevant" (e.g., "reservoir pressure is irrelevant") leads to a suboptimal decision. Any procedure or process that doesn't consider the needs of the reservoir is suboptimum. Any facility that puts an artificial constraint on the reservoir is inappropriate.

10.2.3 You get what you measure

In 1981, John Sweringin, CEO of Standard Oil (Indiana) was asked "Can you tell us why you consider stock analysts irrelevant to your operating decisions?" and his reply was "(We) are in the business of finding, developing, producing, transporting, processing, and marketing hydrocarbon products. If we do that very well, we will be very successful and people investing in Standard Oil (Indiana) will make a lot of money, how do stock analysts add value to this?" I took this statement as an endorsement of the concept that a company should be measured on its results, and the same is absolutely true for individuals.

Someone who does a great job at some piece of work that is not part of her job description will be quite unlikely to be recognized for that performance. A field tech whose incentive pay is based on the number of safety observations that he submits will step over production-enhancing opportunities to write safety observations. Someone whose incentive pay is based on meeting a rig schedule will be reluctant to add rig time to completely clean out the well bore before rigging down. In short, you will get the performance that you have measures for.

The POD project deployed oil-flooded screw compressors as part of the overall plan. These machines were foreign to the compressor operators and the learning curve was very steep. In the first year, the project experienced excessive employee turnover (entire staff was replaced 1.5 times) and the team tried to determine the cause and solutions. This analysis determined that a successful incentive program must

- Be based on objective parameters.
- Have key performance parameters that are capable of being directly influenced by the individual.
- Provide quick feedback.
- Encourage performance that positively impacts the company's goals.
- Encourage employment stability.
- Establish confidence that payouts will happen.

The team looked at all of the industry key performance indicators and rejected them all (Table 10.2) because they did not encourage behavior that was consistent with the company's goals (i.e., maximize production and retain trained staff). Using production as a compressor-operator target always begs the question "why should I be punished for well problems?" and the way that the team addressed this was

- Calculate a target production rate for each well, each month using an accessible algorithm that allowed the operator to verify the value used.
- Remove any well (both target and actual) from the calculation that was shut-in for more than one-fourth of the month, and remove any well that had had compression installed for less than 3 months.
- Sum the target and actual for each remaining well in the operator's area.
- For each month, calculate and publish (but do not pay) each operator's incentive pay.

Table 10.2 Traditional compressor KPI

KPI	Pro	Con
Run time	Easy to measure	No clear tie from run time to gas sales
Mechanical availability	Easy to measure	Creates incentives to leave compressors that went down because of well or gathering system conditions out of service
Utilized vs installed Hp	Hp weighted	Hp is not necessarily proportional to rate and it is difficult to determine actual Hp
Cost per Hp	Effective for controlling variable costs	The contract for these machines included variable costs as fixed costs in the lease agreement

- If the actual for the area was less than the target, then the incentive pay for that month was zero.
- If the actual for the area was more than 103% of target, then the operator received credit for the maximum incentive for the month.
- In other cases, the incentive was prorated.

- The monthly incentives were accumulated from mid-December of one year through mid-December of the following year and paid out in a lump sum at a nice luncheon.

There were a couple of extra conditions. If an operator had a spill, an environmental incident, or a preventable safety incident, then they forfeited the bonus for that month (but only that month), and if the operator was not currently in the job at the time of the luncheon, they forfeited the entire bonus.

This incentive program was put in place for the compressor operators, the mechanics (the job title for the operator's supervisor), and the superintendent. If an operator lost his bonus in a given month due to health, safety, or environmental issues, then the mechanic and superintendent also lost their bonus for that month.

This program was in place from 1998 through 2003. Some highlights of the program are included in Table 10.3. For the 5 years with data, the net present value at a 15% discount rate was $8 MM.

$$\text{Turnover} = \frac{\text{Number of new hires} \cdot \text{months all new hires worked}}{\text{Total number of positions in program} \cdot 12 \text{ months}} \quad (10.1)$$

The behaviors that were observed are summarized in Table 10.4.

Table 10.3 Incentive program results

Year	Turnover (%)	Sales to target (%)	Benefit to company	Payout
1	250	−2	Before program	Before program
2	18	+1	\$1.9 MM	\$45k
3	12	+3	\$2.6 MM	\$44k
4	22	+4	\$3.1 MM	\$42k
5	14	+5	\$4.0 MM	\$63k
Total			\$11.6 MM	\$194k

Table 10.4 Incentive behaviors

Behaviors	
Before incentives	**After incentives (year 5)**
Each compressor operated in isolation • If a machine on a high rate well blew a turbo charger, the well was down until it could be replaced • Incentives base on run time	Manage the fleet • If a machine on a high-rate well blew a turbo, the mechanic pulled the turbo off a smaller well and left the lower rate well down • Incentives based on production
Wells visited on a rigid schedule	Operator started and ended every day looking at his highest rate well
Mechanics only went to failed compressors	Mechanics spent time-mentoring operators at their big wells
Run-time maximized	Gas-sales maximized
Average time an operator stayed on the job 4 months	Average time an operator stayed on the job 3 years

10.2.4 Major projects

Over the last decade, upstream field development has moved away from "organic field development" to the "engineering, procurement, and construction (EPC) model."

Organic field development implies

- Project designed by company engineers with contractors providing components (e.g., drawings, survey's, etc.) and field personnel providing significant input to the design.
- Procurement done by company personnel (not "supply chain management") with significant ongoing input from the design engineer.
- Company personnel managed construction contractors to build system.
- Design engineer often retained responsibility for operating project.

Under the "EPC model,"

- Company personnel develop a "Pre-front end engineering design (FEED)" project description.
- Supply chain management sends the Pre-FEED to several engineering consultants to bid on developing the "FEED," bids are awarded based on various criteria, none of which have any consideration of the needs of the reservoir.
- An engineering consultant develops a FEED that is typically too big, too expensive, and too complex for anyone within the company to have time to fully understand the ramifications of all the decisions.
- Supply chain management sends the FEED out to "EPC" contractors for bids on the job, which includes something called "detailed engineering" that is often more expensive than the FEED.
- Bids are let and the chosen EPC company.
  - "Finishes" the engineering design, generates engineering drawings, buys/builds stuff for the project, and installs it.
  - Assigns senior plant engineers to direct junior plant engineers in developing the design (zero-field engineering experience is the norm).

It is fairly common for a project being developed under the EPC model to go "smoothly" (i.e., the only contact between the EPC company and the owner is monthly invoices) for 6–18 months before the owner gets concerned about the amount of money they've committed and how little tangible results are on the ground.

On three occasions in 2015, MuleShoe Engineering was contracted to conduct an independent third-party review of the current status of major projects (one each in the United States, Australia, and India). Contractual obligations prevent disclosing either the EPC company or the project owners, but some common threads can be released.

Pre-FEED. The main point of the Pre-FEED is to lay out the "nominal conditions," for example:

- Reservoir will need very low pressures (< 50 kPag (7.3 psig)) on the surface facilities.
- Produced water will be 2000 bbl/day/well (318 m^3/day/well).
- Gas production will be 5 MMSCF/day/well (141 kSCm/day/well).
- Must minimize operating manpower requirements.

The result of these nominal conditions turned out to be:

- FEED defined gathering system MAWP of 350 kPag (50 psig).
- HDPE pipe was selected for all gathering lines.
- Pigging facilities omitted because the plastic pipe isn't subject to normal corrosion and pigging takes a lot of manpower.

- Pro-forma separator too big for average conditions (which are much less than the nominal conditions).
- Automated line drips spotted all over the system.

FEED. FEED is required by supply chain management protocols. Intention is to enter the procurement process with engineering completed and every component specified to a degree that would allow a procurement specialist to send it out for bid **without further Engineering involvement**. The goal sounds laudable when you say it fast, but excluding engineering from procurement decisions hasn't worked out well:

- Key details don't get written into the FEED.
- Alternatives don't get considered (e.g., when a vendor says, "You called for XYZ widget, but this other widget does the same thing and will allow you to eliminate a third widget," the engineer would say, "Let me look at the specs" and supply chain management says "NO!").

The basic concept was naïve, and all the work in the FEED is now redone in EPC, generally with less company input. FEED should go away, but it is now part of the institution.

A field is not a plant. Under organic projects:

- Drawings were limited to pipeline alignment sheets and fabrication isometric drawings.
- No P&ID's were developed (in fact, no wellsite drawings at all were developed).
- Vessels were a collaboration between a vessel fabrication shop and the design engineer.
- Automation was done by analogy (i.e., you put in the same thing as the rest of the field or the last field, the design of that particular wheel was not up for review on every project).
- Pipeline construction done in collaboration, company personnel intimately involved.

Under the EPC model, collaboration is different:

- P&ID is king, and it is common to have many drawings for each wellsite (one project had 107 drawings/well another 109 drawings/well).
- Adjusting equipment location for terrain required modifying dozens of drawings and could shut the work down for weeks.
- Field piping is a shock.
    - One gathering system project had an inspector show up in a small two-door coupe because he knew "he could walk to the pipe rack".

  - Another project shut down for over a week while the head office determined how to lay pipe across a dry wash (the head office wanted to build a pipe bridge).
  - A fence crossing in Colorado caused 10 days delay (and there are fences every km or so) while plant guys decided what to do.
- Wellsite equipment done with nominal values and no input is allowed until the equipment hits the ground, and then it is a change order.

Wellsite vessels. Under the organic model, companies have tended to value field input and lean toward using equipment that the field operators are comfortable with.

Vessel design is a place where the EPC companies show their collective lack of field experience. The EPC companies have some really talented engineers, but there tends to be a lot more plant and academic experience than field-operations experience. Some of the things that are obvious are

- Tendency to design for a nominal value, which:
  - Fails to consider the range of values probable in a field. That is, a field may be expected to produce 200,000 bbl/day (32,000 m^3/day) of water from 100 wells (2000 bbl/day/well (318 m^3/well/day)). The project owner knows that 20 wells will produce 180,000 bbl/day (29,000 m^3/day) and the other 80 wells will produce about 250 bbl/day/well (40 m^3/day/well); but that information does not get into the FEED so every well is designed for 2000 bbl/day (318 m^3/day).
  - Fails to treat gas flow and water flow as related. Wells that make a lot of water tend to not make much gas. Wells that make a lot of gas tend to not make much water. Consequently, a vessel designed for field-average gas and field-average water will be wrong for both.
  - Leads to "clever" solutions (e.g., at 2000 bbl/day (318 m^3/day), a conventional dump valve would operate too often so one designer put a continuous level switch and replaced the dump valve with a centrifugal pump and a variable speed controller, it didn't work very well)
- Rigid adherence to company specifications without critical review (e.g., all of these projects had company-specified liquid retention time for vessels that came into the specifications from oil fields, the EPC used them without ever questioning whether a 3-minute retention time is really necessary for a two-phase separator, it isn't) which leads to very large, very expensive vessels.

- One of the projects was the next phase on a project that had been in production for several years with considerable experience developed regarding corrosion risks. The EPC looked at the gas analysis and saw 6% CO_2 and specified that the separator and all upstream piping and valves would be stainless steel in spite of the fact that the field had not seen CO_2 corrosion.
- The vessels are very expensive. A wellsite vessel for one of the projects that cost $500k was sold at auction for $7k (purchaser claimed that he was going to harvest the valves and electronics for resale and sell the vessel for scrap).

After the review, two of the EPC contracts were canceled less than half finished, and the third was restructured to bring in some field experience. One of the canceled projects was picked back up as an organic project—they still have far too many drawings and far too many nonproductive processes, but their costs are significantly less than the EPC and they are meeting calendar and budget targets. The other canceled project was turned over to a different EPC and preliminary results looked like they were starting down the same path.

This case is not intended to paint the EPC companies or the owner-company management as either incompetent or as crooks. Far from it. For the first 100 years of gas production, gas fields were not engineered, the gas just did not have enough value to warrant the cost of engineering staff. Turning development over to an EPC company is a reasonably logical way to bridge the chasm from no-engineering to some-engineering because the EPC all have experience doing turn-key plants and plants are a lot harder than a field, right? The logic may have not held up as well as many would hope, but it was logical. Hopefully, you've seen from this book that growing a competent field-facilities engineering staff is not a trivial undertaking.

10.3 SIMPSON'S POSTULATES

Several of the chapters in this book have started with a "postulate." These pithy sayings were included in an effort to focus the discussion on the bits of the task that had the best chance of leading to economic success. I've repeated them here to make them easier to find.

Simpson's first postulate: Every activity, joint of pipe, piece of equipment, and facility should have the goal of maximizing reservoir profitability—any activity which ignores that goal is going to result in suboptimum performance.

Simpson's second postulate: A mistake implemented (and corrected) promptly has a much better chance of success than a perfect decision made after months of sober deliberation corollary: If you don't implement (and analyze) mistakes, then you never improve.

Simpson's third postulate: Any process or procedure that inhibits achieving meaningful assessment of wellsite conditions will reduce the profitability of the reservoir.

Simpson's fourth postulate: Gathering systems and compressor stations are "tools of reservoir management." Thinking of them as a "sales tool" will reduce ultimate recovery.

REFERENCES

Mike Rowe, Don't Follow Your Passion, PragerU, Created June 6, 2016, Accessed February 9, 2017, from https://www.prageru.com/courses/life-studies/dont-follow-your-passion.

(Simpson, 1) "Producing Coalbed Methane at High Rates at Low Pressures", SPE 84509, David Simpson, Mike Kutas, SPE Annual Technical Conference and Exhibition, Denver, CO, 5–8 October 2003.

(Simpson. 2) "Facilities Selection Impacts Reservoir Performance", SPE-169837-MS, David Simpson, SPE Hydrocarbon Economics and Evaluation Symposium, Houston, TX, 19–20 May 2014.

EXERCISES

1. In the POD Case study, there is an observation that the project "managed the reservoir from the burner tip." Please describe three things the project did to make that happen.
2. What might an operation do that changes the focus of the operation from backward-looking to "forward-looking" and why might this be undesirable?
3. Why should (or should not) the principles of process safety management be applied to onshore gas wells?

Acronyms and Glossary

Abiotic hydrocarbons Combinations of carbon atoms and hydrogen atoms that derived from chemical reactions that did not involve any living or once-living organisms.

Adiabatic A process that occurs without heat transfer to or from the environment (adiabatic processes also tend to be isentropic and reversible, but these characteristics are not necessary for a process to be adiabatic).

ACF Cubic foot at actual conditions. ACF volumes of gas cannot be added together or subtracted from each other unless they are at the same pressure and temperature.

Acid gas Any gas that has the ability to mix with water and forms an acidic solution. The acid gases of most concern to Oil & Gas are carbon dioxide, hydrogen sulfide, and oxides of sulfur.

AFE Authorization for expenditure

Amine Any of a class of basic organic compounds derived from ammonia by replacement of hydrogen with one or more monovalent hydrocarbon radicals.

Annulus Space between two concentric objects.

ASME American Society of Mechanical Engineers

APB Acid-producing bacteria

API American Petroleum Institute

API gravity A normalized density used to classify crude oil.

Artificial lift Application of external energy to lift a commercial product from reservoir depths to the surface

ASL Above sea level

ASTM ("American Society for Testing and Materials" until a name change in 2001) ASTM International develops voluntary consensus technical standards for materials, products, systems, and services.

atm Atmosphere. This pressure unit is a multiple of "atmospheric pressure," defined as 14.696 psia, 101.325 kPaa, or 760 Torr, 29.92 inHg, or 1.013 bara, too often confused with local atmospheric pressure to be useful outside of very narrow fields of study.

Atmospheric pressure The pressure applied locally by the atmospheric overburden. Values range from around 14.73 psia (101.56 kPaa, or 1.016 bara) at sea level to values around 10.76 psia (74.2 kPa or 0.742 bara) at 8500 ft (2590 m) elevation.

Autoignition temperature The temperature where a flammable mixture will spontaneously ignite (also called "Dieseling"). For methane at atmospheric pressure, autoignition temperature is around 1000°F (538°C), values for elevated pressure are not standardized, and researchers seem to agree that this value drops rapidly at elevated pressure but consistent values for the magnitude of the change are nonexistent.

bbl US Oil Field Barrel (42 US gallons or 0.1589 m^3)

BGL Below ground-level

BHA Bottom-hole assembly

BHP Bottom-hole pressure

BHT Bottom-hole temperature

Biotic hydrocarbons Combinations of carbon and hydrogen that derived from biological activity on once-living material.

BOE Barrel oil equivalent. A generic barrel of crude oil has been arbitrarily assigned an energy content of 5.8 MMBTU (1700 kW h). For gross aggregations of natural gas, the conversion factor is 5.642 MSCF/BOE (0.1569 kSCm/BOE) which implies that natural gas has an averaged energy content of 1028 BTU/SCF (38.302 MJ/SCm) which requires 83.3% methane and 16.7% ethane (or some other mix that includes more methane, less ethane, and small amounts of heavier hydrocarbons, e.g., the conditions are satisfied with 86.6% C1, 10.0% C2, and 3.4% C3).

BPVC Boiler and pressure vessel code, published by ASME

BS&W Bottom sediment and water

BTU British thermal units. In the United States, BTU is defined as the amount of heat required to raise the temperature of 1 lbm of water from 59°F to 60°F (15 to 15.6°C). One US BTU is equivalent to about 1054.7 J. The exact value depends on the temperature range where the experiment started. The Canadian standard is to start at 60°F (15.6°C), and that results in 1054.8 J. If you start at 39°F (3.89°C), then 1 BTU is 1059.67 J. If you average the heat content from the freezing point of water to the boiling point of water, you get 1055.87 J. It is often convenient to define "thousands of BTU" as "MBTU" and "millions of BTU" as "MMBTU" which results in MMBTU/MSCF values for a natural gas clustered around 1.0.

Bulk modulus A measure of a substance's ability to resist changing volume with applied pressure. For liquids, tabulated values refer to the amount of pressure required to reduce the volume of the liquid by 1%.

CapEx Capital expenditure

Casing OCTG used in a wellbore for hole-stabilization and aquifer-protection. Casing is differentiated from "tubing" by being cemented in place.

Cavitation Number A dimensionless parameter that relates vapor pressure minus current pressure to kinetic energy.

CBL Cement bond log

CBM Coalbed methane

CDP or CPD Central delivery point or central point of delivery

CO Carbon monoxide

CO_2 Carbon dioxide

Compressibility A measure of the amount that a given mass of matter will change volume in response to a change in pressure and/or temperature compared to the amount that a theoretical ideal substance would change volume. For gases, compressibility comes out of the equation of state. For liquids, it is the reciprocal of the bulk modulus.

Condensate A mixture of naturally occurring chemicals which contain hydrocarbon species that are gaseous at reservoir conditions and liquid at surface conditions. Condensate generally includes species of hydrocarbons including pentane (C_5H_{12}), hexane (C_6H_{14}), and occasionally heptane (C_7H_{16}) and octane (C_8H_{18}). Lighter hydrocarbon species can be present in condensate, but over time, it will boil off. API gravity (see above) of condensate is generally 45–75°API.

Contactor A vessel designed to facilitate the contact of one fluid with another (e.g., an amine contactor to remove acid gas or a TEG contactor in a dehydrator tower)

Corrosion The wastage of material by the chemical action of the environment

CNG Compressed natural gas

Crude oil A mixture of naturally occurring chemicals which contain hydrocarbons and is liquid at reservoir conditions and remains liquid at temperatures somewhat above ambient at atmospheric pressure.

Cryogenics The study of "very low" temperatures. NIST defines "very low" as less than −300°F (−184°C), GPSA uses −51°F (−46°C) on the basis of application limits for low-temperature carbon steel.

CSG Coal seam gas (used interchangeably with CBM)

CSA Canadian Standards Association

Cum Cumulative production

Deliquification Application of energy to remove an interfering liquid to enhance gas production

Dew Point The temperature at which water vapor begins to condense.

DN "*Diametèr nominal*" or "nominal diameter" pipe sizing. It takes the integer portion of US pipe sizes and multiplies them times 25 (i.e., 4-in. (101.6 mm) steel pipe has an OD of 4.5 in. (114.3 mm) and is designated 100 DN).

DOE US Department of Energy

dP Differential pressure

Dry gas A mixture of gases that does not contain hydrocarbon species that become condensate, but often contains water.

Dynamic pressure Kinetic energy of a flow stream expressed as a pressure. At low velocities (below about 0.3 M), dynamic pressure is low and can usually be disregarded.

Dynamic viscosity (μ) (Also "Shear Viscosity") a measure of a fluid's ability to resist shear forces.

e-NGO Environmental-focused non-government organization.

EA Environmental assessment

EFM or EGM Electronic flow measurement or electronic gas measurement. EFM or EGM refers to replacing an analog pressure signal that is fed into a pen-and-ink chart with pressure transducers feeding electronic recording devices that are further processed by on-site algorithms to determine a flow rate.

EIS Environmental impact statement

EPC Engineering, procurement, and construction

Equation of state An equation showing the relationship between the values of the pressure, volume, and temperature of a quantity of a particular substance.

ESD Emergency shutdown

ESP Electric submersible pump

Erosion Wear on a surface from the viscous forces of a flowing fluid or moving solid.

Euler Number A dimensionless parameter that relates pressure forces to inertial forces.

Empirical equations Mathematical relationships that were developed based on observations of physical phenomena that do not necessarily have a theoretical basis or a closed-form derivation.

FBHP Flowing bottom-hole pressure

FEED Front-end engineering design

FERC The US Federal Energy Regulatory Commission

Filter cake Accumulation of solids on the inflow side of a filter.

FNU Formazin Nephelometric Units (used to quantify turbidity in water, see also FTU)

fps A system of units that uses "feet" for length, "pounds" for mass, and "seconds" for time. This system of units is also called "US Units," "Imperial Units," or even "British Units" even though the United Kingdom largely moved away from a consistent fps system many years ago.

FPSO Floating production, drilling, storage, and offloading vessel.

FRC Fire-resistant clothing. This used to stand for "Fire-Retardant Clothing" but evolution in the language (along with litigation) has changed most manufacturers to "resistant" instead of "retardant."

Froude Number A dimensionless parameter that relates flow inertia to body forces in the flow.

ft^3 Cubic feet, only used in this book for physical volumes like liquid volumes and gas volumes at actual conditions (never for gas volumes at standard conditions).

FTU Formazin turbidity unit (see also "FNU")

FWP Flowing wellhead pressure

g_c Conversion factor to convert lbm to lbf and back.

GIP Gas in place. Original gas in place (see OGIP) less cumulative production.

GLR Gas/liquid ratio

GOR Gas/oil ratio

GPS Global positioning system

GPSA Gas processors suppliers association

Gross heating value (GHV) The amount of heat released by complete oxidation of a specified quantity of a fuel at a specified initial temperature and pressure. Also called "upper heating value" and "higher heating value"

GRP Glass-reinforced plastic (also known as "fiberglass").

GTL Gas to liquids

GWR Gas/water ratio

HAZOP Hazard and operability

HDPE High-density polyethylene. Originally limited to material with a density of 0.941 to 0.965 g/cc, but those limits have blurred with time.

hp Horsepower. One hp is equivalent to 0.7457 kW

HPHT High pressure/high temperature

H_2S Hydrogen sulfide

Hydrostatic pressure The pressure exerted by a fluid on its surroundings due to the weight of the fluid-mass above the reference plane.

Hydrocarbon gases In common practice, hydrocarbon gases are referenced by the number of carbon atoms in a molecule. The common hydrocarbon gases are:

- *C1*: Methane (CH_4)
- *C2*: Ethane (C_2H_6)
- *C3*: Propane (C_3H_8)
- *iC4*: iso-Butane (C_4H_{10})
- *nC4*: normal Butane (C_4H_{10})
- *iC5*: iso-Pentane (C_5H_{12})
- *nC5*: normal Pentane (C_5H_{12})
- *C6*: Hexane (C_6H_{14})
- *C6+*: A "typical" mix of hydrocarbons that are generally liquid at atmospheric pressure and normal ambient temperature ranges. Trace amounts of these compounds are often present in a natural gas sample, and the industry chooses to

lump them into a mixture that represents the combination of them. Each laboratory has its own version of C6 +.

Hydrate (Properly called "clathrate hydrates") a structure of a lattice made up of water molecules with trapped foreign molecules included within the structure. Hydrate formation point is very dependent upon pressure, and at elevated pressures, hydrate can form at temperatures much higher than the freezing point of water.

Hygroscopic Hydrophilic substance that is irreversibly changed by acquiring water.

Hydrophilic Literally "water loving."

ID Inside diameter

IPR Inflow performance relationship

ISO International Organization for Standardization

Isobaric A process that takes place without a change in temperature.

Isentropic A process that takes place without a change in entropy.

Isenthalpic A process that takes place without a change in enthalpy.

Isothermal A process that takes place without a change in temperature.

Kinematic viscosity (ν) (Also called "momentum diffusivity") is the ratio of dynamic viscosity and fluid density.

LACT Lease automatic custody transfer

lbf Pounds force. The force applied if a mass of 1 lbm is subjected to the acceleration of gravity. It is currently popular to define a metric force term used in place of the "Newton" of kgf which is 2.2 lbf.

lbm Pounds mass (2.2 lbm = 1 kg). This term has replaced the "Slug" in common usage for the conversion from mass to weight or other force. On earth 1 lbm weighs 1 lbf.

LCM Lost circulation material

Lean oil A term used for and absorption media used in a contactor to extract natural gas liquids.

LEL or LFL Lower explosive limit or lower flammable limit. The minimum concentration of a flammable gas in air that will burn. LEL for methane in air is 5% at STP (see "STP" in this glossary), higher pressures or higher temperatures decrease this value. Personal hydrocarbon monitors display percentage of LEL so a 100% reading indicates that the atmosphere contains 5% methane by volume.

Liner OCTG used in a wellbore for hole-stabilization. Liner is differentiated from "casing" by having its top terminate within the wellbore instead of being connected to the wellhead.

Line pipe OCTG used outside the wellbore for transportation of hydrocarbon fluids.

LNG Liquefied natural gas

LOE Lease operating expense

LOPA Layers of protection analysis

LWD logging while drilling

m^3 Cubic meter, only used in this book for physical volumes like liquid volumes and gas volumes at actual conditions (never for gas volumes at standard conditions).

Mach Number A dimensionless parameter that expresses speed in terms of a ratio of the actual velocity to the speed of sound. Occasionally, it is abbreviated as "M," but this causes confusion in Oil & Gas where "M" is the Roman numeral for "1000" in traditional oil field units and "million" in metric units.

MAWP or MAOP Maximum Allowable Working Pressure or Maximum Allowable Operating Pressure. The pressure of a pressure-containing system that the designer selected as the maximum pressure that the system has been designed to withstand.

MIC Microbiologically influenced corrosion. This is a kind of corrosion caused by the biological activity of microbes.

mks a system of units that uses "meters" for length, "kilograms" for mass, and "seconds" for time. This is one of the choices for a "metric" system (the other is the cgs system that uses centimeters, grams, and seconds) under the SI system. The mks system uses liters for volumes, so our industry deviates from the standard to keep volume numbers manageable.

MOC Management of change

NACE National Association of Corrosion Engineers

NASA US National Aeronautics and Space Administration

Natural gas liquids (NGL) Hydrocarbon species whose boiling point temperature is near-ambient and cannot be reliably stored at atmospheric pressure in a vented vessel. NGL generally includes ethane (C_2H_6), propane (C_3H_8), normal butane (C_4H_{10}), and iso-butane (C_4H_{10}).

Net heating value (NHV) Gross heating value that has been reduced by the theoretical magnitude of the latent heat of vaporization of the water vapor created in combustion.

NGO Non-government organization

NOAA US National Oceanic and Atmospheric Administration

NPSH Net positive suction head

NPV Net present value, an economic term that brings all future costs back to time zero at a specified interest rate to allow projects to be compared on a consistent basis.

OCD Oil conservation division. Responsible for regulating Oil & Gas operations within the state. This term is used by the governments of several western US states, in places like New Mexico and Colorado the OCD performs the same function as the Railroad Commission in Texas, the Oil & Gas Conservation Division of the Corporation Commission in Oklahoma, and the Department of Environmental Protection in Pennsylvania.

OCTG Oil country tubular goods. Refers to tubing, casing, and line pipe used in "oil country."

OD Outside diameter

OGIP Original gas in place. The total gas that can be contacted by a well-bore over infinite time.

OpEx Operating expenditure

OSHA US Occupational Safety and Health Administration. OSHA is responsible for promulgating regulations to protect health and safety of industrial workers and for enforcing those regulations. Roughly equivalent to the Health and Safety Executive in the United Kingdom, Occupational Health and Safety (OH&S) in Canada, India, and Australia.

P&A Plug and abandon (used for Oil & Gas wells)

PCP Progressing cavity pump. A downhole positive displacement pump.

PD Positive displacement

Permeability A measure of the ability of a porous material to transmit fluids. Common oil field units are "Darcy," mili Darcy (Darcy/1000, symbol mD), micro Darcy (mD/1000, symbol μD), nano Darcy (μD/1000, symbol nD), or pico Darcy (nD/1000, symbol pD)

pH a measure of a substance's free hydrogen. pH is a log scale with 7.0 being neutral. Values less than 7.0 are acidic (i.e. a pH of 6.0 is 10 times more acidic than 7.0).

Values greater than 7.0 are basic or alkaline (i.e., pH of 9.0 is 100 times more alkaline than 7.0)

Pig a device intended to pass through a pipeline to shift liquids, clean pipe walls, and/or evaluate the condition of the pipe.

Piggable able to allow a pipeline pig to pass unmolested.

PLC Program logic control

POD Plan of depletion. This term is used anytime you develop a systematic methodology to develop (or deplete) an oil or gas field. In this book, "The POD" is a project described in SPE 84509.

Porosity A measure of the void space within a reservoir. Presented as a percentage of the reservoir volume that is not occupied by solid material.

ppb Parts per billion. This is usually a mass comparison (μg/kg). It can also be a volume comparison (ppbv). Using "ppb" alone implies a mass comparison.

ppm Parts per million. This is usually a mass comparison (mg/kg). It can also be a volume comparison (ppmv). Using "ppm" alone implies a mass comparison.

Pressure A thermodynamic parameter representing a force applied by a fluid over an area

psia Pounds (force) per square inch absolute—Total pressure at low fluid velocity (i.e., the sum of the local atmospheric pressure and the indicated gauge pressure disregarding dynamic pressure). Absolute pressure is used in scientific calculations and most engineering calculations except as noted below.

psid Pounds (force) per square inch differential. The pressure difference between two points. This is referred to as "psid" or "dP."

psig Pounds (force) per square inch gauge. The static pressure above atmospheric pressure that reads on a pressure gauge. Units are psig, kPag, or barg. Gauge pressure is the differential pressure between system pressure and atmospheric pressure. Gauge pressure is rarely used for scientific or engineering calculations, but many industries (such as "Pressure Vessel" and "Pressure Safety Valves") are concerned about the differential pressure to atmosphere and use gauge pressure in their calculations.

PSD Pressure safety device

PSM Process safety management, specifically referencing US OSHA regulation 1910.119, but used generically around the world as any safety program that largely adheres to the tenets of regulation 1910.119.

PSV or PRV Pressure safety valve or pressure relief valve

QA/QC Quality assurance/quality control. This term is used all over the world, but is rarely defined and it means whatever the speaker wants it to mean.

Rat Hole The area of a well-bore below the producing formation intended to accumulate well-bore debris.

Reversible An action that can be applied and then removed without changing the chemical composition of the fluid acted upon or transferring initial energy out of the system (freezing and thawing water is reversible, energy converted to heat by fluid friction is not reversible).

Reynolds Number A dimensionless parameter that relates friction forces to momentum forces.

RH Relative humidity. Expressed as a percentage, this is the amount of water vapor in a gas stream divided by the theoretical maximum water vapor that the gas can hold.

RO Reverse osmosis. Used to reference any process that overcomes osmotic pressure with externally applied pressure to "reverse" the natural tendency of pure fluids to flow toward less pure fluids.

ROW Right of way. A fiscal right to access physical property. ROW is generally purchased from the property owner.

RTP Reinforced thermoplastic pipe (formerly known as "spoolable composite" pipe).

RVP Reid vapor pressure

San Juan Basin (SJB) CBM Types the San Juan Basin has three distinctly different types of CBM production

- Type I—also known as "Fairway," this is source of the lion's share of CBM production from the San Juan Basin. It is represented by about 550 wells in the center of the field, these wells make very large gas rates, fairly small water rates, and significant CO_2 production. The fairway has been fully developed and no undrilled acreage exists at the time of this writing.
- Type II—also known as "Colorado Type," this area is north of the Fairway and is represented by an increasing number of wells (over 3000 at the time of this writing); these wells make fairly large gas rates, very high water rates, much less CO_2 than Fairway wells.
- Type III—this area is south of the Fairway and is represented by a slowly increasing number of wells (currently over 1000); these wells make very small gas rates, tend to make very little water, and nearly zero CO_2. There is development activity in the Type III, but it is fairly slow.

SAR Sodium absorption ratio

SDR Standard dimension ratio. HDPE is sold by "SDR" number; you can purchase SDR-13.5, SDR-11, SDR-7, etc. The number after the SDR is the ratio of the pipe outside diameter (OD) to the wall thickness. HDPE comes in OD's that are the same as steel (i.e., 8-in. pipe has an OD of 8.625 in.). Using these relationships, 8-inch SDR-7 would have a wall thickness of 1.232 in. (8.625/7). The inside diameter (ID) of that pipe would be

$$ID = OD - 2\cdot\left(\frac{OD}{SDR}\right) = OD\cdot\left(1-\frac{2}{SDR}\right) = 6.161\text{ in. for the example.}$$

SI *le Système international d'unités* or the "International System of Units" also called colloquially "The Metric System."

SIL Safety integrity level

SPE Society of petroleum engineers

Specific gravity Specific gravity is a convenient way to represent a fluid's mass relative to a reference fluid.

- *Liquid:* The specific gravity of a liquid is the density of the liquid divided by the density of pure water at 60°F (15.6°C) (i.e., 62.4 lbm/ft^3 (1000 kg/m^3)). If is known that a liquid's density is 60% of the density of water, then you can input the density of water in units that are convenient for your current calculation and multiply it times 0.60. Liquid specific gravity is not an intrinsic property of the liquid and must be referenced to a specific temperature.
- *Gas:* Specific gravity of a gas is a ratio of the mass per mole of the gas divided by the mass per mole of air (28.9625 lbm/lb mole (28.9625 g/g mole)). Gas specific gravity is an intrinsic property of the gas and is not dependent upon pressure or temperature.

SMYS Specified minimum yield stress. The manufacturer's guaranteed minimum stress that a material can withstand before it will begin to yield (i.e., mechanically deform under stress). This number is always well under the actual yield point (generally material will begin to yield around 125% of SMYS), but the magnitude of the safety factor is unreliable and SMYS must be used in any stress calculations. Companies and regulators have their own ideas about SMYS and often allowable stresses will be capped at 20%–40% of SMYS by fiat.

SRB Sulfur reducing bacteria

SRM The company that holds the basic patents for oil flooded screw compressors, Svenska Rotor Maskiner AB

SRP Sucker rod pump

STP Standard temperature and pressure. There is not really a standard, and the most common "standard" pressures referenced in US regulations and contracts are 14.73 psia, 14.696 psia, or 15.025 psia. STP in the SI unit system is also subject to regulatory and contractual manipulation, but the most common value is 101.325 kPa. Temperature in U.S. customary units is most often given as 60°F. In SI, the most common values are 0, 15, 15.55, and 20°C but there are others used.

Strouhal Number A dimensionless parameter that relates vortex shedding frequency to fluid velocity.

Sweet gas gas without the ability to form an acidic mixture by dissolving in water (either they don't form an acidic mixture or they don't dissolve or both).

T&E Threatened and endangered species

TDS Total dissolved solids. A measure of liquid contamination. Since (by definition) 1 L of pure water weighs 1 kg (1 million mg), this value also works out to ppm so a 10,000 mg/L TDS value is also a 10,000 ppm TDS value.

TEG Triethylene glycol used in dehydration equipment.

Thermocompressor A device that uses one fluid to increase the pressure of a second fluid.

Total Pressure The sum of dynamic pressure, static pressure, and hydrostatic pressure.

TSS Total suspended solids

Turbidity a comparison of the light-refracting characteristics of the sample to the characteristics of a standard Formazine mixture.

Tubing OCTG used in a well-bore for fluids-production. Tubing is differentiated from "casing" by not being cemented in place.

TVD True vertical depth

UEL or UFL Upper explosive limit or upper flammable limit. The maximum concentration of a flammable gas in air that will burn (adding more of the flammable gas will make the mixture "too rich to burn"). UEL for methane in air is 15% at STP (higher pressures or higher temperatures increase this value).

VFD Variable frequency drive (used in AC motors).

VI Volume index. This is a design parameter of a screw compressor and it is the ratio of the inlet volume to the outlet volume.

Viscosity A measure of a fluid's resistance to deformation by externally applied stresses.

VRU Vapor recovery unit

VSD Variable speed drive (used in DC motors)

SCF Standard cubic feet. A surrogate for mass in that a volume stated in SCF can be added or subtracted from any other volume stated in SCF at any other pressure and temperature.

SCm Standard cubic meters. A surrogate for mass in that a volume stated in SCm can be added or subtracted from any other volume stated in SCm at any other pressure and temperature.

Static pressure The force per unit area that a fluid exerts on its surroundings when the fluid is at rest relative to the surroundings.

Weber Number A dimensionless parameter that relates inertial forces to interfacial tension.

Wet gas A mixture of gases that include hydrocarbon species that become liquid with pressure or temperature changes of a magnitude expected in normal gas production.

WGR Water/gas ratio

WI Working interest. The ownership share that is obligated to participate in expenses and capital expenditure.

Wobbe Index A measure of the interchangeability of fuel gases. It is calculated by dividing GHV by the square root of the gas specific gravity.

Unit Conversions

B.1 PREFIXES

Prefixes can be applied to any unit. A mass changes from 1000 g to 1 kg by adding a prefix. Oil field prefixes should never be applied to metric units and vice versa.

B.1.1 Oilfield prefixes

- M → Roman numeral for 1000
- MM → Multiplication of Roman numerals (i.e., in Roman numerals "MM" would be 2000, but in Oil & Gas "MM" is 1 million or 1000 × 1000)
- B → Abbreviation for "billion" or 10^9
- T → Abbreviation for "trillion" or 10^{12}

B.1.2 SI prefixes

- y → Yocto, as a prefix, it is the base value times 10^{-24}
- z → Zepto, as a prefix, it is the base value times 10^{-21}
- a → Atto, as a prefix, it is the base value times 10^{-18}
- f → Femto, as a prefix, it is the base value times 10^{-15}
- p → Pico, as a prefix, it is the base value times 10^{-12}
- n → Nano, as a prefix, it is the base value times 10^{-9}
- μ → Micro, as a prefix, it is the base value times 10^{-6}
- m → Mili, as a prefix, it is the base value times 10^{-3}
- k → Kilo, as a prefix, it is the base value times 10^3
- M→ Mega, as a prefix, it is the base value times 10^6 (this book will try to avoid the use of "M" as an SI unit, but values like MPa are far too ubiquitous and useful to completely forego)
- G → Giga, as a prefix, it is the base value times 10^9
- T → Tera, as a prefix, it is the base value times 10^{12}

- P → Peta, as a prefix, it is the base value times 10^{15}
- E → Exa, as a prefix, it is the base value times 10^{18}
- Z → Zeta, as a prefix, it is the base value times 10^{21}
- Y → Yotta, as a prefix, it is the base value times 10^{24}

B.2 UNIT CONVERSIONS

In all of the tables in this section, the units you have are in the first column. The units you want are in the cells on that row (i.e., in Table B.3 if you have 100 MMSCF, then 100 MMSCF × 28.32 kSCm/MMSCF = 2832 kSCm) (Tables B.1–B.9).

Table B.1 Length and distance units

	in.	ft	mi	μm	mm	m	km
in.	1	0.0833	15.78E − 6	25.4E + 3	25.4	25.40E − 3	25.4E − 6
ft	12.0	1	189.4E − 6	304.8E + 3	304.8	0.3048	304.8E − 6
mi	63.36E + 3	5280	1	1.609E + 9	1.609E + 6	1.609E + 3	1.609
μm	39.37E − 6	3.281E − 6	621.4E − 12	1	1E − 3	1E − 6	1E − 9
mm	39.37E − 3	3.281E − 3	621.4E − 9	1000	1	1E − 3	1E + 6
m	39.37	3.281	621.4E − 6	1E + 6	1E + 3	1	1E − 3
km	39.37E + 3	3.281E + 3	0.6214	1E + 9	1E + 6	1000	1

Table B.2 Mass units conversions

	slug	lbm	oz	grain	ton	kg	tonne
slug	1	32.17	514.8	225.2E + 3	16.09E − 3	14.59	14.59E − 3
lbm	0.03108	1	16	7000	500E − 6	0.4536	453.6E − 6
oz	1.943E − 3	0.0625	1	437.5	31.25E − 6	0.2835	28.35E − 6
grain	4.440E − 6	142.9E − 6	2.286E − 3	1	68.78E − 9	64.80E − 6	64.80E − 9
ton	62.16	2000	32,000	14E + 6	1	907.2	0.9072
kg	0.06852	2.205	35.27	15,430	1.102E − 3	1	0.100
tonne	68.52	2205	35.27E + 3	15.43E + 6	1.102	1000	1

Table B.3 Gas volume unit conversions

	SCF	MSCF	MMSCF	SCm	kSCm
SCF	1	1E − 3	1E − 6	0.0283	0.283E − 3
MSCF	1E3	1	1E − 3	28.32	0.0283
MMSCF	1E6	1E3	1	28,320	28.32
SCm	35.31	0.0353	35.3E − 6	1	1E − 3
kSCm	35,310	35.31	0.0353	1000	1

Note: In this book, when reference temperature and pressure are not stated, 14.73 psia (101.56 kPaa or 1.0156 bara) and 60°F (15.56°C) were used. If your project is expecting a different reference temperature or pressure, then multiply the above table by $\rho_{14.7/60F} \div \rho_{RequiredValues}$

Table B.4 Liquid volume unit conversions

	BBL	Gallon	ft^3	L	$kL(m^3)$
BBL	1	42	5.615	159	0.159
Gallon	23.8E − 3	1	0.134	3.79	3.79E − 3
ft^3	0.178	7.481	1	28.317	0.028
L	6.29E − 3	0.264	0.035	1	1E − 3
$kL(m^3)$	6.290	264	35.315	1000	1

Table B.5 Pressure units

	psi	inH_2O	inHg	kPa	bar	mmH_2O	torr
psia	1	27.707	2.036	6.895	0.068	703.765	51.175
inH_2O	0.036	1	0.073	0.249	0.002	25.400	1.866
inHg	0.491	13.609	1	3.386	0.034	345.656	25.400
kPa	0.145	4.019	0.295	1	0.010	102.072	7.501
bar	14.504	401.860	29.530	100	1	10,207.247	750.064
mmH_2O	1.421E − 3	38.37E − 3	2.983E − 3	9.797E − 3	97.97E − 6	1	73.48E − 3
torr (mmHg)	19.34E − 3	535.8E − 3	39.37E − 3	133.3E − 3	1.333E − 3	13.609	1

Table B.6 Temperature unit conversions

	°F	R	°C	K
°F	1	°F + 459.67	(°F − 32)*5/9	(°F − 32)*5/9 + 273.15
R	R − 459.67	1	(R − 491.67)*5/9	R*5/9
°C	°C*9/5 + 32	°C*9/5 + 491.67	1	°C + 273.15
K	K*9/5 − 459.67	K*9/5	K − 273.15	1

Table B.7 Energy and work units

	MMBTU	therm	ft-lbf	N m	kW h	kJ
MMBTU	1	10	778.2E + 6	1.055E + 9	293.1	1.055E + 6
therm	0.1	1	77.82E + 6	105.5E + 6	29.31	105.5E + 3
ft-lbf	1.285E − 9	12.85E − 9	1	1.356	376.6E − 9	1.356E − 3
J (N m)	947.8E − 12	9.47E − 9	737.6E − 3	1	277.8E − 9	1000
kW h	3.412E − 3	34.12E − 3	2.655E + 6	3.6E + 6	1	3.6E + 3
kJ	947.8E − 9	9.478E − 6	737.6	1000	277.8E − 6	1

Table B.8 Power unit conversions

	hp (British)	BTU/h	hp (metric)	kW	MW
hp (British)	1	2544	1.014	0.7457	745.7E − 6
BTU/h	393.0E−6	1	398.5E − 6	293.1E − 6	293.1E − 9
hp (metric)	0.9863	2510	1	0.7355	735.5E − 6
kW	1.341	3412	1.360	1	0.001
MW	1341	3.412E + 6	1360	1000	1

Table B.9 Viscosity unit conversion

	poise	cP	lbm/ft/s	Pa s
poise	1	0.001	0.0672	0.1
cP	100	1	0.000672	0.001
lbm/ft/s	14.88	1488	1	1.488
Pa s	10	1000	0.672	1

Valve Summary

C.1 GATE VALVE

- *Operating mechanism*: A wedge-shaped gate slides between matching seats. Seal is metal-to-metal. In larger sizes, some manufactures use a gate that is two separate plates separated by springs to hold the gate more firmly on the seat. Other applications use a plate which has a hole that can be lined up with the pipe bore or hidden from the pipe bore.
- *Primary use*: liquids and steam. Used in natural-gas applications where a bubble-tight seal is not required.
- *Throttling characteristics*: very poor.
- *Method of actuation*: actuators are rarely used on gate valves outside of steam plants and then they are configured to simulate turning a valve hand wheel.
- *Flow path*: directly through the valve, generally larger than the pipeline ID.
- *Flow symmetry*: can accommodate flow in either direction.
- *Advantages*: somewhat lower costs.
- *Disadvantages*: extremely tedious to operate (e.g. a 12-in. Grove gate valve requires almost 100 turns of the hand wheel to operate), no provisions for double-block-and-bleed.

C.2 PLUG VALVE

- *Operating mechanism*: a drilled, cone-shaped stem (the plug) rotates in a matching seat. When the valve is opened, the plug is jacked off the seating surface and rotated 90 degrees. When shutting the valve, the plug is rotated and pushed down on the seat.

- *Primary use*: none, it has been sold as an alternative to gate and/or ball valves, but the seating characteristics and the tendency to jam make them a very poor device for field operations.
- *Throttling characteristics*: very poor.
- *Method of actuation*: quarter-turn piston actuators can be used, but they need significantly over-sized torque characteristics to lift the plug off the seat.
- *Flow path*: straight through the valve, but the hole in the plug is smaller than the pipeline diameter so they are not generally piggable.
- *Flow symmetry*: can accommodate flow in either direction.
- *Advantages*: none.
- *Disadvantages*: difficult to operate, poor sealing characteristics, require frequent maintenance.

C.3 BUTTERFLY VALVE

- *Operating mechanism*: a flat plate that pivots about its centerline is placed in the flow. Rotating the plate one-fourth turn toward shut will put the plate against the seating surfaces.
- *Primary use*: on/off in applications where considerable leakage is acceptable.
- *Throttling characteristics*: very poor.
- *Method of actuation*: quarter-turn piston actuators can be used.
- *Flow path*: straight through the valve, but the plate in the flow prevents them from being piggable.
- *Flow symmetry*: can be installed in either direction.
- *Advantages*: very inexpensive.
- *Disadvantages*: poor seal and not piggable.

C.4 GLOBE VALVE

- Includes "motor valves," needle valves, and backpressure valves.
- *Operating mechanism*: a plug-shaped stem seats in a matching seat that is oriented 90 degrees (relative to the flow direction) from the pipe centerline.

- *Primary use*: throttling any fluid.
- *Throttling characteristics*: good across almost half the valve travel.
- *Method of actuation*: actuators in the oil and gas industry are usually diaphragms.
- *Flow path*: up through the seat, across the chamber, and down into the outlet. This flow path causes a pressure drop across even a fully opened valve.
- *Flow symmetry*: although a globe valve can be installed in either direction, the manufacturers generally recommend that they be installed with the upstream flow under the seat to minimize the pressure on the stem-packing. One significant exception to this is actuated on/off valves tend to be flow downward through the seat to allow spring pressure to tend to keep them shut.
- *Advantages:* throttling and easy actuation.

C.5 FLOATING BALL VALVE

- *Operating mechanism*: a drilled ball rotates between seating surfaces. The ball is coupled to the valve body on the top only.
- *Primary use*: low-replacement-cost on/off applications.
- *Throttling characteristics*: poor.
- *Method of actuation*: quarter-turn piston actuators.
- *Flow path*: through the ball. Many floating ball valves have reduced ports so they are often not easily piggable (e.g., a reduced port valve on Tenneco's 36-in. main into New York City requires 3–5 hours for the pig to traverse).
- *Flow symmetry*: can be installed in either direction.
- *Advantages*: are cheaper than trunion-mounted ball valves.
- *Disadvantages*: lack of a body bleed, they have more of a tendency to leak through (because of lateral ball movement as seals wear), and they have poor sealing characteristics in very low dP installations.

C.6 TRUNNION BALL VALVE

- *Operating mechanism*: a drilled ball rotates between seating surfaces. The ball is held rigid top and bottom by trunnion bearings.
- *Primary use*: on/off applications.
- *Throttling characteristics*: poor.
- *Method of actuation*: quarter-turn piston actuators.
- *Flow symmetry*: can be installed in either direction.
- *Advantages*: body bleeds allow a single valve to serve in many double-block-and-bleed applications. Sealing surfaces are very durable. Work well with very low differential pressure.
- *Disadvantages*: purchase cost is about 12% higher than a floating ball valve.

INDEX

Note: Page numbers followed by "*f*" and "*t*" refer to figures and tables, respectively.

D

E

F

G

H

Q

R

S

T

U

V

W

Y

Printed in the United States
By Bookmasters